Discrete Networked Dynamic Systems

Discrete Networked Dynamic Systems

Analysis and Performance

Magdi S. Mahmoud
King Fahd University of Petroleum and Minerals
Systems Engineering Department
Dhahran, Saudi Arabia

Yuanqing Xia
Beijing Institute of Technology
School of Automation
Beijing, China

Academic Press is an imprint of Elsevier
125 London Wall, London EC2Y 5AS, United Kingdom
525 B Street, Suite 1650, San Diego, CA 92101, United States
50 Hampshire Street, 5th Floor, Cambridge, MA 02139, United States
The Boulevard, Langford Lane, Kidlington, Oxford OX5 1GB, United Kingdom

Library of Congress Cataloging-in-Publication Data
A catalog record for this book is available from the Library of Congress

British Library Cataloguing-in-Publication Data
A catalogue record for this book is available from the British Library

ISBN: 978-0-12-823698-7

For information on all Academic Press publications
visit our website at https://www.elsevier.com/books-and-journals

Publisher: Mara Conner
Acquisitions Editor: Sonnini R. Yura
Editorial Project Manager: Chiara Giglio
Production Project Manager: Nirmala Arumugam
Designer: Greg Harris

Typeset by VTeX

To our families.
For the love, the unwavering support, and all the joy they bring.
Each one of you has a special place in our hearts.
With gratitude.

Magdi S. Mahmoud, Yuanqing Xia

Contents

About the authors

Magdi S. Mahmoud obtained his BSc (Honors) in communication engineering, his MSc in electronic engineering, and his PhD in systems engineering, all from Cairo University, in 1968, 1972, and 1974, respectively. He has been a Professor of Engineering since 1984. He is now a Distinguished Professor at King Fahd University of Petroleum and Minerals (KFUPM), Saudi Arabia. He was on the faculty at different universities worldwide, in countries including Egypt (CU, AUC), Kuwait (KU), UAE (UAEU), UK (UMIST), USA (Pitt, Case Western), Singapore (Nanyang), and Australia (Adelaide). He lectured in Venezuela (Caracas), Germany (Hanover), UK (Kent), USA (UoSA), Canada (Montreal), and China (BIT, Yanshan). He is the principal author of 51 books and book chapters and the author/coauthor of more than 600 peer-reviewed papers. He is a fellow of the IEE, a senior member of the IEEE and the CEI (UK), and a registered consultant engineer of information engineering and systems (Egypt). He received the Science State Incentive Prize for outstanding research in engineering (1978, 1986), the State Medal for Science and Art, First Class (1978), and the State Distinction Award (1986), Egypt. He awarded the Abdulhamed Showman Prize for Young Arab Scientists in the field of Engineering Sciences (1986), Jordan. In 1992, he received the Distinguished Engineering Research Award, College of Engineering and Petroleum, Kuwait University (1992), Kuwait. He is cowinner of the Most Cited Paper Award 2009, for "Signal Processing", vol. 86, no. 1, 2006, pp. 140–152. His papers were selected among the 40 best papers in Electrical & Electronic Engineering by the Web of Science ISI in July 2012. He interviewed for "People in Control," IEEE Control Systems Magazine, August 2010. He served as Guest Editor for the special issue "Neural Networks and Intelligence Systems in Neurocomputing" and as Guest Editor for the 2015 International Symposium on Web of Things and Big Data (WoTBD 2015), 18–20 October 2015, Manama, Bahrain. He is a Regional Editor (Middle and East Africa) of the International Journal of Systems, Control and Communications (JSCC), INDERSCIENCE Publishers since 2007, a member of the Editorial Board of the Journal of Numerical Algebra, Control and Optimization (NACO), Australia since 2010, an Associate Editor of the International Journal of Systems Dynamics Applications (IJSDA), since 2011, a member of the Editorial Board of the Journal of Engineering Management, USA since 2012, and an

Academic Member of the Athens Institute for Education and Research, Greece since 2015. Since 2016, He is an Editor of the Journal of Mathematical Problems in Engineering, Hindawi Publishing Company, USA. He is currently actively engaged in teaching and research in the development of modern methodologies to distributed control and filtering, networked control systems, fault-tolerant systems, cyber-physical systems, and information technology.

Yuanqing Xia was born in Anhui Province, China in 1971, and graduated from the Department of Mathematics, Chuzhou University, China in 1991. He received his MSc in Fundamental Mathematics from Anhui University, China, in 1998, and his PhD in Control Theory and Control Engineering from Beijing University of Aeronautics and Astronautics, China in 2001. From 1991 to 1995 he worked as a teacher at Tongcheng Middle-School, China. From January 2002 to November 2003 he was a postdoctoral research associate at the Institute of Systems Science, Academy of Mathematics and System Sciences, Chinese Academy of Sciences, China, where he worked on navigation, guidance, and control. From November 2003 to February 2004 he was a Research Fellow at the National University of Singapore, where he worked on variable structure control. From February 2004 to February 2006 he was a Research Fellow at the University of Glamorgan, UK, where he studied networked control systems. From February 2007 to June 2008 he was a guest professor with Innsbruck Medical University, Austria, where he worked on biomedical signal processing. Since July 2004 he has been with the School of Automation, Beijing Institute of Technology, Beijing, first as an Associate Professor and since 2008 as a Professor. In 2012 he was appointed as a Xu Teli Distinguished Professor at the Beijing Institute of Technology, and in 2016 he became Chair Professor. In 2012 he obtained a grant from the National Science Foundation for Distinguished Young Scholars of China; in 2016 he was honored as the Yangtze River Scholar Distinguished Professor and was supported by the National High Level Talents Special Support Plan ("Million People Plan") by the Organization Department of the CPC Central Committee.

He is now the Dean of the School of Automation, Beijing Institute of Technology. He has published 10 monographs published by Springer, John Wiley, and CRC, and more than 400 papers in international scientific journals. He is a deputy editor of the Journal of Beijing Institute of Technology and an associate editor of Acta Automatica Sinica; Control Theory and Applications; the International Journal of Innovative Computing, Information, and Control; and the International Journal of Automation and Computing. He obtained the Second Award of the Beijing Municipal Science and Technology (No. 1) in 2010 and 2015, the Second National Award for Science and Technology (No. 2) in 2011, and the Second Natural Science Award of the Ministry of Education (No. 1) in 2012 and 2017. His research interests include cloud control systems, networked control systems, robust control and signal processing, active disturbance rejection control, unmanned system control, and flight control.

Preface

Networked dynamic systems (NDSs) are dynamic systems operating over networks and appear quite frequently in many natural and cyber-physical systems. An important class of these systems is discrete NDSs (DNDSs), where the information dynamics and the associated decision making capabilities are processed in a discrete-time framework. A popular representation of such dynamic processes is consensus (agreement), which has been widely incorporated for a variety of control and estimation applications, including robotics and swarm deployment, distributed Kalman filtering, and multiagent systems. In such scenarios, an important question arises naturally: How does the underlying network topology influence the behavior of the dynamic systems operating over networks? In view of the available results, it turns out that research avenues in DNDSs offer great opportunities for further developments from theoretical, simulation, and implementation perspectives.

This book originated from our research activities in control of/over networks and dynamics over networks, both individually and jointly, at King Fahd University of Petroleum and Minerals (KFUPM) and the Beijing Institute of Technology (BIT). The treatment is completely self-contained, with only standard matrix algebra, calculus, and probability as prerequisites. We constantly aim to relate the properties of the information flow (essentially determined by a suitable graph) with the properties of dynamics. Dynamical properties of interest include not only mere convergence but also "performance," broadly intended.

We intend to consider different notions of performance, including the rate of convergence, the accuracy in approximating the average of the initial states, and the robustness against noise and communication errors. In all these cases, the relation between graph and performance is made explicit by the spectral analysis of the update matrix. We develop our approach from the perspective of large-scale networks. Even though our theory is valid for networks of any size, we pay special attention to how dynamical properties depend on the size of the network. Concretely, this leads us to specialize our results to specific families of graphs (for instance, grids) and to take limits where the number of nodes grows to infinity.

Several multiagent models exist in the literature. The proposed book adopts the unique approach of structuring along four dimensions:

- *Agent,*
- *Environment,*
- *Interaction, and*
- *Organization,*

while taking two main viewpoints:

- local (agent-centered) and
- global (system-centered)

From this perspective, the pedagogical objectives of the book are:

1. to introduce a coherent and unified framework for studying DNDSs with particular emphasis on a set-theoretic methodology and relying heavily on deriving our results based on algebraic graph theory and topology;
2. to acquaint students with the system-theoretic background required to read and contribute to the research literature on DNDSs;
3. to contribute to the further development of advanced distributed consensus/control/estimation methods for different classes of DNDSs;
4. to help in expanding the field of consensus/coordination of discrete dynamic systems over graphs, including swarms and multivehicle and swarm robotics; and
5. to provide a modest coverage of new results based on rigorous math tools and implemented in efficient algorithmic procedures.

- **Chapter 1:** (Mathematical background and examples)
  In this chapter, we introduce some representative examples and systems from multiple disciplines to motivate our treatment of a class of linear DNDSs in the following chapters. We focus our attention on DNDSs where the underlying connection topology couples the agents at their outputs. Additionally, we review basic concepts from matrix theory with a special emphasis on the so-called Perron–Frobenius theory. These concepts will be useful when analyzing the convergence of the linear dynamical systems discussed throughout the book. Next, we provide a brief overview on algebraic graph theory with an emphasis on averaging and estimation algorithms defined over graphs.
- **Chapter 2:** (Structural and performance patterns)
  This chapter establishes a general framework for the analysis and performance of a class of linear DNDSs, where the underlying connection topology couples the agents at their outputs. DNDSs with homogeneous agent dynamics are considered. Complete results are provided for agent feedback control using sensed state information and extended to include $\mathbb{H}_2$ performance metrics. It is established that the $\mathbb{H}_2$-norm expression reduces to the Frobenius norm of the underlying connection topology incidence matrix scaled by the $\mathbb{H}_2$-norm of the agents comprising the DNDS. The $\mathbb{H}_2$-norm characterization is then used to synthesize DNDSs with prescribed $\mathbb{H}_2$ performance. Convex analysis is presented to design a local controller for each agent when the underlying topology is fixed and expressed as minimization over linear matrix inequalities.

- **Chapter 3:** (Consensus of systems over graphs)
  In this chapter, we treat different consensus problems for distributed/multiagent systems over graphs. We begin by shedding light on issues pertaining to the consensus problems for distributed/multiagent systems described over graphs. Next, we discuss asymptotic consensus properties over switching graphs and related convergence analysis. The consensus control of distributed systems with time delays is addressed. Finally, we deal with the robustness issues in consensus of distributed/multiagent systems. Several illustrative simulation examples are provided.
- **Chapter 4:** (Energy-based cooperative control)
  This chapter presents a class of energy-based cooperative control problems that have an explicit connection to convex network optimization problems. The main basis is the passivity and dissipativity concepts. The new notion of maximal equilibrium independent passivity is introduced and it is shown that networks of systems possessing this property asymptotically approach the solutions of a dual pair of network optimization problems, namely, an optimal potential and an optimal flow problem. This connection leads to an interpretation of the dynamic variables, such as system inputs and outputs, to variables in a network optimization framework, such as divergences and potentials, and reveals that several duality relations known in convex network optimization theory translate directly to passivity-based cooperative control problems.
- **Chapter 5:** (Performance of consensus algorithms)
  In this chapter, we describe a topology synthesis procedure based on results from combinatorial optimization. We provide a general framework for the analysis and synthesis of a class of relative sensing networks (RSNs) in the context of the peak performance. In an RSN, the underlying connection topology couples each agent at their outputs. A distinction is made between RSNs with homogeneous agent dynamics and RSNs with heterogeneous dynamics. In both cases, explicit graph-theoretic expressions and bounds for the peak performance are derived. The peak performance is structure-dependent and related to the spectral radius of the graph Laplacian. The analysis results are then used to develop synthesis methods for RSNs. Using results from robust semidefinite programming, a synthesis procedure for the design of a robust sensing topology is derived.
- **Chapter 6:** (Event-based coordination control)
  This chapter provides an account of the major works done on spectra of adjacency matrices drawn on networks and the basic understanding attained so far. We divide the results into: (1) extremal eigenvalues, (2) bulk part of the spectrum, and (3) degenerate eigenvalues, based on the intrinsic properties of eigenvalues and the phenomena they capture. We outline a coverage of the works on spectra of various popular model networks, such as the random networks, scale-free networks, one-dimensional lattice, small-world networks,

and various different real-world networks. Additionally, potential applications of spectral properties for natural processes are discussed.

- **Chapter 7:** (Advanced approaches to multiagent coordination)
  In this chapter, we deal with four problems. First, we discuss the synchronization analysis and control problems for a class of nonlinear stochastic complex dynamical networks consisting of identical nodes and account for asymmetric coupling configuration, nonlinear inner coupling structures, and nonidentical exogenous disturbances. Second, the consensus of multiagent systems with linearized dynamics is considered based on an observer-type protocol using the relative outputs of neighboring agents on a directed communication topology. Third, the tracking problem for multiagent systems with a time-varying reference state is investigated and designed based on the event-triggered scheme. Fourth, the robust output regulation of linear multiagent systems under a directed interaction topology is addressed. The digraph is assumed to contain a spanning tree, and every agent or subsystem is identical and uncertain, but subsystems have different external disturbances.
- **Chapter 8:** (State estimation techniques)
  In this chapter, an improved state estimation approach for sensor networks is initially developed. A distinct feature of this approach is towards characterizing the filter's performance limit and weights such that the variance of the estimation errors is minimized. Moreover, certain parameter optimization is alleviated with the application of a particular finite impulse response filter. Next, the important topic of asynchronous multisensor systems is considered, where data losses and delays are likely to occur when performing state estimation at different rates. An algorithm to process both the delayed and missing measurements is developed, under the assumption that the data arrive between two consecutive sampling times. Local estimation is then implemented using a fusion algorithm. Later on, we provide the theory and analytical development of distributed Kalman filtering based on various methods. In one method, the weighted averages, based on the Bayesian point of view, yield the best possible estimate of the state at a particular time instant which can be computed from the conditional probability distribution of state x given all past measurements up to the particular time.
- **Chapter 9:** (Advanced distributed filtering)
  In this chapter, we provide the theory and analytical development of advanced distributed filtering based on various methods. In one method, the weighted averages, based on the Bayesian point of view, yield the best possible estimate of the state at a particular time instant which can be computed from the conditional probability distribution of state x given all past measurements up to the particular time.

Magdi S. Mahmoud
Yuanqing Xia
KFUPM, Saudi Arabia, and BIT, China
January 2020

Acknowledgement

Special thanks are due to the Elsevier team, particularly our Acquisitions Editor Sonnini R. Yura, our Editorial Project Manager Chiara Giglio, and our Production Project Manager Nirmala Arumugam for their support, guidance, and dedication throughout the publishing process. We are grateful to all the anonymous referees for carefully reviewing and selecting the appropriate topics for the final version during this process. Portions of this volume were developed and upgraded while offering the graduate courses **SCE-701-191**, **SCE-555-191**, and **SCE-606-192** at KFUPM, Saudi Arabia. Dr. Mahmoud acknowledges the support afforded by the Deanship of Scientific Research (DSR) at KFUPM through distinguished research project no. **BW 191007**, and Dr. Xia acknowledges the National Natural Science Foundation of China under grant number **61720106010**.

Magdi S. Mahmoud and Yuanqing Xia
February 2020

Chapter 1

Mathematical background and examples

1.1 Introduction

Recent years have witnessed an explosive increase in research activities on developing methodologies for cooperative control and motion coordination. This interest is motivated by the growing possibilities enabled by robotic networks and/or multivehicle systems in the monitoring of natural phenomena and the enhancement of human capabilities in hazardous and unknown environments.

Admittedly, we assert that *networks* are everywhere. We watch TV through the television network; we interact with each other in a closely connected social network; our body itself is a highly complicated biological network. Mathematically, a network can be thought of as a collection of nodes that represent some physical quantities and edges that interconnect different nodes (see Fig. 1.1). On a parallel avenue, we have witnessed the emergence of a discipline of study

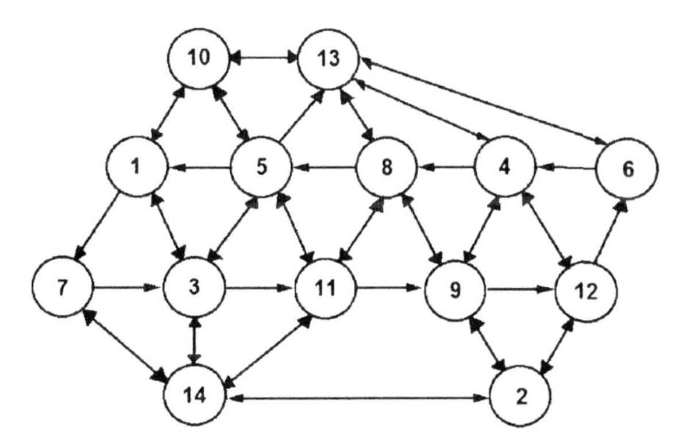

FIGURE 1.1 A network structure.

focused on modeling, analyzing, and designing dynamic phenomena over networks. We refer to such systems as networked dynamic systems; they are also equivalently referred to as multiagent or distributed systems. This emerging discipline, rooted in graph theory, control theory, and matrix analysis, is increasingly relevant because of its broad set of application domains. Discrete networked dynamic systems (DNDSs) appear naturally in

Discrete Networked Dynamic Systems. https://doi.org/10.1016/B978-0-12-823698-7.00009-6

(i) social networks and mathematical sociology,
(ii) electric, mechanical, and physical networks, and
(iii) animal behavior, population dynamics, and ecosystems.

Network systems are designed in the context of networked control systems, robotic networks, power grids, parallel and scientific computation, and transmission and traffic networks, to name a few (see [1]).

This book has been written to be an accurate introduction to the emerging topic of DNDSs. The chief goal is establishing a unified and self-contained theoretical framework that is suitable to the analysis and performance for several related dynamical models that feature time variance, randomness, and heterogeneity with emphasis on the methodological and general aspects of the subject. We will also treat applications to inferential problems in sensor networks, rendezvous of mobile robots, and opinion dynamics in social networks.

The focal point addressed throughout this book is that complex dynamical evolutions originate from the interactions of a large number of simple units. Not only such collective behaviors are evident in biological and social systems, but the digital revolution and the miniaturization in electronics have also made possible the creation of man-made complex architectures of interconnected devices, including computers, sensors, and cameras.

Most of the time, we concentrate on linear discrete-time dynamics, which we identify as the core theoretical issue. We present the fundamental set-theoretic results of dynamic systems over graphs and gather together a unified viewpoint of various models and results scattered in the literature.

To this end, we provide in the sequel some relevant mathematical information.

1.2 Mathematical background

In this chapter, we provide some mathematical notations, present basic notions, collect useful algebraic inequalities and lemmas, and give an introduction to some relevant topics which are quite useful for the book.

At first, we assume that the reader already has basic training in linear algebra, and for a more complete introduction, some familiarity with the numerical software MATLAB® is also encouraged.

In what follows, the notations used are quite standard. We use $\mathbb{Z}_+$ to denote the set of all nonnegative integers. For any positive real number r, $\lfloor r \rfloor$ denotes the largest integer that is less than or equal to r. For any $x \in \Re^n$, x^t is its transpose and $|x|$ its Euclidean norm. For an $n \times m$ matrix A, $|A|$ stands for its induced matrix norm. For any function, including *controls* or *inputs*, $\phi : \mathbb{Z}_+ \to \Re^n$, we denote (with slight abuse of notation) $||\phi|| = \sup\{|\phi| : k \in \mathbb{Z}_+\}$. In the case when ϕ is bounded, this is the standard ℓ_∞-norm. For any $k \in \mathbb{Z}_+$ and any function, $\phi : \mathbb{Z}_+ \to \Re^n$, $\phi_{[k]}$ denotes the truncation of ϕ at k, that is, $\phi^k := \phi - \phi_{[k]}$. We use $\gamma_1 \circ \gamma_2$ to denote the composition of two functions γ_1 and γ_2 which are from $\Re \to \Re$.

Throughout this book, we often consider the same system but with controls restricted to take values in some subset $\Omega \subset \Re^m$; we use $\mathbf{M}_\Omega$ to denote the set of controls taking values in Ω.

In this book, we deal mostly with *finite-dimensional linear spaces*, which are also often called *linear vector spaces*. Seeking generality, we consider the linear space to be **n**-dimensional. Although most of the time we will deal with vectors of real numbers $\Re$, occasionally, we will encounter vectors of complex numbers $\mathbb{C}$. For instance, the eigenvalues or eigenvectors of a real matrix could be complex.

Let $x_j, y_j \in \Re$ (or $\boldsymbol{C}$), $j = 1, 2, ..., n$. Then the **n**-dimensional vectors **x**, **y** are defined by $\mathbf{x} = [x_1\ x_2\ ...\ x_n]^t$, $\mathbf{y} = [y_1\ y_2\ ...\ y_n]^t \in \Re^n$, respectively, where $\Re^n = \Re \times ... \times \Re$. A nonempty set $\mathcal{X}$ of elements **x**, **y** is called the *real (or complex) vector space (or real [complex] linear space)* by defining two algebraic operations, *vector additions and scalar multiplication, in* $x = [x_1\ x_2\ ...\ x_n]^t$.

Given two vector spaces $\mathcal{X}_1$ and $\mathcal{X}_2$ with the same associated scalar field, we use $\mathcal{X}_1 \times \mathcal{X}_2$ to denote the vector space formed by their Cartesian product. Thus every element of $\mathcal{X}_1 \times \mathcal{X}_2$ is of the form

$$(x_1, x_2), \quad \text{where} \quad x_1 \in \mathcal{X}_1 \text{ and } x_2 \in \mathcal{X}_2.$$

A nonempty subset $\mathcal{G} \subset \Re^n$ is called a *linear subspace* of $\Re^n$ if $x + y$ and αx are in $\mathcal{G}$ whenever x and y are in $\mathbb{G}$ for any scalar α. A set of elements $X = \{x_1, x_2, ..., x_n\}$ is said to be a *spanning set* for a linear subspace $\mathbb{G}$ of $\Re^n$ if every element $g \in \mathbb{G}$ can be written as a linear combination of $\{x_j\}$. That is, we have

$$\mathcal{G} = \{g \in \Re : g = \alpha_1 x_1 + \alpha_2 x_2 + \ ...\ \alpha_n x_n \text{ for some scalars } \alpha_1, \alpha_2, ..., \alpha_n\}.$$

Sometimes the shorthand notation $\mathbf{span}\{x_1, x_2, ..., x_n\}$ is used. A spanning set $\mathcal{X}$ is said to be a *basis* for $\mathcal{G}$ if no element x_j of the spanning set X of $\mathcal{G}$ can be written as a linear combination of the remaining elements $x_1, x_2, ..., x_{j-1}, x_{j+1}, ..., x_n$, that is, x_j, $1 \leq i \leq n$ form a linearly independent set. It is common to use $x_j = [0\,0\,...\,0\,1\,0\,...\,0]^t$ as the kth unit vector. The geometric ideas of linear vector spaces have led to the concepts of "*spanning a space*" and a "*basis for a space*."

1.2.1 Basic notions

We begin with some definitions.

Definition 1.1. The **n**-dimensional Euclidean space, denoted throughout this book by $\Re^n$, is a linear vector space equipped with the inner product

$$\langle x, y \rangle = x^t\ y = \sum_{j=1}^{n} x_j y_j$$

and with the *trace inner product* in a matrix space

$$\langle M, N \rangle = \mathbf{trace}(M^t\, N), \quad M, N \in \Re^{n\times},$$
$$\langle M, N \rangle = \mathbf{trace}(M^*\, N), \quad M, N \in \mathbf{C}^{n\times},$$

where $\mathbf{trace} M = \sum_{j=1}^{n} m_{jj}$ and the superscript $*$ means the complex conjugate transpose.

Let $\mathcal{X}$ be a linear space over the *field* $\mathbf{F}$ (typically $\mathbf{F}$ is the field of real numbers $\Re$ or complex numbers $\mathbf{C}$). Then a function

$$||.|| : \mathcal{X} \to \Re$$

that maps $\mathcal{X}$ into the real numbers $\Re$ is a norm on $\mathcal{X}$ if and only if

1. $||x|| \geq 0, \forall x \in \mathcal{X}$ (nonnegativity);
2. $||x|| = 0, \iff x = 0$ (positive definiteness);
3. $||\alpha\, x|| = |\alpha| ||x|| \;\; \forall x \in \mathcal{X}$ (homogeneity with respect to $|\alpha|$);
4. $||x + y|| \leq ||x|| + ||y||, \forall x, y \in \mathcal{X}$ (triangle inequality).

Given a linear space $\mathcal{X}$, there are many possible norms on it. For a given norm $||.||$ on $\mathcal{X}$, the pair $(\mathcal{X}, \; ||.||)$ is used to indicate $\mathcal{X}$ endowed with the norm $||.||$.

Definition 1.2 (A linear space or a vector space). A set (of vectors) V is considered a linear space over the field $\Re$ if its elements, called vectors, are closed under two basic operations: scalar multiplication and vector summation "+." That is, given any two vectors $v_1, \; v_2 \in V$ and any two scalars $\alpha, \; \beta \in \Re$, the linear combination $v = \alpha\, v_1 + \beta\, v_2$ is also a vector in V. Furthermore, the addition is commutative and associative, it has an identity 0, and each element has an inverse, "$-v$," such that $v + (-v) = 0$. The scalar multiplication respects the structure of $\Re$, that is, $\alpha(\beta)v = (\alpha\beta)v$, $1v = v$, and $0\, v = 0$. The addition and scalar multiplication are related by the distributive laws: $(\alpha + \beta)v = \alpha\, v + \beta\, v$ and $\alpha(v + u) = \alpha v + \alpha u$.

For example, $\Re^n$ is a linear space over the field of real numbers $\Re$. To be consistent, we always use a column to represent a vector:

$$\begin{bmatrix} x_1 & x_2 & \cdots & x_n \end{bmatrix}^t = \begin{bmatrix} x_1 \\ x_2 \\ \vdots \\ x_n \end{bmatrix} \in \Re^n, \tag{1.1}$$

where $\begin{bmatrix} x_1 & x_2 & \cdots & x_n \end{bmatrix}^t$ means the (row) vector $\begin{bmatrix} x_1 & x_2 & \cdots & x_n \end{bmatrix}$ transposed. Given two scalars $\alpha, \; \beta \in \Re$ and two vectors $x = \begin{bmatrix} x_1 & x_2 & \cdots & x_n \end{bmatrix}^t \in \Re^n$ and

$y = \begin{bmatrix} y_1 & y_2 & \cdots & y_n \end{bmatrix}^t \in \Re^n$, their linear combination is a component-wise summation weighted by α and β:

$$\begin{aligned}\alpha\, x + \beta\, y &= \alpha \begin{bmatrix} x_1 & x_2 & \cdots & x_n \end{bmatrix}^t + \beta \begin{bmatrix} y_1 & y_2 & \cdots & y_n \end{bmatrix}^t \\ &= \begin{bmatrix} \alpha x_1 + \beta y_1 & \alpha x_2 + \beta y_2 & \cdots & \alpha x_n + \beta y_n \end{bmatrix}^t. \end{aligned} \tag{1.2}$$

In the sequel, if $A \in \mathbb{C}^{n\times n}$, under the symmetry, we mean the conjugate symmetry, that is, $A = A^*$, where A^* is the conjugate transpose of A.

We will now provide a brief review of basic notions and frequently used notation associated with a linear vector space V (that is, $\Re^n$).

Definition 1.3 (Subspace). A subset W of a linear space is called a subspace if the zero vector 0 is in W and $w = \alpha\, w_1 + \beta\, w_2 \in W$ for all $\alpha,\ \beta \in \Re$ and $w_1,\ w_2 \in W$.

Definition 1.4 (Spanned subspace). Given a set of vectors $S = \{v_i\}_{i=1}^m$, the subspace spanned by S is the set of all finite linear combinations $\sum_{i=1}^m \alpha_i\, v_i$ for all $\begin{bmatrix} \alpha_1 & \alpha_2 & \cdots & \alpha_n \end{bmatrix}^t$. This subspace is usually denoted by **span(S)**.

For example, the two vectors $v_1 = \begin{bmatrix} 1 & 0 & 0 \end{bmatrix}^t$ and $v_2 = \begin{bmatrix} 1 & 1 & 0 \end{bmatrix}^t$ span a subspace of $\Re^3$ whose vectors are of the general form $v = \begin{bmatrix} x & y & 0 \end{bmatrix}^t$.

Definition 1.5 (Linear independence). A set of vectors $S = \{v_i\}_{i=1}^m$ is linearly independent if

$$\alpha_1 v_1 + \alpha_2 v_2 + \cdots \alpha_m v_m = 0$$

implies

$$\alpha_1 = \alpha_2 = \cdots = \alpha_m = 0.$$

On the other hand, a set of vectors $\{v_i\}_{i=1}^m$ is said to be linearly dependent if there exist $\begin{bmatrix} \alpha_1 & \alpha_2 & \cdots & \alpha_n \end{bmatrix} \in \Re$ not all zero such that

$$\alpha_1 v_1 + \alpha_2 v_2 + \cdots \alpha_m v_m = 0.$$

Definition 1.6 (Basis). A set of vectors $\mathbf{B} = \{\mathbf{b}_i\}_{i=1}^n$ of a linear space $\mathbf{V}$ is said to be a basis if $\mathbf{B}$ is a linearly independent set and $\mathbf{B}$ spans the entire space $\mathbf{V}$; that is, $\mathbf{V} = \mathbf{span}(\mathbf{B})$.

1.2.2 Signal norms

In the sequel, we use *norms* to measure the size of a signal in the discrete domain and generically symbolized by $||.||$. Four key properties of a norm for signals $x(k)$, $y(k)$, $k \in \mathbb{Z}$ are:

(a) $||x(k)|| > 0$ (positivity);
(b) $||x(k)|| = 0$ if and only if $||x(k) = 0||$ (positive definiteness);
(c) $||\alpha x(k)|| = |\alpha| ||x(k)|| \; \forall$ scalars α (homogeneity);
(d) $||x(k) + y(k)|| \leq ||x(k)|| + ||y(k)||$ (triangle inequality)

for $x(k) \in \mathcal{X} \subset \Re^n$ and $y(k) \in \mathcal{Y} \subset \Re^n$.

Some key signal norms are:

- *Energy of the signal* $x(k)$, 2-norm,

$$||x(k)||_2 = \sqrt{[\sum_{j=\infty}^{\infty} x^2(k)_j]}.$$

- *Maximum value over time*, ∞-norm,

$$||x(k)||_\infty = \max_j \, x(k)_j.$$

It is also called *sup-norm* or ℓ_∞-norm. An important special case is the ℓ_1-norm $||x(k)||_1$, defined by

$$||x||_1 = \sum_{j=1}^{n} |x_j|.$$

- *Average power-norm*,

$$pow[x(k)] = \sqrt{[\lim_{S \to \infty} \frac{1}{2S} \sum_{j=1}^{S} u_j^2]}.$$

1.2.3 Vector norms

By similarity, let $x(k) \in \Re^n$, and let the *vector p-norm* of $x(k)$ be

$$||x(k)||_p := \left(\sum_{j=1}^{n} |x(k)_j|^p \right)^p, \quad \text{for } 1 \leq p \leq \infty.$$

In particular, for $p = 1, 2, \infty$,

(1) $||x(k)||_1 = \sum_{j=1}^{n} |x(k)_j|$;
(2) $||x(k)||_2 = \sqrt{\sum_{j=1}^{n} |x(k)_j|^2}$;
(3) $||x(k)||_\infty = \max_{1 \leq j \leq n} |x(k)_j|$.

Of significant interest is the *Euclidean norm* of a vector $x = [x_1, \ ..., \ x_n]^t$ defined by

$$||x||_2 = \sqrt{x^t x} = \sqrt{[\sum_{j=1}^{n} x_j^2]}.$$

The space $\Re^n$, equipped with this norm, is called a *Euclidean space*. Two important results for the Euclidean norm are the following.

Proposition 1.1 (Pythagorean theorem). *For any two vectors x and y that are orthogonal, we have*

$$||x + y||^2 = ||x||^2 + ||y||^2.$$

Proposition 1.2 (Schwartz inequality). *For any two vectors x and y that are orthogonal, we have*

$$|x^y| \leq ||x||||y||$$

with equality holding if and only if $x = \alpha y$ *for some scalar* α.

1.2.4 Matrix norms

We now introduce the notion of a *norm* for matrices. An $m \times n$ matrix may be viewed as an *operator* on the (finite-dimensional) normed vector space $\mathcal{R}^n$:

$$A^{m \times n} : (\mathcal{R}^m; ||.||) \rightarrow (\mathcal{C}^n; ||.||).$$

Building upon the previous part of vector norms, we define the matrix norm induced by a vector p-norm as follows:

$$||A||_p := \sup_{x \neq 0} \frac{||Ax||_p}{||x||_p}.$$

In particular, the corresponding induced matrix norm for $p = 1, 2, \infty$ can be computed as follows:

(a) $||A||_1 = \max_{1 \leq j \leq n} \sum_{i=1}^{m} |a_{ij}|$ (column sum);
(b) $||A||_2 = \sqrt{\lambda_M (A^t \ A)}$;
(c) $||x(k)||_\infty = \max_{1 \leq i \leq m} \sum_{j=1}^{n} |a_{ij}|$ (row sum);
(d) $||A||_F = \sqrt{trace(A^t \ A)} = \sqrt{\sum_{i=1}^{m} \sum_{j=1}^{n} |a_{ij}|^2}$ (Frobenius norm).

Of particular interest for most of the norms are the following results:

(i) $||AB|| \leq ||A|| \ ||B||$;
(ii) $||A^{-1}|| \geq ||A||^{-1}$;
(iii) $||AB||_F \leq ||A|| \ ||B||_F$ and $||AB||_F \leq ||B|| \ ||A||_F$.

1.2.5 Singular value decomposition

The singular value decomposition (SVD) of a matrix is now presented. The SVD exposes the 2-norm of a matrix, but its value to us goes much further: It enables the solution of a class of matrix perturbation problems that form the basis for the stability robustness concepts introduced later; it solves the so-called total least squares problem, which is a generalization of the least squares estimation problem considered earlier; and it allows us to clarify the notion of conditioning, in the context of matrix inversion. These applications of the SVD are presented at greater length in the next lecture.

Let $A \in \Re^{m \times n}$. There exist unitary matrices

$$U = [u_1, u_2, ..., u_m] \in \Re^{m \times m},$$
$$V = [v_1, v_2, ..., v_n] \in \Re^{n \times n}$$

such that

$$A = U\ \Sigma\ V^t, \quad \Sigma = \begin{bmatrix} \Sigma_1 & 0 \\ 0 & 0 \end{bmatrix},$$

$$\Sigma_1 = \begin{bmatrix} \sigma_1 & 0 & \cdots & 0 \\ 0 & \sigma_1 & \cdots & 0 \\ \vdots & \vdots & \ddots & \vdots \\ 0 & 0 & \cdots & 0 \end{bmatrix},$$

$$\sigma_1 \geq \sigma_2 \geq \cdots \sigma_p \geq 0, \quad p = \min\{m, n\}.$$

It is significant to observe that the singular values $\sigma_1, \sigma_2, \cdots \sigma_p$ are good measures of the "size" of a matrix. Additionally, singular vectors are good indicators of strong/weak input or output directions. Note that

$$A\ v_i = \sigma_{1_i}\ u_i,$$
$$A^t\ u_i = \sigma_{1_i}\ v_i,$$
$$A^t A\ v_i = \sigma_{1_i}^2\ u_i,$$
$$AA^t\ u_i = \sigma_{1_i}^2\ v_i$$

and

$$\overline{\sigma}(A) = \sigma\, 1_M(A) = \sigma_1 = \text{the largest singular value of } A,$$
$$\underline{\sigma}(A) = \sigma\, 1_m(A) = \sigma_p = \text{the smallest singular value of } A.$$

Geometrically, the singular values of a matrix A are precisely the lengths of the semiaxes of the hyperellipsoid E defined by

$$E = \{y\ :\ y = A\,x,\ x \in \mathcal{C},\ ||x|| = 1\}.$$

Thus v_1 is the direction in which $||y||$ is the largest for all $||x|| = 1$, while v_n is the direction in which $||y||$ is the smallest for all $||x|| = 1$. Moreover,

- $v_1(v_n)$ is the highest (lowest) gain input direction;
- $u_1(u_m)$ is the highest (lowest) gain observing direction.

This can be expressed as

$$A = \begin{bmatrix} \cos\theta_1 & -\sin\theta_1 \\ \sin\theta_1 & \cos\theta_1 \end{bmatrix} \begin{bmatrix} \sigma_1 & 0 \\ 0 & \sigma_2 \end{bmatrix} \begin{bmatrix} \cos\theta_2 & -\sin\theta_2 \\ \sin\theta_2 & \cos\theta_2 \end{bmatrix}.$$

Thus A maps a unit disk to an ellipsoid with semiaxes σ_1 and σ_2. Alternatively,

$$\overline{\sigma}(A) = \max_{||x||=1} ||Ax||,$$

$$\underline{\sigma}(A) = \min_{||x||=1} ||Ax||.$$

Suppose that A and Δ are square matrices. Then

(1) $|\underline{\sigma}(A+\Delta) - \underline{\sigma}(A)| \leq \overline{\sigma}(A)$;
(2) $\underline{\sigma}(A\ \Delta) \geq \underline{\sigma}(A)\underline{\sigma}(\Delta)$;
(3) $\overline{\sigma}(A^{-2}) = \frac{1}{\underline{\sigma}(A)}$ if A is invertible.

Let $A \in \Re^{m\times n}$ along with

$$\sigma_1 \leq \sigma_2 \leq \cdots \leq \sigma_r > \sigma_{r+1} = \cdots = 0, \ \ r \leq \min\{m,\ n\}.$$

Then, some useful properties are inferred as follows:

(a) $\mathbf{rank}(A) = r$;
(b) $\mathbf{Ker}(A) = \mathbf{span}\{v_{r+1}, \cdots, v_n\}$ and $\mathbf{Ker}(A)^\dagger = \mathbf{span}\{v_1, \cdots, v_r\}$;
(c) $A \in \Re^{m\times n}$ has a dyadic expansion

$$A = \sum_{i=1}^{r} \sigma_i u_i v_i^t = U - r\ \Sigma_r\ V_r^t,$$

where $U_r = [u_1,\ u_2, ..., u_r]$, $V_r = [v_1,\ v_2, ..., v_n]$, and $\Sigma_r = diag(\sigma_1,\ \sigma_2, ..., \sigma_r)$;

(d) $||A||_F = \sqrt{\sigma_1^2 + \sigma_2^2 + \cdots + \sigma_r^2}$;
(e) $||A|| = \sigma_1$;
(f) $\sigma_j(U_o A V_o) = \sigma_j(A)$, $j = 1, \cdots, p$, for any appropriately dimensional unitary matrices U_o and V_o.

1.2.6 System norms

Consider the system with dynamics $\mathbf{y} = \mathbb{T}(z)\mathbf{u}$, where $\mathbf{u} \in \Re^m$ and $\mathbf{y} \in \Re^m$ are the input and output, respectively, and $\mathbb{T}(z)$ is the transfer function matrix in the $\mathbb{Z}$-domain. Without loss of generality, we assume that $\mathbb{T}(z)$ is a stable linear time-invariant matrix. Let $\mathbf{t}(k)$ be the associated impulse response matrix and hence the pair $\mathbb{T}$, $\mathbf{t}(k)$ is a Z-transform pair.

As in the matrix case, the measure on a system should be induced from the space of signals it operates on. Thus, the size of a system is best measured by the maximum amplification it exerts on a set of signals with unit norm.

The most common system norms are $\mathbb{H}_2$ and $\mathbb{H}_\infty$. In general, $\mathbb{H}_\infty$ is concerned primarily with the *peaks* in the frequency response, and the $\mathbb{H}_2$-norm is concerned with *the overall* response ($\sum_j \sigma_j^2$) over all frequencies.

In the sequel, we summarize some of the key points concerning these norms:

(A) One standard definition of the $\mathbb{H}_2$-norm is given by

$$||\mathbb{T}||_2 = \sqrt{\left(\frac{1}{2\pi}\int_0^{2\pi} \mathbf{trace}[\mathbb{T}(e^{j\theta})]^*[\mathbb{T}(e^{j\theta})]\, d\theta\right)}.$$

For the single-input single-output (SISO) case, $||\mathbb{T}||_2$ is interpreted as the energy in the output $y(k)$ for a unit impulse input $u(k)$.

For the multiinput multioutput (MIMO) case, one can apply an impulsive input separately to each actuator and measure the response z_i; then

$$||\mathbb{T}||_2^2 = \sum_i ||z_i||_2^2.$$

(B) One commonly used definition of the $\mathbb{H}_\infty$-norm is given by

$$||\mathbb{T}||_\infty = \sup_{\theta\in[0,2\pi]} \sigma_{\max}[\mathbb{T}(e^{j\theta})].$$

An interpretation of $||\mathbb{T}||_\infty$ is the "energy gain" from the input $u(k)$ to output $y(k)$,

$$||\mathbb{T}||_\infty = \max_{u(k)\neq 0} \frac{\mathbf{y}^t(k)\mathbf{y}(k)}{\mathbf{u}^t(k)\mathbf{u}(k)}.$$

Using a "worst case" input signal that is essentially a sinusoid at frequency ω^*, the system achieves the maximum gain with input direction that yields $\overline{\sigma}[\mathbb{T}]$ as the amplification, where $*$ stands for the complex-conjugate transpose. One key feature is that the $||\mathbb{T}||_\infty$-norm satisfies the property $||\mathbb{T}\mathbb{S}||_\infty \leq ||\mathbb{T}||_\infty ||\mathbb{S}||_\infty$. However, the same property does not hold for $||\mathbb{T}||_2$.

1.2.7 Brief overview of matrix theory

Throughout the different chapters, we denote by $\Re$ the field of real numbers and by $\Re^{m\times n}$ the space of real $m \times n$ matrices. Let $\Re_+$ stand for the nonnegative real numbers. Let $\mathbb{C}$ be the field of complex numbers and denote $\widehat{\mathbb{C}} = \{s \in \mathbb{C} : \Re(s) < 0\}$, $\mathbf{C} = \{s \in \mathbb{C} : \mathbf{Re}(s) \leq 0\}$.

The transpose of a vector x and a matrix M will be denoted by x^T and M^T, respectively. A matrix $M \in \Re^{n \times n}$ is symmetric if $M = M^T$.

1. Let $M \in \Re^{m \times n}$. Then $M[i : j, k : l]$ denotes a matrix of dimension $(j - i + 1) \times (l - k + 1)$ obtained by extracting rows i to j and columns k to l from the matrix M, with $m \geq j \geq i \geq 1, n \geq k \geq l \geq 1$.
2. The identity matrix of dimension m is denoted by I_m, where $I_m \in \Re^{m \times m}$.
3. Let $\lambda_i(M)$ denote the ith eigenvalue of $M \in \Re^{n \times n}$, $i = 1, \ldots, n$. The spectrum of M will be denoted by $\mathbb{S}(M) = \{\lambda_1(M), \ldots, \lambda_n(M)\}$.

Definition 1.7. Let matrices $A, B \in \Re^{n \times n}$ be symmetric.

1. Matrix A is positive definite if $x^T A x > 0$ for all nonzero $x \in \Re^n$, and A is positive semidefinite if $x^T A x \geq 0$ for all nonzero $x \in \Re^n$. We denote this by $A > 0$ and $A \geq 0$, respectively.
2. Matrix A is negative (semi)definite if $-A$ is positive (semi)definite.
3. Inequalities $A < B$ and $A \leq B$ mean $A - B < 0$ and $A - B \leq 0$, respectively.

Definition 1.8. A matrix $M \in \Re^{n \times n}$ is called Hurwitz (or stable) if all its eigenvalues have a negative real part, that is, $\lambda_i(M) \in \widehat{\mathbb{C}}$, $i = 1, \ldots, n$.

1.2.8 Estimation of spectrum of complex matrices

Generally speaking, there is no analytical formula for calculating eigenvalues of matrices. The following lemmas can be used for the estimation of the spectrum of $A \in \mathbf{C}^{n \times n}$ denoted by $\mathbf{spec}(A) \triangleq \{\lambda_1(A), \cdots, \lambda_n(A)\}$.

The first lemma is usually called the *Gershgorin disc lemma* [2].

Lemma 1.1. *For any $A = [a_{ij}] \in \mathbf{C}^{n \times n}$ we have*

$$\mathbf{spec}(A) \subset \bigcup_{j=1}^{n} D_i, \tag{1.3}$$

$$D_i = \{z \in \mathbf{C} : |z - a_{ii}| \leq \sum_{j=1, j \neq i}^{n} |a_{ij}|\}. \tag{1.4}$$

Observe that the eigenvalues of A are also in the union of the discs:

$$|z - a_{jj}| \leq \sum_{i=1, i \neq j}^{n} |a_{ij}|, \quad j = 1, \cdots, n. \tag{1.5}$$

From (1.4) it follows that $|z| \leq \sum_{j=1}^{n} |a_{ij}|$, which implies that any complex number in the disk D_i has an absolute value less than or equal to the sum of the absolute values of the entries of the same row.

1.2.9 Spectral radius

Given a square matrix $A \in \Re^{n \times n}$, the *spectral radius* of A is the maximum norm of the eigenvalues of A, that is,

$$\varrho(A) := \max\{|\lambda| : \lambda \in \mathbb{S}(A)\},$$

or, equivalently, the radius of the smallest disk in $\mathbb{C}$ centered at the origin and containing the spectrum of A.

The following facts hold:

(i) Matrix A is *convergent* if and only if $\varrho(A) < 1$.
(ii) Matrix A is *semiconvergent* if and only if

- $\varrho(A) < 1$, or
- $\mathbf{1}$ is a semisimple eigenvalue and all other eigenvalues have magnitude less than $\mathbf{1}$.

At this stage, we can rephrase the *Gershgorin disc lemma* in terms of the spectral norm $\varrho(A)$ of matrix A.

Corollary 1.1. *For any matrix $A \in \mathbb{C}^{n \times n}$ we have*

$$\varrho(A) \leq \max_{i \in 1, \cdots, n} \left\{ \sum_{j=1}^{n} |a_{ij}| \right\}. \tag{1.6}$$

Defining

$$\mathbf{H} = diag\{h_j,\ j \in j \in 1, \cdots, n\},\ \ h_j > 0, \tag{1.7}$$

it follows that $\varrho(A) = \varrho(\mathbf{H}A\mathbf{H}^{-1})$. Based on this basic fact, we present the following corollary.

Corollary 1.2. *For any matrix $A \in \mathbb{C}^{n \times n}$ we have*

$$\varrho(A) \leq \min_{h_1, \ldots, h_n > 0} \max_{i \in 1, \cdots, n} \left\{ \sum_{j=1}^{n} \frac{h_i}{h_j} |a_{ij}| \right\}. \tag{1.8}$$

Next, we introduce $\xi \in \mathbb{C}^{n \times 1}$, $\xi^* \xi = 1$. Then, for any eigenvalue λ_i of matrix A, we have $\lambda_i = \lambda_i \xi^* \xi = \xi^* A \xi$. By defining the field of values of A as [2]

$$\mathbf{F}(A) = \{xi^* A \xi : \xi^* \xi = 1\}, \tag{1.9}$$

the next lemma follows.

Lemma 1.2. *For any matrix $A \in \mathbb{C}^{n \times n}$ we have*

$$\varrho(A) \subset \mathbf{F}(A). \tag{1.10}$$

1.2.10 Row-stochastic matrices

Encouraged by the averaging model introduced in the foregoing section, we consider a class of discrete-time linear systems defined by matrices with special properties. Typically, we are looking at matrices with nonnegative entries and whose row-sums are all equal to 1. The square matrix $A \in \Re^{n\times n}$ is

(i) nonnegative (respectively positive) if $a_{ij} \geq 0$ (respectively $a_{ij} > 0$) for all $i, j \in \{1, ..., n\}$;
(ii) row-stochastic if nonnegative and $A\mathbf{1}_n = \mathbf{1}_n$;
(iii) column-stochastic if nonnegative and $A^T\mathbf{1}_n = \mathbf{1}_n$; and
(iv) doubly stochastic if it is row- and column-stochastic.

For a row-stochastic matrix A, the following facts hold:

(i) $\mathbf{1}$ is an eigenvalue, and
(ii) $\mathbb{S}(A)$ is a subset of the unit disk and $\varrho(A) = 1$.

It is readily evident that because $\mathbf{1}$ is an eigenvalue of each row-stochastic matrix A, clearly A is not convergent. But it is possible for A to be semiconvergent.

1.2.11 Products of stochastic matrices

One problem studied which is of particular relevance here is to describe the asymptotic behavior of products of $n\times n$ stochastic matrices of the form

$$\mathbf{S}(j),\ \mathbf{S}(j-1),\ \mathbf{S}(j-2),\ \cdots,\ \mathbf{S}(1),\ \ j \to \infty.$$

This is equivalent to looking at the asymptotic behavior of all solutions to the recursion equation

$$x(j+1) = \mathbf{S}_j\, x(j), \tag{1.11}$$

since any solution $x(j)$ can be written as

$$x(j) = (\mathbf{S}(j)\,\mathbf{S}(j-1)\ \cdots\ \mathbf{S}(1)),\ \ j \geq 1. \tag{1.12}$$

One especially useful idea, which has been extensively used [27], is to consider the behavior of the scalar-valued nonnegative function $V(x) = x_L - x_S$ along solutions to (1.11), where $x = [x_1,\ x_2,\ \cdots,\ x_n]^t$ is a nonnegative n vector and x_L and x_S are its largest and smallest elements, respectively. The key observation is that for any stochastic matrix $\mathbf{S} \in \Re^{n\times n}$, the ith entry of $\mathbf{S}x$ satisfies

$$\sum_{j=1}^{n} \mathrm{s}_{ij} x_j \geq \sum_{j=1}^{n} \mathrm{s}_{ij} x_S = x_S,$$

$$\sum_{j=1}^{n} \mathrm{s}_{ij} x_j \leq \sum_{j=1}^{n} \mathrm{s}_{ij} x_L = x_L.$$

Since these inequalities hold for all rows of $\mathbf{S}x$, it must be true that $[\mathbf{S}x]_S \geq x_S$, that $[\mathbf{S}x]_L \leq x_L$, and, as a consequence, that $V(\mathbf{S}x) \leq V(x)$. These inequalities and (1.11) imply that the sequences

$$x_S(1),\ x_S(2),\ \cdots,\ x_L(1),\ x_L(2),\ \cdots V(x(1)),\ V(x(2)),\ \cdots$$

are each monotone. Thus because each of these sequences is also bounded, the limits

$$\lim_{j\to\infty} x_S(j),\ \cdots,\ \lim_{j\to\infty} x_L(j),\ \cdots \lim_{j\to\infty} V(x(j))$$

all exist. Note that whenever the limit of $V(x(j))$ is zero, all components of $x(j)$ together with $x_S(j)$ and $x_L(j)$ must tend to the same constant value.

1.2.12 Kronecker products and vec

Let $A \in \Re^{m\times n}$, $B \in \Re^{p\times q}$. Then $A \otimes B$ denotes the *Kronecker product* of A and B, given by

$$A \otimes B = \begin{bmatrix} a_{11}B & \cdots & a_{1n}B \\ \vdots & \ddots & \vdots \\ a_{m1}B & \cdots & a_{mn}B \end{bmatrix} \in \Re^{mp\times nq}. \tag{1.13}$$

Some important results on the Kronecker product are presented hereafter.

Theorem 1.1. *[2]: Let $A \in \Re^{m\times n}$ and $B \in \Re^{p\times q}$ each have an SVD of $A = U_A\ \Sigma_A\ V_A^t$ and $B = U_B\ \Sigma_B\ V_B^t$. The SVD of the Kronecker product of A and B is then*

$$A \otimes B = (U_A \otimes U_B)(\Sigma_A \otimes \Sigma_B)(V_A^t \otimes V_B^t).$$

Corollary 1.3. *An immediate consequence of Theorem 1.1 is the following result on the matrix 2-norm, $||A \otimes B||_2 = ||A||_2||B||_2$.*

In case $A \in \Re^{n\times n}$ and $B \in \Re^{n\times n}$, the set of eigenvalues of $C = A \otimes B$ is given by $\lambda_j(A)\lambda_k(B),\ \forall j,k$.

Some properties of Kronecker products are:

(A) for square nonsingular matrices, $(A \otimes B)^{-1} = A^{-1} \otimes B^{-1}$;
(B) $(A \otimes I_n)(I_m \otimes B) = (A \otimes B) = (I_m \otimes B)(A \otimes I_n)$;
(C) $A \otimes (B \otimes C) = (A \otimes B) \otimes C$;
(D) $A \otimes (B + C) = (A \otimes B) + (A \otimes C)$;
(E) $(A \otimes B)(D \otimes E) = (AD \otimes BE)$;
(F) $(A \otimes B)^T = A^T \otimes B^T$;
(G) $(A + B) \otimes C = (A \otimes C) + (B \otimes C)$;
(H) $\alpha \otimes A = (A \otimes \alpha) = \alpha A$, for any scalar α;
(I) $\alpha A \otimes \beta B = \alpha\beta A \otimes B$ for any scalars α, β;

(J) for partitioned matrices, $[A_1,\ A_2] \otimes B = [A_1 \otimes B;\ A_2 \otimes B]$, but $A \otimes [B_1,\ B_2] \neq [A \otimes B_1;\ A \otimes B_2]$;
(K) for matrices $A \in \Re^{n\times n},\ B \in \Re^{m\times m}$, $|A \otimes B| = |A|^n |B|^m$;
(L) $\mathbf{trace}(A \otimes B) = \mathbf{trace}(A) \otimes \mathbf{trace}(B)$;
(M) $\mathbf{rank}(A \otimes B) = \mathbf{rank}(A) \otimes \mathbf{rank}(B)$.

Proceeding further, consider the matrix $A \in \Re^{m\times n}$. The column mn-vector, obtained by stacking column 2 of A after column 1, column 3 after column 2, and so forth, is termed **vec** A.

For M and N matrices for which the product MN can be formed, it turns out that

$$\begin{aligned}\mathbf{vec}(MN) &= [I \otimes M]\,\mathbf{vec}(N)\\ &= [N^t \otimes N]\,\mathbf{vec}(M).\end{aligned}$$

Proposition 1.3. *Consider two matrices $A = \alpha \mathbf{I}_n$ and $B \in \Re^{n\times n}$. Then $\lambda_i(A+B) = \alpha + \lambda_i(B)$, $i = 1, \ldots, n$.*

Proof. Take any eigenvalue $\lambda_i(B)$ and the corresponding eigenvector $v_i \in \mathbb{C}^n$. Then $(A+B)v_i = Av_i + Bv_i = \alpha v_i + \lambda_i v_i = (\alpha + \lambda_i)v_i$. □

Proposition 1.4. *Given $A, C \in \Re^{m\times m}$ and $B \in \Re^{n\times n}$, consider two matrices $\bar{A} = I_n \otimes A$ and $\bar{C} = B \otimes C$, where $\bar{A}, \bar{C} \in \Re^{nm\times nm}$. Then $\mathbb{S}(A + \lambda_i(B)C)$, where $\lambda_i(B)$ is the ith eigenvalue of B.*

Proof. Let $v \in \mathbb{C}^n$ be an eigenvector of B corresponding to $\lambda(B)$, and let $u \in \mathbb{C}^m$ be an eigenvector of $M = (A + \lambda(B)C)$ with $\lambda(M)$ as the associated eigenvalue. Consider the vector $v \otimes u \in \mathbb{C}^{nm}$. Then

$$\begin{aligned}(\bar{A} + \bar{C})(v \otimes u) &= v \otimes Au + Bv \otimes Cu\\ &= v \otimes Au + \lambda(B)v \otimes Cu\\ &= v \otimes (Au + \lambda(B)Cu).\end{aligned}$$

Since $(A + \lambda(B)C)u = \lambda(M)u$, we get $(\bar{A} + \bar{C})(v \otimes u) = \lambda M(v \otimes u)$. □

Some properties of the relationships of vector-operators and Kronecker products are:

(1) $\mathbf{vec}(A\ X\ B) = (B^t \otimes A)\mathbf{vec}(X)$;
(2) $\mathbf{vec}(A\ B) = (I \otimes A)\mathbf{vec}(B) = (B^t \otimes I)\mathbf{vec}(A)$;
(3) $\mathbf{trace}(A\ B\ C) = \mathbf{vec}(A^t)^t (I \otimes B)\mathbf{vec}(C)$;
(4) $\mathbf{trace}(A\ B) = \mathbf{vec}(A^t)^t \mathbf{vec}(B)$;
(5) $\mathbf{vec}(\alpha\beta^t) = (\beta \otimes \alpha)$, for any vectors α and β;
(6) $\mathbf{trace}(A^t\ B\ C\ D^t) = \mathbf{vec}(A)^t (D \otimes B)\mathbf{vec}(C)$.

In the sequel, **Ker(M)** will be used to denote the orthogonal complement of **M**. A block-diagonal matrix with submatrices $\mathbf{X}_1, \mathbf{X}_2, \ldots, \mathbf{X}_p$ on its diagonal will be denoted by diag$\{\mathbf{X}_1, \mathbf{X}_2, \ldots, \mathbf{X}_p\}$. We use $\mathbf{S}^{n\times n}$ to denote the real, symmetric $n \times n$ matrices, and $\mathbf{S}_{+}^{n\times n}$ to denote positive definite matrices. If $\mathbf{M} \in \mathbf{S}^{n\times n}$, then $\mathbf{M} > 0$ ($M \geq 0$) indicates that $\mathbf{M}$ is positive definite (positive semidefinite), and $\mathbf{M} < 0$ $M \leq 0$ denotes a negative definite (negative semidefinite) matrix. Given a symmetric matrix $\mathbb{A}$, its inertia is defined as $\text{In}(\mathbb{A}) = (\pi(\mathbb{A}), \nu(\mathbb{A}), \delta(\mathbb{A}))$, i.e., the numbers of positive, negative, and zero eigenvalues of $\mathbb{A}$. For a temporal and spatial variable $x(t,s)$, $\mathbf{T}$ and $\mathbf{S}$ are the temporal and spatial forward shift operators, defined as

$$\mathbf{Tx}(t,s) = \mathbf{x}(t+1,s), \quad \mathbf{Sx}(t,s) = \mathbf{x}(t,s+1).$$

With slight abuse of notation, the symbols $\mathbf{T}$ and $\mathbf{S}$ are also used to denote the time and spatial differential operators for continuous interconnected systems. The space of square summable functions is denoted by $\mathbb{L}_2$, that is, for any $u \in \mathbb{L}_2$, $\|u\|_2 := [\sum_{k=0}^{\infty}\sum_{s=-\infty}^{\infty} \mathbf{u}^T(k,s)\mathbf{u}(k,s)]^{\frac{1}{2}}$ is finite.

1.2.13 Matrix lemmas

The first lemma relates the inertia of a nonsingular symmetric matrix $\mathbf{N}$ on a subspace with the inertia of its inverse $\mathbf{N}^{-1}$ on the complementary subspace. It can be utilized to derive the dual formulation from its original test. For its proof, readers are referred to [3].

Lemma 1.3. *Suppose that N is symmetric and nonsingular, and $\mathbb{S}$ is a negative subspace of $\mathbf{N}$ with maximal dimension. Then*

$$N^{-1} > 0 \; on \; \mathbb{S}_{\perp}.$$

The second result is basically an extension of the well-known elimination lemma [4,5] to quadratic matrix inequality. It is convenient for eliminating controller parameters from the synthesis conditions. Its proof closely follows the one in [5,6], and is omitted here for lack of space.

Lemma 1.4. *Assume a nonsingular symmetric matrix*

$$\begin{bmatrix} \mathbf{Q} & \mathbf{S} \\ \bullet & \mathbf{R} \end{bmatrix}$$

has its inverse

$$\begin{bmatrix} \tilde{\mathbf{Q}} & \tilde{\mathbf{S}} \\ \bullet & \tilde{\mathbf{R}} \end{bmatrix}$$

with $\text{In}(\tilde{\mathbf{Q}}) = \text{In}(-\mathbf{R})$ *and* $\delta(\mathbf{R}) = 0$. *The quadratic inequality*

$$\begin{bmatrix} \mathbf{I} & \mathbf{B}^T\mathbf{X}^T\mathbb{A} + \mathbf{C}^T \end{bmatrix} \begin{bmatrix} \mathbf{Q} & \mathbf{S} \\ \bullet & \mathbf{R} \end{bmatrix} \begin{bmatrix} \mathbf{I} \\ \mathbb{A}^T\mathbf{XB} + \mathbf{C} \end{bmatrix} < 0 \tag{1.14}$$

has a solution $\mathbf{X}$ *if and only if*

$$[\ \mathbf{B}_{\perp}^T \quad \mathbf{B}_{\perp}^T \mathbf{C}^T\ ] \begin{bmatrix} \mathbf{Q} & \mathbf{S} \\ \bullet & \mathbf{R} \end{bmatrix} \begin{bmatrix} \mathbf{B}_{\perp} \\ \mathbf{C}\mathbf{B}_{\perp} \end{bmatrix} < 0 \tag{1.15}$$

and

$$[\ -\mathbb{A}_{\perp}^T \mathbf{C} \quad \mathbb{A}_{\perp}^T\ ] \begin{bmatrix} \tilde{\mathbf{Q}} & \tilde{\mathbf{S}} \\ \tilde{\mathbf{S}}^T & \tilde{\mathbf{R}} \end{bmatrix} \begin{bmatrix} -\mathbf{C}^T \mathbb{A}_{\perp} \\ \mathbb{A}_{\perp} \end{bmatrix} > 0. \tag{1.16}$$

Finally, a well-known result in matrix theory is the *matrix inversion lemma*, summarized below.

Let $A \in \Re^{n \times n}$ and $C \in \Re^{m \times m}$ be nonsingular matrices. By using the definition of matrix inverse, it can be easily verified that

$$[A + B\ C\ D]^{-1} = A^{-1} - A^{-1}\ B\ [D\ A^{-1} B + C^{-1}]^{-1}\ D A^{-1}. \tag{1.17}$$

1.3 Elements of algebraic graphs

Graph theory plays a crucial role in describing the interconnection topology of multiagent systems. In this section, we only present basic definitions, concepts, and results about graph theory. For a systematic study of graph theory, the reader is referred to [2], [7], [8].

1.3.1 Graph theory

A directed graph (in short, a digraph) $\mathbb{G} = (\mathbb{V}; \mathbb{E}; \mathbb{A})$ of order N is composed of a *vertex set* $\mathbb{V} = \{v_1,\ v_2,\ ...,\ v_N\}$, a set of ordered pairs of vertices called *edges* $\mathbb{E} = \{e_{ij} = (v_i,\ v_j)\} \subseteq \mathbb{V} \times \mathbb{V}$, and a weighted adjacency matrix $\mathbb{A} = [a_{ij}]$ with nonnegative adjacent elements a_{ij}. For emphasis, we denote by $\mathbb{V}(\mathbb{G})$ and $\mathbb{E}(\mathbb{G})$ the vertex set and edge set of graph $\mathbb{G}$, respectively. The node indices belong to a finite index set $\mathbb{I} = \{1, 2, \ldots, n\}$. Moreover, $a_{ij} > 0$ if $(v_i,\ v_j) \in \mathbb{E}$ and $a_{ij} = 0$ if $(v_i,\ v_j) \notin \mathbb{E}$ for all $i = 1, ..., n$. Also, $(v_i,\ v_j) \in \mathbb{E}$ if and only if the ith agent can receive information from the jth agent directly. If a directed graph has the property that $a_{ij} = a_{ji}$ for any $i, j \in \mathbb{I}$, the directed graph is called undirected. In addition, $\mathbb{N} = \{v_j \in \mathbb{V} : (v_i,\ v_j) \in \mathbb{E}\}$ is defined as the set of neighbors of node v_i.

An undirected graph (in short, a graph) consists of a set $\mathbb{V}$ of nodes and of a set $\mathbb{E}$ of unordered pairs of nodes, called edges. For $u;\ v \in \mathbb{V}$ and $u \neq v$, the set $\{u,\ v\}$ denotes an unordered edge.

For an edge $(i,\ j)$, node i is called the *parent* node, j is called the *child* node, and j is neighboring to i. A graph with the property that $(i,\ j) \in \mathbb{E}$ implies $(j,\ i) \in \mathbb{E}$ is said to be *undirected*; otherwise, it is called *directed*. A path on $\mathbb{G}$ from node i_1 to node i_ℓ is a sequence of ordered edges of the form $(i_k,\ i_{k+1})$, $k = 1, ..., \ell - 1$. A *directed graph* has or contains a *directed spanning tree* if there exists a node called *root* such that there exists a *directed path* from this node to every other node in the graph.

1.3.2 Undirected graphs

A digraph $\mathbb{G}(\mathbb{V}', \mathbb{E}')$ is said to be a *subgraph* of a digraph $(\mathbb{V}, \mathbb{E})$ if $\mathbb{V}' \subset \mathbb{V}$ and $\mathbb{E}' \subset \mathbb{E}$. In particular, a digraph $\mathbb{G}(\mathbb{V}', \mathbb{E}')$ is said to be a *spanning subgraph* of a digraph $(\mathbb{V}, \mathbb{E})$ if it is a subgraph and $\mathbb{V}' = \mathbb{V}$. The digraph $(\mathbb{V}', \mathbb{E}')$ is the subgraph of $(\mathbb{V}, \mathbb{E})$ induced by $\mathbb{V}' \subset \mathbb{V}$ if $\mathbb{E}'$ contains all edges in $\mathbb{E}$ between two vertices in $\mathbb{V}'$.

An undirected graph (in short, graph) $\mathbb{G}$ consists of a vertex set $\mathbb{V}$ and a set $\mathbb{E}$ of unordered pairs of vertices. If each edge of the graph $\mathbb{G}$ is given a particular orientation, then we get an oriented graph of $\mathbb{G}$, denoted by $\mathbb{G}^{\rightarrow}$, which is a digraph. Denote by $\mathbb{G}^{\leftarrow}$ the reverse of $\mathbb{G}^{\rightarrow}$. Then, $\mathbb{G} = \mathbb{G}^{\rightarrow} \cap \mathbb{G}^{\leftarrow}$.

For an undirected graph $\mathbb{G}$, the in-neighbor set of any vertex is always equal to the out-neighbor set of the same vertex. Therefore, in the undirected case we simply use the terms neighbor, neighbor set, and degree.

For an undirected graph, if it contains a globally reachable node, then any other vertex is also globally reachable. In that case we simply say that the undirected graph is *connected*.

For an undirected graph, it is said to be a *tree* if it is connected and acyclic. The following results hold.

Theorem 1.2. *$\mathbb{G}(\mathbb{V}, \mathbb{E})$ is a tree if and only if $\mathbb{G}$ is connected and $|\mathbb{E}| = |\mathbb{V}| - 1$. Alternatively, $\mathbb{G}(\mathbb{V}, \mathbb{E})$ is a tree if and only if G is acyclic and $|\mathbb{E}| = |\mathbb{V}| - 1$.*

Theorem 1.3. *A graph is connected if and only if it contains a spanning tree.*

1.3.3 Main graphs

In the sequel, we present the following graphs of dimension n of common use in system and network theory [8]:

(a) *Path graph:* nodes are ordered in a sequence and edges connect subsequent nodes in the sequence.
(b) *Cycle (or ring) graph:* all nodes and edges can be arranged as the vertices and edges of a regular polygon.
(c) *Star graph:* edges connect a specific node, called the center, to all other nodes.
(d) *Complete graph:* every pair of nodes is connected by an edge.
(e) *Two-dimensional grid graph:* nodes are ordered in row and column sequences and edges connect subsequent nodes in both horizontal and vertical sequences.
(f) *Petersen graph:* nodes are arranged in a closed, outer pentagon and a closed, inner pentagon.

(g) *Complete bipartite graph:* nodes are divided into two sets and every node of the first set is connected with every node of the second set.

Fig. 1.2 illustrates these graph types. Further details about *bipartite graphs* are now presented.

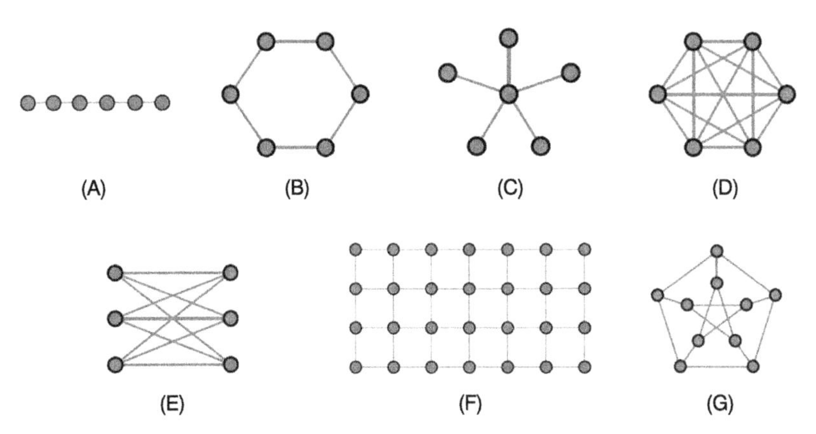

FIGURE 1.2 Graph types: (A) *Path graph* with 6 nodes, denoted by P_6; (B) *Cycle (ring) graph* with 6 nodes, denoted by C_6; (C) *Star graph* with 6 nodes, denoted by S_6; (D) *Complete graph* with 6 nodes, denoted by K_6; (E) *Complete bipartite graph* with $3 + 3$ nodes, denoted by $K_{3,3}$; (F) *Two-dimensional grid graph* with 4×7 nodes, denoted by $G_{4,7}$; (G) *Petersen graph.*

A graph $\mathbb{G}$ is a bipartite graph with vertex classes $\mathbb{V}_1$ and $\mathbb{V}_2$ if $\mathbb{V}(\mathbb{G})$ is a direct sum of $\mathbb{V}_1$ and $\mathbb{V}_2$, $\mathbb{V} = \mathbb{V}_1 \oplus \mathbb{V}_2$, which implies that $\mathbb{V} = \mathbb{V}_1 \cup \mathbb{V}_2$ and $\mathbb{V} = \mathbb{V}_1 \cap \mathbb{V}_2$, and every edge joins a vertex of $\mathbb{V}_1$ to a vertex of $\mathbb{V}_2$. It is also said that $\mathbb{G}$ has bipartition $(\mathbb{V}_1,\ \mathbb{V}_2)$.

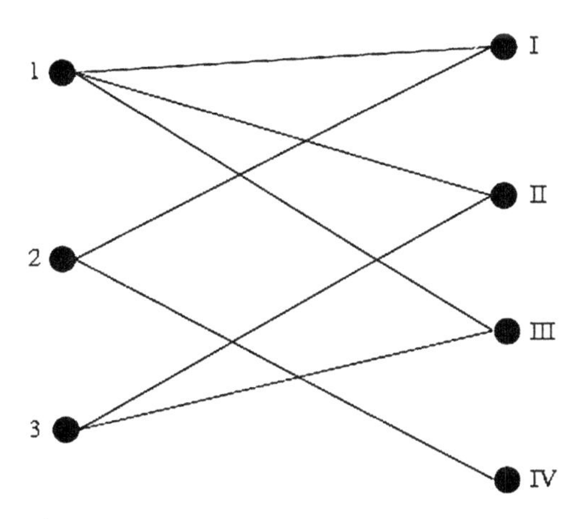

FIGURE 1.3 A bipartite graph.

Fig. 1.3 shows an example of a bipartite graph, where the vertex set $\mathbb{V}$ is a direct sum of $\mathbb{V}_1 = \{v_1, v_2, v_3\}$ and $\mathbb{V}_2 = \{v_I, v_{II}, v_{III}, v_{IV}\}$. Each vertex in $\mathbb{V}_1$ has neighbors only in $\mathbb{V}_2$, and vice versa.

The following result holds.

Theorem 1.4. *A graph is bipartite if and only if it does not contain an odd cycle.*

1.3.4 Graph operations

In a multiagent system, each agent can be considered as a vertex in a digraph, and the information flow between two agents can be regarded as a directed path between the vertices in the digraph. Thus, the interconnection topology of a multiagent system can be described by a digraph. However, differing from the classic signal-flow graph [9], in this book and in many other references on distributed control/multiagent systems, the direction of an edge in the digraph does not mean the direction of an information flow. Let us consider the digraph shown in Fig. 1.4 for instance. We denote by $x_i \in \Re$, $i = 1, ..., 5$, the state of agent i associated with vertex i. The existence of edge e_{ij} implies that agent i gets the state information x_j from agent j. For example, agent 1 gets information from agent 2.

In principle, we can construct new graphs from old ones by graph operations.

For two graphs $\mathbb{G}_1 = (\mathbb{V}_1; \mathbb{E}_1)$ and $\mathbb{G}_2 = (\mathbb{V}_2; \mathbb{E}_2)$, the *intersection* and *union* of $\mathbb{G}_1$ and $\mathbb{G}_2$ are defined by

$$\mathbb{G}_1 \cap \mathbb{G}_2 := (\mathbb{V}_1 \cap \mathbb{V}_2, \ \mathbb{E}_1 \cap \mathbb{E}_2),$$
$$\mathbb{G}_1 \cup \mathbb{G}_2 := (\mathbb{V}_1 \cup \mathbb{V}_2, \ \mathbb{E}_1 \cup \mathbb{E}_2).$$

For a digraph $\mathbb{G} = (\mathbb{V}; \mathbb{E})$, the *reverse digraph* of $\mathbb{G}$ is a pair $rev(\mathbb{G}) = (\mathbb{V}; rev(\mathbb{E}))$, where $rev(\mathbb{E})$ consists of all edges in $\mathbb{E}$ with reversed directions.

If $\mathbb{W} \subset \mathbb{G}(\mathbb{G})$, then $\mathbb{G} - \mathbb{W} = \mathbb{G}[\mathbb{V} \backslash \mathbb{W}]$ is the subgraph of $\mathbb{G}$ obtained by deleting the vertices in $\mathbb{W}$ and all edges incident with them. Obviously, $\mathbb{G} - \mathbb{W}$ is the subgraph of $\mathbb{G}$ induced by $\mathbb{V} \backslash \mathbb{W}$. Similarly, if $\mathbb{E}' \subset \mathbb{E}$, then $\mathbb{G} - \mathbb{E}' = (\mathbb{V}(\mathbb{G}), \ \mathbb{E}(\mathbb{G}) \setminus \mathbb{E}')$. If $\mathbb{W}$ or $(\mathbb{E}')$ contains a single vertex w (or a single edge xy), respectively, the notion is simplified to $\mathbb{G} - w$ or $\mathbb{G} - xy$, respectively. Similarly, if x and y are nonadjacent vertices of $\mathbb{G}$, then $\mathbb{G} + xy$ is obtained from $\mathbb{G}$ by joining x to y.

1.3.5 Basic properties

For a graph $\mathbb{G}$ with m nodes, the row-stochastic matrix $\mathbb{D} \in \Re^{m \times m}$ is defined with $d_{ii} > 0$, $d_{ij} > 0$ if $(j, i) \in \mathbb{E}$ but 0 otherwise, and $\sum_{j=1}^{m} d_{ij} = 1$. It follows from the foregoing consideration that all of the eigenvalues of $\mathbb{D}$ are either in the open unit disk or equal to 1, and furthermore, 1 is a simple eigenvalue of $\mathbb{D}$ if and only if graph $\mathbb{G}$ contains a directed spanning tree. For an undirected graph, $\mathbb{D}$ is symmetric.

Let Γ_m denote the set of all directed graphs with m nodes such that each graph contains a directed spanning tree, and let $\Gamma_{\leq\delta}$ $(0 < \delta < 1)$ denote the set of all directed graphs containing a directed spanning tree, whose eigenvalues different from one lie in the disk of radius δ centered at the origin.

A *path* in a digraph is an ordered sequence of vertices such that any ordered pair of vertices appearing consecutively in the sequence is an edge of the digraph. A path is simple if no vertices appear more than once in it, except possibly for the initial and final vertices. The *length* of a path is defined as the number of consecutive edges in the path. For a simple path, the path length is less than the number of vertices contained in the path by unity.

A vertex v_i in digraph $\mathbb{G}$ is said to be *reachable* from another vertex v_j if there is a path in $\mathbb{G}$ from v_i to v_j. A vertex in the digraph is said to be *globally reachable* if it is reachable from every other vertex in the digraph. A digraph is *strongly connected* if every vertex is globally reachable. In Fig. 1.4, v_1, v_2, v_3, v_4 are globally reachable vertices. But the digraph is not strongly connected because v_3 is unreachable from the other vertices. A *cycle* is a simple path that

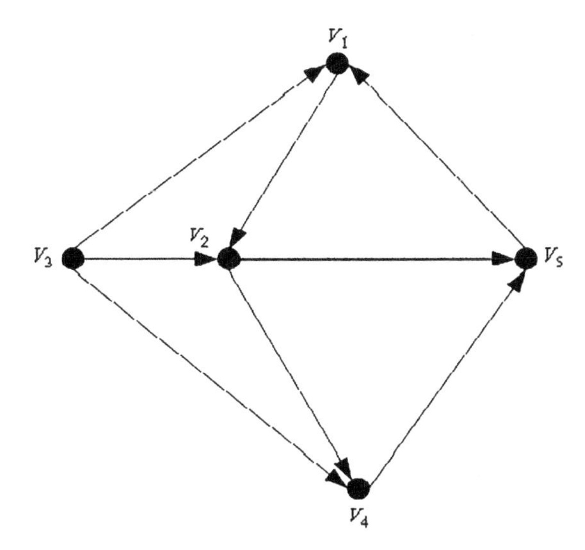

FIGURE 1.4 A digraph.

starts and ends at the same vertex. A cycle containing only one vertex is called a *self-cycle (or self-loop)*. The *length* of a cycle is defined as the number of edges contained in the cycle. A cycle is *odd (even)* if its length is odd (even). If a vertex in a cycle is globally reachable, then any other vertex in the cycle is also globally reachable. In Fig. 1.4, the path (v_1, v_2, v_5, v_1) is a cycle. The path $\{v_2, v_4, v_5, v_2\}$ and the path $\{v_1, v_2, v_4, v_5, v_1\}$ are also cycles. This digraph has no self-cycle. A digraph with self-cycle is shown in Fig. 1.5. A digraph is *acyclic* if it contains no cycles. An acyclic digraph is called a *directed tree* if it satisfies the following property: There exists a vertex, called the *root*, such that

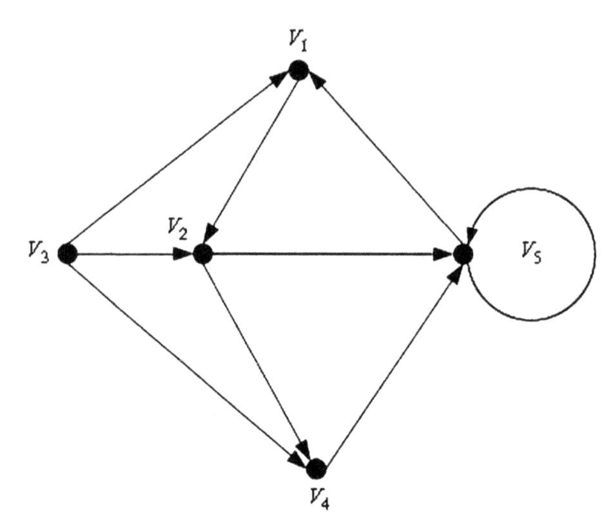

FIGURE 1.5 A digraph with a self-cycle.

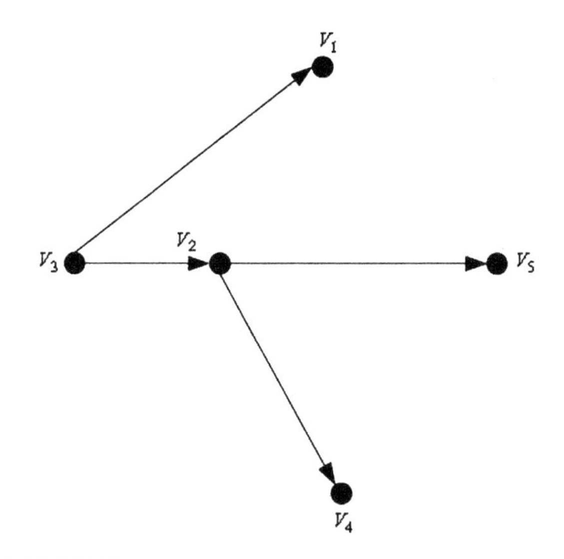

FIGURE 1.6 A directed tree.

any other vertex of the digraph can be reached by one and only one path starting at the root. A *directed spanning tree* of a digraph is a spanning subgraph that is a directed tree.

The digraph shown in Fig. 1.6 is a directed tree. Obviously, it is a directed spanning tree of both the digraph in Fig. 1.4 and the digraph in Fig. 1.5.

The degree matrix $\mathbb{D}(\mathbb{G}) = diag\{d_1, d_2, ..., d_n\}$ is a diagonal matrix, whose diagonal elements are given as $d_i = deg_{out}(v_i)$. A directed path is a sequence of edges in a directed graph of the form (v_1, v_2), (v_2, v_3), ..., $v_i \in \mathbb{V}$. A directed graph contains a directed spanning tree if there exists at least one agent which is called root node that has a directed path to every other agent. The $\mathbb{E}(\mathbb{G})$ defines the incidence matrix for a graph with arbitrary orientation. The columns of $\mathbb{E}(\mathbb{G})$ are indexed by the edges, and the ith row entry takes the value one if it is the initial node of the corresponding edge, negative one if it is the terminal node, and zero otherwise. The adjacency matrix $A(\mathbb{G})$, is the symmetric $|\mathbb{V}| \times |\mathbb{V}|$ matrix with zero on the diagonal and one in the ijth position if node i is adjacent to node j.

The (graph) Laplacian of $\mathbb{G}$ is a rank-deficient, symmetric, and positive semidefinite matrix defined by

$$\begin{aligned}\mathbb{L}(\mathbb{G}) &:= \mathbb{E}(\mathbb{G})\mathbb{E}(\mathbb{G})^t = \Delta(\mathbb{G}) - A(\mathbb{G}) \\ &:= [\ell_{ij}],\ \ell_{ij} = -a_{ij},\ \ell_{ii} = \sum_{j=1}^{n} a_{ij}.\end{aligned}$$

Lemma 1.5 ([10]). *If the graph $\mathbb{G}$ has a spanning tree, then its Laplacian L has the following properties:*

1. *Zero is a simple eigenvalue of $\mathbb{L}$, and $\mathbf{1_n}$ is the corresponding eigenvector, that is, $\mathbb{L}\mathbf{1_n} = \mathbf{0}$.*
2. *The remaining $n-1$ eigenvalues all have positive real parts. In particular, if the graph $\mathbb{G}$ is undirected, then all these eigenvalues are positive and real.*

Lemma 1.6 ([11]). *Consider a directed graph $\mathbb{G}$. Let $\mathbb{D} \in \Re^{n\times|\mathbb{E}|}$ be the 01-matrix with rows and columns indexed by the nodes and edges of $\mathbb{G}$, and let $\mathbb{E} \in \Re^{|\mathbb{E}|\times n}$ be the 01-matrix with rows and columns indexed by the edges and nodes of $\mathbb{G}$, such that*

$$D_{uf} = \begin{cases} 1 & \text{if the node } u \text{ is the tail of the edge } f, \\ 0 & \text{otherwise,} \end{cases}$$

$$E_{fu} = \begin{cases} 1 & \text{if the node } u \text{ is the head of the edge } f, \\ 0 & \text{otherwise,} \end{cases}$$

where $|\mathbb{E}|$ is the number of the edges. Let $\mathbb{Q} = diag\{q_1, q_2, \ldots, q_{|\mathbb{E}|}\}$, where $q_p (p = 1, \ldots, |\mathbb{E}|)$ is the weight of the pth edge of G (i.e., the value of the adjacency matrix on the pth edge). Then the Laplacian of $\mathbb{G}$ can be transformed into $\mathbb{L} = (\mathbb{D}\mathbb{Q}\mathbb{D}^T - \mathbb{E})$.

1.3.6 Connectivity properties of digraphs

There are four useful connectivity notions for a digraph $\mathbb{G}$:

(1) $\mathbb{G}$ is strongly connected if there exists a directed path from any node to any other node;

(2) $\mathbb{G}$ is weakly connected if the undirected version of the digraph is connected;

(3) $\mathbb{G}$ possesses a globally reachable node if one of its nodes can be reached from any other node by traversing a directed path; and

(4) $\mathbb{G}$ possesses a directed spanning tree if one of its nodes is the root of directed paths to every other node.

1.3.7 Properties of adjacency matrices

Note that in case of *undirected* graphs, the adjacency matrix is symmetric. In general, given the adjacency matrix depicted in Fig. 1.7, the adjacency matrix is given by

$$\mathbf{A} = \begin{bmatrix} 0 & 3.7 & 2.6 & 0 & 0 \\ 8.9 & 0 & 0 & 1.2 & 0 \\ 0 & 0 & 0 & 1.9 & 2.3 \\ 0 & 0 & 0 & 0 & 0 \\ 4.4 & 0 & 0 & 2.7 & 4.4 \end{bmatrix}.$$

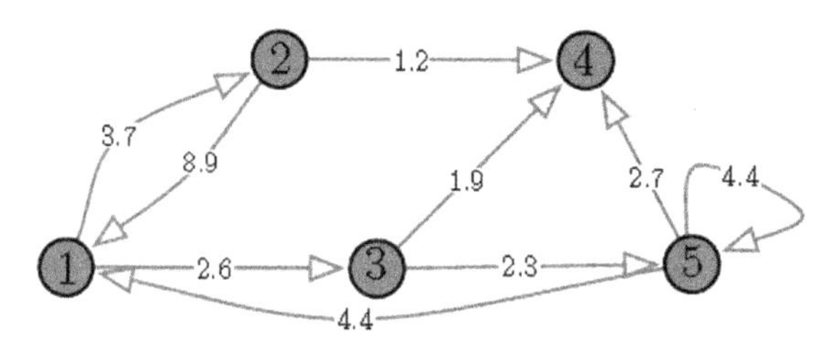

FIGURE 1.7 A weighted graph.

In the sequel, we focus on the case of binary adjacency matrices. We denote $\mathbb{A}(\mathbb{G})$ by the (0, 1) adjacency matrix of the graph $\mathbb{G}$. Let $\mathbb{A}_{ij} \in \Re$ be its i, j element. Then $\mathbb{A}_{i,i} = 0, \forall i = 1, \ldots, N$, $\mathbb{A}_{i,j} = 0$ if $(i, j) \notin \mathcal{A}$ and $\mathbb{A}_{i,j} = 1$ if $(i, j) \in \mathbb{A}, \forall i, j = 1, \ldots, N, i \neq j$. If the in-degree equals the out-degree for all nodes $i \in N$, the graph is said to be *balanced.*

Let $\mathbb{S}(\mathbb{A}(\mathbb{G})) = \{\lambda_1(\mathbb{A}(\mathbb{G})), \ldots, \lambda_N(\mathbb{A}(\mathbb{G}))\}$ be the spectrum of the adjacency matrix associated with an undirected graph $\mathbb{G}$ arranged in nondecreasing semiorder.

1) ***Property 1:*** $\lambda_N(\mathbb{A}(\mathbb{G})) \leq d_{\max}(\mathbb{G})$.

This property together with Proposition 1.1 implies the following.

2) ***Property 2:*** $\gamma_i \geq 0, \forall \gamma_i \in \mathbb{S}(d_{\max} I_N - \mathbb{A})$.

In what follows we let $\mathbb{G}$ be a weighted digraph and $\mathbb{A}$ its weighted adjacency matrix or, equivalently, we let $\mathbb{A}$ be a nonnegative matrix and $\mathbb{G}$ its associated weighted digraph (that is, the digraph with nodes $\{1, ..., n\}$ and with weighted

TABLE 1.1 A table of correspondence.

Digraph $\mathbb{G}$	Nonnegative matrix $\mathbb{A}$ (adjacency of $\mathbb{G}$)
$\mathbb{G}$ is undirected	$\mathbb{A} = \mathbb{A}^t$
$\mathbb{G}$ is weight-balanced	$\mathbb{A}\mathbf{1}_n = \mathbb{A}^t\mathbf{1}_n$, that is, $\mathbb{D}_{out} = \mathbb{D}_{in}$
(no self-loops) node i is a sink	(zero diagonal) ith row-sum of $\mathbb{A}$ is zero
(no self-loops) node i is a source	(zero diagonal) ith column-sum of $\mathbb{A}$ is zero
each node has weighted out-degree equal to 1 ($\mathbb{D}_{out} = I_n$)	$\mathbb{A}$ is a row-stochastic
each node has weighted out-degree equal to 1 ($\mathbb{D}_{in} = \mathbb{D}_{out} = I_n$)	$\mathbb{A}$ is a doubly stochastic

adjacency matrix $\mathbb{A}$). To summarize, we give some straightforward statements, organized as a table of correspondences in Table 1.1.

1.3.8 Laplacian spectrum of graphs

This section is a concise review of the relationship between the eigenvalues of a Laplacian matrix and the topology of the associated graph. We list a collection of properties associated with undirected graph Laplacians and adjacency matrices, which will be used in subsequent sections of the chapter.

A graph $\mathbb{G}$ is defined as

$$\mathbb{G} = (\mathbb{V}, \mathbb{A}), \tag{1.18}$$

where $\mathbb{V}$ is the set of nodes (or vertices) $\mathbb{V} = \{1, \ldots, N\}$ and $\mathbb{A} \subseteq \mathbb{V} \times \mathbb{V}$ is the set of edges (i, j) with $i \in \mathbb{V}$, $j \in \mathbb{V}$. The degree d_j of a graph vertex j is the number of edges which start from j. Let $d_{\max}(\mathbb{G})$ denote the maximum vertex degree of the graph $\mathbb{G}$.

We define the Laplacian matrix of a graph $\mathbb{G}$ in the following way:

$$\mathbb{L}(\mathbb{G}) = \mathbb{D}(\mathbb{G}) - \mathbb{A}(\mathbb{G}), \tag{1.19}$$

where $\mathbb{D}(\mathbb{G})$ is the diagonal matrix of vertex degrees d_i (also called the valence matrix). Eigenvalues of Laplacian matrices have been widely studied by graph theorists. Their properties are strongly related to the structural properties of their associated graphs. Every Laplacian matrix is a singular matrix. By Gershgorin's theorem [13], the real part of each nonzero eigenvalue of $\mathbb{L}(\mathbb{G})$ is strictly positive.

For undirected graphs, $\mathbb{L}(\mathbb{G})$ is a symmetric, positive semidefinite matrix, which has only real eigenvalues. Let $\mathbb{S}(\mathbb{L}(\mathbb{G})) = \{\lambda_1(\mathbb{L}(\mathbb{G})), \ldots, \lambda_N(\mathbb{L}(\mathbb{G}))\}$ be the spectrum of the Laplacian matrix $\mathbb{L}$ associated with an undirected graph $\mathbb{G}$ arranged in nondecreasing semiorder. Then we have the following.

3) ***Property 3:***

1. $\lambda_1(\mathbb{L}(\mathbb{G})) = 0$ with corresponding eigenvector of all ones, and $\lambda_2(L(\mathbb{G}))$ if and only if $\mathbb{G}$ is connected. In fact, the multiplicity of 0 as an eigenvalue of $L(\mathbb{G})$ is equal to the number of connected components of $\mathbb{G}$.
2. The modulus of $\lambda_i(L(\mathbb{G}))$, $i = 1, \ldots, N$, is less than N.

The second smallest Laplacian eigenvalue $\lambda_2(\mathbb{L}(\mathbb{G}))$ of graphs is probably the most important information contained in the spectrum of a graph. This eigenvalue, called the algebraic connectivity of the graph, is related to several important graph invariants, and it has been extensively investigated.

Let $\mathbb{L}(\mathbb{G})$ be the Laplacian of a graph $\mathbb{G}$ with N vertices and with maximal vertex degree $d_{\max}(\mathbb{G})$. Then properties of $\lambda_2(\mathbb{L}(\mathbb{G}))$ include the following.

4) ***Property 4:***

1. $\lambda_2(\mathbb{L}(\mathbb{G})) \leq (N/(N-1)) \min\{d(v), v \in \mathbb{V}\}$;
2. $\lambda_2(\mathbb{L}(\mathbb{G})) \leq v(\mathcal{G}) \leq \eta(\mathbb{G})$;
3. $\lambda_2(\mathbb{L}(\mathbb{G})) \geq 2\eta(\mathcal{G})(1 - \cos(\pi/N))$;
4. $\lambda_2(\mathbb{L}(\mathbb{G})) \geq 2(\cos\frac{\pi}{N} - \cos 2\frac{\pi}{N})\eta(\mathbb{G}) - 2\cos\frac{\pi}{N}(1 - \cos\frac{\pi}{N})d_{\max}(\mathbb{G})$.

Here, $v(\mathbb{G})$ is the vertex connectivity of the graph $\mathbb{G}$ (the size of a smallest set of vertices whose removal renders $\mathbb{G}$ disconnected) and $\eta(\mathbb{G})$ is the edge connectivity of the graph $\mathbb{G}$ (the size of a smallest set of edges whose removal renders $\mathbb{G}$ disconnected) [14].

Further relationships between the graph topology and Laplacian eigenvalue locations are discussed in [12] for undirected graphs. Spectral characterization of Laplacian matrices for directed graphs can be found in [13].

A lemma about Laplacian $\mathbb{L}$ associated with a balanced digraph $\mathbb{G}$ is given hereafter.

Lemma 1.7. *If $\mathbb{G}$ is balanced, then there exists a unitary matrix*

$$\mathbf{V} = \begin{pmatrix} \frac{1}{\sqrt{n}} & * & \ldots & * \\ \frac{1}{\sqrt{n}} & * & \ldots & * \\ \vdots & \vdots & & \vdots \\ \frac{1}{\sqrt{n}} & * & \ldots & * \end{pmatrix} \in \mathbb{C}^{m \times n} \tag{1.20}$$

such that

$$\mathbf{V}^* \mathbb{L} \mathbf{V} = \begin{pmatrix} 0 & \\ & \mathbf{H} \end{pmatrix} = \Lambda \in \mathbb{C}^{n \times n}, \qquad \mathbf{H} \in \mathbb{C}^{(n-1)\times(n-1)}. \tag{1.21}$$

Moreover, if $\mathbb{G}$ has a globally reachable node, $\mathbf{H} + \mathbf{H}^*$ is positive definite.

Proof. Let $\mathbf{V} = [\zeta_1, \zeta_2, \ldots, \zeta_n]$ be a unitary matrix where $\zeta_i \in \mathbb{C}^n$ $(i = 1, \ldots, n)$ are the column vectors of V and

$$\zeta_1 = (1/\sqrt{n})1 = (1/\sqrt{n}, 1/\sqrt{n}, \ldots, 1/\sqrt{n})^T.$$

Note that if G is balanced, this implies that $\zeta_1^* L = 0$. Then we have

$$\begin{aligned}
\mathbf{V}^*\mathbb{L}\mathbf{V} &= \mathbf{V}^*\mathbb{L}[\zeta_1, \zeta_2, \ldots, \zeta_n] \\
&= \begin{pmatrix} \zeta_1^* \\ \zeta_2^* \\ \vdots \\ \zeta_n^* \end{pmatrix} [0_n, \mathbb{L}\zeta_2, \ldots, \mathbb{L}\zeta_n] \\
&= \begin{pmatrix} 0 & 0_{n-2}^T \\ \bullet & \mathbf{H} \end{pmatrix}.
\end{aligned}$$

Furthermore, if G has a globally reachable node, then $\mathbb{L} + \mathbb{L}^t$ is positive semidefinite (see Theorem 7 in [20]). Hence, $\mathbf{V}^*(\mathbb{L} + \mathbb{L}^t)\mathbf{V}$ is also positive semidefinite. From Lemma 1.5, zero is a simple eigenvalue of $\mathbb{L}$ and, therefore, $\mathbf{H} + \mathbf{H}^*$ is positive definite. □

Lemma 1.8. *[15] Suppose that $\mathcal{G}$ is strongly connected. Let $\xi = [\xi_1, \xi_2, \ldots, \xi_N]^T$ be the positive left eigenvector of $\mathbb{L}$ associated with zero eigenvalue. Then, $\Xi\mathbb{L} + \mathbb{L}^t\Xi \geq 0$, where $\Xi = diag(\xi_1, \xi_2, \ldots, \xi_N)$.*

Lemma 1.9. *[16] For a strongly connected graph $\mathbb{G}$ with Laplacian matrix $\mathbb{L}$, define its generalized algebraic connectivity as*

$$\mathbf{a}(\mathbb{L}) = \min_{\xi^t x = 0, x \neq 0} \frac{x^t(\Xi\mathbb{L} + \mathbb{L}^t\Xi)}{x^t\,\Xi x},$$

where ξ and Ξ are defined as in Lemma 1.8. Then, $\mathbf{a}(\mathbb{L}) > 0$.

Let the symbol $d_{\max}(\mathbb{A})$ denote a maximal in-degree of $\mathbb{G}(\mathbb{A})$. In correspondence with the Gershgorin theorem [13], we can deduce another important property of the Laplacian $\mathbb{L}$.

All eigenvalues of the matrix $\mathbb{L}(\mathbb{A})$ have a nonnegative real part and belong to the circle with center on the real axis at the point $(0, d_{\max}(\mathbb{A}))$ and with a radius which equals $d_{\max}(\mathbb{A})$.

Let $\{\lambda_1, \cdots, \lambda_N\}$ denote eigenvalues of the matrix $\mathbb{L}(\mathbb{A})$. We arrange them in ascending order of real parts:

$$0 \leq \mathbf{Re}(\lambda_1) \leq \mathbf{Re}(\lambda_2) \leq \cdots \leq \mathbf{Re}(\lambda_N).$$

By virtue of Lemma 1.5, if the graph has a spanning tree, then $\lambda_1 = 0$ is a simple eigenvalue and all other eigenvalues of $\mathbb{L}$ are in the open right half of the complex plane.

The second eigenvalue λ_2 of $\mathbb{L}$ is important for analysis in many applications. It is usually called *Fiedler eigenvalue*. For undirected graphs it was shown in [1] that

$$\mathbf{Re}(\lambda_2) \leq \frac{N}{N-1} \min_{i \in N} d_i(\mathbb{A}) \tag{1.22}$$

and for the connected undirected graph $\mathbb{G}(\mathbb{A})$

$$\mathbf{Re}(\lambda_2) \geq \frac{1}{\mu\theta}, \tag{1.23}$$

where μ is the longest distance between two nodes and $\theta = \sum_{i \in N} d_i(\mathbb{A})$.

1.4 Lyapunov stability

In the sequel, we lay down some preliminary information about Lyapunov stability and related issues.

1.4.1 Preliminaries

Consider the discrete-time system

$$x(k+1) = f[x(k)], \tag{1.24}$$

where the states are $x(k) \in \Re^n$ and control values are $u(k) \in \Re^m$, for some n and m and for each time instant $k \in \mathbb{Z}_+$, and the map $f : \Re^n \to \Re^n$ can be generally nonlinear, nonsmooth, or even uncertain. Starting from an initial condition x_0, *what can be said about the asymptotic behavior of the state $x(k)$ as $k \to \infty$?* Questions of this type play a central role in control theory and engineering, as well as sciences including ecology, biology, and economics.

Hereinafter, we focus on the notion of *global asymptotic stability (GAS)*. Taking the unique equilibrium point of (1.24) to be the origin, that is, $f[0] = 0$, we have the following formal definition and fundamental theorem.

Definition 1.9. The origin is a globally asymptotic stable equilibrium of (1.24) if:

- $\forall \varepsilon > 0, \exists \delta > 0$ such that

$$||x(0)|| < \delta \Rightarrow ||x(k)|| < \varepsilon, \ \forall k;$$

- $\lim_{k \to \infty} x(k) = 0, \ \forall x(0) \in \Re^n$.

Theorem 1.5. *Consider system (1.24). If there exists a continuous radially unbounded function* $\mathbf{V} : \Re^n \to \Re$ *such that* $\mathbf{V}(x) > 0$, $\forall x \neq 0$, $\mathbf{V}(0) = 0$ *and*

$$\mathbf{V}(x+1) < \mathbf{V}(x),$$

then the origin is a GAS equilibrium of (1.24).

The significance of Theorem 1.5 is that it allows stability of the system to be verified without explicitly solving the difference equation.

Next, we proceed further and summarize some pertinent results on the input-to-state stability (ISS) property for discrete-time nonlinear systems of the general form

$$x(k+1) = f[x(k), u(k)], \tag{1.25}$$

where the states are $x(k) \in \Re^n$ and control values are $u(k) \in \Re^m$, for some n and m and for each time instant $k \in \mathbb{Z}_+$. We assume that $f : \Re^n \times \Re^m \to \Re^n$ is a continuous function, where $\mathbb{Z}_+$ denotes the set of all nonnegative integers.

For each $\xi \in \Re^n$ and each input u, we denote by $x(., \xi, u)$ the sequence of system (1.25) with initial state $x(0) = \xi$ and the input u. Clearly such a sequence is defined uniquely on $\mathbb{Z}_+$.

Recall that a function $\gamma : \Re_{\geq 0} \to \Re_{\geq 0}$ is a $\mathbb{K}$-function if it is continuous and strictly increasing and $\gamma(0) = 0$; it is a $\mathbb{K}_\infty$-function if it is a $\mathbb{K}$-function and also $\gamma(s) \to \infty$ as $s \to \infty$; and it is a *positive definite function* if $\gamma(s) > \infty$ for all $s > 0$ and $\gamma(0) = 0$. A function $\beta : \Re_{\geq 0} \times \Re_{\geq 0} \to \Re_{\geq 0}$ is a $\mathbb{KL}$-function if, for each $k \geq 0$, the function $\beta(., k)$ is a $\mathbb{K}$-function, and for each $s \geq 0$, the function $\beta(s, .)$ is decreasing with $\beta(s, k) \to 0$ as $k \to \infty$.

Hereinafter, we are motivated by the corresponding ISS notion, which was originally developed in [17] for continuous-time nonlinear systems. Simply stated, *the ISS property concerns the continuity of state trajectories on the initial states and the inputs. Roughly speaking, a system is ISS if every state trajectory corresponding to a bounded control remains bounded, and the trajectory eventually becomes small if the input signal is small no matter what the initial state is.*

ISS turns out to be a very natural stability property and, indeed, has been successfully employed in the stability analysis and control synthesis of nonlinear systems with complex structure [18].

1.4.2 Input-to-state stability properties

In the sequel, the main concern is to expose the dependence of state trajectories on the magnitude of inputs for system (1.25) where control inputs $u(k)$ are functions from $\mathbb{Z}_+$ to $\Re^m$ and $f(0, 0) = 0$, implying that the origin 0 is an equilibrium of the 0-input system.

Definition 1.10. System (1.25) is globally ISS if there exist a $\mathbb{KL}$-function $\beta : \Re_{\geq 0} \times \Re_{\geq 0} \rightarrow \Re_{\geq 0}$ and a $\mathbb{K}$-function γ such that, for each input $u \in \ell_{\infty}^{m}$ and each $\xi \in \Re^{n}$, we have

$$|x(k, \xi, u)| \leq \beta(|\xi|, k) + \gamma(||u||)$$

for each $k \in \mathbb{Z}_{+}$ and

$$|x(k, \xi, u)| \leq \beta(|\xi|, k) + \gamma(||u_{[k-1]}||)$$

for every $k \in \mathbb{Z}_{+} \{0\}$.

Remark 1.1. It can be seen from Definition 1.10 that the ISS property implies that the 0-input system $x(k+1) = f[x(k), 0]$ is globally asymptotically stable (GAS) and that system (1.25) is "converging-input converging-state," that is, every sequence $x(k, \xi, u)$ goes to 0 if $u(k)$ goes to 0 as $k \rightarrow \infty$. However, the converse is not true.

Definition 1.11. A continuous function $\mathbf{V} : \Re^{n} \rightarrow \Re_{\geq 0}$ is called an ISS-Lyapunov function for system (1.25) if the following holds:

A) There exist $\mathbb{K}_{\infty}$-functions α_1, α_2 such that

$$\alpha_1(|\xi|) \leq \mathbf{V}(\xi) \leq \alpha_2(|\xi|). \tag{1.26}$$

B) There exist a $\mathbb{K}_{\infty}$-function α_3 and a $\mathbb{K}$-function σ such that

$$\mathbf{V}(f(\xi, \mu) - \mathbf{V}(\xi) \leq -\alpha_3(|\xi|) + \sigma(|\mu|) \tag{1.27}$$

for all $\xi \in \Re^{n}$ and for all $\mu \in \Re^{m}$.

It should be noted that a smooth ISS-Lyapunov function is one which is smooth. Evidently, if $\mathbf{V}$ is an ISS-Lyapunov function for system (1.25), then $\mathbf{V}$ is a Lyapunov function for the 0-input system $x(k+1) = f[x(k), 0]$. It can be proved that a system is ISS if and only if it admits an ISS-Lyapunov function.

1.4.3 More on the lemma of Lyapunov

The standard *lemma of Lyapunov* for discrete-time systems states that for a matrix $\mathbf{C} > 0$, there exists a unique matrix $\mathbf{P} > 0$ such that $A^t\mathbf{P}A - \mathbf{P} + \mathbf{C} = 0$ if and only if $|\lambda_j(A)| < 1$.

Let $\mathbf{C} = \mathbf{S}\mathbf{S}^t$ with $(A, \mathbf{S})$ being completely observable. The first strengthening states that if $(A, \mathbf{S})$ is completely observable, then there exists a unique $\mathbf{P} > 0$ such that $A^t\mathbf{P}A - \mathbf{P} + \mathbf{S}\mathbf{S}^t = 0$ if and only if $|\lambda_j(A)| < 1$.

The second strengthening states that if $(A, \mathbf{S})$ is completely detectable, there exists a unique $\mathbf{P} \geq 0$ such that $A^t\mathbf{P}A - \mathbf{P} + \mathbf{S}\mathbf{S}^t = 0$ if and only if $|\lambda_j(A)| < 1$.

In all cases, $\mathbf{P}$ exists and is defined by

$$\mathbf{P} = \sum_{m=0}^{\infty} A^m \mathbf{S}\mathbf{S}^t A^{tm}.$$

1.5 Minimum mean square estimate

We now present a brief overview of the minimum mean square estimate (MMSE), which is quite useful in estimation methods to be discussed in later chapters.

We start by presenting a definition.

Definition 1.12. Consider a random variable $\mathbf{Y}$ that depends on another random variable $\mathbf{X}$. Of interest is the *minimum mean square error estimate* (MMSEE), which, simply stated, is $\hat{\mathbf{X}}$, the estimate of $\mathbf{X}$, such that the mean square error given by

$$\mathbb{E}[\mathbf{X} - \hat{\mathbf{X}}]^2$$

is minimized, where the expectation is taken over the random variables $\mathbf{X}$ and $\mathbf{Y}$.

In the following, some of the standard results are given.

1.5.1 Standard results

Proposition 1.5. *The MMSEE is given by the conditional expectation* $\mathbb{E}[\mathbf{X}|\mathbf{Y} = \mathbf{y}]$.

Proof. Consider the functional form of the estimator as $g(\mathbf{Y})$. Let $f_{\mathbf{X},\mathbf{Y}}(\mathbf{x}, \mathbf{y})$ denote the joint probability density function of $\mathbf{X}$ and $\mathbf{Y}$. Then the cost function is given by

$$\begin{aligned}
\mathbb{C} := \mathbb{E}\left[\mathbf{X} - \hat{\mathbf{X}}\right]^2 &= \int_x \int_y (\mathbf{x} - g(\mathbf{y}))^2 f_{\mathbf{X},\mathbf{Y}}(\mathbf{x}, \mathbf{y})\, dx\, dy \\
&= \int_y dy\, f_{\mathbf{Y}}(\mathbf{y}) \int_x (\mathbf{x} - g(\mathbf{y}))^2 f_{\mathbf{X}|\mathbf{Y}}(\mathbf{x}|\mathbf{y})\, dx.
\end{aligned}$$

Taking the derivative of the cost function with respect to the function $g(\mathbf{y})$,

$$\begin{aligned}
\frac{\partial \mathbb{C}}{\partial g(\mathbf{y})} &= \int_y dy\, f_{\mathbf{Y}}(\mathbf{y}) \int_x 2(\mathbf{x} - g(\mathbf{y})) f_{\mathbf{X}|\mathbf{Y}}(\mathbf{x}|\mathbf{y})\, dx \\
&= 2 \int_y dy\, f_{\mathbf{Y}}(\mathbf{y}) \left(g(\mathbf{y}) - \int_x \mathbf{x} f_{\mathbf{X}|\mathbf{Y}}(\mathbf{x}|\mathbf{y})\, dx \right) \\
&= 2 \int_y dy\, f_{\mathbf{Y}}(\mathbf{y}) \left(g(\mathbf{y}) - \mathbb{E}[\mathbf{X}|\mathbf{Y} = \mathbf{y}] \right).
\end{aligned}$$

Therefore the only stationary point is $g(\mathbf{y}) = \mathbb{E}[\mathbf{X}|\mathbf{Y} = \mathbf{y}]$ and it can be easily verified that it is a minimum. □

Remark 1.2. It is noted that the result established in Proposition 1.5 holds for vector random variables as well. Observe that MMSE are important because for Gaussian variables, they coincide with the maximum likelihood (ML) estimates. It is a standard result that for Gaussian variables, the MMSE is linear in the state value.

In what follows, we will assume zero mean values for all the random variables with $\mathbf{R_X}$ being the covariance of $\mathbf{X}$ and $\mathbf{R_{XY}}$ being the cross-covariance between $\mathbf{X}$ and $\mathbf{Y}$.

Proposition 1.6. *The best linear MMSE of* $\mathbf{X}$ *given* $\mathbf{X} = \mathbf{y}$ *is*

$$\hat{\mathbf{x}} = \mathbf{R_{XY}} \mathbf{R_Y}^{-1} \mathbf{y}$$

with the error covariance

$$\mathbf{P} = \mathbf{R_X} - \mathbf{R_{XY}} \mathbf{R_Y}^{-1} \mathbf{R_{XY}}.$$

Proof. Let the estimate be $\hat{\mathbf{x}} = \mathbf{K}\mathbf{y}$. Then the error covariance is

$$\begin{aligned} \mathbf{P} &:= \mathbb{E}\big[(\mathbf{x} - \mathbf{K}\mathbf{y})(\mathbf{x} - \mathbf{K}\mathbf{y})^t\big] \\ &= \mathbf{R_X} - \mathbf{K}\mathbf{R_{YX}} - \mathbf{R_{XY}}\mathbf{K}^t + \mathbf{K}\mathbf{R_Y}\mathbf{K}^t. \end{aligned}$$

Differentiating $\mathbf{P}$ with respect to $\mathbf{K}$ and setting it equal to zero yields

$$-2\mathbf{R_{XY}} + 2\mathbf{K}\mathbf{R_Y}^{-1}.$$

The result follows immediately. □

Extending Proposition 1.5 to the case of linear measurements $\mathbf{y} = \mathbf{H}\mathbf{x} + \mathbf{v}$, we have the following standard result.

Proposition 1.7. *Let* $\mathbf{y} = \mathbf{H}\mathbf{x} + \mathbf{v}$*, where* $\mathbf{H}$ *is a constant matrix and* $\mathbf{v}$ *is a zero mean Gaussian noise with covariance* $\mathbf{R_V}$ *independent of* $\mathbf{X}$*. Then the MMSE of* $\mathbf{X}$ *given* $\mathbf{Y} = \mathbf{y}$ *is*

$$\hat{\mathbf{X}} = \mathbf{R_X}\mathbf{H}^T(\mathbf{H}\mathbf{R_X}\mathbf{H}^T + \mathbf{R_V})^{-1}\mathbf{y}$$

with the corresponding error covariance

$$\mathbf{P} = \mathbf{R_X} - \mathbf{R_X}\mathbf{H}^T(\mathbf{H}\mathbf{R_X}\mathbf{H}^T + \mathbf{R_V})^{-1}\mathbf{H}\mathbf{R_X}.$$

1.6 Motivating problems

A sensor network is a collection of interconnected nodes that are deployed in a geographic area to perform monitoring tasks. Each node is equipped with sensing and computing capability. Sensor networks consisting of a large collection of nodes are currently under development or envisioned for the near future [19], [21]. Usually, each node can communicate with only a small subset of the remaining nodes. These communication constraints define a graph whose vertices are the nodes and whose edges are the communication links. In typical situations, a node may lack knowledge of certain attributes such as its own position in a global reference frame. However, nodes might be capable of measuring relative values with respect to nearby nodes. In this scenario, it is desirable to use relative measurements to estimate global attributes. We describe three scenarios that motivate these problems.

P1. Consider the problem of localization, where a sensor does not know its position in a global coordinate system but can measure its position relative to a set of nearby nodes. These measurements can be obtained, for example, from range data and bearing (that is, angle) data (see Fig. 1.8). In particular, two nearby sensors u and v located in a plane at positions p_u and p_v, respectively, have access to the measurement

$$\xi_{uv} = p_u - p_v + \epsilon_{u,v}, \tag{1.28}$$

where $\epsilon_{u,v}$ denotes measurement error. The problem of interest is to use the ξ_{uv}s to estimate the positions of all the nodes in a common coordinate system whose origin is fixed arbitrarily at one of the nodes.

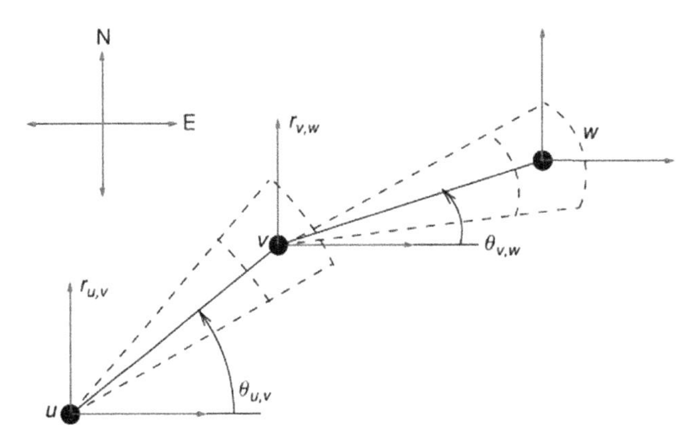

FIGURE 1.8 Relative position measurement in a Cartesian reference frame.

P2. The second scenario involves the time synchronization problem, in which the sensing nodes are part of a multihop communication network. Each node has a local clock, but each pair of clocks differs by a constant offset. However, nodes that communicate directly can estimate the difference between their local clocks by exchanging "hello" messages that are time-stamped with local clock times. For example, suppose that nodes u and v can communicate directly with each other and have clock offsets t_u and t_v with respect to a reference clock. By passing messages back and forth, the nodes can measure the relative clock offset $t_u - t_v$ with the measurement

$$\xi_{uv} = t_u - t_v + \epsilon_{u,v} \in \Re, \tag{1.29}$$

where $\epsilon_{u,v}$ denotes measurement error (see Fig. 1.9). The task is now to estimate the clock offsets with respect to the global time, which is defined as the local time at some reference node.

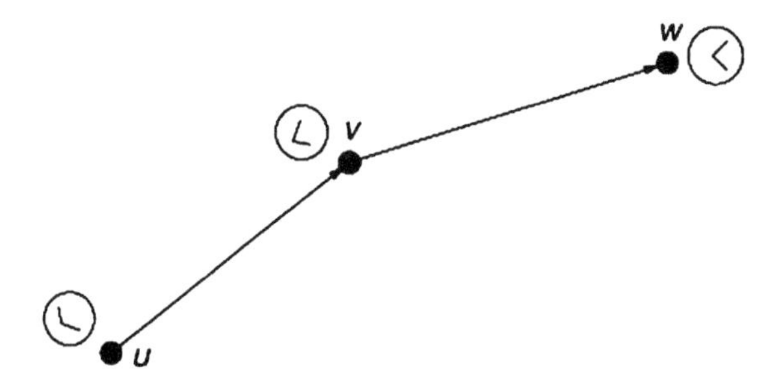

FIGURE 1.9 Measurement of differences in local times.

P3. The third scenario is a motion consensus problem, in which each agent wants to determine its velocity with respect to the velocity of a leader using only measurements of its relative velocities with respect to nearby agents. These measurements can be obtained, for example, by using vision-based sensors. In particular, two nearby agents u and v moving with velocities $\dot{p}_u$ and $\dot{p}_v$, respectively, have access to the measurement

$$\xi_{uv} = \dot{p}_u - \dot{p}_v + \epsilon_{u,v} \in \Re, \tag{1.30}$$

where $\epsilon_{u,v}$ denotes measurement error (see Fig. 1.9). The task is to determine the velocity of each agent with respect to the leader based solely on the available relative velocities between pairs of neighboring agents.

In this section, we present a coherent introduction to a particular class of networked dynamic systems, namely, discrete-time networked dynamic systems (DNDSs). This emerging discipline sits at the intersection of different areas such as distributed algorithms, parallel processing, control, and estimation. Our objective is to provide a self-contained, broad exposition of the notions and tools from these areas that are relevant in cooperative control problems.

These concepts include graph-theoretic notions (connectivity, adjacency, and Laplacian matrices), distributed algorithms from computer science (leader election, basic tree computations) and from parallel processing (averaging algorithms, convergence rates), and geometric models and optimization.

Needless to emphasize that graph theory provides key concepts to model, analyze, and design network systems and distributed algorithms; the language of graphs pervades modern science and technology and is therefore essential.

1.6.1 Wireless sensor networks

A wireless sensor network (WSN) is a collection of spatially distributed devices capable of measuring physical and environmental variables (including temperature, vibrations, sound, and light), performing local computations, and transmitting information to neighboring devices and, in turn, throughout the network (including, possibly, an external operator).

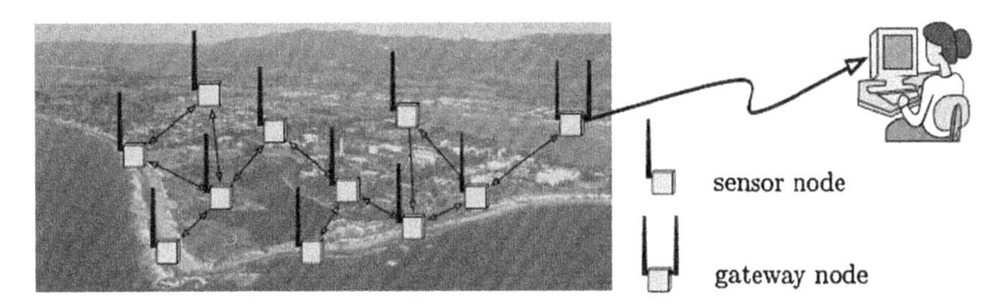

FIGURE 1.10 A wireless sensor network.

With reference to Fig. 1.10, a WSN is composed of a collection of spatially distributed sensors in a field and a gateway node to carry information to an operator. The nodes are meant to measure environmental variables, such as temperature, sound, and pressure, and cooperatively filter and transmit the information to an operator.

Now suppose that each node in a WSN has measured a scalar environmental quantity, say, x_j. Consider the following simple distributed algorithm, based on the concepts of linear averaging. Each node repeatedly executes

$$x_j^* := \textbf{average}\Big(x_j,\ \{x_k,\ \text{for all neighbor nodes } k\}\Big), \tag{1.31}$$

where x_j^* represents the new value of x_j and **average** denotes the average operation. For example, for the graph in Fig. 1.11, one can easily write

$$x_1^* := (x_1 + x_4 + x_6 + x_7)/4, \ \ x_2^* := (x_2 + x_4 + x_5 + x_6)/4,$$

and so forth. In effect, the algorithm proceeds by

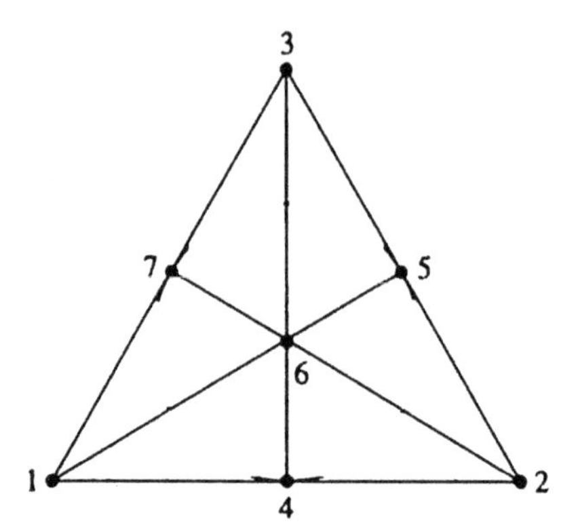

FIGURE 1.11 A representative graph.

$$x^* := \begin{bmatrix} 1/4 & 0 & 0 & 1/4 & 0 & 1/4 & 1/4 \\ 0 & 1/4 & 0 & 1/4 & 1/4 & 1/4 & 0 \\ 0 & 0 & 1/4 & 0 & 1/4 & 1/4 & 1/4 \\ 1/4 & 1/4 & 0 & 1/4 & 0 & 1/4 & 0 \\ 0 & 1/4 & 1/4 & 0 & 1/4 & 1/4 & 0 \\ 1/7 & 1/7 & 1/7 & 1/7 & 1/7 & 1/7 & 1/7 \\ 1/4 & 0 & 1/4 & 0 & 0 & 1/4 & 1/4 \end{bmatrix} x = A_{sn}\, x, \qquad (1.32)$$

where A_{sn} in (1.32) is a row-stochastic matrix (has nonnegative entries and unit row sums).

It is readily seen that one aspect of modeling WSNs can be cast into the general framework of a *discrete dynamic system (DDS)* defined by

$$x(k+1) := A\, x(k), \quad x(0) = x_0, \qquad (1.33)$$

where the matrix $A \in \Re^{n \times n}$ and k is the discrete-time variable. It follows from (1.33) that

$$x(k) = A^k\, x_0,$$

where the sequence $\{x(k)\}_{k \in \mathbb{Z}_{\geq 0}}$ is called the solution, trajectory, or evolution of the discrete system with $\mathbb{Z}_{\geq 0}$ denoting the space of nonnegative integers.

To study the evolution of system (1.33), we recall from linear matrix algebra [2] that for all $j \in \{1, ..., \mathbf{N}\}$,

$$\begin{aligned} \max_j (Ax)_j &= \max_j \sum_{m=1}^{\mathbf{N}} a_{jm} x_m \leq \max_j \sum_{m=1}^{\mathbf{N}} a_{jm} (\max_s x_s) \\ &= (\max_j \sum_{m=1}^{\mathbf{N}} a_{jm})(\max_s x_s) = 1 \times (\max_s x_s), \end{aligned}$$

$$\min_j (Ax)_j = \min_j \sum_{m=1}^{\mathbf{N}} a_{jm} x_m \geq \min_j \sum_{m=1}^{\mathbf{N}} a_{jm} (\min_s x_s)$$
$$= (\min_j \sum_{m=1}^{\mathbf{N}} a_{jm})(\min_s x_s) = 1 \times (\min_s x_s).$$

This leads to the conclusion that all initial conditions $x(0) := x_0$ are bounded in the following manner:

$$\min_j \ x_j(0) \leq \min_j \ x_j(k) \leq \min_j \ x_j(k+1)$$
$$\leq \max_j \ x_j(k+1) \leq \max_j \ x_j(k) \leq \max_j \ x_j(0).$$

In what follows, we proceed to examine the behavior of a solution from an arbitrary initial condition $x(0) \in \Re^n$ having an asymptotic limit as time progresses and to what value the solution converges. Formally, we present this property as follows.

- It is *semiconvergent* if $\lim_{k \to \infty} A^k = A_\infty$ exists. This implies that $\lim_{k \to \infty} x(k) = A_\infty x(0)$.
- It is *convergent* if it is *semiconvergent* and $\lim_{k \to \infty} A^k = 0_{n \times n}$.

Extending (1.33) to the controlled case, we have

$$x(k+1) := A\, x(k) + B\, u(k), \quad x(0) = x_0, \tag{1.34}$$

where the matrix $A \in \Re^{n \times n}$ is a convergent matrix and $B \in \Re^{n \times m}$, $u \in \Re^m$. Let x^* be an equilibrium point. It follows that x^* exists if and only if $x^* = A\, x^*(k) + B\, u(k)$, which implies that $x^* = [I_n - A]^{-1}\, B\, u(k)$. Next, we introduce a new sequence $y(k) = x(k) - x^*$, $k \in k \in \mathbb{Z}_{\geq 0}$, for an appropriate x^*. It is readily seen that

$$y(k+1) := x(k+1) - x^* = A\, x(k) + B\, u(k) - x^*$$
$$:= A\, [x(k) - x^*] = A\, y(k), \tag{1.35}$$

which means that every solution to $y(k+1) = Ay(k)$ asymptotically converges to zero. This leads to the conclusion that

$$\lim_{k \to \infty} x(k) = [I_n - A]^{-1}\, B\, u(k) \ \ \forall\, x(0) \in \Re^n.$$

1.6.2 Distributed parameter estimation

Following ideas from [22], the goal is to estimate an unknown parameter $\alpha \in \Re^m$ via the measurements taken by an **N**-node sensor network. Each node $j \in \{1, ..., \mathbf{N}\}$ measures

$$z_j = M_j \alpha + \varepsilon_j, \tag{1.36}$$

where $z_j \in \Re^{m_j}$, M_j is a known matrix, and ε_j is random measurement noise. Two assumptions are made:

A1) The noise vectors $\varepsilon_1, ..., \varepsilon_{\mathbf{N}}$ are independent jointly Gaussian variables with zero mean $\mathbb{E}[\varepsilon_j] = 0_{m_j}$ and positive definite covariance $\mathbb{E}[\varepsilon_j \varepsilon_j^T] = \Sigma_j = \Sigma_j^T$, for $j \in \{1, ..., \mathbf{N}\}$.

A2) The measurement parameters satisfy $\sum_j m_j \geq m$, and

$$\begin{bmatrix} M_1 \\ \vdots \\ M_{\mathbf{N}} \end{bmatrix}$$

is full rank.

Given the measurements $z_1, ..., z_{\mathbf{N}}$, it is of interest to compute a least square estimate of α, say, $\hat{\alpha}$, that is, *an estimate $\hat{\alpha}$ that minimizes a least square error. Specifically, we aim to minimize the following weighted least square error:*

$$\min_{\hat{\alpha}} \sum_{j=1}^{\mathbf{N}} ||z_j - M_j\hat{\alpha}||^2_{\Sigma_j^{-1}} = \sum_{j=1}^{\mathbf{N}} [z_j - M_j\hat{\alpha}]^T \Sigma_j^{-1} [z_j - M_j\hat{\alpha}]. \quad (1.37)$$

Note in (1.37) that the individual errors are weighted by their corresponding inverse covariance matrices so that an accurate (respectively, inaccurate) measurement corresponds to a high (respectively, low) error weight. To this end, the least square estimate coincides with the so-called *maximum likelihood estimate*. Now under Assumptions **A1** and **A2**, the optimal solution is expressed as

$$\hat{\alpha}^o := \left(\sum_{j=1}^{\mathbf{N}} M_j^T \Sigma_j^{-1} M_j \right)^{-1} \sum_{j=1}^{\mathbf{N}} M_j^T \Sigma_j^{-1} z_j. \quad (1.38)$$

This formula is quite standard in the field of parameter estimation and it is easy to implement by a single processor with all the information about the problem.

To make advantage of sensor (and processor) networks, we can compute $\hat{\alpha}^o$ in two steps:

S1: We run two distributed algorithms in parallel to compute the average of the quantities $M_j^T \Sigma_j^{-1} M_j$ and $M_j^T \Sigma_j^{-1} z_j$.

S2: We compute the optimal estimate via

$$\begin{aligned} \hat{\alpha}^o &:= H_a \, H_c, \\ H_a &:= \mathbf{average}\left(M_1^T \Sigma_1^{-1} M_1, ..., M_{\mathbf{N}}^T \Sigma_{\mathbf{N}}^{-1} M_{\mathbf{N}} \right)^{-1}, \\ H_c &:= \mathbf{average}\left(M_1^T \Sigma_1^{-1} z_1, ..., M_{\mathbf{N}}^T \Sigma_{\mathbf{N}}^{-1} z_{\mathbf{N}} \right). \end{aligned} \quad (1.39)$$

The key lesson to be learned from these examples is that it is beneficial to deploy distributed algorithms whenever sensor networks are used. Several issues arise concerning the design of computational algorithms, the properties that the graph needs to have in order for such an algorithm to exist, and the need to design an algorithm with fastest convergence.

All of these concerns and others will be addressed in later sections of the book.

1.6.3 Population dynamics

This continuous model represents the spread of HIV-1 infection inside the human organism. Here $y(k)$ represents the number of susceptible cells which are present at time in a unit of plasma. The process of infection of a cell is divided into several sequential stages; therefore, is the number of infected cells at time at stage [23].

A large class of discrete models concerning the dynamics of an infection in an organism or in a host population [24] is described by

$$
\begin{aligned}
y(k+1) &= \sigma + (1-\beta)y(k) - \sum_{j=1}^{n} \psi_j(y_j(k+1))x_j(k+1), \quad k \geq 0, \\
x_1(k+1) &:= (1-a_1)x_1(k) - \phi_1(y(k))x_L(k), \quad 1 \leq L \leq n, \\
x_j(k+1) &:= (1-a_j)x_j(k) - \phi_j(y(k))x_{j-1}(k), \quad j = 2, 3, ...n,
\end{aligned}
\tag{1.40}
$$

where $\psi_j(x),\ \phi_j(x) \in \mathbf{C}^o\Re,\ 1 \leq j \leq n$. Moreover, $\phi_j,\ 1 \leq j \leq n$, and at least one of $\psi,\ 1 \leq j \leq L$, are strictly monotone increasingly functions satisfying $y\psi(y) \geq 0,\ \forall y \in \Re,\ j = 1, ..., n$.

The model (1.40) involves typically two populations: *susceptible* individuals represented by y and *infective* ones represented by one of the sequences x_I, $1 \leq I \leq L$.

The following are two cases of special interest:

- The spread of HTLV-I infection in a human organism [25] is a special case of model (1.40) by setting $n = 2,\ I = L = 2,\ \psi_1(y) = 0,\ \phi_1(y) = \psi_2(y) = c\ y$, $\phi_2(y) = g,\ a_1 = d_B,\ \sigma = a,\ a_1 = d_A,\ \beta = b$.
- The spread of HIV-I infection in a human organism [26] is a special case of model (1.40) by setting $n = 2,\ I = 1,\ L = 2,\ \psi_1(y) = 0,\ \phi_1(y) = \psi_2(y) = c\ y,\ \phi_2(y) = g,\ a_1 = d_B,\ \sigma = \lambda,\ a_1 = d_A,\ \beta = b$.

Recalling that functions y and x_j $(j = 1, ..., n)$ represent populations following the work in [24], we can infer the dynamic properties of positivity and boundedness by using very natural hypotheses.

Lemma 1.10. *Assume that*

F1 $\sigma > 0$;
F2 $0 < \beta < 1$;
F3 $0 < a_j < 1$ $(j = 1, ..., n)$;
F4 $\phi_j(x)$ *is not decreasing and* $\phi_j(0) \geq 0,\ j = 1, ..., m$;
F5 $y\psi_j(y) \geq 0,\ \forall y \in \Re,\ j = 1, ..., m$;
F6 $\exists \hat{j}$ *such that* $1 \leq \hat{j} \leq L$ *and* $\psi(y)$ *is strictly increasing.*

Then we have $y(k) > 0,\ x_j(k) > 0,\ k \geq 0,\ j = 1, ..., n$.

Proof. From **F3**, **F4**, and $x_j(0) > 0$, we have $x_j(1) > 0$. Let $y(1) \leq 0$. From **F5**, **F6**, we get $\sum_{j=1}^{n} \psi_j(y(1))x_j(1) \leq 0$ and from **F1**, **F2**, and model (1.40), we obtain $y(1) > 0$, which is a contradiction. By induction in the same way, we can establish the remainder of Lemma 1.10. □

Looking at these models, it can be easily cast into the standard nonlinear discrete models:

$$\begin{aligned} z(k) &:= [z_s(k) \quad z_f(k)]^t, \\ z(k+1) &:= \mathbf{F}[z_s(k), z_f(k)], \quad k \geq 0, \quad z(0) = z_0, \end{aligned} \tag{1.41}$$

for two populations: *susceptible* individuals represented by $z_s(k)$ and *infective* ones represented by $z_f(k)$.

1.6.4 Distributed control systems

This section concerns some common characteristics of distributed control systems, such as distributive interconnections, local control rules, and scalability. Basic notions and results of algebraic graph theory are introduced as a theoretic foundation for modeling interconnections of subsystems (agents) in distributed control systems. Coordination control systems and end-to-end congestion control systems are also introduced as two typical kinds of distributed control systems.

Network-based distributed control system

What kinds of control systems can be called distributed control systems? Are there any differences between a distributed control system and a so-called large-scale control system or a decentralized control system? What is the advantage of distributed control over centralized control? We will not attempt to answer all questions; rather, we will try to grasp some common features of distributed control systems.

Cooperation among agents

Cooperation is perhaps the soul of a distributed control system. Since there is no centralized control unit, the only way through which all agents in the system

can work in harmony is to conduct some kind of cooperation. This is governed effectively by three simple rules:
(1) collision avoidance,
(2) agreement on velocity, and
(3) approaching center, which actually implies approaching any neighbor.

Nowadays, distributed control mechanisms are widely used in engineering systems, such as aircraft traffic control, multirobot control, coverage control of sensor networks, and formation control of unmanned aerial vehicles, to name a few.

Recall that in a conventional feedback control system, the controller gets the output signal of the plant, compares it with some reference signal, and makes decisions on how to act. Such a system also serves as one of basic units (subsystems) in a distributed control system. However, to cooperate with other units in the entire system, it should be equipped with a sensor/communication (S/M) module besides the classic decision/action module (controller), as shown in Fig. 1.12. Such a subsystem is sometimes called an "agent." In a distributed

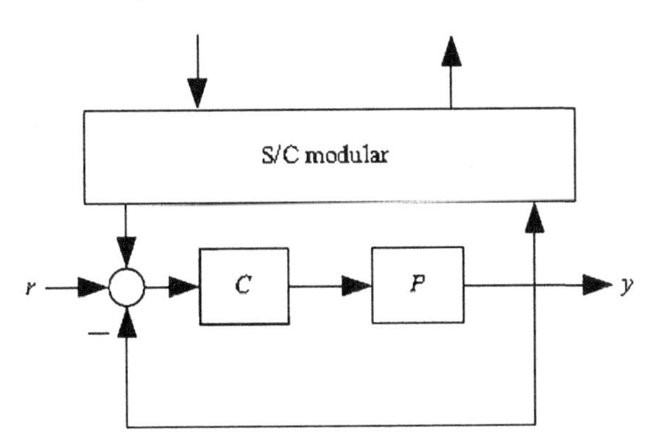

FIGURE 1.12 An agent in a distributed control system.

control system, therefore, each subsystem gets not only the information of the output of itself but also the information of some other subsystems via sensor or/and communication networks (see Fig. 1.13). To this end, a distributed control system interconnected with multiple agents is also called a multiagent system.

For many distributed control systems, such as the Internet, power grids, traffic control systems, etc., subsystems (agents) are often distributed across multiple computational units in an immense space and connected through long-distance packet-based communications. In these systems, packet loss and delay are unavoidable, and hence, computational and communication constraints cannot be ignored. A new formalism to ensure stability, performance, and robustness is required in the analysis and design of distributed control systems.

A key feature is that a distributed control law should be subject to the *agent control rule*, which implies that in most cases there is no kind of centralized

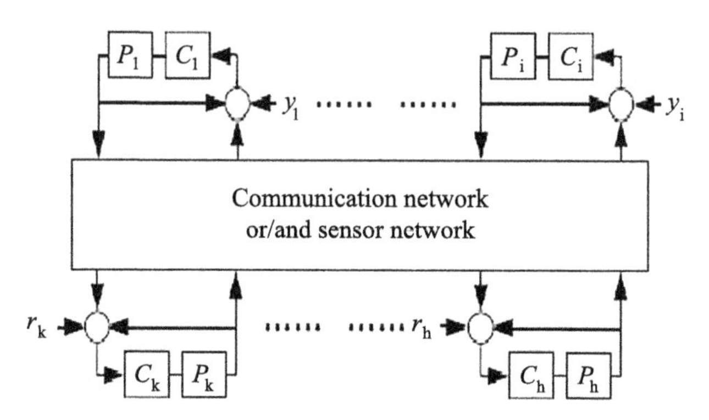

FIGURE 1.13 A network-based distributed control system.

supervision or control unit in the system, and each agent makes decisions based only on the information received by its own sensor or from its neighboring agents through communication. The agent control rule is the most important feature of distributed control systems and distinguishes distributed control from the decentralized control of large-scale systems.

A remarkable advantage of the agent control rule over the nonagent control rules is its higher fault tolerance capability. This is extremely important for many large-scale engineering systems which must operate continuously even when some individual units fail.

Scalability is another very important reason why distributed control systems prefer the agent control rule. In the sequel the scalability of a distributed control system implies that the controller of the system and its maintenance utilize only local information around each agent and rarely depend on the scale of the system. In other words, by scalability we mean that not only the control law but also most important properties of the system rely on local information.

For example, in checking the stability of the entire distributed control system, scalability requires that the stability criterion does not need to use information about the global interconnection topology of the system because such global information is usually unavailable to individual agents. A scalable system allows new applications to be designed and deployed without requiring changes to the underlying system.

1.6.5 Bipartite distributed control systems

There are many distributed control systems in which information exchanges are conducted through forward and backward channels. The interconnection of agents in such systems can be described by the weighted bipartite digraph shown in Fig. 1.14. We call these systems bipartite systems, as depicted in Fig. 1.15. We denote by (S, L) and A the vertex bipartition and the weighted adjacency

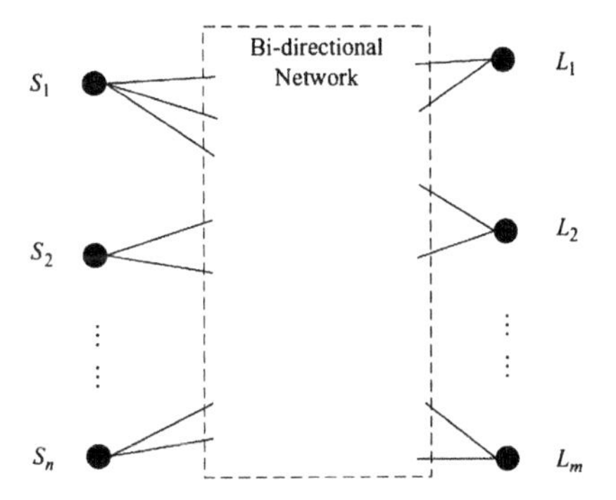

FIGURE 1.14 Bipartite graph for distributed control systems.

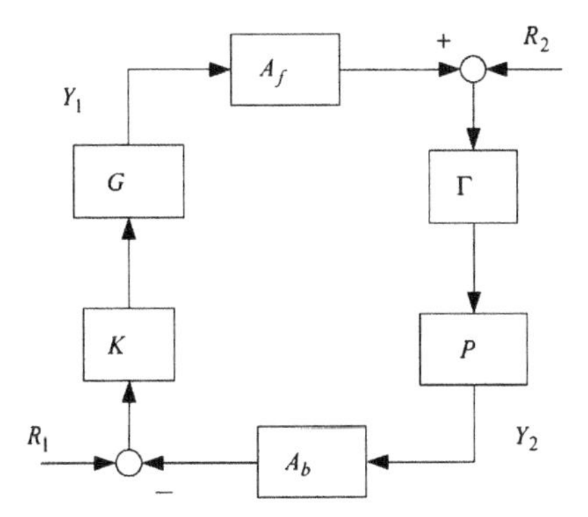

FIGURE 1.15 Bipartite distributed control system.

matrix of the graph, respectively, where

$$\mathsf{S} = \{\mathsf{S}_1, \cdots, \mathsf{S}_n\},$$
$$\mathsf{L} = \{\mathsf{L}_1, \cdots, \mathsf{L}_n\},$$
$$\mathsf{A} = \begin{bmatrix} 0 & \mathsf{A}_{12} \\ \mathsf{A}_{12}^t & 0 \end{bmatrix}. \tag{1.42}$$

Let the agents associated with vertex class S be described by transfer functions $\mathsf{G}_i(.)$, $i \in 1, \cdots, n$, and let the agents associated with L be described by transfer functions $\mathsf{P}_j(.)$, $j \in 1, \cdots, m$. Agent $i \in \mathsf{S}$ receives output information from its neighbors in L, and agent $j \in \mathsf{L}$ receives output information from its neighbors in S, that is,

$$z_i(k) = \sum_{\ell \in \mathbb{N}_i} a_{i\ell}^b y_\ell(k - \tau_{\ell i}^b), \quad i \in \mathsf{S},$$
$$z_\ell(k) = \sum_{i \in \mathbb{N}_i} a_{\ell i}^f y_\ell(k - \tau_{\ell i}^f), \quad \ell \in \mathsf{L}, \tag{1.43}$$

where $\tau^b_{\ell i}$, $a^b_{i\ell}$ are the communication delay and the weight of the backward channel from agent $\ell \in \mathsf{L}$ to agent $i \in \mathsf{S}$, respectively, and $\tau^f_{\ell i}$, $a^f_{\ell i}$ are the communication delay and the weight of the forward channel from agent $i \in \mathsf{S}$ to agent $\ell \in \mathsf{L}$. Considering (1.42), the following relationship should hold:

$$a^b_{i\ell} = a^f_{\ell i}. \tag{1.44}$$

The transfer function matrix from $R_1(.) \to Y_1(.)$ is given by

$$\mathsf{W}_1(.) = [I + \mathsf{G}(.)\mathsf{K}(.)\mathsf{A}_b(.)\mathsf{P}(.)\Gamma(.)\mathsf{A}_f(.)]^{-1}\mathsf{G}(.)\mathsf{K}(.) \tag{1.45}$$

and the transfer function matrix from $R_2(.) \to Y_2(.)$ is given by

$$\mathsf{W}_2(.) = [I + \mathsf{P}(.)\Gamma(.)\mathsf{A}_f(.)\mathsf{GKA}_b(.)]^{-1}\mathsf{P}(.)\Gamma(.). \tag{1.46}$$

Denote $\hat{\mathsf{A}}_1(.) = \mathsf{A}_b(.)\mathsf{P}(.)\Gamma(.)\mathsf{A}_f(.)$ and $\hat{\mathsf{A}}_2(.) = \mathsf{A}_f(.)\mathsf{GKA}_b(.)$. Then, the open-loop transfer function of $\mathsf{W}_1(.)$ or $\mathsf{W}_2(.)$ has the same structure as that of the conventional transfer function of multivariable systems. More importantly, it consists of a diagonal matrix corresponding to the agents' dynamics and a square matrix corresponding to the topology graph. This suggests that it is possible to deal with the stability problem of two kinds of distributed control systems in a unified framework.

1.7 Notes

This section proposed some ideas for future research investigations in distributed or multiagent systems. The primary goal in distributed coordination control is mainly to have systems that can coordinate using available resources and knowledge of system-wide objectives to reach an agreement over a particular task. These problems are solved almost on a daily basis by biological systems. Much of the research in distributed coordination assumes a fixed consensus protocol over which multiagent system objectives can be achieved. However, real-life scenarios are subject to numerous uncertainties that could affect perception and communication in multiagent systems. In this regard, suggestions which could stimulate new research interests are classified under the following categories: autonomous multiagent systems, decision-based coordination, evolutionary graph theory, and bio-inspired algorithms.

Resilient and bipartite consensus problems are just recently being discussed by researchers. There is still a wealth of open topics to be studied under these areas, especially considering input and network communication delays. Also, resilient group consensus and resilient bipartite consensus problems still remain unexplored.

Control protocols for distributed/multiagent systems are not fully autonomous; rather, they are designed based on rules described a priori. In real-life coordination, coordination rules are dynamic and formed based on the interest of agents and the overall group objective. Hence, new algorithms should be designed where the multiagent system network is fully autonomous and interactions between agents are formed based on immediate environment, state-of-resources, and collective objectives.

Distributed control algorithms could be designed based on limited states of resources of each agent in networked distributed/multiagent systems; hence, decisions to coordinate should be informed by its state of resources and perception of the immediate environment. In applications of distributed/multiagent systems where teams of robots are deployed for surveillance, decision-based control may be useful for each agent to determine whether or not to interact in a distributed/multiagent framework, especially in cases where it may be potentially dangerous to do so, for example if an agent is compromised or its immediate environment is too dangerous to act.

In designing fully distributed multiagent systems to solve real-world problems, there is quite a lot to learn from nature. For problems of distributed/multiagent systems appearing in team robotics, bio-inspired control algorithms could be designed which allow teams of robots to demonstrate intelligent coordination behaviors, as demonstrated in fish and bird swarms.

Useful collections of example networks are freely available online, and we briefly mention below some examples.

(A) The *Koblenz Network Collection*, available at

http://konect.uni-koblenz.de,

contains model graphs in easily accessible MATLAB format (as well as a MATLAB toolbox for network analysis).

(B) A broad range of example networks is available online at the *Stanford Large Network Dataset Collection*,

http://snap.stanford.edu/data.

(C) The *Suite Sparse Matrix Collection* (formerly known as the University of Florida Sparse Matrix Collection), available at

http://suitesparse.com,

contains a large and growing set of sparse matrices and complex graphs arising in a broad range of applications.

(D) The *UCI Network Data Repository*, available at

http://networkdata.ics.uci.edu,

facilitates the scientific study of networks.

Looking for software libraries for network analysis and visualization, there are some freely available libraries online.

(1) *Network X*, available at

http://networkx.github.io,

is a Python library for network analysis.

(2) *Mathematica* provides functionality for modeling, analyzing, synthesizing, and visualizing graphs and networks – beside the ability to simulate dynamical systems; the interested reader can consult more information by searching for "graphs and networks" at

http://www.wolfram.com.

(3) *Graphviz*, available at

http://www.graphviz.org/,

is open source graph visualization software which is also compatible with MATLAB.

(4) *Gephi*, available at

https://gephi.org,

is an interactive visualization and exploration platform for all kinds of networks, complex systems, and dynamic and hierarchical graphs.

(5) *Cytoscape*, available at

http://www.cytoscape.org,

is an open source software platform for visualizing complex networks and integrating them with attribute data.

References

[1] F. Fagnani, P. Frasca, Introduction to Averaging Dynamics over Networks, Lecture Notes in Control and Information Sciences, vol. 472, Springer International Publishing AG, 2018.

[2] R.A. Horn, C. Johnson, Topics in Matrix Analysis, Cambridge University Press, 1991.

[3] C.W. Scherer, LPV control and full block multiplier, Automatica 37 (2001) 361–375.

[4] A.K. Packard, Gain scheduling via linear fractional transformations, Syst. Control Lett. 22 (2) (1994) 79–92.

[5] R. D'Andrea, G.E. Dullerud, S. Lall, Convex $\mathcal{L}_2$ synthesis for multidimensional systems, in: Proc. 37th IEEE Conf. Decision and Control, Tampa, FL, December 1998, pp. 1883–1888.

[6] R. D'Andrea, Extension of Parrott's theorem to non-definite scalings, IEEE Trans. Autom. Control 45 (5) (2001) 937–940.

[7] C. Godsil, G. Royle, Algebraic Graph Theory, Springer, 2001.

[8] F. Bullo, Lectures on Network Systems, ed. 1.3, Kindle Direct Publishing, 2019.

[9] M.S. Mahmoud, Y. Xia, Applied Control Systems Design: State-Space Methods, Springer-Verlag, UK, May 2012.

[10] W. Ren, R.W. Beard, Consensus seeking in multiagent systems under dynamically changing interaction topologies, IEEE Trans. Autom. Control 50 (5) (2005) 655–661.

[11] P. Lin, Y. Jia, L. Lin, Distributed robust $\mathcal{H}_\infty$ consensus control in directed networks of agents with time-delay, Syst. Control Lett. 57 (8) (2008) 643–653.

[12] R.S. Smith, F.Y. Hadaegh, Control of deep-space formation-flying spacecraft; relative sensing and switched information, J. Guid. Control Dyn. 8 (1) (2005) 106–114.
[13] D. Angeli, P.-A. Bliman, Convergence speed of unsteady distributed consensus: decay estimate along the settling spanning-trees, SIAM J. Control Optim. 48 (1) (2009) 1–32.
[14] J. Zhou, Q. Wang, Convergence speed in distributed consensus over dynamically switching random networks, Automatica 45 (6) (2009) 1455–1461.
[15] Z. Qu, Cooperative Control of Dynamical Systems: Applications to Autonomous Vehicles, Springer-Verlag, London, 2009.
[16] W. Yu, G. Chen, M. Cao, J. Kurths, Second-order consensus for multiagent systems with directed topologies and nonlinear dynamics, IEEE Trans. Syst. Man Cybern., Part B, Cybern. 40 (3) (2010) 881–891.
[17] E.D. Sontag, Further facts about input to state stabilization, IEEE Trans. Autom. Control 35 (1990) 473–476.
[18] D. Kazakos, J. Tsinias, The input to state stability conditions and global stabilization of discrete-time systems, IEEE Trans. Autom. Control 39 (1994) 2111–2113.
[19] D. Estrin, D. Culler, K. Pister, G. Sukhatme, Connecting the physical world with pervasive networks, IEEE Pervasive Comput. 1 (1) (2002) 59–69.
[20] R. Olfati-Saber, R.M. Murray, Consensus problems in networks of agents with switching topology and time delays, IEEE Trans. Autom. Control 49 (9) (2004) 1520–1533.
[21] P. Barooah, J.P. Hespanha, Estimation on graphs from relative measurements, IEEE Control Syst. Mag. 27 (4) (2007) 57–74.
[22] F. Garin, L. Schenato, A survey on distributed estimation and control applications using linear consensus algorithms, in: A. Bemporad, M. Heemels, M. Johansson (Eds.), Networked Control Systems, in: LNCIS, Springer, 2010, pp. 75–107.
[23] Z. Grossman, M. Feinberg, V. Kuznetsov, D. Dimitrov, W. Paul, HIV infection: how effective is drug combination treatment?, Immunol. Today 19 (11) (1998) 528–532.
[24] G. Izzo, Y. Muroya, A. Vecchio, A general discrete time model of population dynamics in the presence of an infection, Discrete Dyn. Nat. Soc. 2009 (2009) 143019, 15 pp.
[25] D.S. Callaway, A.S. Perelson, HIV-1 infection and low steady state viral loads, Bull. Math. Biol. 64 (1) (2002) 29–64.
[26] L. Wang, M.Y. Li, D. Kirschner, Mathematical analysis of the global dynamics of a model for HTLV-I infection and ATL progression, Math. Biosci. 179 (2) (2002) 207–217.
[27] J.N. Tsisiklis, Problems in Decentralized Decision Making and Computation, PhD thesis, MIT, Cambridge, MA, 1984.

Chapter 2

Structural and performance patterns

2.1 Introduction

Discrete networked dynamic systems (DNDSs) are a collection of multiple dynamic systems coupled together through a network. These types of systems are found in a range of applications, including the coordination of multiple space, air, and land vehicles [1], [2], [3]. Recent interests in NDSs focused on the foundation of graph theory [4], [5] and examined graph-theoretic constructs in the context of networked systems, particularly the interplay between the agreement protocol (also known as the consensus algorithm) and graph theory. DNDSs are a particular class of NDSs when the information is processed in a discrete-time environment. A basic premise in network science is that the structure and attributes of the network influence the dynamical properties exhibited at the system level. For linear and discrete time-invariant models, it is known that all the essential system-theoretic properties can be derived from the quadruple system matrices (A, B, C, D). When considering multiagent systems, the underlying connection topology, $\mathbb{G}$, can typically be embedded into the system matrices. It is then enlightening to consider how certain properties of the system explicitly depend on that topology. Therefore, when studying linear DNDSs, one should consider the quintuple $(A, B, C, D, \mathbb{G})$ and explicitly describe the dependence of the underlying topology on the system properties.

In this chapter, the focus is on a general class of linear discrete-time dynamic systems interconnected over an information network, exchanging relative state measurements or output measurements. It should also be noted that the design of the underlying topology in the context of system-theoretic properties, such as the $\mathbb{H}_2$-norm, has received little attention in the literature.

2.2 Discrete networked dynamic systems

DNDSs are a collection of multiple dynamic systems that are coupled together through a network (or graph) and where the information is processed in discrete-time format. These types of systems are found in a range of applications that involve, for example, the coordination of multiple space, air, and land vehicles [1], [6], [7], [8]. Studying system-theoretic notions from the perspective of the underlying topology can lead to interpretations that explicitly characterize the effects of the network on the behavior of the system.

Discrete Networked Dynamic Systems. https://doi.org/10.1016/B978-0-12-823698-7.00010-2

2.2.1 Linear time-invariant systems

To begin with, we start with linear and time-invariant (LTI) systems, in which all the essential system-theoretic properties can be derived from the quadruple system matrices (A, B, C, D). When considering multiagent systems, the underlying connection topology, $\mathbb{G}$, can typically be embedded into the system matrices. It is then enlightening to consider how certain properties of the system explicitly depend on that topology. Therefore, when studying linear NDSs, one should consider the quintuple $(A, B, C, D, \mathbb{G})$ and explicitly describe the dependence of the underlying topology on the system properties. Recent examples of such network-centric analysis include relating closed-loop stability properties of DNDSs to the spectral properties of the graph Laplacian [9], relating controllability in consensus seeking systems to graph symmetry [10], and graph-centric observability properties of relative sensing DNDSs [11].

We start by presenting an initial consideration.

2.2.2 Preliminary view

Let $x_i(k) \in \Re$, $i = 1, \cdots, n$, be the state of agent i associated with vertex i, where k represents the time instant. By our stipulation of the physical meaning of the edge direction in a digraph, the existence of e_{ij} implies that agent i gets the state information x_j from agent j. Hence, the weight a_{ij} associated with e_{ij} can be regarded as the gain for the information flow. To this end, for a system with $\mathbb{G}(\mathbb{V}, \mathbb{E}, \mathbb{A})$ as its interconnection topology digraph, the existence of e_{ij} implies that agent i gets the amplified state information $a_{ij}x_j(k)$ from agent j. Let the measurement $y_i(k)$ of agent i be equal to the sum of all the amplified states at the present time received by agent i, that is,

$$y_i(k) = a_{ii}x_i(k) + \sum_{j \in \mathbb{N}_i} a_{ij}x_j(k). \tag{2.1}$$

Such a measurement will be also referred to as aggregated measurement. Sometimes, each agent can get only some relative measurement, which can be expressed as

$$y_i(k) = a_{ii}x_i(k) - \sum_{j \in \mathbb{N}_i} a_{ij}x_j(k). \tag{2.2}$$

We denote the state vector by $x(k) = [x_1(k), \cdots, x_n(k)]^t$ and the measurement vector by $y(k) = [y_1(k), \cdots, y_n(k)]^t$. Then, the matrix form of the aggregated measurement (2.1) is

$$y(k) = \bar{A}x(k). \tag{2.3}$$

Note that (2.3) provides an interpretation of the generalized adjacency matrix $\bar{A}$ from a viewpoint of system theory: *it can be considered as a measurement matrix.*

Suppose the state formation of each agent is updated by the following local law:

$$x(k+1) = \mathbf{K}y(k), \tag{2.4}$$

where $\mathbf{K} = \mathbf{diag}\{\kappa_i,\ i = 1,\ \cdots,\ n\}$. It follows that

$$x(k+1) = \mathbf{K}\bar{A}x(k). \tag{2.5}$$

Therefore, with the updating law (2.4) is a state-transfer matrix of a closed-loop system, as shown in Fig. 2.1.

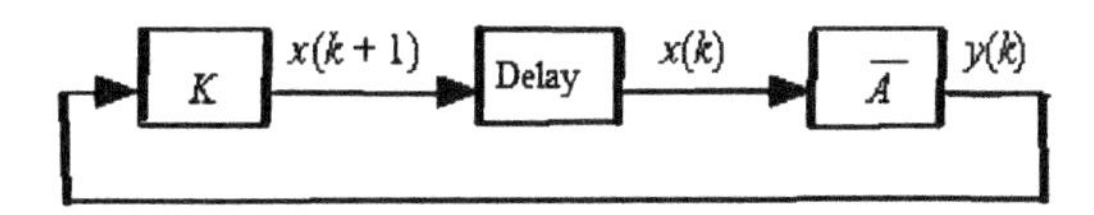

FIGURE 2.1 A closed-loop system.

In this section, we focus on a class of linear DNDSs where the underlying connection topology couples the agents at their outputs. Such systems are prevalent in formation flying applications where relative sensing is used to measure interagent distances [12].

In the sequel, we adopt the view that a DNDS consists of two system layers:

1) The first can be considered as the local agent layer corresponding to the dynamics of the individual agents in the ensemble.
2) The second layer is a global DNDS layer that represents the complete interconnected system.

In what follows, we develop new results for a general linear model of DNDSs that includes both the local and global layers and establish some interesting set-theoretic properties.

2.2.3 Problem statement

We will work with the following class of DNDSs:

$$\begin{aligned} x_i(k+1) &= A_i x_i(k) + B_i u_i(k) + \Gamma_i \omega_i(k), \\ y_i(k) &= C_i x_i(k) + D_i u_i(k) + \Phi_i \omega_i(k), \\ z_i(k) &= M_i x_i(k). \end{aligned} \tag{2.6}$$

We define the vertex set $\mathbb{N} = \{1, ..., \mathbf{N}\}$. We use $\{A_o, ..., \Phi_o\}$ to imply generic system matrices and $\{A_{oj}, ..., \Phi_{oj},\ j \in \mathcal{N}\}$ to represent the respective values at the vertices. In system (2.6), $x(k) \in \Re^n$ is the state vector, $\omega(k) \in \Re^q$ is the disturbance input which belongs to $\ell_2[0, \infty)$, $y(k) \in \Re^p$ is the measured output, and $z(k) \in \Re^q$ is the performance output. The matrices $A_o \in \Re^{n\times n}$, $B_o \in \Re^{n\times m}$, $C_o \in \Re^{p\times n}$, $D_o \in \Re^{p\times n}$, $M_o \in \Re^{q\times n}$, $\Gamma_o \in \Re^{n\times q}$, and $\Phi_o \in \Re^{p\times q}$

are real and known constant matrices. In the sequel, we assume that the pairs (A_i, B_i)–(A_i, C_i) are stabilizable-detectable.

The parallel interconnection of all agents is described with the following state-space description of DNDSs:

$$\begin{aligned} \mathbf{x}(k+1) &= \mathbf{A}\mathbf{x}(k) + \mathbf{B}\mathbf{u}(k) + \Gamma\omega(k), \\ \mathbf{y}(k) &= \mathbf{C}\mathbf{x}(k) + \mathbf{D}\mathbf{u}(k) + \Phi\omega(k), \\ \mathbf{z}(k) &= \mathbf{M}\mathbf{x}(k), \end{aligned} \tag{2.7}$$

with $\mathbf{x}$, $\mathbf{u}$, $\mathbf{y}$, $\mathbf{z}$, and ω denoting, respectively, the concatenated state vector, control vector, measured output vector, controlled vector, and exogenous disturbance input at time k of all the agents in the DNDS. The matrices $\mathbf{A}$, $\mathbf{B}$, $\mathbf{C}$, $\mathbf{D}$, $\mathbf{M}$, Γ, and Φ are the block-diagonal aggregation of each agent's state-space matrices. In the sequel, when working with homogeneous NDSs, we omit the subscript for all state-space and operator representations of the system.

Remark 2.1. We incorporate a feedforward term of the control $u_i(k)$ and added noises in the measurements as well. At instants when we seek a minimal realization for each agent with the outputs of each agent being compatible, system outputs correspond to the same physical quantity, and we can set $D_i \equiv 0$ and $\Phi_i \equiv 0$ for all agents.

The global DNDS layer we examine in this section is motivated by the relative sensing processing problem and focuses on *sensed state information*, where each node has the capacity to sense the state information of its adjacent nodes.

The global layer is visualized in the diagram of Fig. 2.2.

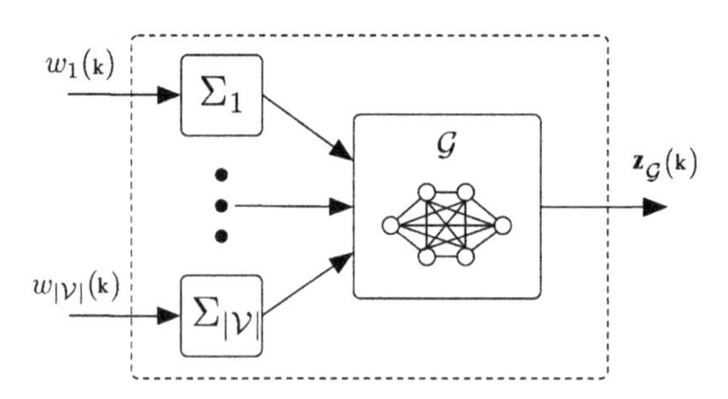

FIGURE 2.2 Block diagram of global DNDSs.

When considering the analysis of the global layer, we are interested in studying the map from the agent's exogenous inputs to the DNDS sensed output, which we denote by the operator $T_{hom}^{\omega \to \mathbb{G}}$ for homogeneous DNDSs, where it is assumed that each dynamic agent in the DNDS is described by the same set of linear state-space dynamics, that is, $\{A_{oi}, ..., \Gamma_{oi}\}, = \{A_{oj}, ..., \Gamma_{oj}, \forall\, i, j \in \mathcal{N}\}$. Hence, we omit the subscript for all state-space and operator representations of the system. Using the above notations and the Kronecker properties, we can express the homogeneous dynamics as

$$\begin{aligned}
\mathbf{x}(k+1) &= \hat{\mathbf{A}}\mathbf{x}(k) + \hat{\mathbf{B}}\mathbf{u}(k) + \hat{\Gamma}\omega(k), \\
\mathbf{y}(k) &= \hat{\mathbf{C}}\mathbf{x}(k) + \hat{\mathbf{D}}\mathbf{u}(k) + \hat{\Phi}\omega(k), \\
\mathbf{z}(k) &= \hat{\mathbf{M}}\mathbf{x}(k), \\
\mathbf{z}_{\mathbb{G}}(k) &= (\mathbb{E}(\mathbb{G})^t \otimes \mathbf{M})\mathbf{x}(k),
\end{aligned} \tag{2.8}$$

where

$$\begin{aligned}
&\hat{\mathbf{A}} = (I_{|\mathbb{V}|} \otimes \mathbf{A}), \ \hat{\mathbf{B}} = (I_{|\mathbb{V}|} \otimes \mathbf{B}), \\
&\hat{\mathbf{\Gamma}} = (I_{|\mathbb{V}|} \otimes \Gamma), \ \hat{\mathbf{M}} = (I_{|\mathbb{V}|} \otimes \mathbf{M}), \\
&\hat{\mathbf{C}} = (I_{|\mathbb{V}|} \otimes \mathbf{C}), \ \hat{\mathbf{D}} = (I_{|\mathbb{V}|} \otimes \mathbf{D}), \\
&\hat{\mathbf{\Phi}} = (I_{|\mathbb{V}|} \otimes \Phi).
\end{aligned} \tag{2.9}$$

2.3 System properties

In the sequel, we will review the basic information about system properties: stability, controllability, and observability.

2.3.1 Glimpse of stability

Intuitively, stability means that, without inputs, a system's response will converge to some equilibrium. Consider a general nonlinear system

$$x(k+1) = A[x(k)],$$

where $x \in \Re^n$ are the state variables and $A : \Re^n \to \Re^n$ is a (nonlinear) function. Assume $A(0) = 0$, and the equilibrium point $x(0) = 0$ is asymptotically stable if there exists a neighborhood of $x(0) = 0$ such that if the system starts in the neighborhood, then its sequence converges to the equilibrium point $x(0) = 0$ as k increases indefinitely. It is known that determining stability of a system is not easy if the system is nonlinear. One standard approach often used is the *Lyapunov approach*, which can be explained as follows.

Given a system, let us define some suitable "energy" function of the system. This function must have the property that it is zero at the origin and positive elsewhere. Assume further that the system dynamics are such that the energy of the system is monotonically decreasing with time and hence eventually reduces to zero. Then the sequences of the system have no other place to go but the origin. Therefore, the system is asymptotically stable.

This generalized energy function is called a *Lyapunov function*. The Lyapunov approach will be used in deriving much of the subsequent results.

2.3.2 Controllability and observability

Controllability and observability represent two major concepts of modern control system theory. For an introduction into these concepts, see the review in [13]. They can roughly be described as follows.

- **Controllability:** *In order to be able to do whatever we want with the given dynamic system under control input, the system must be* **controllable**. The importance of controllability is due to the fact that if a system is controllable, then we can move or place its poles or eigenvalues in arbitrary places within the unit disc by using state feedback.
- **Observability:** *In order to see what is going on inside the system under observation, the system must be* **observable**. It turns out that an LTI system is observable if the initial state $x(0)$ can be uniquely deduced from the knowledge of the input sequence $u(k)$ and output sequence $y(k)$ over the interval $k \in [0, \Delta]$ for some $\Delta > 0$. It is important to note that the initial state $x(0)$ is arbitrary: We can find $x(0)$ no matter where the system starts.

2.3.3 Stabilizability and detectability

Stabilizability is related to both stability and controllability. We learned that if a system is controllable, then we can use a state feedback to move its poles or eigenvalues to any locations within the unit disc in the complex plane. Therefore, we can always use state feedback to stabilize a controllable system: We just need to move the eigenvalues to within the unit disc in the complex plane. However, if the system is not controllable, then we may not be able to stabilize a system using state feedback.

Consider the LTI case

$$\begin{aligned} x(k+1) &= \mathbf{A}x(k) + \mathbf{B}u(k), \\ y(k) &= \mathbf{C}x(k) + \mathbf{D}u(k). \end{aligned} \tag{2.10}$$

Formally, $\lambda_j \in \mathbb{S}(\mathbf{A})$ is unstable if $|\lambda_j| > 1$. We say that an eigenvalue λ_j is not controllable if it cannot be moved by state feedback.

The LTI system (2.10) is *stabilizable* if all unstable eigenvalues are controllable. The following two sufficiency conditions can be easily obtained from the definition of stabilizability:

1. If a system is stable, then it is stabilizable.
2. If a system is controllable, then it is stabilizable.

Necessary and sufficient conditions for checking stabilizability are more complex. We first need to find all the eigenvalues of the system and then determine if these eigenvalues are controllable or not.

Lemma 2.1 ([13]). *An eigenvalue $\lambda_j \in \mathbb{S}(A) = \mathbb{S}(A^t)$ of an LTI system (2.10) is controllable if and only if its corresponding eigenvector v_j of A^t satisfies the condition $v_j^t B \neq 0$.*

Checking stabilizability is more complex than checking controllability. Since controllability implies stabilizability, we can first check controllability. If the system is controllable, then we know it is stabilizable.

In a similar way, the notion of "detectability" is presented. The LTI system (2.10) is *detectable* if all unstable eigenvalues are observable. The dual of Lemma 2.1 is provided by the following lemma.

Lemma 2.2 ([13]). *An eigenvalue $\lambda_j \in \mathbb{S}(A)$ of system (2.10) is observable if and only if its corresponding eigenvector v_j of A satisfies the condition $Cv_j^t \neq 0$.*

In the sequel, we will work with the following class of DNDSs:

$$\begin{aligned} x_i(k+1) &= A_i x_i(k) + B_i u_i(k) + \Gamma_i \omega_i(k), \\ y_i(k) &= C_i x_i(k) + D_i u_i(k) + \Phi_i \omega_i(k), \\ z_i(k) &= M_i x_i(k). \end{aligned} \tag{2.11}$$

We define the vertex set $\mathbb{N} = \{1, ..., \mathbf{N}\}$. We use $\{A_o, ..., \Phi_o\}$ to imply generic system matrices and $\{A_{oj}, ..., \Phi_{oj},\ j \in \mathcal{N}\}$ to represent the respective values at the vertices. In system (2.6), $x(k) \in \Re^n$ is the state vector, $\omega(k) \in \Re^q$ is the disturbance input which belongs to $\ell_2[0, \infty)$, $y(k) \in \Re^p$ is the measured output, and $z(k) \in \Re^q$ is the performance output. The matrices $A_o \in \Re^{n\times n}$, $B_o \in \Re^{n\times m}$, $C_o \in \Re^{p\times n}$, $D_o \in \Re^{p\times n}$, $M_o \in \Re^{q\times n}$, $\Gamma_o \in \Re^{n\times q}$, and $\Phi_o \in \Re^{p\times q}$ are real and known constant matrices. In the sequel, we assume that the pairs $(A_i,\ B_i)$–$(A_i,\ C_i)$ are controllable-observable and matrix A_i is stable, or in short, stabilizable-detectable.

2.4 Controllability Gramian

We proceed to provide a description of the $\mathbb{H}_2$ system norm by considering the controllability Gramian for an individual agent (from the exogenous input channel) based on the dynamics in (2.11), defined as

$$\mathbb{X}_c^i = \sum_{m=o}^{\infty} \mathbf{A}_i{}^m \Gamma_i \Gamma_i^t \mathbf{A}_i^{tm}. \tag{2.12}$$

The controllability Gramian $\mathbb{X}_c^i$ can be calculated by solving the corresponding Lyapunov equation

$$\mathbf{A}_i \mathbb{X}_c^i \mathbf{A}_i^t - \mathbb{X}_c^i + \Gamma_i \Gamma_i^t = 0. \tag{2.13}$$

It is known that the $\mathbb{H}_2$-norm of each agent from the exogenous input channel to the measured output can be expressed in terms of the Gramian as

$$||T_i^{\omega\to z}||_2 = \sqrt{\mathbf{trace}(\mathbf{G}_i \mathbb{X}_c^i \mathbf{G}_i^t)}, \tag{2.14}$$

where **trace**(.) is the trace operator. Using the above description we can begin to clarify how the underlying network topology influences the system norm. For simplicity in exposition, we separate our analysis into the homogeneous and heterogeneous cases.

2.4.1 $\mathbb{H}_2$-norm of homogeneous discrete networked dynamic systems

The $\mathbb{H}_2$-norm of the homogeneous DNDS described in (2.8) can be written in terms of the controllability Gramian. We consider the map $T_{hom}^{\omega \to \mathbb{G}}$ for examining the global NDS layer. Therefore, the expression for the controllability Gramian of the global DNDS layer in (2.8) is

$$\begin{aligned} \mathbb{X}_c &= \sum_{m=o}^{\infty} \hat{\mathbf{A}}_i^{t^m} (\mathbb{E}(\mathbb{G})^t \otimes \mathbf{G})^t (\mathbb{E}(\mathbb{G})^t \otimes \mathbf{G}) \hat{\mathbf{A}}_i^m \\ &= \mathbb{L}(\mathbb{G}) \otimes \mathbb{X}, \end{aligned} \tag{2.15}$$

where $\mathbb{X} := [\mathbb{X}^1, ..., \mathbb{X}^N]$ represents the aggregate controllability Gramian of agents in the network. On utilizing (2.14), we have the following characterization of the $\mathbb{H}_2$-norm:

$$\begin{aligned} ||T_{hom}^{\omega \to \mathbb{G}}||_2 &= \sqrt{\mathbf{trace}(I_N \otimes \Gamma)^t (\mathbb{L}(\mathbb{G}) \otimes \mathbb{X})(I_N \otimes \Gamma)} \\ &= ||\mathbb{E}(\mathbb{G})||_F ||T^{\omega \to z}||_2, \end{aligned} \tag{2.16}$$

where $||R||_F$ denotes the Frobenius norm of the matrix R. It is noted that the expression in (2.16) gives an explicit characterization of how the network affects the overall gain of the DNDS. In the homogeneous case, we can focus our attention on how the Frobenius norm of the incidence matrix changes with the addition or removal of an edge. Recall that the Frobenius norm of a matrix can be expressed as the sum of the vector 2-norm of each column.

In the case of the incidence matrix, each column, representing a single edge of the graph, always has the same structure. Therefore, the Frobenius norm of the incidence matrix can be expressed in terms of the number of edges in the graph, $|\mathbb{E}|$, as $\mathbb{E}(\mathbb{G})_F = \sqrt{2|\mathbb{E}|}$.

It follows that the DNDS $\mathbb{H}_2$-norm is only dependent on the number of edges in the graph rather than the actual structure of the topology.

Considering only connected graphs, we have immediate lower and upper bounds on the $\mathbb{H}_2$-norm of the system,

$$||T_{hom}^{\omega \to \mathbb{G}}||_2^2 \geq ||T_{hom}^{\omega \to y}||_2^2 (|\mathbb{V}| - 1). \tag{2.17}$$

The lower bound is attained with equality whenever the underlying graph is a spanning tree. Assuming that all graphs are simple, that is, they do not have multiple edges between a single pair of nodes, the upper bound for the system norm is achieved by the complete graph,

$$||T_{hom}^{\omega\to\mathbb{G}}||_2^2 \leq ||T_{hom}^{\omega\to y}||_2^2|\mathbb{V}|(|\mathbb{V}|-1). \tag{2.18}$$

2.4.2 Observability Gramian

Following [13], we know that the observability of the pair (A_i, C_i) and the stability of matrix A_i guarantee that the unique solution of the corresponding Lyapunov equation

$$A_i^t \mathbb{X}_o^i A_i - \mathbb{X}_o^i + C_i^t C_i = 0 \tag{2.19}$$

is positive definite and is given by

$$\mathbb{X}_o^i = \sum_{m=o}^{\infty} A_i^{tm} C_i^t C_i A_i^m, \tag{2.20}$$

where $\mathbb{X}_o^i$ is called the observability Gramian.

2.4.3 Homogeneous and heterogeneous discrete networked dynamic systems

Henceforth, we will be referring to two classes of DNDS:

1) those with homogeneous dynamics and
2) those with heterogeneous dynamics.

For both cases, we will work with LTI systems for each agent, driven by a generalized disturbance

$$\begin{aligned} x_i(k+1) &= A_i x_i(k) + B_i \omega_i(k), \\ y_i(k) &= C_i x_i(k). \end{aligned} \tag{2.21}$$

Each agent is indexed by the subscript i, and the subscript is omitted for the homogeneous case.

In the homogeneous case, it is assumed that each dynamic agent in the NDS is described by the same set of linear state-space dynamics (e.g., $(A_i, B_i, C_i) = (A_j, B_j, C_j), \ \forall i, j$). In the heterogeneous case, each agent is assumed to have different linear dynamics.

For both cases, we will assume a stable, strictly proper system ($D_i = 0$) with a minimal realization. Finally, we also assume that each agent has compatible outputs (e.g., system outputs correspond to the same physical quantity) and dimensions.

The observability Gramian of a dynamic system is an important operator that will be used throughout this chapter. The observability Gramian for an individual agent based on the dynamics in (1) is defined by (2.20). As each agent is assumed to be minimal, the Gramian is a positive definite matrix and can be expressed in terms of its singular value decomposition, $\mathbb{X}_o^i = U_i \Sigma_i U_i^T$. We denote, respectively, the largest and smallest singular values of $\mathbb{X}_o^i$ as $\overline{\sigma}(\mathbb{X}_o^i)$ and $\underline{\sigma}(\mathbb{X}_o^i)$.

The Gramian can be calculated by solving a system of linear equations, described by the Lyapunov equation (2.19).

Additionally, the Gramian can be used as a quantitative way to compare the relative observability of different modes in the system, as $\|y(t)\|_2 = \|Y_o^{1/2} x(0)\|_2$.

In both cases, we will represent the parallel interconnection of all the agents with the following state-space description:

$$
\begin{aligned}
\mathbf{x}(k+1) &= \mathbf{A}\mathbf{x}(k) + \mathbf{B}\omega(k), \\
\mathbf{y}(k) &= \mathbf{C}\mathbf{x}(k),
\end{aligned}
\tag{2.22}
$$

with $\mathbf{x}(k)$, $\mathbf{w}(k)$, and $\mathbf{y}(k)$ denoting, respectively, the concatenated state vector, the generalized disturbance vector, and the output vector of all the agents in the NDS. The dimensions of each agent's state, control, and output vectors are $x_i(k) \in \Re^n$, $u_i(k) \in \Re^m$, and $y_i(k) \in \Re^r$, respectively. In the heterogeneous case, the dimension of the state and the control need not be the same, but for notational convenience we only examine the above case. The matrices $\mathbf{A}$, $\mathbf{B}$, and $\mathbf{C}$ are the block-diagonal aggregation of each agent's state-space matrices.

The model we examine hereinafter is motivated by the relative sensing problem. The sensed output of the NDS is a vector $\mathbf{y}_{\mathcal{G}}(t) \in \mathbb{R}^{r|\mathcal{E}|}$ containing the relative information of each agent and its neighbors. For example, the output sensed across an edge $e = (i, i')$ would be of the form $y_i(t) - y_{i'}(t)$.

This can be compactly written as

$$
\mathbf{y}_{\mathbb{G}}(k) = (\mathbb{E}(\mathbb{G})^t \otimes I_r)\mathbf{y}(k). \tag{2.23}
$$

Here, "$\otimes$" denotes the Kronecker product and I_r is the $r \times r$ identity matrix. An important result on the singular value decomposition of Kronecker products will prove useful in subsequent discussions.

Theorem 2.1. *[14] Let $A \in \mathbb{R}^{m \times n}$ and $B \in \mathbb{R}^{p \times q}$ each have a singular value decomposition of $A = U_A \Sigma_A V_A^T$ and $B = U_B \Sigma_B V_B^T$. The singular value decomposition of the Kronecker product of A and B is*

$$
A \otimes B = (U_A \otimes U_B)(\Sigma_A \otimes \Sigma_B)(V_A^T \otimes V_B^T). \tag{2.24}
$$

In subsequent sections, we will refer to homogeneous and heterogeneous DNDSs by $\Sigma_{hom}(\mathcal{G})$ and $\Sigma_{het}(\mathcal{G})$, respectively. We also refer to $\Sigma_{hom}(\mathcal{G})$ and

$\Sigma_{het}(\mathcal{G})$ in an operator context. Using the above notations, we have the following compact descriptions for homogeneous and heterogeneous NDSs:

$$\Sigma_{hom}(\mathcal{G}) \begin{cases} \mathbf{x}(k+1) = (I_N \otimes A)\mathbf{x}(k) + (I_N \otimes B)\omega(k), \\ \mathbf{y}_{\mathbb{G}}(k) = (\mathbb{E}(\mathbb{G})^t \otimes C)\mathbf{x}(k), \end{cases} \tag{2.25}$$

$$\Sigma_{het}(\mathcal{G}) \begin{cases} \mathbf{x}(k+1) = \mathbf{A}(x) + \mathbf{B}\omega(k), \\ \mathbf{y}_{\mathbb{G}}(k) = (\mathbb{E}(\mathbb{G})^t \otimes I_r)\mathbf{C}x(k). \end{cases} \tag{2.26}$$

2.5 Observability properties of discrete networked dynamic systems

Examination of observability properties of a system is an important tool in the analysis of dynamical systems. It can be used to characterize, for example, the $\mathbb{H}_2$-norm of a system, as well as to provide answers to the existence of a state estimator. In this section we present observability analysis for the homogeneous and heterogeneous linear NDS separately. The results of both are valuable, noting that each agent can be individually compensated to achieve homogeneous or heterogeneous dynamics. Furthermore, it will be shown that the homogeneous system is a specialization of the heterogeneous case.

We also include in this discussion expressions for the observability Gramian for both the homogeneous and heterogeneous cases. As mentioned in Section 2.4.3, the observability Gramian of a dynamic system can give additional insight about the observability properties of the system. In the context of NDSs, the Gramian leads to an explicit characterization of how the underlying topology affects the observability properties.

2.5.1 Homogeneous systems

For the homogeneous case, we have the following result on its observability properties.

Proposition 2.1. *The homogeneous networked dynamic system (2.25) is unobservable.*

Proof. Using the PHB test for observability of a linear system, it suffices to show that we can construct a nonzero vector q such that

$$(I_n \otimes A)\mathbf{q} = \lambda\mathbf{q} \text{ and } (\mathbb{E}(\mathbb{G})^t \otimes C)\mathbf{q} = 0,$$

where λ is an eigenvalue of $(I_n \otimes A)$.

Let $\tilde{q}$ be an eigenvector of A such that $A\tilde{q} = \lambda\tilde{q}$. To construct an eigenvector for $I_n \otimes A$, we only need to concatenate n versions of $\tilde{q}$ into one vector. Thus, $q = 1 \otimes \tilde{\mathbb{G}}$.

Exploiting properties of the Kronecker product reveals that q is in the null space of $\mathbb{E}(\mathbb{G})^t \otimes C$, and

$$(\mathbb{E}(\mathbb{G})^t \otimes C)(1 \otimes \tilde{q}) = (\mathbb{E}(\mathbb{G})^T 1) \otimes (C\tilde{q}) = 0$$

renders the proof complete. □

It is also beneficial to discuss how the observable and unobservable subspaces of the NDS relate to the structure of the network. One way to examine these subspaces is to find a transformation matrix S that separates the system into its observable and unobservable components. There are many ways to construct such a transformation, and we will do so by exploiting properties of the incidence matrix associated with the underlying graph.

First, we define a partition of the network into a tree and its cycles. Denote $\mathbb{E}_\tau$ as the incidence matrix corresponding to any spanning tree subgraph of $\mathbb{G}$. The remaining edges necessarily complete the cycles in $\mathbb{G}$, and $\mathbb{E}_c$ denotes the incidence matrix for those edges. Therefore, with an appropriate permutation of the columns of $\mathbb{E}(\mathbb{G})$, we can always write $\mathbb{E}(\mathbb{G}) = [\mathbb{E}_\tau \;\; \mathbb{E}_c]$.

One important property of $\mathbb{E}_\tau$ is that its columns are linearly independent. We can construct the transformation matrix as $\mathbb{S} = ([\mathbb{E}_\tau \;\; 1]) \otimes I_n$.

Now, we define the new state $\mathbf{z}(k)$ such that $\mathbb{S}\mathbf{z}(k) = \mathbf{x}(k)$. The transformed state-space is thus

$$\begin{aligned} \mathbf{z}(k+1) &= \mathbb{S}^{-1}(I_n \otimes A)\mathbb{S}\mathbf{z}(k) + \mathbb{S}^{-1}(I_n \otimes B)\omega(k). \\ \mathbf{y}_{\mathbb{G}}(k) &= (\mathbb{E}^t \otimes C)S\mathbf{z}(k). \end{aligned} \tag{2.27}$$

Using properties of the Kronecker product, we note the following simplifications:

$$\mathbb{S}^{-1}(I_t \otimes A)\mathbb{S} = I_n \otimes A, \tag{2.28}$$

$$(\mathbb{E}(\mathbb{G})^t \otimes C)\mathbb{S} = \left[\left(\begin{bmatrix} \mathbb{E}_\tau^t \mathbb{E}_\tau \\ \mathbb{E}_c^t \mathbb{E}_\tau \end{bmatrix} \otimes C \right) \; 0 \right]. \tag{2.29}$$

This transformation clearly shows that the unobservable subspace is spanned by the $\mathbf{1}$ vector. Physically, this corresponds to a rigid body-type motion of the NDS. That is, we are not able to observe the inertial position of the formation. For the estimation problem, if the objective is to estimate the relative states between each agent, then we can accept the unobservable subspace. However, if we require an estimate of the inertial states, then we must effectively anchor one of the nodes to reconstruct the states of all the other nodes. We should also note that based on our earlier assumption that each agent is minimal and stable, we are at least guaranteed that the unobservable mode is stable.

An expression for the observability Gramian of the entire NDS in (2.25) is

$$\mathbb{X}_{\mathbf{o}} = \sum_{0}^{\infty}(I_N \otimes A)^{k^t}(\mathbb{E}^t \otimes C)^t(\mathbb{E}^t \otimes C)(I_N \otimes A)^k$$
$$= \mathbb{L}(\mathbb{G}) \otimes \mathbb{X}_o, \tag{2.30}$$

where $\mathbb{X}_o$ represents the observability Gramian of a single agent in the network (described in model (2.20)).

The form of (2.30) explicitly shows how the network structure affects the observability Gramian. In fact, (2.30) can be used as an alternative proof to Proposition 2.1 by invoking Theorem 2.1. Since zero is an eigenvalue of $L(\mathcal{G})$, it must also be an eigenvalue of $\mathbf{Y}_o$ with multiplicity $|\mathcal{V}|$, resulting in a positive semidefinite Gramian. This is equivalent to the system being unobservable.

2.5.2 Heterogeneous systems

We give conditions for when (2.26) is observable or unobservable.

Proposition 2.2. *The heterogeneous networked dynamic system (2.26) is observable if there is no eigenvalue of* $\mathbf{A}$ *that is an eigenvalue for each* A_i

Proof. We must show that we cannot construct a nonzero vector q that satisfies $\mathbf{A}q = \lambda q$ and $(\mathbb{E}^t \otimes I_r)\mathbf{C}_q = 0$. We assume that no agents share the same eigenvalues. The proof is similar if a subset of agents does share an eigenvalue. In this case, an eigenvector of $\mathbf{A}$ must have the form

$$q = \begin{bmatrix} 0_n^t & 0_n^t & \cdots & \tilde{q}_i^t & 0_n^t & \cdots & 0_n^t \end{bmatrix}^t,$$

where q_i is the eigenvector for A_i.

We now check to see if $0 \in \mathcal{N}\{(\mathbb{E}(\mathcal{G})^1 \otimes I_r)\mathbf{C}\}$

$$(\mathbb{E}(\mathbb{G})^t \otimes I_r)\mathbf{Cq} = (\mathbb{E}(\mathbb{G})^t \otimes I_r)\begin{bmatrix} 0_n^t & \cdots & (C_i\tilde{q}_i)^t & 0_n^t & \cdots \end{bmatrix}^t.$$

By assumption, $C_i\tilde{q}_i \neq 0$, and (2.26) is observable. □

Proposition 2.2 only provides a sufficient condition for observability. In order for a heterogeneous system to be unobservable, not only does each agent need to share a common eigenvalue, but the outputs of each agent associated with a certain direction must be indistinguishable. This is characterized in the following proposition.

Proposition 2.3. *The heterogeneous networked dynamic system (2.26) is unobservable if the following conditions are met:*

1. *there exists an eigenvalue,* λ^*, *of* $\mathbf{A}$ *that is common to each* A_i;
2. $C_i q_i = C_j q_j$ $\forall i, j$ *with* $A_i q_i = \lambda^* q_i$ $\forall i$.

Proof. By assumption, there exists a λ^* that is an eigenvalue for each A_i. We can construct an eigenvector for $\mathbf{A}$ as $q = [q_1^t, \cdots, q_{|\mathcal{V}|}^t]^t$, with $A_i q_i = \lambda^* q_i$. By condition 2, we have $\mathbf{C}_q = 1 \otimes r$, where $r = C_i\, q_i \neq 0$ for all i.

Using properties of the Kronecker product we have

$$(\mathbb{E}(\mathbb{G})^t \otimes I_r)\mathbf{C}\, q = (\mathbb{E}(\mathbb{G})^t 1 \otimes r) = 0.$$

This shows the system is unobservable. □

It is readily evident that Proposition 2.3 is a generalization of the homogeneous case.

The advantage of an observable heterogeneous system is the ability to reconstruct the inertial states of each agent using an observer (given that the conditions of Proposition 2.3 are not met). However, a heterogeneous system introduces another degree of complexity. For the homogeneous case, the assignment of an agent to a certain position in the network topology does not change the observability properties. In the heterogeneous case, the assignment of an agent to a certain position can have a dramatic effect on the system observability.

As in the homogeneous case, we can derive an expression for the observability Gramian of the heterogeneous NDS. Using the definition directly, we have

$$\mathbb{X}_{\mathbf{o}} = \sum_{0}^{\infty} \mathbf{A}^t \mathbf{C}^t (\mathbb{L}(\mathbb{G}) \otimes I_r)\mathbf{C}\mathbf{A}.$$

The above form, however, is not as satisfying as the form derived for the homogeneous case. The reason is that the graph structure does not decouple cleanly from the expression. However, with some manipulation, the expression can be derived to highlight the role of the network in a more transparent way.

We begin by first noting that

$$\mathbf{C}\mathbf{A}^k = \sum_{i=1}^{|\mathbb{V}|} (e_i e_i^t \otimes C_i A_i^k),$$

where $e_i \in \Re^{|\mathbb{V}|}$ is the ith unit coordinate basis vector for $\Re^{|\mathbb{V}|}$.

It can also be verified that

$$\mathbb{L}(\mathcal{G}) = \sum_{i=1}^{|\mathbb{V}|} \sum_{j=1}^{|\mathbb{V}|} e_i e_i^t \mathbb{L}(\mathbb{G}) e_j e_j^t.$$

Using these results, the expression for the observability Gramian can be further simplified to

$$\mathbb{X}_o = \sum_{i=1}^{|\mathbb{V}|} \sum_{j=1}^{|\mathbb{V}|} \sum_{0}^{\infty} e_i e_i^t \mathbb{L}(\mathbb{G}) e_j e_j^t \otimes \left(A_i^{k^t} C_i^t C_j A_j^k \right).$$

We now can introduce a notational simplification by defining the observability operator and its adjoint:

$$\Psi_i(x) = C_i A_i^k x \quad \text{and} \quad \Psi_i^*(y(k)) = \sum_{0}^{\infty} A_i^{k^t} C_i^t y(k).$$

Each agent is assumed to be stable and minimal, so we have $\Psi_i : \Re^n \mapsto L_2^m[0, \infty)$ and the adjoint $\Psi_i^* : L_2^m[0, \infty) \mapsto \Re^n$. We also note that the composition of Ψ_i^* with its adjoint, as in $\Psi_j^* \Psi_i$, is precisely equal to the observability Gramian of agent i, $\mathbb{X}_o^{(i)}$. More generally, $\mathbb{X}_{ij} = \Psi_i^* \Psi_j$ can be calculated by solving the Sylvester equation

$$A_i^t \mathbb{X}_o^i A_i - \mathbb{X}_o^i + C_i^t C_i = 0. \tag{2.31}$$

All the results above can be used to derive the following expression for the observability Gramian of a heterogeneous DNDS:

$$\mathbb{X}_o^i = (\mathbb{L}(\mathbb{G}) \otimes J_n) \circ (\Psi^* \Psi), \tag{2.32}$$

where J_n is the $n \times n$ matrix of all ones, $\Psi[\Psi_1 \quad \cdots \quad \Psi_{|\mathcal{V}|}]$, and $A \circ B$ denotes the Hadamard product of A and B.

The form of (2.32) is appealing in how it separates the role of the network from each agent. A precise characterization of the eigenvalues of (2.32) is nontrivial, but we can construct bounds on those values, as presented in [14]. In particular, since both terms in the Hadamard product are positive semidefinite matrices, we can apply the Schur theorem to obtain the following bound:

$$\underline{d}\, \underline{\sigma}(\Psi^* \Psi) \leq \underline{\sigma}(\mathbb{X}_o^i) \leq \overline{\sigma}(\mathbb{X}_o^i) \leq \bar{d}\, \overline{\sigma}(\Psi^* \Psi), \tag{2.33}$$

where $\underline{d} = \min_i [L(\mathcal{G}) \otimes J_n]_{ii}$ and $\bar{d} = \max_x [\mathbb{L}(\mathbb{G}) \otimes J_n]_{ii}$. These correspond, respectively, to the minimum and maximum degree vertices of the underlying graph.

The Gramian expression (2.32) can be represented alternatively as a node-weighted Laplacian. Consider scalar weights W_i on each node collected together in a diagonal matrix $\mathbf{W} = \mathbf{diag}\{w_1, \ldots, w_{|\mathcal{V}|}\}$. The node-weighted Laplacian can be defined as

$$\hat{\mathbb{L}}(\mathbb{G}) = \mathbf{W} \mathbb{L}(\mathbb{G}) \mathbf{W} = \mathbb{L}(\mathbb{G}) \circ w w^t. \tag{2.34}$$

This can be generalized to $n \times n$-block matrix weights, and (2.34) can be equivalently written as

$$\hat{\mathbb{L}}(\mathbb{G}) = \mathbf{W}(\mathbb{L}(\mathbb{G}) \otimes I_n)\mathbf{W}^t. \tag{2.35}$$

Using (2.35) leads to a new interpretation of the expression in (2.32). Each node in the graph is weighted by the observability operator of the agent assigned to that node, and we have

$$\mathbb{X}_o^i = \mathbf{diag}\{\Psi^*\}(\mathbb{L}(\mathbb{G}) \otimes I_n)\mathbf{diag}\{\Psi\}. \tag{2.36}$$

2.5.3 Necessary and sufficient conditions

The results of the previous sections provide necessary and sufficient conditions for the observability of an NDS. These conditions do not depend on the homogeneity of the NDS, as the general conditions capture both scenarios. We combine the results into the following theorem.

Theorem 2.2. *Consider an NDS composed of homogeneous or heterogeneous dynamics that are individually observable. The NDS is unobservable if and only if the following conditions are met:*

1. *there exists an eigenvalue of* $\mathbf{A}$, λ^* *that is common to each* A_i;
2. $C_i q_i = C_j q_j \ \forall i, j$ *with* $A_i q_i = \lambda^* q_i \ \forall i$.

Proof. The proof follows immediately from the proofs of Propositions 2.1, 2.2, and 2.3. □

2.6 Index of homogeneity and heterogeneity

The previous section only provides a definite answer ("yes" or "no") to the question of observability in an NDS. As discussed earlier, the singular values of the observability Gramian can be used to give a quantitative comparison of the relative observability between different modes of the system. In the context of a single agent, the symmetry of the observability ellipsoid could be considered as a description of the homogeneity of that agent's initial condition to output map. As an example, the ellipsoid in Fig. 2.3A is symmetric, which corresponds to the output energy being independent of the direction of the initial condition of the system. On the other hand, the ellipsoid in Fig. 2.3B shows the output energy is strongly dependent on the direction of the initial condition.

The shape of the ellipsoid, of course, corresponds to the relative magnitude of the singular values of the observability Gramian.

This notion can be extended for DNDSs to answer the following questions:

1. How does the structure of the underlying network topology affect the relative observability of the DNDS?
2. How does the placement of agents in the network affect the relative observability of the NDS?

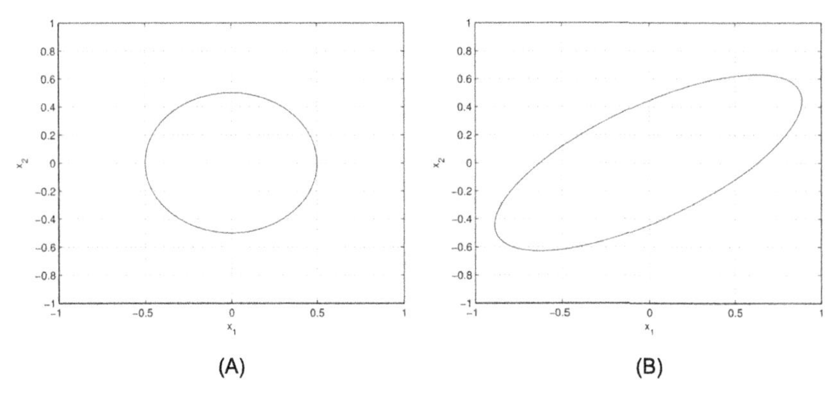

FIGURE 2.3 Visualization of observability Gramian ellipsoids for a "symmetric" system and a "stretched" system. (A) Symmetric ellipsoid; (B) Stretched ellipsoid.

More fundamentally, these questions suggest that certain topologies in a homogeneous system might be "more homogeneous" than others. Similarly, placing heterogeneous agents in different locations of a DNDS might result in a "more heterogeneous" DNDS. This would correspond to a symmetry, or lack thereof, of the observability ellipsoid of the DNDS.

This section aims to develop an index of homogeneity for homogeneous DNDSs and an index of heterogeneity for heterogeneous DNDSs that can be used to answer these questions. It is natural that these measures should somehow relate to the observability Gramian of the DNDS. As in Section 2.5, we separate the discussion into the homogeneous and heterogeneous settings.

2.6.1 DNDS index of homogeneity

In the homogeneous case, as indicated by (2.30), we recognize that the network topology has a direct effect on the observability Gramian. Furthermore, the statement of Theorem 2.1 shows that the eigenvalues of $\mathbb{X}_{\mathbf{o}}$ are the eigenvalues of $\mathbb{X}_o$ scaled by the eigenvalues of the graph Laplacian, $\mathbb{L}(\mathbb{G})$. The index of homogeneity should capture the effect of the network on the overall observability properties. Using the symmetry analogy developed earlier, a more homogeneous NDS should correspond to a more symmetric observability Gramian.

The index of homogeneity will be denoted as $\rho(\Sigma_{hom}(\mathbb{G}))$. One choice for this index is

$$\rho(\Sigma_{hom}(\mathbb{G})) = \left(\frac{\lambda_2(\mathbb{G})}{\lambda_{|\mathbb{V}|}(\mathbb{G})}\right)\frac{\underline{\sigma}(\mathbb{X}_o)}{\overline{\sigma}(\mathbb{X}_o)}, \tag{2.37}$$

where $\lambda_2(\mathbb{G})$ and $\lambda_{|\mathbb{V}|}(\mathbb{G})$ denote, respectively, the second smallest and largest eigenvalue of the graph Laplacian.

Using this index for characterizing the relative observability properties of the homogeneous DNDS leads to some interesting observations. First, note that whenever the graph is disconnected, $\rho(\Sigma_{hom}(\mathbb{G})) = 0$. This corresponds to the intuitive result that a disconnected graph should somehow be "less homogeneous" than a connected one. In terms of this specific index, the homogeneity of the NDS is lower-bounded by 0, and is indistinguishable from any disconnected graph on $|\mathbb{V}|$ nodes.

This index is also upper-bounded by $\underline{\sigma}(\mathbb{X}_o)/\overline{\sigma}(\mathbb{X}_o)$. This upper bound is achieved whenever the underlying graph is complete. The complete graph is the only graph where $\lambda_2(\mathbb{G}) = \lambda_{|\mathbb{V}|}(\mathbb{G})$.

Finally, we note the set of graphs that are cospectral with respect to the graph Laplacian will all result in the same index of homogeneity. This property could prove to be useful if reconfiguration of the connection topology is required.

The motivation for choosing such a function has a more intuitive explanation relating to the symmetry arguments of the observability Gramian.

The term containing the ratio of the smallest and largest singular values of Y_o corresponds loosely to a measure of the eccentricity of the Gramian ellipsoid. The closer this ratio is to the value 1, the more symmetric the ellipsoid is. Conversely, as this ratio approaches 0, the ellipsoid becomes more elongated (along one plane). We have considered a minimal realization for the system dynamics in order to guarantee that this ratio will always be strictly positive.

Next, we consider the observability Gramian of the parallel configuration of homogeneous DNDSs, corresponding to the system in (2.22). The Gramian can be written as $\tilde{\mathbb{X}}_o = I_N \otimes \mathbb{X}_o$.

In the N-agent case, the ellipsoid of agent i is oriented orthogonally to the ellipsoid of agent j. This is illustrated by the solid lines in Fig. 2.4. In this example, we look at the Gramian for a 4-agent homogeneous system with two states. The Gramian for each agent is the same, and its two-dimensional projection is plotted for each pair of state variables.

When the parallel DNDS is coupled by a network, say, a path graph, the ellipsoid becomes scaled and rotated. This is visualized by the dotted lines in Fig. 2.4. We immediately notice that one ellipsoid is scaled by the 0 eigenvalue of graph Laplacian. Using the statement of Theorem 2.1, we see that $\underline{\sigma}(\mathbb{X}_o) = \lambda_2 \underline{\sigma}(\mathbb{X}_o)$ and $\overline{\sigma}(\mathbb{X}_o) = \lambda_N \overline{\sigma}(\mathbb{X}_o)$ are, respectively, the minimum and maximum nonzero singular values of $\mathbb{X}_o$. We thus have the following relationship:

$$0 < \lambda_2 \underline{\sigma}(\mathbb{X}_o) \leq \lambda_N \overline{\sigma}(\mathbb{X}_o). \tag{2.38}$$

In the homogeneous DNDS, $\mathbb{X}_o$ represents a fixed property of the system, determined by the agent dynamics. Thus, in terms of the symmetry argument, a more homogeneous NDS should preserve as closely as possible the shape of the Gramian. Scaling the eigenvalues of $\mathbb{X}_o$ by $\lambda_{|v|}(\mathbb{G})\overline{\sigma}(\mathbb{X}_o)$ is effectively normalizing the observability Gramian singular values to 1.

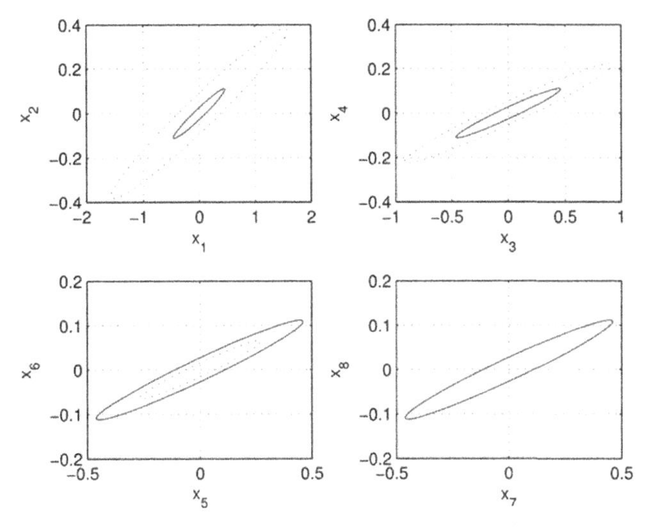

FIGURE 2.4 Visualization of observability Gramian ellipsoids.

2.6.2 DNDS index of heterogeneity

In the heterogeneous case, we aim to characterize how the topology affects the observability properties as well as how the placement of agents within that topology affects the observability of the DNDS. Contrary to the homogeneous case, the interplay between the graph Laplacian eigenvalues and the eigenvalues of the DNDS Gramian is less straightforward. A nice property of the index of homogeneity is that it can be computed by studying – independently – the spectral properties of the graph and the observability properties of the homogeneous agents. Finding an analogous approach for the index of heterogeneity reduces to understanding the spectral properties of (2.32) or (2.36), which requires further examination.

An index of heterogeneity can be developed using the numerical evaluation of the Gramian. The index of heterogeneity will be denoted as $\rho(\Sigma_{het}(\mathcal{G}))$. One choice for this index is

$$\rho(\Sigma_{het}(\mathbb{G})) = \left(\min_{\sigma_i(\mathbb{X}_o)\neq 0} \sigma_i(\mathbb{X}_o) \right)^{-1} \overline{\sigma}(\mathbb{X}_o), \tag{2.39}$$

where $\mathbb{X}_o$ is given in (2.36).

Although not as transparent as the index of homogeneity, some useful observations can be made about this choice of index. It can be seen that the index is upper-bounded by 1, which corresponds to an upper bound on the homogeneity of the DNDS. It is interesting to note that this upper bound can be achieved by a homogeneous DNDS with a complete graph topology, and with the agent Gramian ellipsoid being completely symmetric.

In fact, if all the agents in the NDS are homogeneous, then the index of heterogeneity reduces to (2.37). It might be natural to assume that the observed properties of (2.37) also apply to the heterogeneous case. Unfortunately, this is not the case, as is best illustrated with a simple example.

We consider a heterogeneous NDS with 4 agents and three different topologies. The topologies used are a star graph, a path graph, and the complete graph, which are shown in Fig. 2.5.

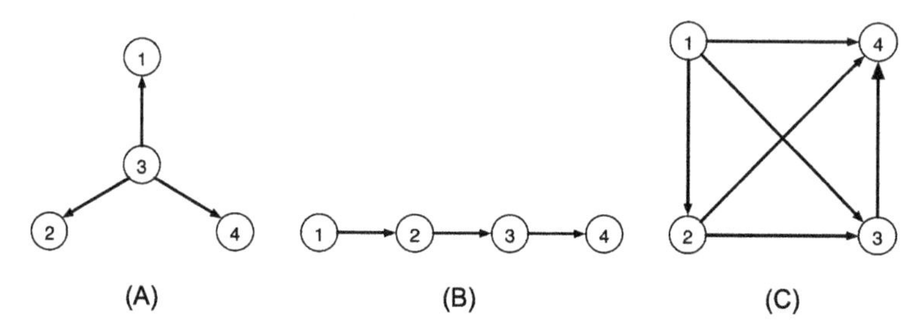

FIGURE 2.5 Three topologies on 4 nodes. (A) Star; (B) Path; (C) Complete.

Note that there are only 4 unique node assignments for the star graph, 12 unique assignments for the path graph, and 1 for the complete graph. For each permutation of the agent's position, the index of heterogeneity was calculated and plotted in Fig. 2.6. As indicated in the above discussion, larger values of $\rho(\Sigma_{het}(\mathbb{G}))$ correspond to the DNDS being "more homogeneous." The important point to notice in the figure is that the topology alone is not sufficient to determine which systems are more homogeneous. Furthermore, it can be seen that the complete graph does not correspond to the most homogeneous system.

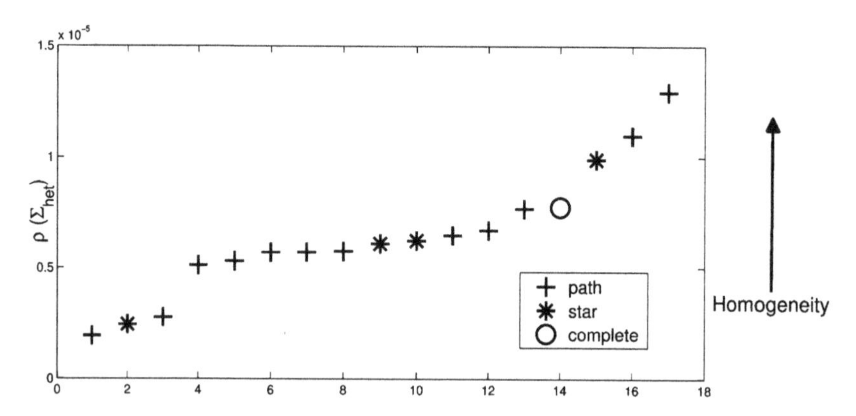

FIGURE 2.6 Index of heterogeneity.

2.7 Agent feedback stability

In what follows, we treat the closed-loop feedback stability based on sensed state information. At the agent layer, the stability protocol for each agent is distributed and only depends on the information of the agent itself and its neighbors due to agent limited capability. In what follows, we adopt the stability protocol

$$u_j(k) = K_s \sum_{m \in \mathbb{N}_j} a_{jm}[\mathbf{x}_j(k) - \mathbf{x}_m(k)], \;\; j = 1, 2, .., \mathrm{N}, \tag{2.40}$$

as determined by the information each node can sense from its adjacent nodes, which is the state measurement or the output measurement. Here, $\mathbb{N}_j$ denotes the neighbors of agent j, a_{jm} is an adjacent element of $\mathbb{A}$ in the graph $\mathbb{G} = (\mathbb{V},\ \mathbb{E},\ \mathbb{A})$, and K_s is the associated controller gain of the protocol.

Now, in terms of system model (2.8) at time instant k together with (2.40), we have

$$\begin{aligned} \mathbf{x}(k+1) &= \hat{\mathbf{A}}\,\mathbf{x}(k) + \hat{\mathbf{B}}\,\mathbf{u}(k) + \hat{\mathbf{\Gamma}}\,\omega(k), \\ \mathbf{y}(k) &= \hat{\mathbf{C}}\,\mathbf{x}(k) + \hat{\mathbf{D}}\,\mathbf{u}(k) + \hat{\mathbf{\Phi}}\,\omega(k), \\ \mathbf{u}(k) &= \hat{K}_s\,\mathbf{x}(k) = (I \otimes K_s)\,\mathbf{x}(k). \end{aligned} \tag{2.41}$$

Since the Laplacian matrix $\mathbb{L}$ of the undirected network is symmetric, positive semidefinite, by the spectral decomposition we have $\mathbb{L} = \mathbb{T}\,\Lambda\,\mathbb{T}^{-1}$, where $\mathbb{T} \in \Re^{N \times N}$ is an orthogonal matrix formed from the eigenvectors of $\mathbb{L}$ and $\Lambda = diag\{\lambda_1,\ \lambda_2, \cdots,\ \lambda_N\}$ is the matrix of the eigenvalues of $\mathbb{L}$. Now consider an orthogonal state transformation $x \longmapsto (\mathbb{T}^{-1} \otimes \mathbf{I})\xi$. Then, in the new coordinates, system (2.41) with (2.40) can be expressed as

$$\begin{aligned} \xi(k+1) &= [\hat{\mathbf{A}} + \Lambda \otimes \hat{\mathbf{B}}\hat{K}_s]\xi(k) + \hat{\mathbf{\Gamma}}\,\omega(k), \\ \zeta(k) &= [\hat{\mathbf{C}} + \Lambda \otimes \hat{\mathbf{D}}\hat{K}_s]\xi(k). \end{aligned} \tag{2.42}$$

For the purpose of computing the $\mathbb{H}_2$-norm, we denote the open-loop map of systems (2.6), (2.41), and (2.42) as $\mathbf{T}_j$ $(j = 1,\ 2,\ ...,\ N)$, $\widetilde{\mathbf{T}}$, and $\widehat{\mathbf{T}}$, respectively. This will pave the way to transfer the linear-quadratic regulator (LQR) control of the multiagents into the LQR control of the single-agent system. After that, we analyze the LQR performance region with respect to the eigenvalues of the Laplacian matrix $\mathbb{L}$.

2.7.1 Computation of the $\mathbb{H}_2$-norm

We have the following initial result.

Theorem 2.3. *The maps* $\mathbf{T}_j$ $(j = 1,\ 2,\ ..,\ N)$, $\widetilde{\mathbf{T}}$, *and* $\widehat{\mathbf{T}}$ *are equal, that is,*

$$||\widetilde{\mathbf{T}}||_2^2 = ||\widehat{\mathbf{T}}||_2^2 = \sum_{j=1}^{N} \lambda_j\, ||\mathbf{T}_j||_2^2,\ \ 1,\ 2,\ ...,\ N. \tag{2.43}$$

Proof. We have

$$
\begin{aligned}
||\widetilde{\mathbf{T}}||_2^2 &= ||\hat{\mathbf{C}}(zI-\hat{\mathbf{A}})^{-1}\hat{\mathbf{B}}+\hat{\mathbf{D}}||_2^2 \\
&= ||(I\otimes C)(zI-I\otimes A)^{-1}(L\otimes B)+(\mathbb{L}\otimes D)||_2^2 \\
&= ||(I\otimes C)(I\otimes (zI-A)^{-1}(L\otimes B))+(\mathbb{L}\otimes D)||_2^2 \\
&= ||\mathbb{L}\otimes[\hat{\mathbf{C}}(zI-\hat{\mathbf{A}})^{-1}\hat{\mathbf{B}}+\hat{\mathbf{D}}]||_2^2.
\end{aligned}
\tag{2.44}
$$

It is evident that

$$
\begin{aligned}
||\widetilde{\mathbf{T}}||_2^2 &= ||(\mathbb{T}^{-1}\otimes I_n)\widetilde{\mathbf{T}}(\mathbb{T}\otimes I)||_2^2 \\
&= ||\widehat{\mathbf{T}}||_2^2.
\end{aligned}
\tag{2.45}
$$

Additionally,

$$
\begin{aligned}
||\widetilde{\mathbf{T}}||_2^2 &= ||(\mathbb{T}^{-1}\otimes I)\widetilde{\mathbf{T}}(\mathbb{T}\otimes I_n)||_2^2 \\
&= ||(\mathbb{T}^{-1}\mathbb{L}\mathbb{T})\otimes[\hat{\mathbf{C}}(zI-\hat{\mathbf{A}})^{-1}\hat{\mathbf{B}}+\hat{\mathbf{D}}]||_2^2 \\
&= ||\Lambda\otimes[\hat{\mathbf{C}}(zI-\hat{\mathbf{A}})^{-1}\hat{\mathbf{B}}+\hat{\mathbf{D}}]||_2^2 \\
&= ||[I\otimes(\hat{\mathbf{C}}(zI-\hat{\mathbf{A}})^{-1})(\Lambda\otimes B)+(\Lambda\otimes D)]||_2^2 \\
&= ||diag\begin{bmatrix}\Pi_1 & \cdots & \Pi_j & \cdots & \Pi_N\end{bmatrix}||_2^2 \\
&= \sum_{j=1}^{N}\lambda_j\,||G_j||_2^2,
\end{aligned}
\tag{2.46}
$$

$$
\Pi_j = C(zI-A)^{-1}(\lambda_j B)+(\lambda_j D).
$$

This completes the proof. □

From Theorem 2.1, one can easily conclude the following corollary.

Corollary 2.1. *System (2.41) is asymptotically stable and achieves the LQR performance specification if and only if the N subsystems are simultaneously asymptotically stable and achieve the desired LQR performance.*

Remark 2.2. From Corollary 2.1, we can conclude that minimizing $||\widetilde{\mathbf{T}}||_2^2$ equals minimizing $||\mathbf{T}_j||_2^2$, $j=1,\ 2,\ \ldots,\ N$, simultaneously. This means that the LQR control of the whole multiagent system (MAS) can be decoupled into the LQR control of N subsystems. Considering the particular structure of the state feedback gain matrix $\widetilde{K}_s$, the control objective is to find the common feedback gain K_s such that the distributed subsystems achieve a suboptimal performance specification. In the sequel, we examine this issue and an associated region of the LQR performance region.

2.7.2 Agent state feedback control

We consider an LQR control problem for system (2.41), with the following cost function coupling the dynamic behavior of the systems:

$$\begin{aligned}\mathbb{J} &:= ||\mathbf{y}||_2^2 \\ &= \sum_{k=0}^{\infty} \left[\mathbf{x}^t(k)\hat{\mathbf{C}}^t\hat{\mathbf{C}}\mathbf{x}(k) + \mathbf{u}^t(k)\hat{\mathbf{D}}^t\hat{\mathbf{D}}\mathbf{u}(k)\right].\end{aligned} \tag{2.47}$$

Remark 2.3. The cost function (2.47) is used to describe the period and difference among various node dynamics before stability is achieved. It represents a suitable measure of both swiftness and vibration of an agent's stability. Here, the term vibration means the maximum amplitude of the state signals among all nodes of the network before stability is achieved.

2.7.3 Linear matrix inequality-based formulation

In what follows, we present a linear matrix inequality (LMI)-based formulation to the static feedback control of system (2.41) while minimizing the quadratic cost (2.47). By adopting Lyapunov theory, we select a Lyapunov functional $\mathbf{V}(\mathbf{x}(k))$ of the form $\mathbf{V}(\mathbf{x}(k)) = \mathbf{x}^t(k)\mathbf{P}\mathbf{x}(k)$, $\mathbf{P} > 0$, that satisfies

$$\begin{aligned}\Delta\mathbf{V}(\mathbf{x}(k)) &= [\mathbf{V}(\mathbf{x}(k+1)) - \mathbf{V}(\mathbf{x}(k))] \\ &\leq -\mathbf{x}^t(k)[\hat{\mathbf{C}}^t\hat{\mathbf{C}} + \hat{K}_s{}^t\hat{\mathbf{D}}^t\hat{\mathbf{D}}\hat{K}_s]\mathbf{x}(k).\end{aligned} \tag{2.48}$$

The following theorem provides an LMI-based LQR design.

Theorem 2.4. *System (2.41) with $\omega(k) \equiv 0$ and the stabilizing control $\mathbf{u}(k) = \widetilde{K}_s\mathbf{x}(k)$ is asymptotically stable and $\mathbb{J}_\infty \leq \mathbf{V}(x_o)$, if there exist matrices $\mathbf{W} > 0$, $\mathbf{S}$ such that*

$$\begin{bmatrix} -\mathbf{W} & \mathbf{S}\hat{\mathbf{D}}^t & \mathbf{W}\hat{\mathbf{A}}^t + \mathbf{S}\hat{\mathbf{D}}^t & \mathbf{W}\hat{\mathbf{C}}^t \\ \bullet & -\mathbf{I} & 0 & 0 \\ \bullet & \bullet & -\mathbf{W} & 0 \\ \bullet & \bullet & \bullet & -\mathbf{I} \end{bmatrix} < 0 \tag{2.49}$$

has a feasible solution, then the feedback gain matrix is $\hat{K}_s = \mathbf{S}^t\mathbf{W}^{-1}$.

Proof. Considering system (2.41) with $\omega(k) \equiv 0$, it follows from (2.47) with $\mathbf{P} > 0$ that

$$\begin{aligned}
\mathbb{J} &= \sum_{k=0}^{\infty} \mathbf{x}^t(k)[\hat{\mathbf{C}}^t\hat{\mathbf{C}} + \hat{K_s}^{t}\hat{\mathbf{D}}^t\hat{\mathbf{D}}\hat{K_s}]\mathbf{x}(k) \\
&\leq \sum_{k=0}^{\infty} \mathbf{x}^t(k)[\hat{\mathbf{C}}^t\hat{\mathbf{C}} + \hat{K_s}^{t}\hat{\mathbf{D}}^t\hat{\mathbf{D}}\hat{K_s}]\mathbf{x}(k) \\
&+ \sum_{k=0}^{\infty} \mathbf{x}^t(k)[\widetilde{\mathbf{A}}_s^t\mathbf{P}\widetilde{\mathbf{A}}_s]\mathbf{x}(k), \qquad (2.50)
\end{aligned}$$

$$\widetilde{\mathbf{A}}_s = (\hat{\mathbf{A}} + \hat{\mathbf{B}}\widetilde{K_s}). \qquad (2.51)$$

If there exist matrices $\mathbf{P} > 0$, $\widetilde{K_s}$ such that

$$\hat{\mathbf{C}}^t\hat{\mathbf{C}} + \hat{K_s}^{t}\hat{\mathbf{D}}^t\hat{\mathbf{D}}\hat{K_s} + \widetilde{\mathbf{A}}_s^t\mathbf{P}\widetilde{\mathbf{A}}_s - \mathbf{P} < 0, \qquad (2.52)$$

then, as $\mathbf{x}(k) \to 0$, $k \to \infty$, we get

$$\mathbb{J} \leq \sum_{k=0}^{\infty} \mathbf{x}^t(k)\mathbf{P}\mathbf{x}(k) = \mathbf{x}^t(0)\mathbf{P}\mathbf{x}(0). \qquad (2.53)$$

Using the Schur complements, applying the congruence transformation $\mathbb{C} = diag[\mathbf{W},\ I,\ \mathbf{W},\ I]$, $\mathbf{W} = \mathbf{P}^{-1}$ to (2.52), and introducing the linearization $\mathbf{S} = \mathbf{W}\hat{K_s}^{t}$ for some matrix $\mathbf{S}$, we readily obtain LMI (2.49) and therefore the proof is completed. □

2.8 Synthesis schemes of discrete networked dynamic systems

The results of the foregoing section can be used to develop a performance metric for the synthesis of DNDSs. The objective is to design a local controller K_i for each agent in the ensemble that minimizes some local performance objective, $||T_i^{\omega \to y}||_2$, while additionally minimizing the global DNDS objective, $||T_{hom}^{\omega \to \mathbb{G}}||_2$ (see Fig. 2.7). It should be noted that this procedure is analogous to an inner-loop control design, whereas additional performance involving the network would be likened to the outer-loop design of the system. To this end, we consider designing the local control for each agent when the underlying topology and the placement of agents within that topology are given and fixed. A semidefinite programming solution is developed to solve this problem.

2.8.1 Local agent control for fixed topology

For this problem we will consider a heterogeneous NDS with a given and fixed topology, $\mathbb{E}(\mathbb{G})$. Each agent, Σ_i, is also assigned a fixed location within the

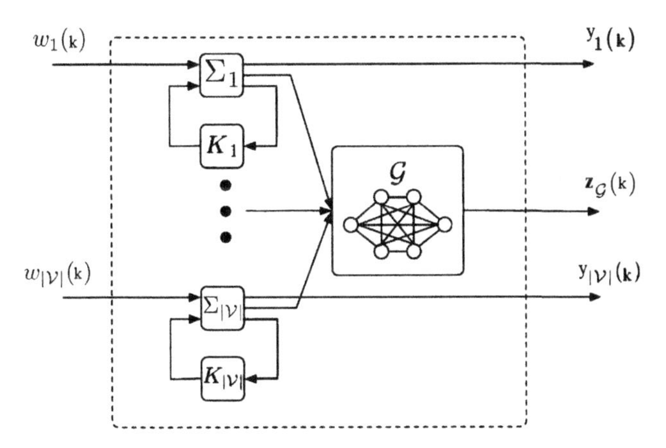

FIGURE 2.7 Synthesis of DNDSs.

network. Looking at it in this light, each agent behaves independently and does not use information from the DNDS for its control. For simplicity in exposition, we consider the state sensing case, that is, setting $M_i = 0$. Further, we assume that the global DNDS output corresponds to a relative position measurement and the DNDS output $\mathbf{z}_{\mathbb{G}}(k)$ will be described as

$$\mathbf{z}_{\mathbb{G}}(k) = \mathbb{E}(\mathbb{G})^t \otimes [\mathbf{1}^t 0...0] = \mathbb{E}(\mathbb{G})^t \otimes M_p, \tag{2.54}$$

where we have assumed the states corresponding to position are the first p states of the state vector $\mathbf{x}(k)$.

To proceed further, we now provide design of $\mathbb{H}_2$ controllers to guarantee a stabilizing system (2.41) with prescribed performance. We start with the design of the $\mathbb{H}_2$ controller.

2.8.2 $\mathbb{H}_2$ control design

Based on Lyapunov theory, we select a Lyapunov functional $\mathbf{V}(\mathbf{x}(k))$ of the form $\mathbf{V}(\mathbf{x}(k)) = \mathbf{x}^t(k)\mathbb{P}\mathbf{x}(k)$, $\mathbb{P} > 0$. The objective of the $\mathbb{H}_2$ controller is to ensure the stability of the closed-loop system and to keep the $\mathbb{H}_2$-norm of the transfer function $\widetilde{\mathbf{G}}$ from ω to $\mathbf{y}$ as small as possible. Using the sensed state information model (2.41) with control law (2.41), the control system can be cast into the following form:

$$\begin{aligned}
\mathbf{x}(k+1) &= \widetilde{\mathbf{A}}_c\, \mathbf{x}(k) + \widetilde{\mathbf{\Gamma}}\, \omega(k), \\
\mathbf{y}(k) &= \widetilde{\mathbf{C}}_c\, \mathbf{x}(k) + \widetilde{\mathbf{\Phi}}\, \omega(k), \\
\widetilde{\mathbf{A}}_c &= \hat{\mathbf{A}} + \hat{\mathbf{B}}(I \otimes K_s), \\
\widetilde{\mathbf{C}}_c &= \hat{\mathbf{C}} + \hat{\mathbf{D}}(I \otimes K_s).
\end{aligned} \tag{2.55}$$

It is known that the closed-loop system (2.55) is internally asymptotically stable with $\omega(k) \equiv 0$ if

$$\mathbb{P} - \widetilde{\mathbf{A}}_c^t \mathbb{P} \widetilde{\mathbf{A}}_c > 0. \tag{2.56}$$

Then the square of the $\mathbb{H}_2$-norm of the transfer function $\widetilde{\mathbf{G}}$ can be expressed in terms of the solution of a Lyapunov equation (controllability Gramian) such that the corresponding minimization problem with respect to the controller gain $\widetilde{K}_s$ is given by

$$\min \left\{ \mathbf{trace}[\widetilde{\mathbf{C}}_c \mathbf{P}_c \widetilde{\mathbf{C}}_c^t] : \mathbf{P}_c - \widetilde{A}_c^t \mathbf{P}_c \widetilde{\mathbf{A}}_c + \widetilde{\mathbf{\Gamma}} \widetilde{\mathbf{\Gamma}}^t = 0 \right\}. \tag{2.57}$$

Since $\mathbf{P}_c < \mathbb{P}$ for any $\mathbb{P}$ satisfying

$$\mathbb{P} - \widetilde{\mathbf{A}}_c^t \mathbb{P} \widetilde{\mathbf{A}}_c + \widetilde{\mathbf{\Gamma}} \widetilde{\mathbf{\Gamma}}^t < 0, \tag{2.58}$$

it is readily verified that $||\widetilde{\mathbf{G}}||_2^2 = \mathbf{trace}[\widetilde{\mathbf{C}}_c \mathbf{P}_c \widetilde{\mathbf{C}}_c^t] < \nu$ with $\widetilde{\Phi} \equiv 0$ if and only if there exists $\mathbb{P} > 0$ satisfying (2.58) and $\mathbf{trace}[\widetilde{C}_c \mathbb{P} \widetilde{C}_c^t] < \nu$.

Introducing an auxiliary parameter $\mathbb{Z}$, the following design result is obtained.

Theorem 2.5. *System (2.55) is stabilizable by the controller* $\mathbf{u} = (I \otimes K_s)\, \mathbf{x}(k)$ *and* $||\widetilde{\mathbf{T}}||_2^2 < \nu$ *for a prescribed* ν *if and only if there exist matrices* $\mathbb{P} > 0$, $\mathbb{Q}$, $\mathbb{Z} > 0$ *such that*

$$\mathbf{trace}(\mathbb{Z}) < \nu, \quad \begin{bmatrix} \mathbb{Z} & \hat{\mathbf{C}}\mathbb{P} + \hat{\mathbf{D}}\mathbb{Q} \\ \bullet & \mathbb{P} \end{bmatrix} > 0,$$

$$\begin{bmatrix} \mathbb{P} & \hat{\mathbf{A}}\mathbb{P} + \hat{\mathbf{B}}\mathbb{Q} & \widetilde{\mathbf{\Gamma}} \\ \bullet & \mathbb{P} & 0 \\ \bullet & \bullet & I \end{bmatrix} > 0. \tag{2.59}$$

Moreover, the feedback gain is given by $(I \otimes K_s) = \mathbb{Q}\mathbb{P}^{-1}$.

Proof. It follows from standard convex analysis similar to [15]. □

The following remarks stand out:

- *From the foregoing convex analysis and minimization over LMIs, we can infer that* $||T_i^{\omega \to y}||_2^2 = \mathbf{Tr}(Z_i)$. *Here, we note that in the above SDP, the matrix* $\mathbb{X}_i$ *actually corresponds to the controllability Gramian of the closed-loop system for agent* i. *That is, it is the controllability Gramian for a realization of the system* $||T_i^{\omega \to y}||_2$.
- *The adopted convex analysis and minimization over LMIs, however, does not incorporate the global DNDS performance objective into the problem. While each agent can generate a solution to the foregoing state or output feedback schemes independently of each other, the addition of the global NDS layer couples the design of each agent's controller.*

To elaborate the previous remark, we should examine the map $||T_{het}^{\omega\to\mathbb{G}}||_2$ in the context of Fig. 2.7. Keeping in mind the fact that the variables in system (2.7) are block-diagonal expansion of those of system (2.6), we will treat the DNDS output $\mathbf{z}_{\mathbb{G}}(k)$ as an additional performance variable and proceed to define

$$\boldsymbol{\xi}(k) = \begin{bmatrix} \mathbf{y}(k) \\ \mathbf{z}_{\mathbb{G}}(k) \end{bmatrix}, \Xi = \begin{bmatrix} \hat{\mathbf{C}} \\ (\mathbb{E}(\mathbb{G})^t \otimes \mathbf{M}) \end{bmatrix}, \Pi = \begin{bmatrix} \hat{\mathbf{D}} \\ 0 \end{bmatrix}. \tag{2.60}$$

Then, we express the dynamics of the augmented system as

$$\begin{aligned} \mathbf{x}(k+1) &= \hat{\mathbf{A}}\mathbf{x}(k) + \hat{\mathbf{B}}\mathbf{u}(k) + \hat{\Gamma}\omega(k), \\ \boldsymbol{\xi}(k) &= \Xi\mathbf{x}(k) + \Pi\mathbf{u}(k), \\ \mathbf{z}(k) &= \hat{\mathbf{M}}\mathbf{x}(k). \end{aligned} \tag{2.61}$$

Using the sensed state information model (2.61) with control law $\mathbf{u}(k) = (I \otimes \mathbf{K}_t)\,\mathbf{x}(k) = \widetilde{\mathbf{K}}_t\,\mathbf{x}(k)$, the control system can be cast into the following form:

$$\begin{aligned} \mathbf{x}(k+1) &= \widetilde{\mathbf{A}}\,\mathbf{x}(k) + \widetilde{\Gamma}\,\omega(k), \\ \boldsymbol{\xi}(k) &= \mathbb{N}\mathbf{x}(k), \\ \widetilde{\mathbf{A}} &= \hat{\mathbf{A}} + \hat{\mathbf{B}}\widetilde{\mathbf{K}}_t, \\ \mathbb{N} &= \begin{bmatrix} \hat{\mathbf{C}} + \hat{\mathbf{D}}\widetilde{\mathbf{K}}_t \\ (\mathbb{E}(\mathbb{G})^t \otimes \mathbf{M}) \end{bmatrix}. \end{aligned} \tag{2.62}$$

Observe from the closed-loop system (2.62) that

$$\begin{aligned} ||\mathbf{T}^{\omega\to\xi}||_2^2 &= \mathbf{trace}\left\{\begin{bmatrix} \hat{\mathbf{C}} + \hat{\mathbf{D}}\widetilde{\mathbf{K}}_t \\ (\mathbb{E}(\mathbb{G})^t \otimes \mathbf{M}) \end{bmatrix} \mathbb{X}_t \begin{bmatrix} \hat{\mathbf{C}} + \hat{\mathbf{D}}\widetilde{\mathbf{K}}_t \\ (\mathbb{E}(\mathbb{G})^t \otimes \mathbf{M}) \end{bmatrix}\right\} \\ &= \sum_j^{|\mathbb{V}|} [\mathbf{trace}(\mathbb{Y}_j) + \mathbf{trace}(\mathbb{S}_j)], \\ \sum_j^{|\mathbb{V}|} [\mathbf{trace}(\mathbb{S}_j)] &= \mathbf{trace}\left\{(\hat{\mathbf{C}} + \hat{\mathbf{D}}\widetilde{\mathbf{K}}_t)\mathbb{X}_t(\hat{\mathbf{C}} + \hat{\mathbf{D}}\widetilde{\mathbf{K}}_t)^t\right\}, \\ \sum_j^{|\mathbb{V}|} [\mathbf{trace}(\mathbb{Y}_j)] &= \mathbf{trace}\left\{(\mathbb{E}(\mathbb{G})^t \otimes \mathbf{M})\mathbb{X}_t(\mathbb{E}(\mathbb{G})^t \otimes \mathbf{M})^t\right\}. \end{aligned} \tag{2.63}$$

From (2.63), the first term corresponds to the $\mathbb{H}_2$-norm of system (2.7) under the state feedback control $\mathbf{u}(k) = \widetilde{\mathbf{K}}_t\mathbf{x}(k)$ and the second term accounts for the

$\mathbb{H}_2$-norm of $||T_{het}^{\omega \to \mathbb{G}}||_2$ of the homogeneous DNDS. This can be simplified to

$$
\begin{aligned}
||T_{het}^{\omega \to \mathbb{G}}||_2^2 &= \mathbf{trace}\left\{ (\mathbb{E}(\mathbb{G})^t \otimes \mathbf{M})\mathbb{X}_t(\mathbb{E}(\mathbb{G})^t \otimes \mathbf{M})^t \right\} \\
&= \sum_{j=1}^{|\mathbb{V}|} d_j \, ||T_i^{\omega \to y}||_2^2 \\
&= \sum_{j=1}^{|\mathbb{V}|} d_j \, \mathbf{trace}\{\mathbf{M}\mathbb{X}_j\mathbf{M}^t\}.
\end{aligned} \tag{2.64}
$$

Considering the $\mathbb{H}_2$ controller scheme, the objective is to ensure the stability of the closed-loop system and to keep the $\mathbb{H}_2$-norm $\mathbf{T}^{\omega \to \xi}$ as small as possible. The design result is summarized by the following theorem.

Theorem 2.6. *System (2.62) can be stabilized by the local state feedback controller* $\mathbf{u}_j = \widetilde{\mathbf{K}}_j \, \mathbf{x}_j(k), \; j = 1, ...|\mathbb{V}|$, *while minimizing the* $\mathbb{H}_2$*-norm* $||\mathbf{T}^{\omega \to \xi}||_2^2$ *if and only if there exist matrices* $\mathbb{X}_j > 0, \; \mathbb{Q}_j, \; \mathbb{S}_j \; \mathbb{Z}_j > 0, \; \mathbb{Y}_j > 0$ *such that*

$$
\min_{\mathbb{X}_j, \mathbb{S}_j, \mathbb{Z}_j, \mathbb{Y}_j} \sum_{j}^{|\mathbb{V}|} [\mathbf{trace}(\mathbb{S}_j) + \mathbf{Tr}(\mathbb{Y}_j)] \tag{2.65}
$$

subject to

$$
\begin{bmatrix} \mathbb{X}_j & \mathbb{X}_j\widetilde{A}_j^t + \mathbb{Q}_j^t\widetilde{B}^t & \widetilde{\Gamma} \\ \bullet & \mathbb{X}_j & 0 \\ \bullet & \bullet & I \end{bmatrix} > 0, \tag{2.66}
$$

$$
\begin{bmatrix} \mathbb{Z}_j & \mathbb{X}_j\hat{C}_j^t + \mathbb{Q}_j^t\hat{D}_j^t \\ \bullet & \mathbb{S}_j \end{bmatrix} > 0, \tag{2.67}
$$

$$
\begin{bmatrix} \mathbb{X}_j & \mathbb{X}_j\mathbf{M}_j^t \\ \bullet & (1/d_i)\mathbb{Y}_j \end{bmatrix} > 0. \tag{2.68}
$$

Moreover, the feedback gain is given by $\widetilde{\mathbf{K}}_j = \mathbb{Q}_j\mathbb{X}_j^{-1}$.

Proof. To guarantee stability, we select a Lyapunov functional $\mathbf{V}(\mathbf{x}(k))$ of the form $\mathbf{V}(\mathbf{x}(k)) = \mathbf{x}^t(k)\hat{\mathbb{P}}\mathbf{x}(k), \; \hat{\mathbb{P}} > 0$. The closed-loop system (2.62) is internally asymptotically stable with $\omega(k) \equiv 0$ if

$$
\hat{\mathbb{P}} - \widetilde{\mathbf{A}}^t\hat{\mathbb{P}}\widetilde{\mathbf{A}} > 0. \tag{2.69}
$$

Following parallel lines to Theorem 2.5, the square of the $\mathbb{H}_2$-norm of the transfer function $\widetilde{\mathbf{T}}$ can be expressed in terms of the solution of a controllability

Gramian equation such that the corresponding minimization problem with respect to the controller gain $\widetilde{\mathbf{K}}_t$ becomes

$$\min \left\{ \mathbf{trace}[\mathbb{N}\mathbf{P}_t\mathbb{N}^t] : \mathbf{P}_t - \widetilde{\mathbf{A}}_t^t \mathbf{P}_t \widetilde{\mathbf{A}}_t + \widetilde{\mathbf{\Gamma}}\widetilde{\mathbf{\Gamma}}^t = 0 \right\}. \tag{2.70}$$

Since $\mathbf{P}_t < \hat{\mathbb{P}}$ for any $\hat{\mathbb{P}}$ satisfying

$$\hat{\mathbb{P}} - \widetilde{\mathbf{A}}^t \hat{\mathbb{P}} \widetilde{\mathbf{A}} + \widetilde{\mathbf{\Gamma}}\widetilde{\mathbf{\Gamma}}^t < 0, \tag{2.71}$$

convexifying (2.71) using $\mathbb{X} = \hat{\mathbb{P}}^{-1}$ and the linearization $\mathbb{Q}_j = \widetilde{\mathbf{K}}_j \mathbb{X}_j$ yields inequality (2.66). It is readily verified from (2.64) that minimizing $||\mathbf{T}^{\omega \to \xi}||_2^2$ and employing the fact that inequality $d_j \{\mathbf{M}\mathbb{X}_j\mathbf{M}^t < \mathbb{Y}_j\}$ corresponds to inequality (2.68) based on d_j $\mathbf{trace}\{\mathbf{M}\mathbb{X}_j\mathbf{M}^t\} < \mathbf{trace}\{\mathbb{Y}_j\}$. LMI (2.67) follows by parallel arguments, which completes the proof. □

Remark 2.4. A significant feature of the LMIs (2.65)–(2.68) lies in their structure. Observe that the global DNDS layer couples each agent; however, the coupling can be removed via the formulation of the $\mathbb{H}_2$-norm. The minimization problem over LMIs is therefore separable across each of the agents, which has implications for the parallel arrangement of the computation and decision making process.

2.9 $\mathbb{H}_\infty$ performance and robust topology design

Many applications in multiagent systems rely on relative sensing to achieve their team objectives. These include space applications relying on relative sensing, such as spacecraft constellations for studying the structure of the heliopause, stereographic imaging and tomography for space physics, and space-borne optical interferometry for probing the origins of the cosmos and identifying Earth-like planets.

At the core of all of these systems is the implicit presence of a "network." The exchange of information between each agent in a relative sensing network describes an underlying connection topology, which can have profound implications for the performance and design of decentralized schemes for estimation and control. It is becoming more important to examine notions of system performance from the perspective of the underlying connection topology describing the interactions of each agent.

In this section we focus on systems that rely on relative sensing to achieve their mission objectives. We refer to this class of systems as relative sensing networks (RSNs). In RSNs, the underlying connection topology couples the agents at their outputs. Such systems are prevalent in formation flying applications where relative sensing is employed to measure interagent distances. More fundamentally, these types of networks are relevant for applications involving distributed sensing for purposes of estimation and control.

This section is an extension of the work of the foregoing chapter, which examined the $\mathbb{H}_2$ performance of RSNs and described a topology synthesis procedure based on results from combinatorial optimization. In this work, we consider a graph-centric characterization of the system's $\mathbb{H}_\infty$ performance of RSNs for both analysis and synthesis purposes. A distinction is made between RSNs with homogeneous agent dynamics and RSNs with heterogeneous agent dynamics. Although homogeneous RSNs can be considered a subset of heterogeneous RSNs, it is more illuminating to consider these cases separately due to the algebraic simplicity of the former case. Understanding how properties of the interconnection graph lead to improvements in system performance can give insight into the design of sensing topologies. In particular, we consider how characterization of the $\mathbb{H}_\infty$ performance leads to a robust topology design formulation.

In the sequel, a matrix and/or a vector that consists of all zero entries will be denoted by $\mathbf{0}$, whereas, "0" will simply denote the scalar zero. Similarly, the vector $\mathbf{1}$ denotes the vector of all ones.

From the fundamental theories of graphs and matrices, we recall the incidence matrix, $\mathbb{E}(\mathbb{G})$, for a graph with arbitrary orientation. The incidence matrix is a $\{0, \pm 1\}$-matrix with rows and columns indexed by the vertices and edges of $\mathbb{G}$ such that $[\mathbb{E}(\mathbb{G})]_{ik}$ has the value "+1" if node i is the initial node of edge e_k, "−1" if it is the terminal node, and "0" otherwise. The degree of vertex i, d_i, is the cardinality of the set of vertices adjacent to it; we define the degree matrix as $\Delta(\mathbb{G}) = diag\{d_1, \cdots, d_{|\mathbb{V}|}\}$. The adjacency matrix of a graph, $\mathbb{A}(\mathbb{G})$, is the symmetric $|\mathbb{V}| \times |\mathbb{V}|$ matrix such that $[\mathbb{A}(\mathbb{G})]_{ij}$ takes the value "+1" if node i is connected to node j, and "0" otherwise.

The (graph) Laplacian of $\mathcal{G}$,

$$\mathbb{L}(\mathbb{G}) := E(\mathcal{G})\mathbb{E}(\mathbb{G})^T = \Delta(\mathbb{G}) - \mathbb{A}(\mathbb{G}), \tag{2.72}$$

is a rank-deficient positive semidefinite matrix. The eigenvalues of the graph Laplacian are real and will be ordered and denoted as $0 = \lambda_1(\mathbb{G}) \leq \lambda_2(\mathbb{G}) \leq \cdots \leq \lambda_{|\mathbb{V}|}(\mathbb{G})$.

In order to apply the framework to specific graphs, we will work with the complete graph and its generalization in terms of k-regular graphs, which are defined as follows. The complete graph on n nodes, K_n, is the graph where all possible pairs of vertices are adjacent, or equivalently, if the degree of all vertices is $|\mathcal{V}| - 1$. Fig. 2.8A depicts K_{10}, the complete graph on 10 nodes. When every node in a graph with n nodes has the same degree $k \leq n - 1$, it is called a k-regular graph. Fig. 2.8B shows a 4-regular graph.

2.9.1 Relative sensing network model

In this section we derive a general plant model for relative sensing networks. An RSN, in its most general setting, is comprised of individual sensing agents

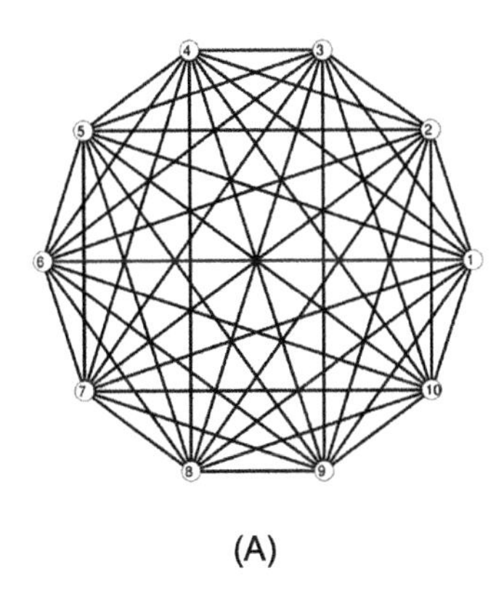

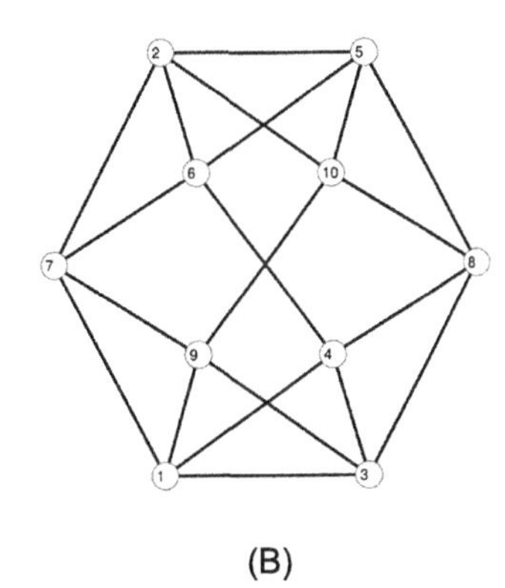

(A) (B)

FIGURE 2.8 Example of regular graphs. (A) K_{10} graph; (B) A 4-regular graph.

which, for this work, are assumed to contain LTI dynamics. The agents are coupled to other agents through their sensed outputs; the output coupling is defined by the connection topology describing their interactions. We identify two classes of RSNs:

1) homogeneous RSNs and
2) heterogeneous RSNs.

For both cases, we will work with a group of n dynamic systems (the "agents"), with state-space representation

$$\Sigma_i : \begin{cases} x_i(k+1) = A_i x_i(k) + B_i u_i(k) + \Gamma_i w_i(k), \\ z_i(k) = C_i^z x_i(k) + D_i^{zu} u_i(k) + D_i^{zw} w_i(k), \\ y_i(k) = C_i^y x_i(k) + D_i^{yw} w_i(k), \end{cases} \tag{2.73}$$

where each agent is indexed by the subscript i. Here, $x_i(k)$ represents the state, $u_i(k)$ the control, $w_i(k)$ an exogenous input (that is, disturbances and noises), $z_i(k)$ the controlled variable, and $y_i(k)$ the locally measured output.

We denote the pulse transfer function representation of Σ_i as $\hat{\Sigma}_i$,

$$\begin{aligned} \begin{bmatrix} Z_i(z) \\ Y_i(z) \end{bmatrix} &= \begin{bmatrix} H_i^{zu}(z) & H_i^{zu}(z) \\ H_i^{yu}(z) & H_i^{yu}(z) \end{bmatrix} \begin{bmatrix} U_i(z) \\ W_i(z) \end{bmatrix} \\ &= \hat{\Sigma}_i \begin{bmatrix} U_i(z) \\ W_i(z) \end{bmatrix}, \end{aligned} \tag{2.74}$$

with

$$\begin{aligned} H_i^{zu}(z) &= C_i^z (zI - A_i)^{-1} B_i + D_i^{Zu}, \\ H_i^{zw}(z) &= C_i^z (zI - A_i)^{-1} \Gamma_i + D_i^{zw}, \\ H_i^{yu}(z) &= C_i^y (zI - A_i)^{-1} B_i, \\ H_i^{yw}(z) &= C_i^y (zI - A_i)^{-1} \Gamma_i + D_i^{yw}. \end{aligned}$$

In the homogeneous case, it is assumed that each dynamic agent in the RSN is described by the same set of linear state-space dynamics (that is, $\Sigma_i = \Sigma_j$ for all i, j). When working with a homogeneous RSN, we omit the subscript for all state-space and operator representations of the system. We will also assume no feedforward terms of the control to the measured output. Additionally, we assume a minimal realization for each agent with compatible outputs for all agents, that is, system outputs will correspond to the same physical quantity. It should be noted that in a heterogeneous system, the dimension of each agent need not be the same; however, using a "padding argument," it can be assumed that all agents have identical dimensions for their respective state space.

The parallel interconnection of all the agents has a state-space description

$$\begin{aligned} \mathrm{x}(k+1) &= \mathbf{Ax(k)} + \mathbf{Bu(k)} + \mathbf{w(k)}, \\ \mathbf{z}(k) &= \mathbf{C^z x(k)} + \mathbf{D^{zu} u(k)} + \mathbf{D^{zw} w(k)}, \\ \mathbf{y}(k) &= \mathbf{C^y x(k)} + \mathbf{D^{yw} w(k)}, \end{aligned} \tag{2.75}$$

with $\mathbf{x}(k)$, $\mathbf{u}(k)$, $\mathbf{w}(k)$, $\mathbf{z}(k)$, and $\mathbf{y}(k)$ denoting, respectively, the concatenated state vector, control vector, exogenous input vector, controlled vector, and output vector of all the agents in the RSN. The boldface matrices represent the block-diagonal aggregation of each agent's state-space matrices, that is, $\mathbf{A} = \mathrm{diag}\{\mathbf{A}_1, \cdots, \mathbf{A}_n\}$.

The global RSN layer we examine in this section is motivated by the relative sensing problem. The sensed output of the RSN is the vector $\mathbf{y}_{\mathcal{G}}(t)$ containing relative state information of each agent and its neighbors. The incidence matrix of a graph naturally captures state differences and will be the algebraic construct used to define the relative outputs of RSNs. For example, the output sensed between agent i and agent j would be of the form $y_i(k) - y_j(k)$. This can be compactly written using the incidence matrix for the entire RSN as

$$\mathrm{y}_{\mathbb{G}}(k) = (\mathbb{E}(\mathbb{G})^t \otimes I)\mathrm{y}(k). \tag{2.76}$$

Here, $\mathbb{G}$ is the graph that describes the connection topology of the RSN; the node set is given as $\mathbb{V} = \{1, \cdots, n\}$. The global layer can be visualized as in the block diagram shown in Fig. 2.9.

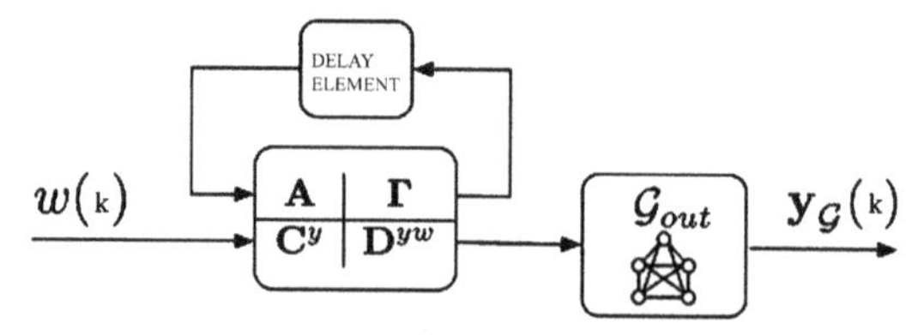

FIGURE 2.9 Global RSN layer block diagram; the feedback connection represents an upper fractional transformation.

When considering the analysis of the global layer, we are interested in studying the map from the agent's exogenous inputs to the RSN sensed output. Using the above notations we can express the heterogeneous RSN in a compact form,

$$\Sigma_{het}(\mathbb{G}): \begin{cases} \mathbf{x}(k+1) = \mathbf{A}\mathbf{x}(t) + \mathbf{B}\mathbf{u}(t) + \Gamma\mathbf{w}(t), \\ \mathbf{z}(t) = \mathbf{C}^z\mathbf{x}(t) + \mathbf{D}^{zu}\mathbf{u}(t) + \mathbf{D}^{zw}\mathbf{w}(t), \\ \mathbf{y}(t) = \mathbf{C}^y\mathbf{x}(t) + \mathbf{D}^{yw}\mathbf{w}(t), \\ \mathbf{y}_{\mathcal{G}}(t) = (E(\mathcal{G})^T \otimes I)\mathbf{C}^y\mathbf{x}(t). \end{cases} \tag{2.77}$$

The homogeneous RSN, $\Sigma_{hom}(\mathcal{G})$, can be expressed using the Kronecker product. For example, $\mathbf{A} = I \otimes A$ and $(E(\mathcal{G}) \otimes I)\mathbf{C}^y = E(\mathcal{G})^t \otimes C^y$.

Similarly, the pulse transfer function representation is written as

$$\hat{\Sigma}_{het} = \begin{bmatrix} \mathbf{H}^{zu}(z) & \mathbf{H}^{zw}(z) \\ \mathbf{H}^{yu}(z) & \mathrm{H}^{yw}(z) \\ (E(\mathcal{G})^t \otimes I)\mathbf{H}^{yu}(z) & (E(\mathcal{G})^t \otimes I)\mathbf{H}^{yw}(z) \end{bmatrix}, \tag{2.78}$$

where, as in the state-space model, boldface transfer functions denote the block-diagonal aggregation of each agent's corresponding transfer function, e.g., $\mathbf{H}^{zu}(s) = \mathbf{diag}\{H_1^{zu}(s), \ldots, H_n^{zu}(s)\}$. The homogeneous system, $\hat{\Sigma}_{hom}$, can also be written using the Kronecker product in a similar manner as described above.

For notational simplicity, we denote $T_{hom}^{w \mapsto \mathcal{G}}$ and $T_{het}^{w \mapsto \mathcal{G}}$ as the map from the exogenous inputs to the RSN sensed output for homogeneous and heterogeneous systems, respectively, e.g., $T_{hom}^{w \mapsto \mathcal{G}} = E(\mathcal{G})^T \otimes H^{yw}(s)$. We also use transfer function and state-space representations interchangeably noting the appropriate realization can be inferred by context. For example, H_i^{yw} will be used to represent both the state-space and transfer function representation of the open-loop map from the exogenous inputs to the measurement of agent i.

2.9.2 Graph-theoretic bounds on $\mathbb{H}_\infty$ performance

In this section we explore a graph-theoretic characterization of the $\mathbb{H}_\infty$ performance of the RSN model presented in Section 2.2. The main goal is to highlight the role of the underlying connection topology on the system norms mapping the exogenous inputs $\mathbf{w}(t)$ to the relative sensed output $\mathbf{y}_{\mathcal{G}}(t)$, $T^{w \mapsto \mathcal{G}}$. We assume that the observation matrix for the sensed output is the same as for the local measurement; that is, $\mathbf{C} = \mathbf{C}^y$ and $\mathbf{H}^{\mathbb{G}w} = \mathbf{H}^{yw}$, as in (2.77) and (2.78). Additionally, we assume throughout this section that the underlying connection graph $\mathbb{G}$ is connected and $\mathbb{V} = \{1, \ldots, n\}$. For analysis, we finally assume that each agent has stable dynamics.

We first recall that the $\mathcal{H}_\infty$-norm for a dynamic system captures how a measurable signal with finite energy, i.e., a signal in $\mathcal{L}_2$, is amplified at the monitored output of the system. Moreover, this norm has implications for robustness, disturbance rejection, and uncertainty management for dynamic systems. Specifically, the $\mathbb{H}_\infty$-norm of a linear system with transfer function representation

$H(s)$ is characterized as

$$\|H(j\omega)\|_\infty = \sup_{\omega}\{\overline{\sigma}[H(j\omega)]\}$$
$$= \sup_{\|U(j\omega)\|_2=1} \|H(j\omega)U(j\omega)\|_{\mathbb{L}_2}, \tag{2.79}$$

where $\bar{\sigma}[A]$ denotes the largest singular value of the matrix A. The induced-norm description allows us to state the submultiplicative property of the $\mathcal{H}_\infty$-norm for two operators as $\|H(j\omega)P(j\omega)\|_\infty \leq \|H(j\omega)\|_\infty \|P(j\omega)\|_\infty$.

In the context of RSNs, therefore, the $\mathbb{H}_\infty$ system norm can be used to capture how disturbances and finite energy exogenous signals, including reference signals, result in the asymptotic deviation of the sensed output of the network. This section aims to explicitly characterize the effect of the network on the $\mathbb{H}_\infty$-norm of the system. We separate our analysis into the homogeneous and heterogeneous cases.

2.9.3 Homogeneous RSN $\mathbb{H}_\infty$ performance

Given the transfer function representation of the homogeneous RSN in (2.78), we can write the map from the disturbances to the networked output as

$$\mathrm{Y}_{\mathbb{G}}(z) = (Pi(\mathbb{G})^t \otimes H^{yw})\mathrm{U}(z). \tag{2.80}$$

Theorem 2.7. *The $\mathbb{H}_\infty$-norm of the homogeneous RSN (2.77) is given as*

$$\|T_{hom}^{w\mapsto\mathbb{G}}\|_\infty = \|\mathbb{E}(\mathbb{G})\| \, \|H^{yw}\|_\infty. \tag{2.81}$$

Proof. The norm expression follows directly from the definition in (2.79) and the Kronecker product property

$$\|A \otimes B\| = \|A\|\,\|B\|. \quad \square$$

The expression (2.81) states that the overall ℓ_2-gain of the system is proportional to the matrix 2-norm of the incidence matrix. In fact, since $\|E(\mathcal{G})\| = \sqrt{\|\mathbb{L}(\mathbb{G})\|} = \lambda_n^{1/2}$, the behavior of the largest eigenvalue of the graph Laplacian is of particular interest. Moreover, an important observation is that certain graph structures will naturally lead to a smaller $\mathbb{H}_\infty$-norm. If we restrict our topology to spanning trees we can state a set of stronger results.

Corollary 2.2. *When the underlying topology is a spanning tree, the path graph is the topology resulting in the smallest $\mathbb{H}_\infty$-norm for the homogeneous RSN (2.77). Moreover, the star graph is the topology resulting in the largest $\mathbb{H}_\infty$-norm for the homogeneous (2.77) among all spanning trees.*

Proof. It can be easily shown that the path graph has the smallest spectral norm for the graph Laplacian among all spanning trees. In addition, the star graph has the largest spectral norm for the graph Laplacian among all spanning trees. $\square$

2.9.4 Bounds on the heterogeneous RSN $\mathbb{H}_\infty$ performance

We follow a similar procedure for the heterogeneous case. Using the transfer function representation of the heterogeneous RSN in (2.78) we can write the map from the disturbances to the networked output as

$$\mathbf{Y}_{\mathbb{G}}(z) = (\mathbb{E}(\mathbb{G})^t \otimes I)\mathbf{H}^{yw}(z)\mathrm{U}(z). \tag{2.82}$$

Calculating the $\mathbb{H}_\infty$-norm involves finding the singular values of the transfer function

$$T_{het}^{w\mapsto\mathbb{G}} = (\mathbb{E}(\mathbb{G})^t \otimes I)\mathbf{H}^{yw}(z). \tag{2.83}$$

In general, an analytic expression for the singular values of the system in (2.83) is difficult to obtain. However, it is possible to generate bounds on the system norm, leading to the following result.

Theorem 2.8. *The $\mathbb{H}_\infty$-norm of the homogeneous RSN (2.77) is bounded as*

$$\|T_{het}^{w\mapsto\mathbb{G}}\|_\infty \leq \|\mathbb{E}(\mathbb{G})^t Q\| \leq \|\mathbb{E}(\mathbb{G})^t\| \max_i \|H_i^{yw}\|_\infty, \tag{2.84}$$

where $Q = \mathbf{diag}\{\|H_1^{yw} 1_\infty, \cdots, \|H_\eta^{yw}\|_\infty\}$.

Proof. The upper bound immediately arises from the submultiplicative property of the matrix 2-norm as $\|\mathbb{E}(\mathbb{G})^t Q\| \leq \|\mathbb{E}(\mathbb{G})^t\|\|Q\|$. Since Q is a diagonal matrix we conclude that $\|Q\| = \max_i \|H_i^{yw}\|_\infty$. To show the lower bound we follow the following chain of inequalities:

$$\begin{aligned}
\|T_{het}^{w\mapsto\mathbb{G}}\|_\infty^2 &= \sup_{\|U(j\omega)\|_{\mathbb{L}_2}=1} \|(\mathbb{E}(\mathbb{G})^t \otimes I)\mathbf{H}^{yw}(j\omega)U(j\omega)\|_{\mathbb{L}_2}^2 \\
&= \sup_{\|U\|_{\mathbb{L}_2}=1} \int_{-\infty}^{\infty} U^*(\mathbf{H}^{yw})^*(\mathbb{L}(\mathbb{G}) \otimes I)\mathbf{H}^{yw} U d\omega \\
&= \sup_{\|U\|_{\mathbb{L}_2}=1} \int_{-\infty}^{\infty} \mathbf{trace}[UU^*(\mathbf{H}^{yw})^*(\mathbb{L}(\mathbb{G}) \otimes I)\mathbf{H}^{yw}]d\omega \\
&\leq \sup_{\|U\|_{\mathbb{L}_2}=1} \int_{-\infty}^{\infty} \mathbf{trace}(U^*)\mathbf{trace}[(\mathrm{H}^{yw})^*(\mathbb{L}(\mathbb{G}) \otimes I)\mathbf{H}^{yw}]d\omega \\
&\leq \|Q\mathbb{E}(\mathbb{G})\|,
\end{aligned} \tag{2.85}$$

where the second to last inequality follows from the property that for Hermitian matrices, M and N, $\mathbf{trace}[M\ N] \leq \mathbf{trace}[M]\mathbf{trace}[N]$, and the last identity follows from the property that the positive definite ordering $\mathbf{H}^{yw}(j\omega)^*(\mathbb{L}(\mathbb{G}) \otimes I)\mathbf{H}^{yw}(j\omega) \leq (Q \otimes I)(\mathbb{L}(\mathbb{G}) \otimes I)(Q \otimes I)$ holds for all ω. □

Corollary 2.3. *When each agent in (2.77) is a single-input single-output (SISO) system, the norm bound in (2.84) is tight.*

An interesting implication of the norm bounds developed in the proof relates the $\mathbb{L}_2$-gain of a heterogeneous RSN to that of a homogeneous RSN. Consider an ordering of each agent in a heterogeneous RSN by the value of the $\mathbb{H}_\infty$-norm of each agent,

$$\|H_{k(1)}^{yw}\|_\infty \leq \cdots \leq \|H_{k(n)}^{yw}\|_\infty, \tag{2.86}$$

where $k : \{1, \ldots, n\} \mapsto \{1, \ldots, n\}$ maps the old index set to the norm-ordered one. The $\mathcal{H}_\infty$-norm of the heterogeneous system $T_{het}^{w \mapsto \mathcal{G}}$ can be bounded from above and below by homogeneous systems as

$$\|E(\mathcal{G})\| \|H_{k(1)}^{yw}\|_\infty \leq \|T_{het}^{w \mapsto \mathbb{G}}\|_\infty \leq \|\mathbb{E}(\mathbb{G})\| \|H_{k(n)}^{yw}\|_\infty.$$

This inequality suggests that in addition to the structure of the underlying topology, one can consider the dynamic differences between agents as an important factor in the performance of the overall system.

2.9.5 Robust synthesis of relative sensing networks

We now consider the synthesis of the underlying connection topology, as shown in Fig. 2.10. As we are only considering the topology, we use the following heterogeneous state-space model for the RSN:

$$T_{het}^{w \mapsto \mathbb{G}} : \begin{cases} \mathrm{x}(k+1) = \mathbf{A}\mathbf{x}(k) + \Gamma \mathbf{w}(k), \\ \mathrm{y}_{\mathbb{G}}(k) = (\mathbb{E}(\mathbb{G})^t \otimes C_p)\mathbf{x}(k). \end{cases} \tag{2.87}$$

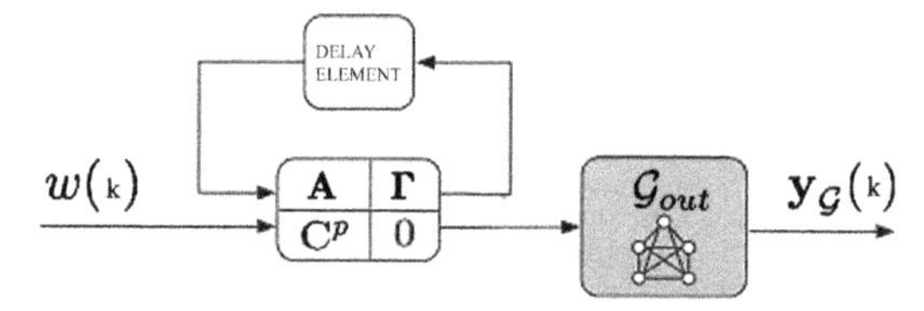

FIGURE 2.10 Topology design: the feedback connection represents an upper fractional transformation.

We would like to find topologies that minimize the effect of disturbances entering each agent on the relative sensed output of the entire system, that is, minimizing the performance objective $\|T_{het}^{w \mapsto \mathbb{G}}\|_\infty$. This can be considered a problem in combinatorial optimization, as the decision to include an edge in the graph is binary. The general synthesis problem can be written as

$$\begin{aligned} \min_{\mathbb{G}} \quad & \|T_{het}^{w \mapsto \mathbb{G}}\|_\infty \\ \text{s.t.} \quad & \mathbb{G} \text{ is connected.} \end{aligned} \tag{2.88}$$

The challenge, therefore, is to find numerically tractable algorithms to solve (2.88). In what follows, we solve a variation of (2.88) that minimizes the robust performance of a weighted version of (2.87) with uncertainty on the edge weights.

Motivated by the foregoing results, we find that (2.88) reduces to the minimization of the spectral norm of the weighted incidence matrix, $\| Q\mathbb{E}(\mathbb{G})\|$, where Q was defined in Theorem 2.8. Minimization of this objective can be formulated as a mixed-integer semidefinite program. For reasonably sized problem instances this can be solved using, for example, branch-and-bound algorithms [16].

While topology design is an important application, the $\mathbb{H}_\infty$ framework allows us to consider the robustness of certain topologies. In this direction, we consider a variation of (2.88) that aims to minimize the robust performance of the RSN in (2.87). For such an analysis, we adjust the RSN model to allow for uncertainty in the sensing protocol. Specifically, we introduce the notion of a weighted edge for the sensed output. This model might be used to capture the fidelity of a relative measurement. We have

$$T_{het}^{w\mapsto\mathbb{G}}(W_o): \begin{cases} \mathbf{x}(k+1) = \mathbf{A}\mathbf{x}(k) + \Gamma\mathbf{w}(k), \\ \mathbf{y}_{\mathbb{G}}(k) = (W_o\mathbb{E}(\mathbb{G})^t \otimes C_p)\mathrm{x}(k). \end{cases} \tag{2.89}$$

In (2.89), each diagonal entry of $W_o = \mathbf{diag}\{w_1, \ldots, w_{|\mathbb{E}|}\}$ represents the nominal weights on each edge in the graph. A weight of zero corresponds to the absence of an edge. We will also assume all the weights are nonnegative. The model (2.89) relates to (2.87) through the output as $T_{het}^{w\mapsto\mathcal{G}}(W_o) = (W_o \otimes I)T_{het}^{w\mapsto\mathbb{G}}$.

Using (2.89), we can introduce a structured uncertainty on each edge weight. The uncertainty set is defined as

$$\Delta = \{\mathrm{diag}\{\delta_1, \ldots, \delta_{|\mathbb{E}|}\} : \delta_i \in \Re, |\delta_i| \leq 1\}. \tag{2.90}$$

The true edge weight can thus be written as $W = W_o + \Delta, \, for \Delta \in \Delta_w$. This can be considered as an output-multiplicative uncertainty, as shown in Fig. 2.11.

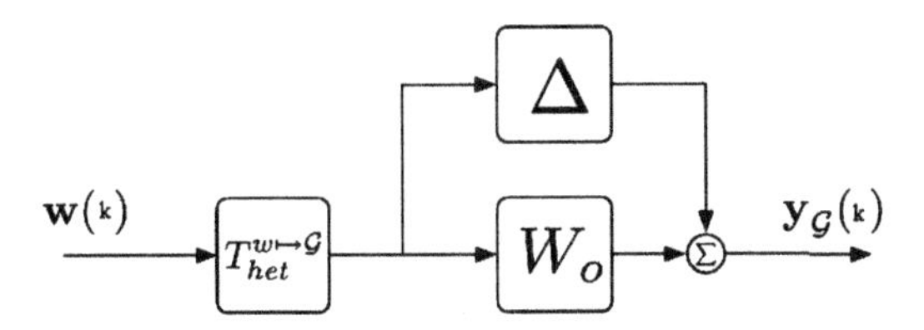

FIGURE 2.11 Multiplicative uncertainty for NDSs.

The problem (2.88) can now be restated as the robust optimization problem [17]

$$\min_{W_0} \max_{\|\Delta\|\leq 1} \|Q\mathbb{E}(\mathbb{G})(W_o+\Delta)\| \tag{2.91}$$

s.t. $\mathbb{G}$ is connected in the presence of edge weight uncertainty.

This problem can be solved as a semidefinite program, the procedure of which is outlined in [17]. To apply these results, we must express the objective and constraints of (2.91) as a perturbed LMI in the form

$$F(x,\delta) = F_0(x) + \sum_{i=1}^{l} \delta_i F_i(x), \tag{2.92}$$

where each $F_i(x)$ is a symmetric matrix and affine in the variable x.

First, we scalarize the objective function by introducing a new variable γ and note that $\|Q\mathbb{E}(\mathbb{G})(W_o+\Delta)\| \leq \gamma$ can be written (via the Schur complement) as the LMI

$$\begin{bmatrix} \gamma I & QF_i(\mathbb{G})(W_o+\Delta) \\ \bullet & I \end{bmatrix} \geq 0. \tag{2.93}$$

Defining the matrices $S_i \in \Re^{|\mathbb{E}|\times|\mathbb{E}|}$ and $\mathbb{V}(\gamma)$ as

$$[S_i]_{kl} = \begin{cases} 1 & k=l=i, \\ 0 & \text{otherwise}, \end{cases} \qquad V(\gamma) = \begin{bmatrix} \gamma I & 0 \\ \bullet & I \end{bmatrix}, \tag{2.94}$$

we can express (2.93) in the form (2.92) as

$$\begin{aligned} F_1(w,\delta) = V(\gamma) + \sum_{i=1}^{|\mathbb{E}|}(w_i+\delta_i)\begin{bmatrix} 0 & Q\mathbb{E}(\mathbb{G})S_i \\ \bullet & 0 \end{bmatrix} \\ \geq 0. \end{aligned} \tag{2.95}$$

Similarly, the robust connectivity constraint can also be expressed in the form (2.92). Recall that for a connected graph, $\lambda_2(\mathcal{G}) > 0$, and the eigenvector associated with $\lambda_1(\mathcal{G}) = 0$ is the vector of all ones, $\mathbf{1}$. Defining the matrix $\mathbf{P}$ such that $\mathbf{IM}\{\mathbf{P}\} = \mathbf{span}\{\mathbf{1}^{\perp}]$, we obtain

$$F_2(w,\delta) = \sum_{i=1}^{|\mathbb{E}|}(w_i+\delta_i)\mathbf{P}^t(e_i e_i^T)\mathbf{P} > 0. \tag{2.96}$$

Using (2.95) and (2.96) we define

$$F_0^1(w) = \begin{bmatrix} \gamma I & QF_j(\mathbb{G})W_o \\ \bullet & I \end{bmatrix}, \tag{2.97}$$

$$F_0^2(w) = \mathbf{P}^t \mathbb{E}(\mathbb{G}) W_o \mathbb{E}(\mathbb{G})^t \mathbf{P}, \tag{2.98}$$

$$F_i^1 = \begin{bmatrix} 0 & Q\mathbb{E}(\mathbb{G})S_i \\ \bullet & 0 \end{bmatrix}, \quad F_i^2 = \mathbf{P}^t e_i e_i^t \mathbf{P}. \tag{2.99}$$

The expressions in (2.97) and (2.99) can now be applied to obtain the following SDP:

$$\begin{aligned} &\min_{w, S_i, T_i} \quad \gamma \\ &\text{s.t.} \quad \begin{bmatrix} S_i & F_1^i & \cdots & F_{|\mathcal{E}|}^i \\ F_1^i & T_i & & \\ \vdots & & \ddots & \\ F_{|\mathcal{E}|}^i & & & T_i \end{bmatrix} \geq 0, \; i = 1, 2, \\ &\quad S_i + T_i \leq 2F_0^i, \; i = 1, 2, \; \sum_i w_i = \alpha, \\ &\quad 0 \leq w_i \leq w_{\max}, \; i = 1, \ldots, |\mathbb{E}|, \end{aligned} \tag{2.100}$$

where the last constraints constrain the aggregate edge weight sum and edge weight range.

To illustrate this procedure, we consider an RSN with $n = 10$ heterogeneous and SISO systems (generated randomly in MATLAB®). The input graph is the complete graph, K_n, allowing the program in (2.100) to select the optimal weights on every possible edge combination. For $\alpha = n - 1$ and $w_{\max} = 2$, (2.100) was solved using *SeDuMi* in MATLAB. The resulting topology is shown in Fig. 2.12. Note that every edge has a positive weight; however, only edges with $w_i > 0.1$ were drawn. The thickness of the line indicates a larger weight.

Remark 2.5. While the problem formulation presented above concerns static edge weight uncertainty, the principle can be extended to include dynamic edge weights. For example, each relative sensor may be characterized by a frequency-dependent weight, $w_i(s)$, and the corresponding uncertainty can be considered as an unstructured norm-bounded uncertainty.

Remark 2.6. The SDP (2.100) presents an analytic framework for solving the robust topology design problem. However, it should be noted that due to the auxiliary variables defined, the size of this problem can grow very large with the number of nodes (for the complete graph on n nodes, there are $n(n-1)/2$ edges). While interior-point methods offer polynomial-time algorithms, for excessively large problem instances (2.100) might lead to numerical problems. This points to the need to consider specialized solution methods or alternative problem formulations.

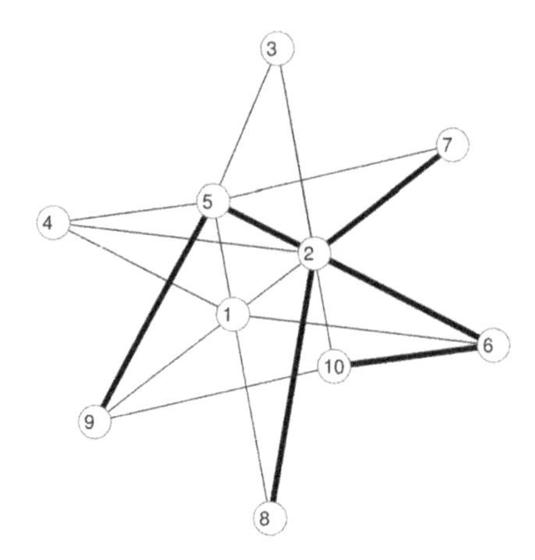

FIGURE 2.12 Optimal topology.

2.10 Notes

In this chapter, we have been concerned with the analysis and performance of a class of linear DNDSs based on a relative sensing model. The results highlight an important connection between certain graph-theoretic concepts and system-theoretic properties. Complete results pertaining to dynamic representation, local feedback stability, and linear controller design using either sensed state or output information and performance evaluation of $\mathbb{H}_2$ scheme are provided.

The contribution of this chapter is twofold. We first introduce two distinct classes of DNDS consisting of homogeneous and heterogeneous agent dynamics. Although the homogeneous case can be considered as a subset of the heterogeneous case, it turns out to be enlightening, whether through algebraic simplicity or development of intuition, to consider both cases separately. The second contribution is a rigorous observability analysis of NDSs based on the relative sensing problem. Specifically, this chapter highlights how both the qualitative and quantitative notions of observability change as the dynamics of each agent become more homogeneous. In this direction, an index of homogeneity and an index of heterogeneity are introduced to quantitatively capture how the dynamics of each agent in the ensemble and the underlying connection topology affect the overall observability properties of the overall system.

Then, we focused on the development of graph-theoretic performance bounds and synthesis techniques for distinct classes of RSNs. The results of this chapter highlight an important connection between certain graph-theoretic concepts and system-theoretic properties. In particular, the $\mathbb{H}_\infty$ performance depends on the spectral norm of a node-weighted incidence matrix, which is strongly dependent on the actual structure of the graph.

Synthesis methods for RSNs were also presented. Using methods from robust semidefinite programming, a synthesis procedure was then developed that aims to minimize the $\mathbb{H}_\infty$ performance of an RSN with uncertainty on the edge weights. This work also suggests that the relationship between system-theoretic properties and graph properties in RSNs can be examined further in the systems and control community. In fact, we believe that developing efficient solution methods for the synthesis of such systems will involve further interpreting results from graph theory and combinatorial optimization in a system-theoretic context.

By using tools from Lyapunov stability analysis and algebraic graph theory, the $\mathbb{H}_\infty$ consensus problem of a group of linear multiagent systems with strongly connected directed communication graph has been studied in this chapter. To achieve consensus, a distributed consensus protocol based only on the relative states of the neighboring agents has been proposed and employed. Some sufficient conditions have been provided to achieve $\mathbb{H}_\infty$ consensus with a guaranteed $\mathbb{H}_\infty$ performance for a group of linear agents subject to disturbances. Another theorem and algorithm have been given to design a protocol which can achieve consensus with a guaranteed $\mathbb{H}_2$ performance index. Future work will focus on the design of a distributed output feedback control protocol for achieving the $\mathbb{H}_\infty$ consensus problem of linear multiagent systems under communication delays with fixed or switching directed communication topologies.

References

[1] W. Ren, R.W. Beard, A survey of consensus problems in multiagent coordination, in: Proc. ACC, Portland, OR, 2005.

[2] R.R. Negenborn, B. De Schutter, H. Hellendoorn, Multiagent model predictive control of transportation networks, in: Proc. IEEE Int. Conference on Networking, Sensing and Control (ICNSC 2006), Ft. Lauderdale, FL, 2006, pp. 296–301.

[3] T.C. Yang, Networked control systems: a brief survey, IEE Proc., Control Theory Appl. 152 (2006) 403–412.

[4] P. Neumann, Communication in industrial automation—what is going on?, Control Eng. Pract. 15 (2007) 1332–1347.

[5] L. Magni, R. Scattolini, Stabilizing decentralized model predictive control of nonlinear systems, Automatica 42 (2006) 1231–1236.

[6] P.K.C. Wang, F.Y. Hadaegh, Coordination and control of multiple micro-spacecraft moving in formation, J. Astronaut. Sci. 44 (1996) 315–355.

[7] J.A. Fax, R.M. Murray, Information flow and cooperative control of vehicle formations, IEEE Trans. Autom. Control 49 (9) (2004) 1465–1476.

[8] M. Mesbahi, F.Y. Hadaegh, Formation flying control of multiple spacecraft via graphs, matrix inequalities, and switching, J. Guid. Control Dyn. 24 (2) (2001) 369–377.

[9] M.S. Mahmoud, Networked control systems analysis and design: an overview, Arab. J. Sci. Eng.: Invited Paper 41 (3) (March 2016) 711–758.

[10] A. Rahmani, M. Mesbahi, Pulling the strings on agreement: anchoring, controllability, and graph automorphisms, in: Proc. American Control Conference, NY, 2007.

[11] D. Zelazo, M. Mesbahi, On the observability properties of homogeneous and heterogeneous networked dynamic systems, in: IEEE CDC, Cancun, Mexico, 2008.

[12] R.S. Smith, F.Y. Hadaegh, Control of deep space formation flying spacecraft; relative sensing and switched information, J. Guid. Control Dyn. 28 (1) (2005).

[13] M.S. Mahmoud, Y. Xia, Applied Control Systems Design: State-Space Methods, Springer-Verlag, UK, May 2012.

[14] R.A. Horn, C. Johnson, Topics in Matrix Analysis, Cambridge University Press, 1991.
[15] M.S. Mahmoud, Resilient L_2/L_∞ filtering of polytopic systems with state-delays, IET Control Theory Appl. 1 (1) (2007) 141–154.
[16] B.H. Korte, J. Vygen, Combinatorial Optimization: Theory and Algorithms, Springer-Verlag, Berlin, 2000.
[17] A. Bental, L. El Ghaoui, A. Nemirovski, Robust Semidefinite Programming, Springer-Verlag, New York, 2000.

Chapter 3

Consensus of systems over graphs

3.1 Dynamic consensus protocol

In recent years, the consensus problem of distributed/multiagent systems has received increasing attention from scientific communities, for its broad applications in areas such as satellite formation flying, cooperative unmanned air vehicles, scheduling of automated highway systems, and air traffic control.

Consider a network of n identical agents with linear or linearized dynamics, where the dynamics of the ith agent are described by

$$\begin{aligned} x_i(k+1) &= Ax_i(k) + Bu_i(k), \\ y_i(k) &= Cx_i(k), \end{aligned} \tag{3.1}$$

where $x_i \in \Re^n$ is the state, $u_i \in \Re^p$ is the control input, $y_i \in \Re^q$ is the measured output, and A, B, and C are constant matrices with compatible dimensions.

The communication topology among agents is represented by a directed graph $\mathbb{G} = (\mathbb{V}, \mathbb{E})$, where $\mathbb{V} = \{1, \ldots, n\}$ is the set of nodes (i.e., agents) and $\mathbb{E} \subset \mathbb{V} \times \mathbb{V}$ is the set of edges. An edge $(i,\ j)$ in graph $\mathbb{G}$ means that agent j can obtain information from agent i, but not conversely.

At each time instant, the information available to agent i is the relative measurements of other agents with respect to i itself, given by

$$\zeta_i(k) = \sum_{j=1}^{N} d_{ij}[y_i(k) - y_j(k)], \tag{3.2}$$

where $\mathbb{D} := [d_{ii}] \in \Re^{n \times n}$ is the row-stochastic matrix associated with graph $\mathbb{G}$. The consensus protocol takes the following observer-type form:

$$\begin{aligned} v_i(k+1) &= (A+BK)v_i(k) + L\left(\sum_{j=1}^{n} d_{ij}C[v_i(k) - v_j(k)] - \zeta_i(k)\right), \\ u_i(k) &= Kv_i(k), \end{aligned} \tag{3.3}$$

where $v_i \in \Re^n$ is the protocol state, $i = 1, 2, \ldots, n$ denotes the coupling strength, and $L \in \Re^{q \times n}$ and $K \in \Re^{p \times n}$ are feedback gain matrices to be determined.

Discrete Networked Dynamic Systems. https://doi.org/10.1016/B978-0-12-823698-7.00011-4

The term $\sum_{j=1}^{n} d_{ij} C[v_i(k) - v_j(k)]$ denotes the information exchanges between the protocol of agent i and those of its neighboring agents. It is observed that the protocol in (3.3) maintains the same communication topology of the agents in (3.1). Introducing

$$z_i = [x_i^t,\ v_i^t]^t, \quad z = [x_1^t,\ \cdots,\ v_n^t]^t,$$

systems (3.1) and (3.3) together can be written as

$$z(k+1) = [I_n \otimes \mathcal{A} + (I_n - \mathbb{D}) \otimes \mathcal{H}] z(k), \tag{3.4}$$

where

$$\mathcal{A} = \begin{bmatrix} A & BK \\ 0 & A + BK \end{bmatrix}, \quad \mathbf{H} = \begin{bmatrix} 0 & 0 \\ -LC & LC \end{bmatrix}.$$

3.1.1 Preliminaries

We start by introducing a concept of dynamic consensus.

Definition 3.1. Given agents (3.1), the protocol (3.3) is said to solve the dynamic consensus problem if the states of system (3.4) satisfy

$$\lim_{k \to \infty} \|z_i(k) - z_j(k)\| = 0, \quad \forall i, j = 1, 2, \ldots, n. \tag{3.5}$$

Now, let $r \in \Re^n$ be such that $r^t L = 0$ and $r^t \mathbf{1} = \mathbf{1}$.

Remark 3.1. We introduce a new variable

$$\begin{aligned} \delta(k) &= z(k) - [(\mathbf{1} r^t) \otimes I_{2n}] z(k) \\ &= [(I_n - \mathbf{1} r^t) - I_{2n}] z(k), \end{aligned} \tag{3.6}$$

where $\delta \in \Re^{2Nn \times 2Nn}$ satisfies

$$(r^t \otimes I_{2n}) \delta = 0.$$

By similarity to [1], δ is referred to as the disagreement vector. It is easy to see that 0 is a simple eigenvalue of $I_N - \mathbf{1} r^t$ with $\mathbf{1}$ as the right eigenvector, and 1 is another eigenvalue with multiplicity $N - 1$.

It follows from (3.6) that $\delta = 0$ if and only if $z_1 = \cdots = z_n$, that is, the dynamic consensus problem can be recast into the asymptotic stability of vector δ, which evolves according to the following dynamics:

$$\delta(k+1) = [I_n \otimes \mathcal{A} + (I_n - \mathbb{D}) \otimes \mathbf{H}] \delta(k). \tag{3.7}$$

In the sequel, we provide a decomposition approach to the dynamic consensus problem, following [2].

Theorem 3.1. *For the communication topology $\mathbb{G} \in \Gamma_n$ containing a directed spanning tree, protocol (3.3) solves the dynamic consensus problem if and only if all the matrices*

$$A + BK,\ \ A + (1 - \lambda_i)LC,\ i = 2, 3, \cdots, n,$$

are Schur stable, where the eigenvalues are located in the open unit disk.

Proof. Let

$$F \in \Re^{n\times(n-1)},\ J \in \Re^{(n-1)\times n},\ Q = \Re^{n\times n}$$

such that

$$Q = \begin{bmatrix} \mathbf{1} & F \end{bmatrix},\ Q^{-1} = \begin{bmatrix} r^t \\ J \end{bmatrix},\ Q^{-1}(I_n - \mathbb{D})Q = \begin{bmatrix} 0 & 0 \\ 0 & \Theta \end{bmatrix} = \Phi,$$

where $\Theta \in \Re^{(n-1)\times(n-1)}$ in the upper triangular with diagonal entries are the nonzero eigenvalues of $(I_n - \mathbb{D})$. Defining the state transformation

$$\sigma(k) \triangleq (Q^{-1} \otimes I_{2n})\delta(k),\quad \sigma = [\sigma_1^t, \cdots, \sigma_n^t]^t,$$

it follows that (3.7) can be cast into the form

$$\sigma(k+1) = [I_n \otimes \mathcal{A} + \Phi \otimes \mathcal{H}]\sigma(k). \tag{3.8}$$

Note that $\sigma_1(k)$ can be inferred from (3.6) as

$$\sigma_1(k) = (r^t \otimes I_{2n})\sigma(k) \equiv 0. \tag{3.9}$$

It is important to observe from (3.8) that the elements of the state matrix are either block-diagonal or block upper diagonal. This leads to $\sigma_j(k)$, $j = 2, \cdots, n$, converging asymptotically to zero if and only if the $n-1$ subsystems along the diagonal

$$\sigma_j(k+1) = [\mathcal{A} + (1 - \lambda_j)\mathbf{H}]\sigma_j(k),\ j = 2, \cdots, n, \tag{3.10}$$

are Schur stable. It is easy to verify that matrices $\mathcal{A} + \lambda_j \mathbf{H}$ are similar to

$$\begin{bmatrix} A + (1 - \lambda_j)LC & 0 \\ -(1 - \lambda_j)LC & A + BK \end{bmatrix},\ j = 2, \cdots, n.$$

It is concluded that the Schur stability of the matrices

$$A + BK, A + (1 - \lambda_j)LC,\ j = 2, \cdots, n,$$

is equivalent to that the state σ of (3.7) converges asymptotically to zero, implying that consensus is achieved. □

Remark 3.2. Essentially, Theorem 3.1 converts the consensus problem of arbitrary large multiagent networks under the observer-type protocol (3.3) to the stability assessment of a set of matrices with the same dimension as a single agent. To this end, it significantly reduces the computational complexity. The directed communication topology $\mathbb{G}$ is only assumed to have a directed spanning tree. The effects of the communication topology on the consensus problem are characterized by the eigenvalues of the corresponding row-stochastic matrix $\mathcal{D}$, which may be complex, rendering the matrices complex-valued in Theorem 3.1.

Remark 3.3. The observer-type consensus protocol (3.3) can be seen as an extension of the traditional observer-based controller for a single system to one for multiagent systems. The "separation principle" of the traditional observer-based controllers still holds in this multiagent setting. Moreover, the protocol (3.3) is based only on relative output measurements between neighboring agents.

One more result is presented here.

Theorem 3.2. *Consider the closed-loop multiagent system (3.4) with a communication topology $\mathbb{G} \in \Gamma_n$. If protocol (3.3) satisfies Theorem 3.1, then*

$$x_j(k) \to \mu(k) \triangleq (r^t \otimes A^k) \begin{bmatrix} x_1(0) \\ \vdots \\ x_n(0) \end{bmatrix},$$

$$v_j(k) \to 0, \quad j = 1, \cdots, n, \quad as\ k \to \infty, \tag{3.11}$$

where $r \in \Re^n$ satisfies $r^t(I_n - \mathbb{D})$ and $r^t\mathbf{1} = 1$.

Proof. It follows from (3.4) that

$$\begin{aligned} z(k) &= [I_n \otimes \mathcal{A} + (I_n - \mathbb{D}) \otimes \mathbf{H}]^k z(0) \\ &= (Q \otimes I)[I_n \otimes \mathcal{A} + (I_n - \mathbb{D}) \otimes \mathbf{H}]^k (Q^{-1} \otimes I) z(0) \\ &= (Q \otimes I) \begin{bmatrix} \mathcal{A}^k & 0 \\ 0 & [I_{n-1} \otimes \mathcal{A} + \Theta \otimes \mathbf{H}]^k \end{bmatrix} (Q^{-1} \otimes I) z(0). \end{aligned}$$

By Theorem 3.1, it follows that matrix

$$[I_{n-1} \otimes \mathcal{A} + \Theta \otimes \mathbf{H}]$$

is Schur stable. Hence,

$$\begin{aligned} z(k) &\to (\mathbf{1} \otimes I)\mathcal{A}^k (r^t \otimes I) z(0) \\ &= (\mathbf{1} r^t) \otimes \mathcal{A}^k z(0), \quad k \to \infty. \end{aligned}$$

In turn, this implies that

$$z_j(k) \to (\mathbf{1} \otimes I)\mathcal{A}^k (r^t \otimes I) z(0)$$
$$= (r^t \otimes \mathcal{A}^k z(0), \; k \to \infty, \; j = 1, \cdots, n. \qquad (3.12)$$

Since $A + BK$ is Schur stable, condition (3.12) guarantees the desired result. □

Remark 3.4. It is important to recall that if A is Schur stable, then $\mu(k) \to 0$, $k \to \infty$. In the case that matrix A in (3.1) has eigenvalues located outside the open unit circle, the consensus value $\mu(k)$ reached by the agents will tend to infinity exponentially. Alternatively, if A has eigenvalues within the closed unit circle, then the agents in (3.1) may reach consensus nontrivially, which means that some states of each agent might approach a common nonzero value. This typically takes place in the commonly studied first-, second-, and high-order integrators.

3.1.2 Discrete consensus region

It follows from Theorem 3.1 that the consensus of the given agents (3.1) under protocol (3.3) depends on the feedback gain matrices K, L and the eigenvalues λ_i of matrix $\mathcal{D}$ associated with the communication graph $\mathbb{G}$. In the sequel, we proceed to analyze the correlated effects of matrix L and graph $\mathbb{G}$ on consensus. For this purpose, the notion of consensus region is introduced.

Definition 3.2. Suppose that matrix K has been designed such that $A + BK$ is Schur stable. The region $\boldsymbol{\mathcal{S}}$ of the parameter $\eta \subset \mathbb{C}$, such that matrix $A + (1 - \eta)LC$ is Schur stable, is called the *discrete-time consensus region (DTCR)* of network (3.4).

Based on Theorem 3.1, the following result holds.

Corollary 3.1. *The agents network (3.1) reaches consensus under protocol (3.3) if $\lambda_i \in \boldsymbol{\mathcal{S}}$, $i = 2, ..., n$, where $\{\lambda_i\}$ are the eigenvalues of $\mathbb{D}$ located in the open unit disk.*

The following observations are made:

(1.) For an undirected communication graph, it turns out that the consensus region of network (3.4) is a bounded interval or a union of several intervals on the real axis.

(2.) For a directed graph where the eigenvalues of $\mathcal{D}$ are generally complex numbers, the consensus region $\boldsymbol{\mathcal{S}}$ is either a bounded region or a set of several disconnected regions in the complex plane.

(3.) Due to the fact that the eigenvalues of the row-stochastic matrix $\mathcal{D}$ lie in the unit disk, bounded consensus regions exist for the discrete-time consensus considered here before.

The following example has a disconnected consensus region.

3.1.3 Simulation example 3.1

The agent dynamics and the consensus protocol are given by (3.1) and (3.4), respectively, with

$$A=\begin{bmatrix} 0 & 1 \\ -1 & -1.02 \end{bmatrix},\quad B=\begin{bmatrix} 1 \\ 0 \end{bmatrix},\quad C=\begin{bmatrix} 1 & 0 \\ 0 & 1 \end{bmatrix},$$
$$L=\begin{bmatrix} 0 & -1 \\ 1 & 0 \end{bmatrix},\quad K=\begin{bmatrix} -0.5 & -0.5 \end{bmatrix}.$$

It is easy to see that matrix $A+BK$ is Schur stable. To simplify the illustration, assume that the communication graph $\mathbb{G}$ is undirected here. It follows that the consensus region is a set of intervals on the real axis and the characteristic equation of matrix $A+(1-\eta)LC$ is

$$\mathbf{det}(zI - A - \eta LC) = z^2 - 1.02\,z + \eta^2 = 0. \tag{3.13}$$

Invoking the bilinear transformation $z = s + 1/s - 1$, (3.13) becomes

$$(\eta^2 - 0.02)\,s^2 + (1-\eta^2)\,s + 2.02 + \eta^2 = 0. \tag{3.14}$$

It is well known that, under the bilinear transformation, (3.13) has all roots within the unit disk if and only if the roots of (3.14) lie in the open left half-plane (LHP). According to the fundamental Hurwitz criterion [3], (3.14) has all roots in the open LHP if and only if $0.02 < \eta^2 < 1$. Therefore, the DTCR in this case is

$$\mathcal{S} = (-1,\ -0.1414) \cup (0.1414,\ 1),$$

a union of two disconnected intervals. For the communication graph shown in Fig. 3.1, the corresponding row-stochastic matrix is

$$\mathcal{D}=\begin{bmatrix} 0.3 & 0.2 & 0.2 & 0.2 & 0 & 0.1 \\ 0.2 & 0.6 & 0.2 & 0 & 0 & 0 \\ 0.2 & 0.2 & 0.6 & 0 & 0 & 0 \\ 0.2 & 0 & 0 & 0.4 & 0.4 & 0 \\ 0 & 0 & 0 & 0.4 & 0.2 & 0.4 \\ 0.1 & 0 & 0 & 0 & 0.4 & 0.5 \end{bmatrix},$$

whose eigenvalues are $\{1,\ -0.2935,\ 0.164,\ 0.4,\ 0.4624,\ 0.868\}$, which all belong to the DTCR $\mathcal{S}$. It follows from Corollary 3.1 that network (3.4) with graph given in Fig. 3.1 can achieve consensus. To examine the reliability of the closed-loop agent network under the observer-based protocol, we address two special cases.

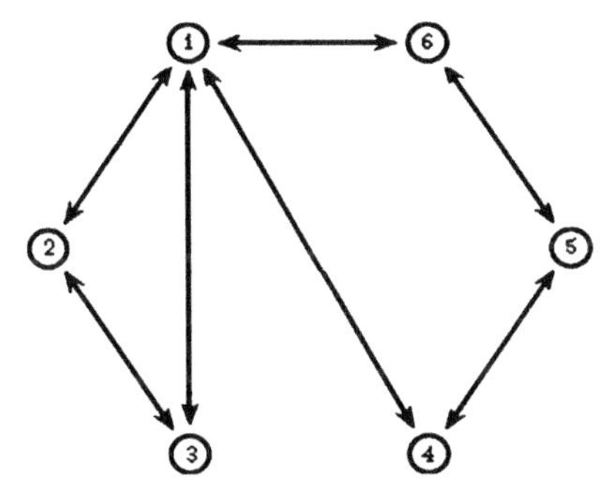

FIGURE 3.1 The communication topology I.

Case 1: An edge is added between nodes 1 and 5, thus more information exchange will exist inside the network. Then, the row-stochastic matrix $\mathcal{D}$ becomes

$$\mathcal{D} = \begin{bmatrix} 0.2 & 0.2 & 0.2 & 0.2 & 0.1 & 0.1 \\ 0.2 & 0.6 & 0.2 & 0 & 0 & 0 \\ 0.2 & 0.2 & 0.6 & 0 & 0 & 0 \\ 0.2 & 0 & 0 & 0.4 & 0.4 & 0 \\ 0.1 & 0 & 0 & 0.4 & 0.2 & 0.3 \\ 0.1 & 0 & 0 & 0 & 0.4 & 0.5 \end{bmatrix},$$

whose eigenvalues are $\{1, -0.2346, 0.0352, 0.4, 0.4634, 0.836\}$. Clearly, the eigenvalue 0.0352 does not belong to the DTCR $\mathcal{S}$, that is, consensus can not be achieved in this case.

Case 2: The edge between nodes 5 and 6 is removed. The row-stochastic matrix $\mathcal{D}$ becomes

$$\mathcal{D} = \begin{bmatrix} 0.3 & 0.2 & 0.2 & 0.2 & 0 & 0.1 \\ 0.2 & 0.6 & 0.2 & 0 & 0 & 0 \\ 0.2 & 0.2 & 0.6 & 0 & 0 & 0 \\ 0.2 & 0 & 0 & 0.4 & 0.4 & 0 \\ 0.1 & 0 & 0 & 0.4 & 0.6 & 0 \\ 0.1 & 0 & 0 & 0 & 0 & 0.9 \end{bmatrix},$$

whose eigenvalues are $\{1, -0.0315, 0.2587, 0.4, 0.8676, 0.9052\}$. In this case, the eigenvalue -0.0315 does not belong to the DTCR $\mathcal{S}$, that is, consensus cannot be achieved either.

These sample cases imply that, for disconnected consensus regions, consensus can be quite "fragile" to the variations of the network's communication topology. Therefore, the consensus protocol should be designed to have a sufficiently large bounded consensus region in order to be “robust” with respect to the communication topology. This will be explored later on.

3.1.4 Networks with neutrally stable agents

In this subsection, a special case where matrix A is neutrally stable is considered. First, the following result is recalled.

Lemma 3.1. *For matrix $Q = Q^* \in \mathbb{C}^{n\times n}$, consider the following Lyapunov equation:*

$$A^* X A - X + Q = 0.$$

If $X > 0$, $Q \geq 0$, and (Q, V) is observable with $VV^t = I$, then matrix A is Schur stable.

Proposition 3.1. *For matrices $Q \in \Re^{n\times n}$, $V \in \Re^{m\times n}$, $\eta \in \mathbb{C}$, where Q is orthogonal, $VV^t = I$, and (Q, V) is observable. If $|\eta| < 1$, then the matrix $Q - (1-\eta)QVV^t$ is Schur stable.*

Proof. It follows that

$$\begin{aligned}
&(Q - (1-\eta)QVV^t)^*(Q - (1-\eta)QVV^t) - I \\
&= Q^t Q - (1-\eta)Q^t QV^t V - (1-\eta^*)V^t V Q^t Q \\
&+ |1-\eta|^2 V^t V Q^t Q V^t V - I \\
&= [-2\,\mathbf{Re}(1-\eta) + |1-\eta|^2]V^t V \\
&= [|\eta|^2 - 1]V^t V.
\end{aligned} \tag{3.15}$$

Since (Q, V) is observable, it is straightforward to check that

$$(Q - (1-\eta)QVV^t),\ V^t V$$

is also observable. It follows by Lemma 3.1 that $(Q - (1-\eta)QVV^t)$ is Schur for any $|\eta| < 1$. □

Next, a constructive algorithm for protocol (3.3) is presented, which will be used in the sequel.

Algorithm 3.1. Given that $A \in \Re^{n\times n}$ is neutrally stable and that the triple (A, B, C) is stabilizable and detectable, the dynamic protocol (3.3) can be constructed as follows:

1) Select gain matrix K such that $A + BK$ is Schur stable.
2) Choose matrices $\mathbf{U} \in \Re^{n\times n_1}$ and $\mathbf{W} \in \Re^{n\times(n-n_1)}$, satisfying

$$\begin{bmatrix} \mathbf{U} & \mathbf{W} \end{bmatrix}^{-1} A \begin{bmatrix} \mathbf{U} & \mathbf{W} \end{bmatrix} = \begin{bmatrix} \mathbf{M} & 0 \\ 0 & \mathbf{X} \end{bmatrix}, \tag{3.16}$$

where matrix $\mathbf{M} \in \Re^{n_1 \times n_1}$ is orthogonal and matrix $\mathbf{X} \in \Re^{(n-n_1)\times(n-n_1)}$ is Schur stable.

3) Choose $\mathbf{V} \in \Re^{m \times n_1}$ such that $\mathbf{V}\mathbf{V}^t = I_m$ and $\mathbf{range}(\mathbf{V}^t) = \mathbf{range}(\mathbf{U}^t C^t)$.

4) Define $L = -\mathbf{U}\mathbf{M}\mathbf{V}^t(C\mathbf{U}\mathbf{V}^t)^{-1}$.

Note in the above algorithm that matrices **U** and **W** can be derived by rendering matrix A into the real Jordan canonical form [4].

Theorem 3.3. *Let matrix $A \in \Re^{n \times n}$ be neutrally stable and that the triple $(A,\ B,\ C)$ is stabilizable and detectable. The protocol (3.3) constructed via Algorithm 3.1 has the open unit disk as its bounded consensus region. Thus, such a protocol solves the consensus problem for (3.1) with respect to Γ_n, the set of all the communication topologies containing a spanning tree.*

Proof. Initially, we recall the related variables defined in Algorithm 3.1. Without loss of generality, we assume that matrix $C\,\mathbf{U}$ is of full row rank. In view of the fact that $\mathbf{V}^t\mathbf{V}$ is an orthogonal projection onto $\mathbf{range}(\mathbf{V}^t) = \mathbf{range}(\mathbf{U}^t C^t)$, matrix $C\mathbf{U}\mathbf{V}^t$ is invertible and $\mathbf{V}^t\mathbf{V}\mathbf{U}^t C^t = \mathbf{U}^t C^t$, it follows that $V = (C\mathbf{U}\mathbf{V}^t)^{-1}C\mathbf{U}$, and hence

$$L\,C\,\mathbf{U} = -\mathbf{U}\mathbf{M}V^t V.$$

Additionally, the detectability of $(A,\ C)$ implies that $(\mathbf{M},\ \mathbf{V})$ is observable.

Let $\mathbf{U}^\dagger \in \Re^{n_1 \times n}$ and $\mathbf{W}^\dagger \in \Re^{(n-n_1)\times n}$ be such that

$$\begin{bmatrix} \mathbf{U}^\dagger \\ \mathbf{W}^\dagger \end{bmatrix} = \begin{bmatrix} \mathbf{U} & \mathbf{W} \end{bmatrix}^{-1},$$
$$\mathbf{U}^\dagger\mathbf{U} = I,\ \ \mathbf{W}^\dagger\mathbf{W} = I,\ \ \mathbf{U}^\dagger\mathbf{W} = 0,\ \ \mathbf{W}^\dagger\mathbf{U} = 0.$$

Then,

$$\begin{aligned} &\begin{bmatrix} \mathbf{U} & \mathbf{W} \end{bmatrix}^{-1} [A + (1-\eta)LC] \begin{bmatrix} \mathbf{U} & \mathbf{W} \end{bmatrix} \\ &= \begin{bmatrix} \mathbf{M} + (1-\eta)\mathbf{U}^\dagger LC\mathbf{U} & (1-\eta)\mathbf{U}^\dagger LC\mathbf{W} \\ (1-\eta)\mathbf{W}^\dagger LC\mathbf{U} & \mathbf{X} + (1-\eta)\mathbf{W}^\dagger LC\mathbf{W} \end{bmatrix} \\ &= \begin{bmatrix} \mathbf{M} - (1-\eta)\mathbf{M}V^t V & -(1-\eta)\mathbf{U}^\dagger LC\mathbf{W} \\ 0 & \mathbf{X} \end{bmatrix}. \end{aligned} \tag{3.17}$$

By Lemma 3.1, matrix $\mathbf{M} - (1-\eta)\mathbf{M}V^t V$ is Schur stable for any $|\eta| < 1$. Therefore, (3.17) implies that matrix $[A + (1-\eta)LC]$ with L given by Algorithm 3.1 is Schur stable for any $|\eta| < 1$, that is, the protocol (3.3) constructed via Algorithm 3.1 has a bounded consensus region in the form of the open unit disk. Since the eigenvalues of any communication topology containing a spanning tree lie in the open unit disk, except eigenvalue 1, it follows from Corollary 3.1 that this protocol solves the consensus problem with respect to Γ_n. □

Had we considered the consensus problem of the following coupled network:

$$x_i(k+1) = Ax_i(k) + LC\sum_{j=1}^{n} d_{ij}(x_i - x_j),\ i = 1,\ 2, \cdots,\ n, \tag{3.18}$$

where $\mathbb{D} := [d_{ii}] \in \Re^{n\times n}$ is the row-stochastic matrix associated with graph $\mathbb{G}$ and matrix L is to be designed, we would have arrived at the following results.

Corollary 3.2. *There exists a matrix L such that network (3.18) has the open unit disk as its consensus region, that is, the network can reach consensus with respect to Γ_n if and only if the pair $(A,\ C)$ is detectable. Such a matrix L can be constructed via Algorithm 3.1.*

Proof. For any communication topology $\mathbb{G} \in \Gamma_n$, it follows from Theorem 3.1 that there exists a matrix L such that network (3.18) achieves consensus if and only if matrices $A + (1 - \eta_i)LC, i = 2,\ \cdots,\ n$, are Schur stable. The necessity is straightforward and the sufficiency follows readily from Theorem 3.3. □

3.1.5 Simulation example 3.2

Consider a network of agents described by (3.1), with

$$A = \begin{bmatrix} 0.2 & 0.6 & 0 \\ -1.4 & 0.8 & 0 \\ 0.7 & 0.2 & -0.5 \end{bmatrix},\quad B = \begin{bmatrix} 0 \\ 1 \\ 0 \end{bmatrix},\quad C = \begin{bmatrix} 1 & 0 & 1 \end{bmatrix}.$$

The eigenvalues of matrix A are $\{0.5,\ 0.5 \pm j\ 0.866\}$, $j = \sqrt{1}$, thus A is neutrally stable. In protocol (3.3), we choose $K = [1.2\quad -0.9\quad -0.2]$ such that $A + BK$ is Schur stable. The matrices

$$\mathbf{U} = \begin{bmatrix} 0.1709 & -0.4935 \\ 0.7977 & 0 \\ -0.0570 & -0.2961 \end{bmatrix},\quad \mathbf{W} = \begin{bmatrix} 0 \\ 1 \\ 0 \end{bmatrix}$$

satisfy (3.16) with

$$\mathbf{M} = \begin{bmatrix} 0.5 & 0.886 \\ -0.866 & 0.5 \end{bmatrix},\quad \mathbf{X} = -0.5.$$

This leads to $\mathbf{U}^t\ C^t = [0.1139\quad -0.7896]^t$. Taking $\mathbf{V} = [0.1428\quad -0.9898]$ such that $\mathbf{V}\mathbf{V}^t = 1$ and $\mathbf{range}(\mathbf{V}^t) = \mathbf{range}(\mathbf{U}^t\ C^t)$, by Algorithm 3.1, one obtains the gain $L = [-0.2143\quad 0.7857\quad -0.2857]^t$. In light of Theorem 3.3, the agents considered in this example will reach consensus under the protocol (3.3), with K and L given as above, with respect to all the communication topologies containing a spanning tree.

3.1.6 Networks with unstable agents

This subsection considers the general case where matrix A is not neutrally stable, that is, A is allowed to have eigenvalues outside the unit circle or has at least one eigenvalue with unit magnitude whose corresponding Jordan block is larger than 1.

For this purpose, we introduce the following modified algebraic Riccati equation (MARE):

$$\begin{aligned} &\mathbf{P} = A\mathbf{P}A^t - (1-\delta^2)A\mathbf{P}C^t(C\mathbf{P}C^t + I)^{-1}C\mathbf{P}A^t + \mathbf{Q}, \\ &\mathbf{P} \geq 0, \ \ \mathbf{Q} > 0, \ \ \delta \in \Re. \end{aligned} \tag{3.19}$$

For $\delta = 0$, the MARE (3.19) is reduced to the commonly used discrete-time Riccati equation (see [3]).

The following lemma concerns the existence of solutions for the MARE (3.19).

Lemma 3.2. *[3] Let the pair $(A,\ C)$ be detectable. Then, the following statements hold.*

1) *Suppose that the matrix A has no eigenvalues with magnitude larger than 1. Then, the MARE (3.19) has a unique positive definite solution $\mathbf{P}$ for any $0 < \delta < 1$.*
2) *For the case where A has at least one eigenvalue with magnitude larger than 1 and the rank of B is one, the MARE (3.19) has a unique positive definite solution $\mathbf{P}$, if*

$$0 < \delta < \frac{1}{\coprod_i |\lambda_i^u(A)|}.$$

3) *If the MARE (3.19) has a unique positive definite solution $\mathbf{P}$, then*

$$\mathbf{P} = \lim_{k\to\infty} \mathbf{P}_k$$

for any initial condition $\mathbf{P}_0 \geq 0$, where $\mathbf{P}_k$ satisfies

$$\mathbf{P}_{k+1} = A\mathbf{P}_k A^t - (1-\delta^2)A\mathbf{P}_k C^t(C\mathbf{P}_k C^t + I)^{-1}C\mathbf{P}_k A^t + \mathbf{Q}.$$

Based thereon, the following results is established.

Proposition 3.2. *Suppose that the pair $(A,\ C)$ is detectable. Then, for the case where A has no eigenvalues with magnitude larger than 1, the matrix $[A + (1-\pi)LC]$ with $L = -A\mathbf{P}C^t(C\mathbf{P}C^t + I)^{-1}$ is Schur stable for any $|\pi| \leq \delta$, $0 < \delta < 1$, where $\mathbf{P} > 0$ is the unique solution to the MARE (3.19).*

Moreover, for the case where A has at least one eigenvalue with magnitude larger than 1 and B is of rank one, $[A+(1-\pi)LC]$ with

$$L=-A\mathbf{P}C^t(C\mathbf{P}C^t+I)^{-1}$$

is Schur stable for any

$$|\pi|\leq\delta,\ 0<\delta<\frac{1}{\coprod_i|\lambda_i^u(A)|}.$$

Proof. Considering the discrete-time Lyapunov inequality $[A+(1-\pi)LC]\mathbf{P}[A+(1-\pi)LC]^*-\mathbf{P}$ and manipulating using the matrix inversion lemma, we get

$$\begin{aligned}
&[A+(1-\pi)LC]\mathbf{P}[A+(1-\pi)LC]^*-\mathbf{P}\\
&=A\mathbf{P}_kA^t-2\mathbf{Re}(1-\pi)A\mathbf{P}C^t(C\mathbf{P}C^t+I)^{-1}C\mathbf{P}A^t-\mathbf{P}\\
&+|1-\pi|^2A\mathbf{P}C^t(C\mathbf{P}_kC^t+I)^{-1}C\mathbf{P}A^t\\
&=A\mathbf{P}_kA^t+(2\mathbf{Re}(1-\pi)+|1-\pi|^2)A\mathbf{P}C^t(C\mathbf{P}C^t+I)^{-1}C\mathbf{P}A^t-\mathbf{P}\\
&+|1-\pi|^2A\mathbf{P}C^t(C\mathbf{P}C^t+I)^{-1}[-I+(C\mathbf{P}C^t+I)^{-1}]C\mathbf{P}A^t \qquad (3.20)\\
&=A\mathbf{P}_kA^t+(|\pi|^2-1)A\mathbf{P}C^t(C\mathbf{P}_kC^t+I)^{-1}C\mathbf{P}A^t-\mathbf{P}\\
&-|1-\pi|^2A\mathbf{P}C^t(C\mathbf{P}_kC^t+I)^{-1}C\mathbf{P}A^t\\
&\leq A\mathbf{P}_kA^t-(1-\delta^2)A\mathbf{P}C^t(C\mathbf{P}C^t+I)^{-1}C\mathbf{P}A^t-\mathbf{P}\\
&=-\mathbf{Q}<0.
\end{aligned}$$

The assertion follows directly from Lemma 3.2. □

Algorithm 3.2. Assuming the triple $(A,\ B,\ C)$ is stabilizable and detectable, the dynamic protocol (3.3) can be constructed as follows:

1) Select gain matrix K such that $A+BK$ is Schur stable.
2) Select $L=-A\mathbf{P}C^t(C\mathbf{P}C^t+I)^{-1}$, where $\mathbf{P}>0$ is the unique solution of (3.19)

By using Algorithm 3.2, it follows that a sufficient and necessary condition for the existence of the consensus protocol is the stabilizability and detectability of the triple $(A,\ B,\ C)$ for the case where matrix A has no eigenvalues with magnitude larger than 1. Alternatively, δ has to satisfy

$$\delta<\frac{1}{\coprod_i|\lambda_i^u(A)|}$$

for the case where A has at least one eigenvalue outside the unit circle and matrix B is of rank one.

On combining Theorem 3.2 and Proposition 3.2, the following result holds.

Theorem 3.4. *Suppose that the triple* (A, B, C) *is stabilizable-detectable. Then, the protocol prescribed by Algorithm 3.2 has a bounded consensus region in the form of an origin-centered disk of radius* δ*, that is, this protocol solves the consensus problem for networks with agents (3.4) with respect to* $\Gamma_{\geq\delta}$*, where* δ *satisfies* $0 < \delta < 1$ *for the case where matrix* A *has no eigenvalues with magnitude larger than* 1 *and satisfies*

$$0 < \delta < \frac{1}{\coprod_i |\lambda_i^u(A)|}$$

for the case where matrix A *has a least one eigenvalue outside the unit circle and matrix* B *is of rank one.*

3.1.7 Simulation example 3.3

Consider the network of agents (3.1) described by

$$A = \begin{bmatrix} 1 & 1 \\ 0 & 1 \end{bmatrix}, \quad B = \begin{bmatrix} 0 \\ 1 \end{bmatrix}, \quad C = \begin{bmatrix} 1 & 0 \end{bmatrix}.$$

We choose $K = [-0.5 \quad -1.5]$ so that matrix $A + BK$ is Schur stable. Solving the MARE (3.19) with $\delta = 0.95$ gives

$$\mathbf{P} = 10^4 \times \begin{bmatrix} 1.1780 & 0.0602 \\ 0.0602 & 0.0062 \end{bmatrix}.$$

By Algorithm 3.2, one obtains $L = [-1.051 \quad -0.051]^t$. It follows from Theorem 3.4 that the agents (3.1) reach consensus under protocol (3.3) with K and L given as above with respect to $\Gamma_{\geq 0.95}$. Consider that the communication topology $\mathbb{G}$ is given as in Fig. 3.2, and the row-stochastic matrix $\mathcal{D}$ becomes

$$\mathcal{D} = \begin{bmatrix} 0.4 & 0 & 0 & 0.1 & 0.3 & 0.2 \\ 0.5 & 0.5 & 0 & 0 & 0 & 0 \\ 0.3 & 0.2 & 0.5 & 0 & 0 & 0 \\ 0.5 & 0 & 0 & 0.5 & 0 & 0 \\ 0 & 0 & 0 & 0.4 & 0.4 & 0.2 \\ 0 & 0 & 0 & 0 & 0.3 & 0.7 \end{bmatrix},$$

whose eigenvalues, other than 1, are $\lambda_i = 0.5,\ 0.5565,\ 0.2217 \pm j0.2531$. Clearly, $|\lambda_i| < 0.95$, $i = 2, \cdots, 6$. Fig. 3.3 depicts the state trajectories of network (3.4) for this example, which shows that consensus is actually achieved.

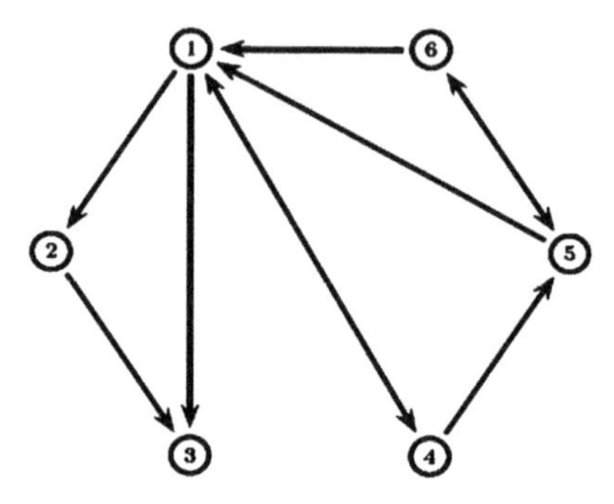

FIGURE 3.2 The communication topology II.

3.1.8 Application to formation control

In this section, the consensus algorithms are modified to solve the formation control problems of multiagent systems.

Let $\tilde{H} = [h_1, h_2, \cdots, h_n] \in \Re^{n \times n}$ describe a constant formation structure of the agent network in a reference coordinate frame, where $h_i \in \Re^n$ is the formation variable corresponding to agent i. For example, $h_1 = [0 \;\; 0]^t$, $h_2 = [0 \;\; 1]^t$, $h_3 = [1 \;\; 0]^t$, and $h_4 = [1 \;\; 1]^t$ represent a unit square. To this end, the variable $h_i - h_j$ denotes the relative formation vector between agents i and j, which is independent of the reference coordinate.

A distributed formation protocol is proposed as

$$v_i(k+1) = (A + BK)v_i(k) + L\left(\sum_{j=1}^{n} d_{ij}C[v_i(k) - v_j(k)] - \omega_i(k)\right),$$
$$u_i(k) = K v_i(k), \tag{3.21}$$

where

$$\omega_i = \sum_{j=1}^{n} d_{ij}C[v_i(k) - v_j(k) - C(h_i - h_j)], \tag{3.22}$$

and the rest of the variables are the same as in (3.3). It should be noted that (3.21) reduces to the consensus protocol (3.3) when $h_i - h_j = 0$ for all $i, \; j = 1, 2, \cdots, n$.

Definition 3.3. The agents (3.3) under protocol (3.21) achieve a given formation $\tilde{H} = [h_1, h_2, \cdots, h_n]$ if

$$||x_i(k) - h_j - x_j(k) + h_j|| \rightarrow 0, \;\; k \rightarrow \infty, \;\; \forall i, \; j = 1, 2, \cdots, n. \tag{3.23}$$

Theorem 3.5. *For any $\mathbb{G} \in \Gamma_n$, the agents (3.1) reach the formation $\tilde{H}$ under protocol (3.21) if all the matrices $A + BK$, $A + (1 - \lambda_i)LC$, $i = 2, \cdots, n$, are*

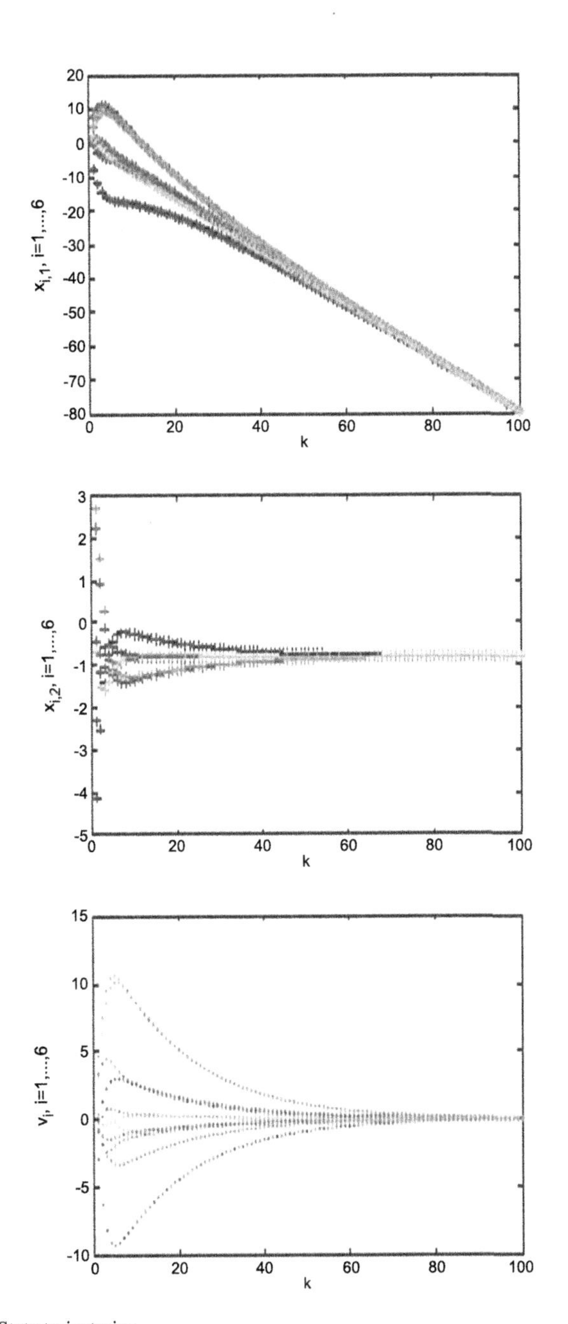

FIGURE 3.3 State trajectories.

Schur stable, and $(A - I)(h_i - h_j) = 0$, $\forall i,\ j = 1, 2, \cdots, n$, *where* λ_i, $i = 1, 2, \cdots, n$, *denote the eigenvalues of* $\mathcal{D}$ *located in the open unit disk.*

Proof. Let $e_{x_i} = x_i - h_i - x_1 + h_1$ and $e_{v_i} = v_i - v_1$ for $i = 2, \cdots, n$. Then, the agents (3.1) can reach the formation $\tilde{H}$ if and only if $e_{x_i} \to 0$, $k \to \infty$, $\forall i,\ j = 1,\ 2,\ \cdots, n$. By invoking $(A - I)(h_i - h_j) = 0$, $\forall i,\ j = 1,\ 2,\ \cdots, n$, it follows from (3.1) and (3.21) that

$$\begin{aligned} e_{x_i}(k+1) &= Ae_{x_i}(k) + BKe_{v_i}, \\ e_{v_i}(k+1) &= (A + BK)e_{v_i}(k) \\ &\quad + LC\Bigg(\sum_{j=2}^{n} d_{ij}[e_{v_i}(k) - e_{v_j}(k)] - \sum_{j=2}^{n} d_{1j}e_{v_j}(k) \\ &\quad - \sum_{j=2}^{n} d_{ij}[e_{x_i}(k) - e_{x_j}(k)] + \sum_{j=2}^{n} d_{1j}e_{x_j}(k)\Bigg), \quad i = 2, \cdots, n. \end{aligned} \tag{3.24}$$

Proceeding further, we let $e_i = [e_{x_i}^t,\ e_{v_i}^t]^t$, $i = 2, \cdots, n$, and $e = [e_2^t, \cdots, e_n^t]^t$. Then, one has

$$e(k+1) = [(I_{n-1} \otimes \mathcal{A} + (I_{n-1} - \mathbb{D})_2 + \mathbf{1}_{n-1}\alpha) \otimes \mathbf{H}]e(k), \tag{3.25}$$

where $\mathcal{A}$, $\mathbf{H}$ are defined in (3.4) and

$$\begin{aligned} \alpha &= \begin{bmatrix} d_{12} & d_{13} & \cdots & d_{1n} \end{bmatrix}, \\ \mathbb{D}_2 &= \begin{bmatrix} d_{22} & d_{23} & \cdots & d_{2n} \\ d_{32} & d_{33} & \cdots & d_{3n} \\ \vdots & \vdots & \ddots & \vdots \\ d_{n2} & d_{n3} & \cdots & d_{nn} \end{bmatrix}, \\ \mathcal{T} &= \begin{bmatrix} 1 & 0 \\ \mathbf{1}_{n-1} & \mathbf{1}_{n-1} \end{bmatrix}. \end{aligned}$$

It can be readily seen that

$$\mathcal{T}(I_n - \mathbb{D})\mathcal{T} = \begin{bmatrix} 0 & \alpha \\ 0 & (I_{n-1} - \mathbb{D}_2) + \mathbf{1}_{n-1}\alpha \end{bmatrix}.$$

Hence, the nonzero eigenvalues of $(I_n - \mathbb{D})$ are all the eigenvalues of $(I_{n-1} - \mathbb{D}_2) + \mathbf{1}_{n-1}\alpha$. Following the procedure of proving Theorem 3.1, one concludes that system (3.25) is asymptotically stable if and only if all the matrices $A + BK$, $A + (1 - \lambda_i)LC$, $i = 2, \cdots, n$, are Schur stable, which completes the proof. □

3.1.9 Simulation example 3.4

In this example, we consider a network of six double integrators, described by

$$\begin{aligned} x_i(k+1) &= x_i(k) + w_i, \\ w_i(k+1) &= w_i(k) + u_i, \\ y_i(k) &= x_i(k),\ i = 1,\ 2,\ \cdots,\ 6, \end{aligned}$$

where $x_i \in \Re^2$, $w_i \in \Re^2$, $y_i \in \Re^2$, and $u_i \in \Re^2$ are the position, the velocity, the measured output, and the acceleration input of agent i, respectively.

The objective is to design a protocol (3.21) such that the agents will evolve to a regular hexagon with edge length 8.

In this case, we choose

$$\begin{aligned} h_1 &= \begin{bmatrix} 0 & 0 & 0 & 0 \end{bmatrix}^t, \ h_2 = \begin{bmatrix} 8 & 0 & 0 & 0 \end{bmatrix}^t, \\ h_3 &= \begin{bmatrix} 12 & 4\sqrt{3} & 0 & 0 \end{bmatrix}^t, \ h_4 = \begin{bmatrix} 8 & 8\sqrt{3} & 0 & 0 \end{bmatrix}^t, \\ h_5 &= \begin{bmatrix} 0 & 8\sqrt{3} & 0 & 0 \end{bmatrix}^t, \ h_6 = \begin{bmatrix} -4 & 4\sqrt{3} & 0 & 0 \end{bmatrix}^t. \end{aligned}$$

As in Example 3.3, we take $K = [-0.5I_2 \quad -1.5I_2]$ and $L = [-1.051I_2 \quad -0.051I_2]$ in protocol (3.21). In this case, the agents will form a regular hexagon with respect to $\Gamma_{\leq 0.95}$. The state trajectories of the six agents are depicted in Fig. 3.4 based the communication topology given in Fig. 3.2.

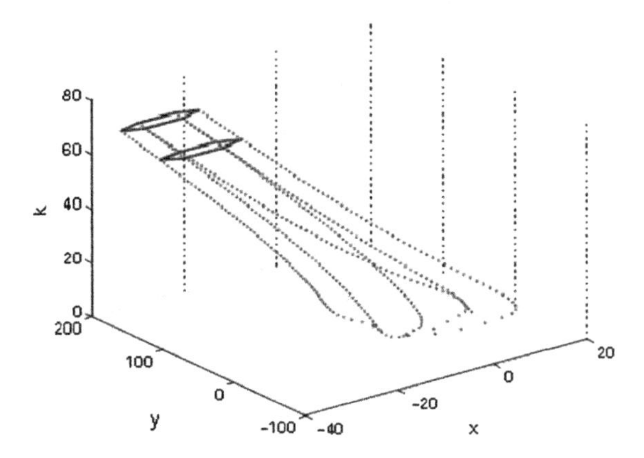

FIGURE 3.4 The agents form a regular hexagon.

3.2 Multiagent systems with diverse time delays

Consensus problems for multiagent systems have attracted great attention in many fields, such as biology, physics, robotics, and control engineering.

Roughly speaking, consensus means that multiple agents reach agreement on a common value which might be the attitude in multispacecraft alignment, the heading direction in flocking behavior, or the average in distributed computation.

3.2.1 Introduction

In the 2010s, numerous studies have been conducted on the consensus problem for agents with first-order dynamics [7], [6], [11], [13], [15], [16], [17]. Nonlinear discrete-time multiagent systems with time-dependent communication links investigated and introduced a novel method based on the notion of convexity [15]. In [16], the authors investigated a systematical framework of consensus problems with directed interconnection graphs or time delays by a Lyapunov-based approach. Moreover, Ren et al. extended the results of [11] and [16] and presented some more relaxable conditions for consensus of information under dynamically changing interaction topologies [17].

In [7], [6], the authors investigated state consensus problems for discrete-time multiagent systems with arbitrary bounded nonuniform time delays and changing communication graphs that may not have spanning trees. Recently, more attention has been paid to consensus-related problems for agents with second-order dynamics [9], [12], [14]. For example, in [12], the authors investigated a method for decentralized stabilization of vehicle formations. An asynchronous discrete-time formulation with fixed topology and derived conditions under which a multiagent system achieves cohesiveness in the presence of sensing delays, sensing errors, and sensing topology was considered in [14].

In this section, we first introduce a linear consensus protocol using the local velocity information and the distributed relative state information. Then by model transformations, we turn the original system into an equivalent nondelayed system whose system matrix is stochastic. Based on the obtained equivalent system, we derive sufficient conditions for state consensus of the system. The most important feature of the results obtained hereafter is that we do not impose restrictive conditions on the communication topologies and the communication time delays, and allow not only for arbitrary bounded nonuniform time delays but also for dynamically changing directed graphs that may not have spanning trees.

3.2.2 Preliminaries

In this section, we introduce some preliminary knowledge of graph theory and matrix theory for the following analysis (referring to [8] and [10]). Let $\mathbb{G}(\mathbb{V}, \mathbb{E}, \mathbb{A})$ be a directed graph of order n, where $\mathbb{V} = \{s_1, \ldots, s_n\}$ is the set of nodes, $\mathbb{E} \subseteq \mathbb{V} \times \mathbb{V}$ is the set of edges, and $\mathbb{A} = [a_{ij}]$ is a weighted adjacency matrix. The node indices belong to a finite index set $\mathbb{I} = \{1, 2, \ldots, n\}$. An edge of G is denoted by $e_{ij} = (s_i, s_j)$, where the first element s_i of e_{ij} is said to be the tail of the edge and the other s_j to be the head. Then the set of neighbors of node s_i is denoted by $N_i = \{s_j \in \mathbb{V} : (s_i, s_j) \in \mathbb{E}\}$. The adjacency elements associated with the edges are positive, i.e., $e_{ij} \in \mathbb{E} \Leftrightarrow a_{ij} > 0$. Correspondingly, the Laplacian of the directed graph is defined as $L = \Delta - \mathbb{A}$, where $\Delta = [\Delta_{ij}]$ is a diagonal matrix with $\Delta_{ii} = \sum_{j=1}^{n} a_{ij}$. An important property of L is that all the row sums of L are zero and thus $\mathbf{1}$ is an eigenvector of L associated with the zero eigenvalue. A directed path is a sequence of ordered edges of the form $(s_{i1}, s_{i2}), (s_{i2}, s_{i3}), \ldots$, where $s_{ij} \in \mathbb{V}$. A directed graph is said to be strongly connected, if there is a directed path from every node to every other node. Moreover, a directed graph is said to have spanning trees, if there exists a node such that there is a directed path from every other node to this node, and this node is called the root node of the spanning tree. The union of a collection of directed graphs $\mathbb{G}_{i1}, \mathbb{G}_{i2}, \ldots, \mathbb{G}_{ij}$ with the same node set $\mathbb{V}$ is a directed graph with node set $\mathbb{V}$ and the edge set equal to the union of the edge sets of all of the graphs in the collection.

Given $C = [c_{ij}] \in \Re^{n \times r}$, it is said that $C \geq 0$ (C is nonnegative) if all its elements c_{ij} are nonnegative, and it is said that $C > 0$ (C is positive) if all its elements c_{ij} are positive. Further, $C \geq D$ if $C - D \geq 0$, and $C > D$ if $C - D > 0$. If a nonnegative matrix $C \in \mathbb{R}^{n \times n}$ satisfies $C\mathbf{1} = \mathbf{1}$, then it is said to be stochastic. In addition, a stochastic matrix A is said to be indecomposable and aperiodic (SIA) if $\lim_{k \to +\infty} A^k = \mathbf{1} f^T$, where $f \in \mathbb{R}^n$.

3.2.3 Dynamic model

Suppose that the multiagent system under consideration consists of n agents with discrete-time dynamics. Each agent is regarded as a node in the communication directed graph $\mathbb{G}$. Each edge $(s_j, s_i) \in \mathbb{E}(\mathbb{G}(kT))$ corresponds to an available information channel from agent s_i to agent s_j at time kT, where $k \in \mathbb{Z}_+$, $T > 0$ is the sample time, and $\mathbb{G}(kT)$ denotes the communication topology at time kT. Moreover, each agent updates its current state based upon the information received from its neighbors at time kT. The set of the neighbors of the ith agent at time kT is denoted by $N_i(kT)$. The Laplacian of the graph $\mathbb{G}(kT)$ is denoted by $L(kT)$. Suppose the dynamics of the ith agent ($i \in \mathbb{I}$) are

$$\begin{aligned} x_i((k+1)T) &= x_i(kT) + v_i(kT)T, \\ v_i((k+1)T) &= v_i(kT) + u_i(kT)T, \end{aligned} \tag{3.26}$$

where $x_i \in \mathbb{R}$ is the position, $v_i \in \mathbb{R}$ is the velocity, and $u_i \in \mathbb{R}$ is the control input. To simplify the notation, we replace all "(kT)" by "(k)." It is assumed that $x_i(k) = x_i(0)$ and $v_i(k) = v_i(0) = 0$ for $k < 0$.

We say protocol u_i asymptotically solves the consensus problem, if and only if the states of agents satisfy

$$\lim_{k \to +\infty} (x_i(k) - x_j(k)) = 0, \tag{3.27}$$

for all $i, j \in \mathbb{I}$.

To solve the consensus problem, we introduce the following protocol using the local velocity information and the distributed relative state information:

$$\begin{aligned} u_i(k) &= -p_0 v_i(k) + p_1 \sum_{s_j \in N_i(k)} a_{ij}(k)(x_j(k - \tau_{ij}) - x_i(k)) \\ &+ p_2 \sum_{s_j \in N_i(k)} a_{ij}(k)(v_j(k - \tau_{ij}) - v_i(k)), \end{aligned} \tag{3.28}$$

where $p_1 > 0$, $p_2 > 0$, $a_{ij}(k) > 0$ denotes the edge weight chosen from a finite set $\bar{a}$, $\tau_{ij}(k) \in \mathbb{Z}_+$, $\tau_{ij}(k) \leq \tau_{\max}(i \neq j)$ is the communication time delay from s_j to s_i, and $p_0 > 0$ denotes the velocity damping gain. Here, $\tau_{\max}$ denotes the maximal communication time delay and the weighting factor $a_{ij}(k) > 0$ is allowed to be dynamically changing to stand for possible time-varying relative confidence in the information received or relative reliability of information channels between agents.

Let

$$\begin{aligned} \psi(k) &= [x_1(k),\ v_1(k), \ldots, x_n(k),\ v_n(k)]^T, \\ E &= \begin{bmatrix} 1 & T \\ 0 & 1 - p_0 T \end{bmatrix}, \qquad F = \begin{bmatrix} 0 & 0 \\ p_1 T & p_2 T \end{bmatrix}. \end{aligned}$$

Then, under the protocol (3.28), the network dynamics are

$$\begin{aligned} \psi(k+1) &= [I_n \otimes E - (L_0(k) - \phi_0(k)) \otimes F]\psi(k) \\ &+ (\phi_1(k) \otimes F)\psi(k-1) + \cdots \\ &+ (\phi_{\tau_{\max}}(k) \otimes F)\psi(k - \tau_{\max}), \end{aligned} \tag{3.29}$$

where $L_0(k) = diag(L(k))$, and the ijth element of $\phi_m(k)$ $(m = 0, 1, \ldots, \tau_{\max})$ is either zero or equal to the weight of the edge e_{ij} if $\tau_{ij} = m$. If at some point one agent receives the information of different times from another agent, then the latest information should be taken into consideration. It is clear that $L(k) = L_0(k) - \sum_{m=0}^{\tau_{\max}} \phi_m(k)$.

3.2.4 Design results

In this section, we will investigate the multiagent system (3.29) on dynamically changing topologies. To analyze the stability of such a multiagent system, there are many possible approaches, such as the frequency domain approach [16], the Lyapunov-based approach [9], [13–16], the passivity-based approach [5], and the approach based on the properties of nonnegative matrices [7], [6], [11], [15], [17]. However, the frequency domain approach is limited to the fixed topology case and invalid when the topologies dynamically change, whereas the Lyapunov-based approach and the passivity-based approach are hard to apply to the case of general directed graphs with time delay and switching topologies, especially when the communication graphs have no spanning trees. In addition, when all time delays are equal to zero, some off-diagonal elements of the system matrices of the system (3.29) are negative and the sum of each row is not equal to one. Therefore the properties of nonnegative matrices cannot be used directly to analyze the stability of the system (3.29). In this kind of situation, we perform a model transformation and turn the system (3.29) into an equivalent undelayed one whose system matrix is stochastic. Based on this obtained equivalent system, we apply the properties of nonnegative matrices to the obtained equivalent system and present sufficient conditions under which all agents reach consensus with dynamically changing topologies and arbitrary bounded nonuniform time delays.

Let $\bar{v}_i(k) = x_i(k) + H v_i(k)$, where $H = p_2/p_1$. Then the system (3.29) is equivalent to

$$
\begin{aligned}
x_i(k+1) &= x_i(k) + \frac{\bar{v}_i(k) - x_i(k)}{H} T, \\
\bar{v}_i(k+1) &= \bar{v}_i(k) + \frac{\bar{v}_i(k) - x_i(k)}{H}(1 - Hp_0)T \\
&\quad + p_2 T \sum_{s_j \in N_i(k)} a_{ij}(\bar{v}_j(k - \tau_{ij}) - \bar{v}_i(k)).
\end{aligned}
\tag{3.30}
$$

Denote

$$
\begin{aligned}
\xi(k) &= [x_1(k), \bar{v}_1(k), \ldots, x_n(k), \bar{v}_n(k)]^T, \\
A &= \begin{bmatrix} 1 - \frac{T}{H} & \frac{T}{H} \\ -(1 - Hp_0)\frac{T}{H} & 1 + (1 - Hp_0)\frac{T}{H} \end{bmatrix}, \\
B &= \begin{bmatrix} 0 & 0 \\ 0 & p_2 T \end{bmatrix}.
\end{aligned}
$$

Rewriting (3.30) in matrix form yields

$$
\begin{aligned}
\xi(k+1) &= (\Theta(k) + \phi_0(k) \otimes B)\xi(k) \\
&\quad + (\phi_1(k) \otimes B)\xi(k-1) + \cdots \\
&\quad + (\phi_{\tau_{\max}}(k) \otimes B)\xi(k - \tau_{\max}),
\end{aligned}
\tag{3.31}
$$

where $\Theta(k) = I_n \otimes A - L_0(k) \otimes B$, $L_0(k)$ and $\phi_m(k)$ $(m = 0, 1, \ldots, \tau_{\max})$ are as defined in (3.29).

Different from the system (3.29), the system (3.31) binds the position variable x_i and the velocity variable v_i together as a new variable $\bar{v}_i$, and consequently, its interaction term $(p_2T\sum_{s_j\in N_i(k)}a_{ij}(\bar{v}_j(k-\tau_{ij})-\bar{v}_i(k)))$ only contains the variables $\bar{v}_i$ $(i=1,2,\ldots,n)$, which might simplify the analysis. However, due to the existence of time delay items, it is still hard to perform analysis on the system (3.31). We need to introduce an equivalent augmented system of system (3.31). We define a new state variable $Z(k)=[\xi^T(k),\xi^T(k-1),\ldots,\xi^T(k-\tau_{\max})]^T$. Then the system (3.31) can be equivalently represented by

$$Z(k+1)=\psi(k)Z(k), \tag{3.32}$$

where

$$\psi(k)=\begin{bmatrix}\Theta(k)+\phi_0(k)\otimes B & \phi_1(k)\otimes B & \cdots & \phi_{\tau_{\max}-1}(k)\otimes B & \phi_{\tau_{\max}}(k)\otimes B\\ I & 0 & \cdots & \cdots & 0\\ 0 & I & 0 & \cdots & 0\\ 0 & 0 & \ddots & \ddots & 0\\ 0 & 0 & \cdots & I & 0\end{bmatrix}.$$

Since $L(k)=L_0(k)-\sum_{m=0}^{\tau_{\max}}\phi_m(k)$ and $L(k)\mathbf{1}=0$, it is easy to see $\psi(k)\mathbf{1}=\mathbf{1}$. This property is important and will be used to study the stability of the networks.

Before presenting the main theorem, we first introduce some necessary lemmas.

Lemma 3.3. *Let $P_1,P_2,\ldots,P_k\in\mathbb{R}^{q\times q}$ be a finite set of SIA matrices with the property that for each sequence $P_{i_1},P_{i_2},\ldots,P_{i_j}$ of positive length, the matrix product $P_{i_j}P_{i_{j-1}}\cdots P_{i_1}$ is SIA. Then, for each infinite sequence $P_{i_1},P_{i_2},\ldots$, there exists a vector $f\in\mathbb{R}^q$ such that*

$$\Pi_{j=1}^{\infty}P_{i_j}=\mathbf{1}f^T.$$

Denote $d_{\max}$ as the largest diagonal element of all possible $L(k)$ and $\bar{\mathbb{G}}(\Gamma)$ as a graph whose adjacent matrix is a given square nonnegative matrix Γ. Let

$$1+(1-Hp_0)\frac{T}{H}>p_2Td_{\max},\quad T<H \text{ and } p_0H>1. \tag{3.33}$$

Lemma 3.4. *Under the condition (3.33), $\psi(k)$ is a stochastic matrix.*

Proof. It is obvious that all elements of $\psi(k)$ are nonnegative under the condition (3.33). Also, $\psi(k)\mathbf{1}=\mathbf{1}$. Thus, $\psi(k)$ is a stochastic matrix. □

Lemma 3.5. *Under the condition (3.33), if the union of the graphs $\mathbb{G}(z_1)$, $\mathbb{G}(z_1+1),\ldots,\mathbb{G}(z_2)$ has spanning trees for positive integers z_1, z_2 with $z_2>z_1$, then $\Pi_{j=z_1}^{z_2}\psi(j)$ is SIA.*

Proof. Let $\Lambda(k) = \Theta(k) + \sum_{m=0}^{\tau_{\max}} \phi_m(k) \otimes B = I_n \otimes A - L(k) \otimes B$. Since $\psi(k)$ is nonnegative, it is easy to see $\Lambda(k)$ is also nonnegative. Note that the $2i$th and the $(2i-1)$th nodes of the graph $\bar{\mathbb{G}}(\Lambda(k))$ are strongly connected for any $i \in I$. Also, the union of graphs $\mathbb{G}(z_1), \mathbb{G}(z_1+1), \dots, \mathbb{G}(z_2)$ has spanning trees. Thus the union of graphs $\bar{\mathbb{G}}(\Lambda(z_1)), \bar{\mathbb{G}}(\Lambda(z_1+1)), \dots, \bar{\mathbb{G}}(\Lambda(z_2))$ has spanning trees. Since the edge weights a_{ij} are chosen from a finite set $\bar{\alpha}$, the set of all possible $\Theta(k)$ is finite. Therefore, there must be a positive scalar $0 < \mu < 1$ such that $\Theta(k) + \phi_0(k) \otimes B \geq \Theta(k) \geq \mu I$. □

Theorem 3.6. *Under the condition (3.33), if there exists an infinite sequence of time $k_0, k_1, k_2, \dots$, where $k_0 = 0$, $0 < k_{m+1} - k_m \leq \eta$, $m, \eta \in \mathbb{Z}_+$, such that the union of graphs $\mathbb{G}(k_m), \mathbb{G}(k_{m+1}), \dots, \mathbb{G}(k_{m+1}-1)$ has spanning trees, then the multiagent system (3.26) reaches consensus with the protocol (3.28).*

Proof. For each $k \geq 0$, let m_k be the largest integer such that $k_{m_k} \leq k$. Then,

$$Z(k+1) = \psi(k) \cdots \psi(k_{m_k}) \Pi_{m=0}^{m_k-1} \xi(m) Z(0),$$

where $\xi(m) = \psi(k_{m+1}-1)\psi(k_{m+1}-2) \cdots \psi(k_m)$. Since $k_{m+1} - k_m \leq \eta$ and the union of graphs $\mathbb{G}(km), \mathbb{G}(k_m+1), \dots, \mathbb{G}(k_{m+1}-1)$ has spanning trees, $\xi(m)$ is SIA from Lemma 3.5. Also, the union of the graphs $\mathbb{G}(k_m), \mathbb{G}(k_m+1)$, $\dots, \mathbb{G}(k_{m+j}-1)$ has spanning trees for some positive integer j. Thus, by Lemma 3.5 again, the matrix product $\Pi_{l=m}^{m+j-1} \xi(l)$ is SIA. Since $0 < k_{m+1} - k_m \leq \eta$ and the edge weights $a_{ij}(k)$ are chosen from a finite set $\bar{\alpha}$, it is easy to see that the set of all possible $\xi(j)$ is finite. Hence by Lemma 3.3, we know that

$$\Pi_{i=0}^{+\infty} \xi(i) = \mathbf{1} f^T$$

for some vector $f \in \mathbb{R}^{2(\tau_{\max}+1)n}$.

According to Lemma 3.4, we know that each $\psi(k)$ is a stochastic matrix. Then it follows that

$$\begin{aligned}
\lim_{k\to+\infty} Z(k+1) &= \lim_{k\to+\infty} \psi(k) \cdots \psi(km_k) \Pi_{m=0}^{m_k-1} \xi(m) Z(0) \\
&= \lim_{k\to+\infty} \psi(k) \cdots \psi(km_k) \mathbf{1} f^T Z(0) \\
&= \mathbf{1} f^T Z(0).
\end{aligned}$$

Thus, $\lim_{k\to+\infty} x_i(k) = f^T Z(0)$ for each $i \in I$. That is, consensus is achieved. □

In Theorem 3.6, we derive a condition under which all agents reach consensus. To the best of our knowledge, this is the first result to date capable of dealing with second-order discrete-time multiagent systems with arbitrary bounded nonuniform time delays and dynamically changing graphs that may not have spanning trees.

Corollary 3.3. *Under the assumption of Theorem 3.6, if all communication time delays are equal to zero, then the multiagent system (3.26) reaches consensus with the protocol (3.28).*

Corollary 3.4. *Under the condition (3.33), the multiagent system (3.26) with fixed topology reaches consensus using the protocol (3.28) if and only if the communication topology has spanning trees.*

In Theorem 3.6, each agent is assumed to access both the relative position and the relative velocity information. In fact, consensus can still be reached without relative velocity measurement (i.e., $p_2 = 0$). To demonstrate this, we also need to make a model transformation. Let $\tilde{v}_i(k) = x_i(k) + \frac{2v_i(k)}{p_0}$ and $\tilde{\xi}(k) = [x_1(k), \tilde{v}_1(k), \ldots, x_n(k), \tilde{v}_n(k)]^T$. Then the network dynamics are

$$\begin{aligned}\tilde{\xi}(k+1) = {} & (I_n \otimes \tilde{A} - L_0(k) \otimes \tilde{B} + \phi_0(k) \otimes \tilde{B})\tilde{\xi}(k) \\ & + (\phi_1(k) \otimes \tilde{B})\tilde{\xi}(k-1) + \cdots \\ & + (\phi_{\tau_{\max}}(k) \otimes \tilde{B})\tilde{\xi}(k - \tau_{\max}),\end{aligned} \tag{3.34}$$

where

$$\tilde{A} = \begin{bmatrix} 1 - \frac{p_0 T}{2} & \frac{p_0 T}{2} \\ \frac{p_0 T}{2} & 1 - \frac{p_0 T}{2} \end{bmatrix}, \quad \tilde{B} = \begin{bmatrix} 0 & 0 \\ \frac{2 p_1 T}{p_0} & 0 \end{bmatrix},$$

and ϕ_m are as defined in (3.29).

Observing the form of (3.34) and applying a similar method to that used in Theorem 3.6, we easily obtain that consensus can be achieved for multiagent system (3.26) with arbitrary bounded time delays and dynamically changing topologies if $1 - \frac{p_0 T}{2} > 0$ and $p_0^2 > 4 p_1 d_{\max}$.

3.2.5 Simulation example 3.5

In the sequel, we give an example to illustrate the obtained result. Fig. 3.5 shows four different directed networks, each with 10 nodes. None of the graphs in this figure have spanning trees. The weight of each edge is 0.5. The sample time of the networks is $T = 0.2$ s. Moreover, the time delay for the edges (1, 2) and (3, 4) is T s, for the edges (2, 3), (6, 7), (9, 10), and (1, 10) it is $2T$ s, and for the edges (4, 5), (5, 6), (7, 8), and (8, 9) it is $3T$ s. The protocol parameters are taken as $(p_0, p_1, p_2) = (1, 1, 2)$. In Fig. 3.6, a finite state machine is shown with four states $\{G_a, G_b, G_c, G_d\}$ which denote the states of a network with switching topology and time delay; it starts at G_a, and switches every $5T$ s to the next state. Note that in each time interval of $20T$ s the union of the communication graphs $G_a \cup G_b \cup G_c \cup G_d$ has a spanning tree. Using the protocol (3.28), consensus is achieved as shown in Figs. 3.7 and 3.8, which is consistent with Theorem 3.6, although the state trajectories are not perfectly smooth due to the switching of the network topology.

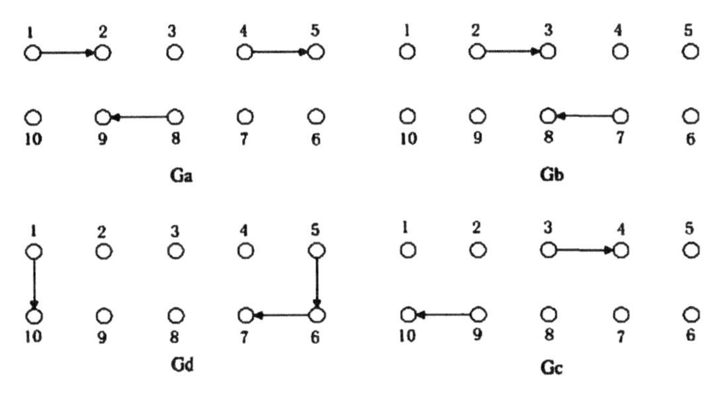

FIGURE 3.5 Four directed graphs.

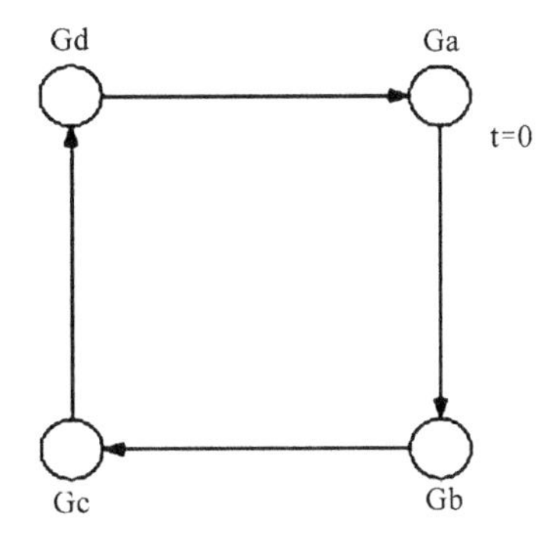

FIGURE 3.6 Finite machine with four states denoting the states of a network with switching topology and time delay.

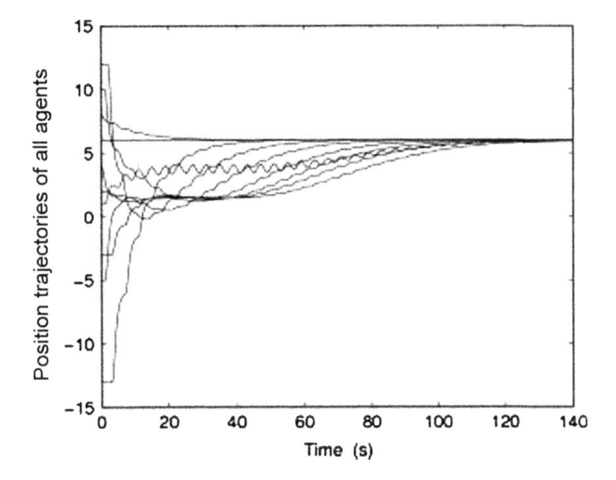

FIGURE 3.7 Position trajectories of all agents.

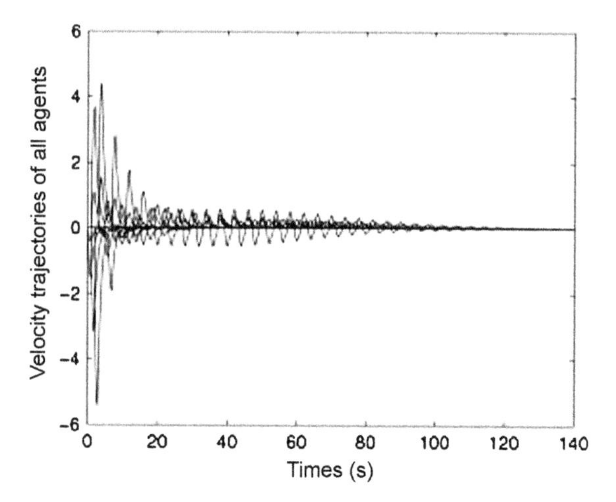

FIGURE 3.8 Velocity trajectories of all agents.

3.3 Decentralized consensus prediction

Using networks to represent individuals and their connections, the consensus problem has broad implications beyond the analysis and design of group collective behavior. Various applications can be cast in this framework, including swarming and flocking, distributed computing, agreement in social networks, or synchronization of coupled oscillators.

In this section, we present a decentralized consensus algorithm that can be embedded as a subalgorithm in a number of distributed algorithms such as distributed Kalman filtering, distributed computation, and the like. The consensus algorithm proposed in this section is much faster than any other algorithm in the literature since it computes the final consensus value in a minimal number of steps.

3.3.1 Formulation of the problem

Let $\mathbb{G}(\mathbb{V}, \mathbb{E}, \mathbb{A})$ be a directed nonweighted graph of order n, where $\mathbb{V} = \{s_1, \ldots, s_n\}$ is the set of nodes, $\mathbb{E} \subseteq \mathbb{V} \times \mathbb{V}$ is the set of edges, and $\mathbb{A} = [a_{ij}]$ is a weighted adjacency matrix with $a_{ij} = 1$ when there is a link from j to i and $a_{ij} = 0$ when there is no link from j to i.

Let $x[i] \in \Re$ denote the state of node i, which might represent the subjective opinion of individual i. Recall that the Laplacian matrix $\mathbb{L} \in \Re^{n \times n}$ induced by the topology $\mathbb{G}$ is defined as $\mathbb{L}(i,i) = \sum_{\ell \neq i}^{n} \mathbb{A}(i, \ell), \forall i = 1, \cdots, n$, and $\mathbb{L}(i,i) = -\mathbb{A}(i,i), \forall i \neq j$.

Hereafter we consider the discrete-time consensus dynamics on a network:

$$x(k+1) = [I_n - \varepsilon\, \mathbb{L}]\, x(k) \stackrel{\Delta}{=} \mathbf{A}\, x(k),$$
$$y(k) = e_r^t\, x(k) \stackrel{\Delta}{=} x_k(r), \tag{3.35}$$

where $x(k) \in \Re^n$ and ε is a prescribed sampling time. Without loss of generality, we focus on the case where the measurable output $y(k) \in \Re$ corresponds to the local state of an arbitrarily chosen individual labeled r.

Let $d_{\max} \stackrel{\Delta}{=} \max_i \mathbb{L}(i,i)$ denote the maximal node in-degree of the graph $\mathbb{G}$. If the network has a rooted directed spanning tree (or is connected in the case of an undirected graph) over time and the sampling time ε is such that $0 < \varepsilon < 1/d_{\max}$, then the discrete-time version of the classical consensus protocol given in (3.35) ensures global asymptotic convergence to consensus in the sense that

$$\lim_{k \to \infty} x(k) = [c^t x(0)]\mathbf{1}_n.$$

Note that the choice of sampling time for each node needs global knowledge about $d_{\max}$ and $\mathbf{1}_n$ is a column vector with all components equal to 1, and c^t is a constant row vector. This implies that the values of all nodes converge asymptotically to the same linear combination of the initial node values $x(0)$.

3.3.2 Distributed asymptotic consensus

When $c^t\ \mathbf{1} \to 1$, the iteration given by (3.35) achieves distributed consensus if and only if:

A.1 Matrix $\mathbf{A}$ has a simple eigenvalue at 1, and all other eigenvalues have a magnitude strictly less than 1.

A.2 The left and right eigenvectors of $\mathbf{A}$ corresponding to the eigenvalue 1 are c^t and 1, respectively.

3.3.3 Finite-time computation

Our starting point is that it is possible to obtain the final value of the consensus dynamics in a finite number of steps and this hinges on the use of the minimal polynomial associated with the consensus dynamics (3.35) in conjunction with the final value theorem.

Definition 3.4 (Minimal polynomial of a matrix). The minimal polynomial of matrix $\mathbf{A} \in \Re^{n \times n}$ is the monic polynomial

$$q(k) = k^{D+1} + \sum_{i=0}^{D} \alpha_i\, k^i$$

with minimal degree $D+1$ that satisfies $q(A) = 0$.

Given the explicit solution of the linear system in (3.35) with initial state $x(0)$, it follows from the definition of the minimal polynomial that the dynamics in (3.35) satisfy the linear regression equation

$$x(k+D+1)+\alpha(D)x(k+D)+\cdots+\alpha(D)x(k+D)$$
$$+\alpha(D)x(k+D)=0,\ \ \forall k\in\mathbb{N}. \tag{3.36}$$

Similarly, the regression equation for $y(k)=x_k(r)$, the measurable output at node r, is determined by the minimal polynomial of the corresponding matrix observability pair $[\mathbf{A},\ e_r^t]$.

Definition 3.5 (Minimal polynomial of a matrix pair). The minimal polynomial associated with the matrix pair $[\mathbf{A},\ e_r^t]$ denoted by

$$q_r(k)=k^{D_r+1}+\sum_{i=0}^{D_r}\alpha(r)_i\ k^i$$

is the monic polynomial of minimal degree $D_r+1\leq D+1$ that satisfies $e_r^t q_r(A)$.

Remark 3.5. The minimal polynomial of a matrix and the minimal polynomial of a matrix pair are unique due to the monic property.

3.4 Performance of agreement protocol

This section presents an $\mathbb{H}_2$ performance analysis of the agreement protocol in the presence of noise. The agreement protocol is first transformed into an equivalent system induced by the dynamics of the relative, or edge, system states. The edge-based representation is used to perform an $\mathbb{H}_2$ analysis of the system highlighting the roles of cycles in this context. For spanning trees and certain k-regular graphs, a characterization of the $\mathbb{H}_2$-norm is given in terms of properties of the graph. These results are used to formulate a semidefinite program for sensor selection and placement. Each sensor has an associated cost and fidelity and the developed SDP determines which sensors to use and where to place them such that the $\mathbb{H}_2$-norm of the system and the sensor costs are minimized.

3.4.1 Introduction

The linear consensus problem has been extensively studied in the dynamic systems and controls community [9], [12]. Applications of linear consensus include distributed computation algorithms [7], sensor fusion [11], [14], and formation flying [13]. The literature includes many variations of the basic setup, including random and switching topologies, stochastic versions, and state-dependent versions [6], [8], [15]. The common analytic theme in all of these approaches relates to the stability and convergent properties of the underlying system.

While these properties describe fundamental aspects for this protocol, they still lack the kind of analysis that is common for more general dynamic systems. For example, the notion of the $\mathbb{H}_2$ performance of the consensus protocol in the context of analysis and synthesis has yet to be examined. This work aims to address this missing system-theoretic component while also providing strong connections between classic systems results and graph-theoretic properties.

At the heart of the analysis in this section lies the edge variant of the consensus problem. Considering the consensus problem from the perspective of states defined over the edges of a graph leads to a transformed system built around an edge variant of the graph Laplacian, which we term the edge Laplacian [16], [17]. A distinct advantage of studying these problems from an edge perspective is that the dominating dynamics – orthogonal to the agreement space – can be isolated while maintaining strong algebraic properties of the graph in the state matrix.

In this direction, we develop an input–output model based on the edge variant of the consensus protocol. We introduce noise as an exogenous input and consider the $\mathbb{H}_2$ performance of the system as a function of the underlying connection topology. The effects of different topologies are examined, emphasizing analytic results for certain classes of graphs including spanning trees and k-regular graphs. The role of cycles in terms of the $\mathbb{H}_2$ performance is also examined. These results are then applied to develop a synthesis procedure for choosing sensors for the consensus protocol. A semidefinite program is developed that determines where sensors of varying fidelity should be placed in the underlying network to achieve the best $\mathbb{H}_2$ performance.

3.4.2 The edge Laplacian

An undirected (simple) graph $\mathbb{G}$ is specified by the vertex set $\mathbb{V}$ and edge set $\mathbb{E}$ whose elements characterize the incidence relation between distinct pairs of elements of $\mathbb{V}$. The cardinalities of the vertex and edge sets of $\mathbb{G}$ will be denoted by $|\mathbb{V}|$ and $|\mathbb{E}|$, respectively. A subgraph of $\mathbb{G}$ is a graph whose vertex and edge sets are subsets of those of $\mathbb{G}$. An orientation of an undirected graph $\mathbb{G}$ is the assignment of directions to its edges, i.e., an edge $e_k \in \mathbb{E}$ is an ordered pair (i, j) such that i and j are, respectively, the initial and the terminal nodes of e_k.

Graphs admit a set of convenient matrix representations. The $|\mathbb{V}| \times |\mathbb{E}|$ incidence matrix $\mathbb{E}(\mathbb{G})$ for an oriented graph $\mathbb{G}$ is a $\{0, \pm 1\}$-matrix with rows and columns indexed by vertices and edges of $\mathbb{G}$, respectively, such that $[\mathbb{E}(\mathbb{G})]_{ik}$ has the value "$+1$" if node i is the initial node of edge e_k, "-1" if it is the terminal node, and "0" otherwise.

From the definition of the incidence matrix it follows that the null space of its transpose, $\mathcal{N}(E(\mathbb{G})^T)$, contains the subspace **span** $\{\mathbf{1}\}$, where $\mathbf{1}$ is the vector with all entries equal to one. The rank of the incidence matrix depends only on $|\mathbb{V}|$ and the number of its connected components, c, with $\mathbf{rank}\,\mathbb{E}(\mathbb{G}) = |\mathbb{V}| - c$ [18].

The degree of a vertex $v_i \in \mathbb{V}$, d_i, is the cardinality of the set of vertices adjacent to it. The degree matrix of $\mathbb{G}$, $\Delta(\mathbb{G})$, is a diagonal matrix with the degree of vertex i at its (i,i) position.

The *complete graph* on n nodes, K_n, is the graph where all possible pairs of vertices are adjacent, or equivalently, if the degree of all vertices is $|\mathbb{V}| - 1$. Fig. 3.9A shows the complete graph on 10 nodes, K_{10}. When every node in a graph has the same degree k, it is called a k-regular graph. The k-regular graph for $k = 2$ is also called the cycle graph, C_n. Fig. 3.9B and Fig. 3.9C show the cycle graph C_{10} and a 4-regular graph.

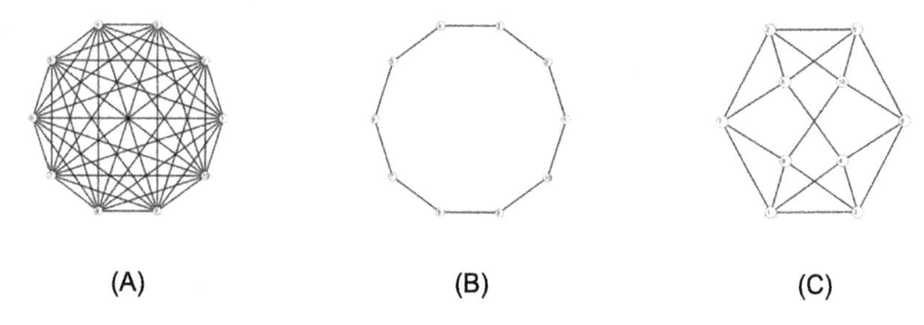

FIGURE 3.9 Example of $3k$-regular graphs. (A) K_{10} graph; (B) C_{10} graph; (C) 4-regular graph.

A sequence of $r + 1$ distinct and consecutively adjacent vertices, starting from vertex i and ending at vertex j, is called a path of length r (form i to j); when $i = j$, we call this path a cycle. We call a graph connected if there exists a path between any pair of vertices. A connected graph without cycles is referred to as a tree. Equivalently, a tree is a connected graph on $|\mathbb{V}|$ vertices with $|\mathbb{V}| - 1$ edges.

Any connected graph $\mathbb{G}$ can be written as the union of two edge-disjoint subgraphs, $\mathbb{G} = \mathbb{G}_\tau \cup \mathbb{G}_c$, where $\mathbb{G}_\tau$ is a spanning tree subgraph and $\mathbb{G}_c$ contains the remaining edges that necessarily complete the cycles in $\mathbb{G}$. Similarly, the columns of the incidence matrix for the graph $\mathbb{G}$ can always be permuted such that $E(\mathbb{G})$ can be written as

$$\mathbb{E}(\mathbb{G}) = \begin{bmatrix} \mathbb{E}(\mathbb{G}_\tau) & \mathbb{E}(\mathbb{G}_c) \end{bmatrix}. \tag{3.37}$$

For notational simplicity, we will use $\mathbb{E}_\tau$ and $\mathbb{E}_c$ to represent the incidence matrix for the tree subgraph and cycle subgraph, respectively.

The cycle edges can be constructed from linear combinations of the tree edges as

$$\mathbb{E}_\tau \mathbb{T}_\tau^c = \mathbb{E}_c, \tag{3.38}$$

where $\mathbb{T}_\tau^c = (\mathbb{E}_\tau^t \mathbb{E}_\tau)^{-1} \mathbb{E}_\tau^t \mathbb{E}_c$. Using (3.38) we obtain the following alternative representation of the incidence matrix:

$$\mathbb{E}(\mathbb{G}) = \mathbb{E}_\tau \begin{bmatrix} I & \mathbb{T}_\tau^c \end{bmatrix} = \mathbb{E}_\tau R(\mathbb{G}). \tag{3.39}$$

We note that the rows of $R(\mathbb{G})$ form a basis for the cut space of $\mathbb{G}$, and the matrix $\begin{bmatrix} -\mathbb{T}_\tau^c & I \end{bmatrix}^t$ forms a basis for the *flow space*;

$$\mathbf{IM}\{E(\mathbb{G})^t\} = \begin{bmatrix} I \\ (\mathbb{T}_\tau^c)^t \end{bmatrix}, \qquad \mathbf{Ker}\{\mathbb{E}(\mathbb{G})\} = \begin{bmatrix} -\mathbb{T}_\tau^c \\ I \end{bmatrix}.$$

The matrix $\mathbb{R}(\mathbb{G})$, which will play an important role in the present work, has a close connection with a number of structural properties of the underlying network. For example, the number of spanning trees in a graph, $\tau(\mathbb{G})$, can be determined from the cut space basis as

$$\tau(\mathbb{G}) = \mathbf{det}[\mathbb{R}(\mathbb{G})\mathbb{R}(\mathbb{G})^t]. \tag{3.40}$$

The graph Laplacian of an oriented graph is defined as

$$\mathbb{L}(\mathbb{G}) := \mathbb{E}(\mathbb{G})\mathbb{E}(\mathbb{G})^T, \tag{3.41}$$

which is independent of a particular orientation of the graph. The graph Laplacian of $\mathbb{G}$ is a rank-deficient positive semidefinite matrix. The eigenvalues are real and will be ordered and denoted as $0 = \lambda_1(\mathbb{G}) \le \lambda_2(\mathbb{G}) \le \ldots \le \lambda_{|\mathbb{V}|}(\mathbb{G})$.

The edge Laplacian is a particular variant of the graph Laplacian [16] and is defined as

$$\mathbb{L}_e(\mathbb{G}) := \mathbb{E}(\mathbb{G})^t E(\mathbb{G}). \tag{3.42}$$

The nonzero eigenvalues of $\mathbb{L}_e(\mathbb{G})$ are equivalent to those of $\mathbb{L}(\mathbb{G})$, and each cycle in $\mathbb{G}$ corresponds to an eigenvalue at 0 in $\mathbb{L}_e(\mathbb{G})$.

Theorem 3.7. *([17]) The graph Laplacian for a connected graph, $\mathbb{L}(\mathbb{G})$, is similar to*

$$\begin{bmatrix} \mathbb{L}_e(\mathbb{G}_\tau)\mathbb{R}(\mathbb{G})\mathbb{R}(\mathbb{G})^t & 0 \\ 0 & 0 \end{bmatrix},$$

where $\mathbb{G}_\tau$ is a spanning tree subgraph of $\mathbb{G}$, and the matrix $\mathbb{R}(\mathbb{G})$ is defined in (3.39).

Proof. We define the transformation matrix

$$\mathbf{S} = \begin{bmatrix} \mathbb{E}_\tau(\mathbb{E}_\tau^t\mathbb{E}_\tau)^{-1} & V \end{bmatrix}, \tag{3.43}$$

where $\mathbb{E}_\tau$ is the incidence matrix of $\mathbb{G}_\tau$ and the columns of $\mathbf{V}$ are a basis for the null space of $\mathbb{L}(\mathbb{G})$, e.g., the vector $\mathbf{1}$. The matrix $\mathbf{S}$ is nonsingular; in fact, its inverse is

$$\mathbf{S}^{-1} = \begin{bmatrix} \mathbb{E}_\tau^t \\ (1/|\mathbb{V}|)\mathbf{1}^t \end{bmatrix}. \tag{3.44}$$

Applying the transformation matrix as

$$\begin{aligned} \mathbf{S}^{-1}\mathbb{L}(\mathbb{G})S &= \begin{bmatrix} \mathbb{E}_\tau^t \mathbb{E}_\tau \\ 0 \end{bmatrix} \mathbb{R}(\mathbb{G})\mathbb{R}(\mathbb{G})^t \begin{bmatrix} I & 0 \end{bmatrix} \\ &= \begin{bmatrix} \mathbb{L}_e(\mathbb{G}_\tau)\mathbb{R}(\mathbb{G})\mathbb{R}(\mathbb{G})^t & 0 \\ 0 & 0 \end{bmatrix} \end{aligned} \tag{3.45}$$

leads to the desired result. □

Note that when $\mathbb{G} = \mathbb{G}_\tau$ (no cycles), then $R(\mathbb{G}) = I$ and we see a tight connection between the graph and edge Laplacians. Furthermore, for a connected graph $\mathbb{G}_\tau$, the edge Laplacian is guaranteed to be invertible as all its eigenvalues are strictly positive.

3.4.3 Performance bounds for consensus

The noise-free consensus problem is comprised of a collection of n first-order dynamic systems of the form

$$\dot{x}_i(t) = u_i(t). \tag{3.46}$$

The dynamic evolution of each agent is coupled through the control input $u_i(t)$ which is defined to be the sum of the differences between states of an individual unit and its neighbors,

$$u_i(t) = \sum_{j \in \mathcal{N}(i)} (x_j(t) - x_i(t)), \tag{3.47}$$

where $\mathcal{N}(i)$ denotes the set of agents that are neighbors of agent i, as defined by the connection topology $\mathbb{G}$. Expressing the dynamic evolution of the resulting system in a compact matrix form with $x(t)^t = [x_1(t), \ldots, x_n(t)]^t$, we arrive at the following autonomous system:

$$\dot{x}(t) = -\mathbb{L}(\mathbb{G})x(t). \tag{3.48}$$

Agreement of the system (3.48) is an asymptotic property where each state approaches the same value. For connected graphs, we have $\lim_{t\to\infty} x(t) = (1/n)Jx(0)$, where J is the matrix of all ones [10].

We now consider a general scenario where noise is introduced at both the process and measurement levels of the consensus protocol. Eq. (3.46) is first modified to include the process noise for each agent, as

$$x_i(k) = u_i(k) + w_i(k). \tag{3.49}$$

We assume that $w_i(k)$ is a zero mean white Gaussian noise with covariance $\mathbf{E}[w(k)w(k)^T] = \sigma_w^2 I$.

The measurement is also corrupted by noise, as

$$y(k) = \mathbb{E}(\mathbb{G})^k x(t) + v(k). \tag{3.50}$$

Here, $v(k) \in \mathbb{R}^{|\mathbb{E}|}$ is also a zero mean white Gaussian noise with covariance $\mathbf{E}[v(k)v(k)^t] = \sigma_v^2 I$.

Eqs. (3.49) and (3.50) can be considered the open-loop consensus model. We denote the open-loop system as Σ_{ol}, i.e.,

$$\Sigma_{ol} : \begin{cases} x(k+1) = u(k) + w(k), \\ y(k) = \mathbb{E}(\mathbb{G})^t x(k) + v(k). \end{cases} \tag{3.51}$$

When the output-feedback control $u(k) = -\mathbb{E}(\mathbb{G})y(k)$ is applied, the system leads to a generalized consensus protocol with noise. The noisy consensus model will be referred to as the Σ model, i.e.,

$$\Sigma : \begin{cases} x(k+1) = -\mathbb{L}(\mathbb{G})x(k) + [I - \mathbb{E}(\mathbb{G})] \begin{bmatrix} w(k) \\ v(k) \end{bmatrix}, \\ z(k) = E(\mathbb{G})^T x(k). \end{cases} \tag{3.52}$$

Here, the variable $z(k)$ is introduced as a performance signal to monitor. Note that as $x(t) \to (1/N)Jx(0)$ we have $z(t) \to \mathbf{0}$. The open-loop system is shown in Fig. 3.10 with the consensus output-feedback law.

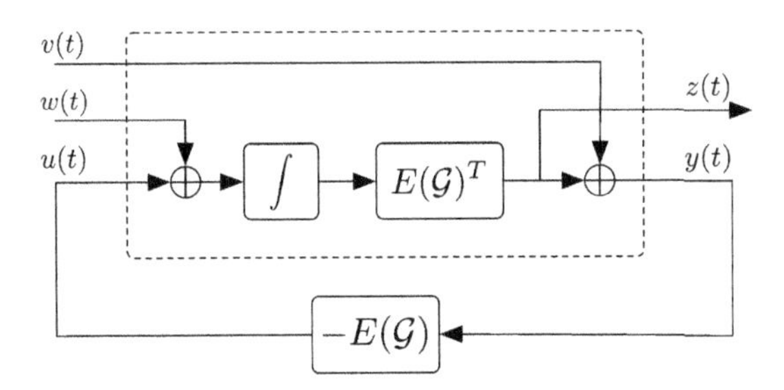

FIGURE 3.10 Open-loop consensus system with output feedback.

The first limiting factor for an $\mathbb{H}_2$ analysis is that for any connected graph, the system Σ has an unbounded $\mathbb{H}_2$-norm due to the presence of the 0 eigenvalue. Therefore, we can consider the performance analysis on the subspace orthogonal to the **1** vector, the eigenvector associated with the 0 eigenvalue.

In this direction, we introduce the coordinate transformation $S\hat{x}(k) = x(k)$, where S is defined in (3.43). Applying this transformation yields

$$\hat{\Sigma} : \begin{cases} \hat{x}(k+1) = \begin{bmatrix} -\mathbb{L}_e^\tau \mathbb{R}\mathbb{R}^t & 0 \\ 0 & 0 \end{bmatrix} \hat{x}(k) + \begin{bmatrix} \mathbb{E}_\tau^t & -\mathbb{L}_e^\tau \mathbb{R} \\ \frac{1}{N}\mathbf{1}^t & 0 \end{bmatrix}, \\ z(k) = \begin{bmatrix} \mathbb{R}^t & \mathbf{0} \end{bmatrix} \hat{x}(k). \end{cases}$$

We use the shorthand notation $\mathbb{L}_e^\tau = \mathbb{L}_e(\mathbb{G}_\tau)$ and $\mathbb{R} = \mathbb{R}(\mathbb{G})$. The benefit of such a transformation is that the algebraic structure of the underlying connection topology is preserved through the edge Laplacian. Furthermore, we note that the new state $\hat{x}(t)$ can be partitioned as $\hat{x}^t(k) = \begin{bmatrix} x_\tau^t(k) & x_\mathbf{1}^t(k) \end{bmatrix}^t$, where $x_\tau(k)$ represents the relative state information across the edges of a spanning tree of $\mathbb{G}$ and $x_\mathbf{1}(k)$ is the mode in the **1** subspace, which corresponds to an unobservable mode of the system.

We can now consider the truncated system containing only the states $x_\tau(t)$ for analysis considerations; in fact, the truncated system represents a minimal realization of (3.52). We refer to this as the Σ_τ system, i.e.,

$$\begin{aligned} \Sigma_\tau : x(k+1) &= -\mathbb{L}_e^\tau \mathbb{R}\mathbb{R}^t x_\tau(k) \\ &\quad + \sigma_w \mathbb{E}_\tau^t \hat{w}(k) - \sigma_v \mathbb{L}_e^\tau \mathbb{R}\hat{v}(k), \\ z(k) &= \mathbb{R}^t x_\tau(k). \end{aligned} \tag{3.53}$$

The signals $\hat{w}(k)$ and $\hat{v}(k)$ are the normalized process and measurement noise signals. The performance variable, $z(k)$, contains information on the tree states in addition to the cycle states. Here we recall that the cycle states are linear combinations of the tree states and we note that $z(k)$ actually contains redundant information. This is highlighted by recognizing that the true states converging to the origin forces the cycle states to do the same.

Consequently, we will consider the system with cycles as well as a system containing only the tree states at the output, which we denote as $\hat{\Sigma}_\tau$. This also allows for a means to quantify the effect of cycles on the performance. We have

$$\begin{aligned} \hat{\Sigma}_\tau : x(k+1) &= -\mathbb{L}_e^\tau \mathbb{R}\mathbb{R}^t x_\tau(k) \\ &\quad + \sigma_w \mathbb{E}_\tau^t \hat{w}(k) - \sigma_v \mathbb{L}_e^\tau \mathbb{R}\hat{v}(k), \\ z(k) &= x_\tau(k). \end{aligned} \tag{3.54}$$

The $\mathbb{H}_2$-norm of Σ_τ (3.37) and $\hat{\Sigma}_\tau$ (3.54) can be computed as

$$\|\Sigma_\tau\|_2^2 = \mathbf{Tr}[\mathbb{R}^t \mathbb{X}^* \mathbb{R}], \qquad \|\hat{\Sigma}_\tau\|_2^2 = \mathbf{Tr}[\mathbb{X}^*], \tag{3.55}$$

where $\mathbb{X}^*$ is the positive definite solution to the Lyapunov equation

$$\mathbb{L}_e^\tau \mathbb{R}\mathbb{R}^t \mathbb{X}^* \mathbb{R}\mathbb{R}^t \mathbb{L}_e^\tau - \mathbb{X}^* + \sigma_w^2 \mathbb{L}_e^\tau + \sigma_v^2 \mathbb{L}_e^\tau \mathbb{R}\mathbb{R}^t \mathbb{L}_e^\tau = 0. \tag{3.56}$$

It is worth noting from the structure of (3.56) that any solution is dependent on certain properties of the graph. A little algebra shows that

$$\sigma_w^2 \mathbb{L}_e^\tau + \sigma_v^2 \mathbb{L}_e^\tau \mathbb{R}\mathbb{R}^t \mathbb{L}_e^\tau = \mathbb{L}_e^\tau \left(\sigma_w^2 (\mathbb{L}_e^\tau)^{-1} + \sigma_v^2 \mathbb{R}\mathbb{R}^t\right) \mathbb{L}_e^\tau.$$

Hence, the solution to (3.56) becomes

$$\mathbb{X}^* = \frac{1}{2}(\sigma_w^2 (\mathbb{R}\mathbb{R}^t)^{-1} + \sigma_v^2 \mathbb{L}_e^\tau), \tag{3.57}$$

and we arrive at the following result.

Theorem 3.8. *The $\mathbb{H}_2$-norm of the Σ_τ system is*

$$\|\Sigma_\tau\|_2^2 = \frac{\sigma_w^2}{2}(n-1) + \sigma_v^2 |\mathbb{E}|. \tag{3.58}$$

The $\mathcal{H}_2$-norm of the $\hat{\Sigma}_\tau$ system is

$$\|\hat{\Sigma}_\tau\|_2^2 = \frac{\sigma_w^2}{2} \mathrm{Tr}[(\mathbb{R}\mathbb{R}^t)^{-1}] + \sigma_v^2 (n-1). \tag{3.59}$$

Proof. The proof follows from (3.57) and noting that $\mathbf{Tr}[\mathbb{L}_e^\tau] = 2(n-1)$, or twice the number of edges in a spanning tree. □

Remark 3.6. We note that $\|\Sigma_\tau\|_2^2$ is a linear function of the number of edges in the graph. This has a clear physical interpretation, as the addition of each edge corresponds to an amplification of the noises. While a general graph-theoretic characterization of (3.59) may be hard to derive, certain graph structures allow for complete characterizations of the solution, which we present hereafter.

3.4.4 Spanning trees

The first case resulting in a simplification of (3.58) arises when $\mathbb{G}$ is a spanning tree. In this case $\mathbb{R} = I$ and (3.58) simplifies to

$$\|\hat{\Sigma}_\tau\|_2^2 = (n-1)\left(\frac{\sigma_w^2}{2} + \sigma_v^2\right). \tag{3.60}$$

An interesting consequence of this result is that all spanning trees result in the same system performance. That is, the choice of spanning tree (e.g., a path or a star) does not affect the performance. As expected, in this scenario $\|\Sigma_\tau\|_2^2 = \|\Sigma_\tau\|_2^2$. We also note that when the noises have different covariance values, this analysis becomes significantly more complicated, as we consider later on.

3.4.5 k-Regular graphs

Regular graphs also lead to a simplification of (3.59). Any connected k-regular graph will contain cycles resulting in a nontrivial expression for $\mathbb{R}\mathbb{R}^t$. The $\mathbb{H}_2$-norm is therefore intimately related to the cut space of the graph. A direct characterization of $\mathbb{R}\mathbb{R}^t$ in terms of basic properties of the graph is challenging for arbitrary k-regular graphs. However, certain k-regular graphs lead to further simplifications, as presented below.

We denote the eigenvalues of $\mathbb{R}\mathbb{R}^t$ by μ_j and note that

$$\mathbf{Tr}[(\mathbb{R}\mathbb{R}^t)^{-1}] = \sum_{j=1}^{n-1} \frac{1}{\mu_j} = \frac{1}{\tau(\mathbb{G})} \sum_{j=1}^{n-1} \prod_{m \neq j}^{n-1} \mu_m. \tag{3.61}$$

The quantity $\prod_{m \neq j}^{n-1} \mu_m$ is recognized as a first minor of the matrix $\mathbb{R}\mathbb{R}^t$.

Lemma 3.6. *The k-regular graph with degree 2 (cycle graph) has n spanning trees and*

$$\mathbf{Tr}[(\mathbb{R}\mathbb{R}^t)^{-1}] = \frac{(n-1)^2}{n}. \tag{3.62}$$

The $\mathbb{H}_2$-norm of the $\hat{\Sigma}_\tau$ system when the underlying graph is the cycle graph C_n is given as

$$\|\hat{\Sigma}_2\|_2^2 = (n-1)\left(\frac{\sigma_w^2(n-1)}{n} + \sigma_v^2\right). \tag{3.63}$$

Proof. Without loss of generality, we consider a directed path graph on n nodes, with initial node v_1 and terminal node v_n as the spanning tree subgraph $\mathbb{G}_\tau$. We index the edges as $e_i = (v_i, v_{i+1})$. The cycle graph is formed by adding the edge $e_n = (v_n, v_1)$. For this graph, we have $\mathbb{T}_\tau^c = \mathbf{1}_{n-1}$ and $\mathbb{R}\mathbb{R}^t = I + J_{n-1}$. From this it follows that $\mathbf{det}[\mathbb{R}\mathbb{R}^t] = N$ and all its first minors have the value $N-1$. Combined with (3.58) yields the desired result. □

Lemma 3.7. *The k-regular graph with degree $n-1$ (complete graph) has n^{n-2} spanning trees and*

$$\mathbf{Tr}[(\mathbb{R}\mathbb{R}^t)^{-1}] = \frac{2(n-1)n^{n-3}}{n^{n-2}} = \frac{2(n-1)}{n}. \tag{3.64}$$

The $\mathbb{H}_2$-norm of the $\hat{\Sigma}_\tau$ system when the underlying graph is the complete graph K_n is given as

$$\|\hat{\Sigma}_\tau\|_2^2 = (n-1)\left(\frac{\sigma_w^2}{n} + \sigma_v^2\right). \tag{3.65}$$

Proof. Without loss of generality, we consider a star graph with center at node v_1 and all edges are of the form $e_k = (v_1, v_{k+1})$. Then the cycles in the graph are created by adding the edges $e = (v_i, v_j)$, $i, j \neq 1$ and $\mathbb{R}\mathbb{R}^t = nI - J_{n-1}$. It then follows that $\mathbf{det}[\mathbb{R}\mathbb{R}^t] = n^{n-2}$ and all the first minors have the value $2n^{n-3}$. Combining this with (3.58) yields the desired result. □

Remark 3.7. In general, one expects the system norm to decrease as the regularity increases. However, it is not clear how the cycle structure directly affects the norm. To illustrate this, 500 random regular graphs of degree 5 were generated in MATLAB®. For each instance, the value $\mathbf{Tr}[(\mathbb{R}\mathbb{R}^t)^{-1}]$ was calculated, sorted, and plotted in Fig. 3.11. In this example, although the degree of each node remains constant, the actual cycle structure, meaning both the number of independent cycles and the length of those cycles, varies greatly.

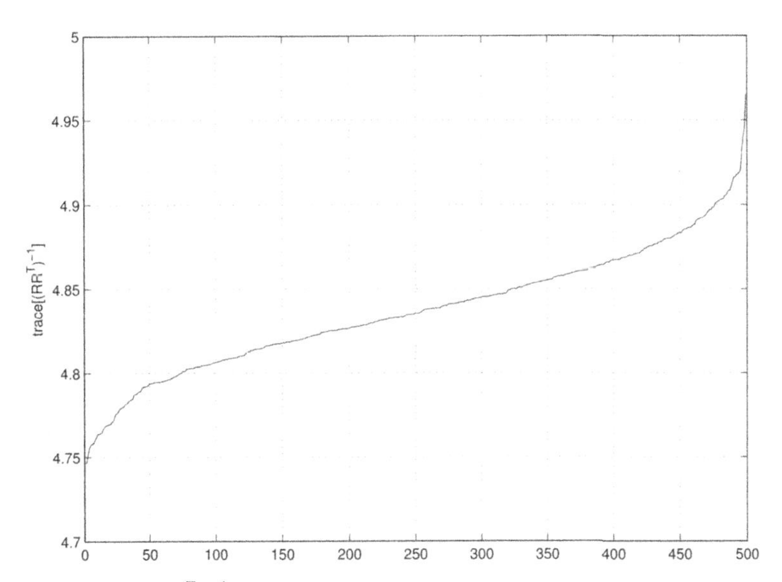

FIGURE 3.11 $\mathbf{Tr}[(RR^T)^{-1}]$ for random 5-regular graphs.

The above analysis suggests that other classes of graphs may exist that lead to a simplification of the expression in (3.58). Ultimately, any such simplification will relate to the matrix $(RR^T)^{-1}$, which is intimately related to the cycles of the graph.

3.4.6 Cycle contributions

Using the above analysis we can begin to quantitatively understand how cycles affect the $\mathcal{H}_2$ performance. For example, examining the ratio

$$\frac{\|\Sigma_\tau(\mathbb{G})\|_2^2}{\|\Sigma_\tau(\mathbb{G}_\tau)\|_2^2}$$

can give a good indication of how the cycles increase the $\mathcal{H}_2$-norm. Recall that $\mathbb{G}$ is in general a graph containing cycles and $\mathbb{G}_\tau \subseteq \mathbb{G}$ is the spanning tree subgraph.

For example, consider the cycle graph C_n and assume unit covariance for both the process and measurement noises. The ratio becomes

$$\frac{\|\Sigma_\tau(C_n)\|_2^2}{\|\Sigma_\tau(\mathbb{G}_\tau)\|_2^2} = \frac{3n-1}{3(n-1)}. \tag{3.66}$$

In this case, we note that as the number of nodes increases, the effect of the cycle (there is only 1) decreases and becomes negligible in the limit.

Similarly, for the complete graph K_n we have

$$\frac{\|\Sigma_\tau(K_n)\|_2^2}{\|\Sigma_\tau(\mathbb{G}_\tau)\|_2^2} = \frac{n+1}{3}. \tag{3.67}$$

Here we see that the norm is amplified linearly as a function of the number of nodes in the graph. It is worth mentioning here that typical performance measures for consensus problems, such as $\lambda_2(\mathbb{G})$, would favor the complete graph over the cycle graph. However, in terms of the $\mathcal{H}_2$ performance we see that there is a penalty to be paid for faster convergence.

Alternatively, insight is also gained by considering the ratio

$$\frac{\|\Sigma_\tau(\mathbb{G})\|_2^2}{\|\hat{\Sigma}_\tau(\mathbb{G})\|_2^2},$$

which highlights the effects of including cycles in the performance variable $z(t)$.

For the cycle graph we have

$$\frac{\|\Sigma_\tau(C_n)\|_2^2}{\|\hat{\Sigma}_\tau(C_n)\|_2^2} = \frac{n(3n-1)}{2(n-1)(2n-1)}. \tag{3.68}$$

Here we note that as $n \to \infty$ the ratio approaches 0.75, which suggests that the effect of including the cycle for performance evaluation does not vary significantly with the size of the graph.

For the complete graph we have

$$\frac{\|\Sigma_\tau(K_n)\|_2^2}{\|\hat{\Sigma}_\tau(K_n)\|_2^2} = \frac{n}{2}. \tag{3.69}$$

As with (3.67) we see that the inclusion of cycles results in a linear function of the number of nodes in the graph.

3.4.7 Sensor placement with $\mathbb{H}_2$ performance

In this section we consider the problem of sensor selection and placement for consensus in the context of its $\mathbb{H}_2$ performance. Consider, for example, a scenario where there are two types of sensors available for the relative measurements in the open-loop consensus problem. One sensor is a high-fidelity and high-cost sensor, with associated noise covariance $\bar{\sigma}_1^2$. The other sensor is cheaper with covariance $\underline{\sigma}_v^2 > \bar{\sigma}_v^2$. When synthesizing the topology for the consensus problem, the designer must consider the tradeoff between the sensor costs and the system performance.

In this direction, we consider a modification of the system in (3.53),

$$\Sigma_\tau : \begin{cases} x(k+1) = -\mathbb{L}_e^\tau \mathbb{R}\mathbb{R}^t x_\tau(t) + \sigma_w \mathbb{E}_\tau^t \hat{w}(k) - \mathbb{L}_e^T \mathbb{R}\Gamma\hat{v}(k), \\ z(k) = \mathbb{R}^t x_\tau(k), \end{cases} \tag{3.70}$$

where $\hat{w}(k)$ and $\hat{v}(k)$ are the normalized noise signals. The matrix Γ is a diagonal matrix with elements σ_i corresponding to the variance of the sensor on edge i.

The most general version of this problem considers a finite set of p sensors each with an associated variance,

$$\Sigma = \{\sigma_i^2, i = 1, 2, \ldots, p\}. \tag{3.71}$$

For each element $\sigma_i^2 \in \Sigma$ there is an associated cost $c(\sigma_i^2)$. The cost function has the property that $c(\sigma_i^2) > c(\sigma_j^2)$ if $\sigma_i^2 < \sigma_j^2$. This yields a mixed-integer program that can then be written as

$$\begin{aligned} &\mathcal{P}_1 \\ &\min_{\mathbb{X},\mathbb{W}} \quad \lambda \mathbf{Tr}[\mathbb{R}^t \mathbb{X}\mathbb{R}] + \sum_{i=1}^{|\mathbb{E}|} c(w_i) \\ &\text{s.t.} \quad W = \mathbf{diag} \quad \{w_1, \ldots, w_{|\mathcal{E}|}\}, \\ &\qquad w_i \in \Sigma, \\ &\qquad \sum_i w_i \le \bar{\sigma}, \\ &\qquad \mathbb{L}_e^\tau \mathbb{R}\mathbb{R}^t \mathbb{X}^* \mathbb{R}\mathbb{R}^t \mathbb{L}_e^\tau - \mathbb{X}^* + \sigma_w^2 \mathbb{L}_e^\tau + \mathbb{L}_e^\tau \mathbb{R}\mathbb{R}^t \mathbb{L}_e^\tau = 0, \end{aligned} \tag{3.72}$$

where λ represents a weighting on the $\mathbb{H}_2$ performance of the solution and $\bar{\sigma}$ represents the maximum allowable aggregate noise. Note that in general

$$|\mathbb{E}| \min_i \sigma_i^2 \le \bar{\sigma} \le |\mathbb{E}| \max_i \sigma_i^2.$$

Remark 3.8. It is important to observe that the problem $\mathcal{P}_1$ is combinatorial in nature, as a discrete decision must be made as to which sensor to use and where to place it in the network. While a tractable solution algorithm is in general difficult, certain relaxations can be made that lead to a more approachable problem.

The first relaxation removes the discrete nature of the set Σ into a simple box-type constraint, as

$$\hat{\Sigma} = [\underline{\sigma}^2, \bar{\sigma}^2]. \tag{3.73}$$

The cost function can now be written as a continuous map $c : \hat{\Sigma} \mapsto \Re$ that is a convex and strictly decreasing function. The simplest version of such a function would be the linear map

$$c(\sigma_i^2) = -\beta \sigma_i^2, \text{ for some } \beta > 0.$$

These relaxations lead to the following modified program, $\mathcal{P}_2$:

$$\begin{aligned} &\mathcal{P}_2 \\ &\min_{X,W} \quad \lambda \mathbf{Tr}[\mathbb{R}^t \mathbb{X} \mathbb{R}] - \beta \mathbf{Tr}[\mathbb{W}] \\ &\text{s.t.} \quad \mathbb{W} = \mathbf{diag}\{w_1, \ldots, w_{|\mathbb{E}|}\}, \\ &\qquad \underline{\sigma}^2 \le w_i \le \bar{\sigma}^2, \\ &\qquad \sum_i w_i \le \bar{\sigma}, \\ &\qquad \mathbb{L}_e^\tau \mathbb{R}\mathbb{R}^t \mathbb{X}^* \mathbb{R}\mathbb{R}^t \mathbb{L}_e^\tau - \mathbb{X}^* + \sigma_w^2 \mathbb{L}_e^\tau + \mathbb{L}_e^\tau \mathbb{R}\mathbb{R}^t \mathbb{L}_e^\tau = 0. \end{aligned} \tag{3.74}$$

As in the foregoing analysis, we would expect that certain topologies lead to a simplification in the above programs. As an example of $\mathcal{P}_2$, we consider the sensor selection for the graph in Fig. 3.12. A random graph on 10 nodes with an edge probability of 0.15 was generated. The graph is connected and contains two independent cycles, resulting in the most general problem instance. The sensor constraints used are $\hat{P} = [0.001, 0.1]$ and $\bar{\sigma}^2 = 0.501$. Finally, the cost function tuning values were chosen as $\beta = 5$ and $\lambda = 1$. It is worth mentioning that the selection of these values is currently a trial and error process. Ideally, they should be chosen in such a way so that one term of the objective function does not overly dominate the other.

Solving $\mathcal{P}_2$ resulted in a nontrivial selection of sensors for each edge. The sensor covariance for each edge is labeled in Fig. 3.12. It is interesting to note that the highest fidelity sensors tend to be concentrated around the node of highest degree. Also, the edge with the lowest fidelity sensor is furthest away from the node of highest degree. It seems rather intuitive to place the lower fidelity sensors in "low-traffic" areas. That is, edges that are adjacent to a low number of other edges.

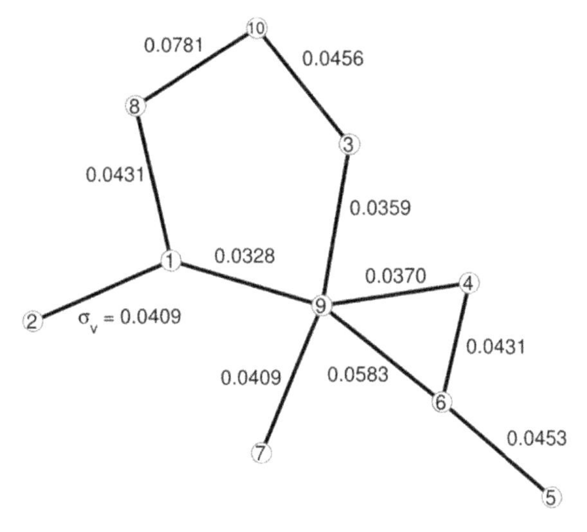

FIGURE 3.12 A graph on 10 nodes with optimal sensor selection.

3.5 Scalable consensus conditions

The design of distributed control systems has "cooperation" as central topic. Evidently, one of the most convenient and efficient approaches to undertaking a cooperative task for the agents is to achieve some agreement (consensus). The purpose of this section is to introduce basic notions in consensus problems, such as consensus protocol, the existence of consensus solutions, and consentability. A unified approach to treating the consensus problem as a stability problem is presented.

3.5.1 Integrator agents over delayless networks

Consider the discrete-time system with consensus protocol

$$x_i(k+1) = x_i(k) + u_i(k), \tag{3.75}$$

$$u_i(k) = \gamma \sum_{j \in n_i} a_{ij}[x_j(k) - x_i(k)]. \tag{3.76}$$

We say a *consensus solution* (or *consensus* in short) is asymptotically achieved in the system if

$$|x_i(k) - x_j(k)| \to 0 \text{ as } k \to \infty, \ \forall i, j \in 1, \cdots n, \ i \neq j. \tag{3.77}$$

We introduce the error variables

$$e_i(k) = x_{i+1}(k) - x_1(k), \ i \in 1, \cdots (n-1). \tag{3.78}$$

Denote $e(k) = [e_1(k), \cdots, e(k)_{n-1}]^t$. Proceeding, the consensus problem for system (3.75) can be converted into the stability problem for the following system:

$$e(k+1) = e(k) - \gamma\,\bar{\mathbb{L}}e(k),$$
$$\bar{\mathbb{L}} = \begin{bmatrix} \ell_{22} - \ell_{12} & \cdots & \ell_{2n} - \ell_{1n} \\ \vdots & \vdots & \vdots \\ \ell_{n2} - \ell_{12} & \cdots & \ell_{nn} - \ell_{1n} \end{bmatrix}. \tag{3.79}$$

One can easily obtain the consensus condition for the system which is summarized in the following theorem.

Theorem 3.9. *System (3.75) with protocol (3.77) reaches a consensus solution asymptotically if and only if the interconnection topology digraph $\mathbb{G}(\mathbb{V}, \mathbb{E}, \mathbb{A})$ has a globally reachable node and the following condition holds:*

$$\gamma\,\mathbf{Re}(\mu_i) < 2, \;\; i \in \{2, \cdots n\}, \tag{3.80}$$

where μ_i are the eigenvalues of the Laplacian matrix $\mathbb{L}(\mathbb{G})$ except for zero.

Invoking *Gershgorin's disc lemma* from Chapter 1, it is easy to get a sufficient but scalable condition for (3.80) as

$$\gamma \sum_{j \in n_i} a_{ij} < 1, \;\; i \in \{2, \cdots n\}. \tag{3.81}$$

3.5.2 Consentability

For a multiagent system, if there exists a consensus protocol such that a consensus can be reached asymptotically, then we say the system is *consentable*. Theorem 3.9 tells us that there always exists a control gain γ such that discrete system (3.75) with protocol (3.76) can reach a consensus solution provided the interconnection topology digraph $\mathbb{G}$ contains a globally reachable node. So, for the system with integrator agents over ideal (delayless) networks the connectivity of the graph serves as the sole condition of consentability.

3.5.3 Effect of input delay

Consider a discrete-time multiagent system with topology graph $\mathbb{G} = (\mathbb{V}, \mathbb{E}, \mathbb{A})$ and integrator agents subject to an input delay described by

$$x_i(k+1) = x_i(k) + u_i(k - \beta_i), \;\; i = 1, \cdots, n. \tag{3.82}$$

Under diverse communication delays, the consensus protocol becomes

$$u_i(k) = \sum_{j \in n_i} a_{ij}[x_j(k - \tau_{ij}) - x_i(k)], \tag{3.83}$$

where τ_{ij} represents the communication delay from agent j to agent i.

Under protocol (3.83), the multiagent system (3.82) is said to achieve a *consensus asymptotically* if

$$\lim_{k \to \infty} x_i(k) = c, \ \ \forall i \in \{1, \cdots, n\}, \tag{3.84}$$

where $c \in \Re$ is a constant. Now, the closed-loop system of (3.82) and (3.83) is

$$\begin{aligned} x_i(k+1) &= x_i(k) \\ &\quad + \sum_{j \in n_i} a_{ij}[x_j(k - \tau_{ij} - \beta_i) - x_i(k - \beta_i)], \\ i &= 1, \ \cdots, \ n. \end{aligned} \tag{3.85}$$

We define

$$\begin{aligned} x(k) &= [x_1(k), \cdots, x_n(k)]^t, \\ d_{m1} &= \tau_{ij} + \beta_i, \ \ m_1 = 1, \cdots, n^2, \\ d_{m2} &= \beta_i, \ \ m_2 = n^2 + 1, \cdots, n(n+1). \end{aligned} \tag{3.86}$$

Using (3.86), we can augment (3.85) to form a time delay system of the following type:

$$x(k+1) = x(k) + \sum_{i=1}^{n_d} A_i \, x(k - d_i), \tag{3.87}$$

where $A_i \in \Re^{n \times n}$ and $n_d = n(n+1)$. It is readily evident that

$$\sum_{i=1}^{n_d} A_i = \mathbb{L},$$

where $\mathbb{L}$ is the Laplacian matrix of the topology graph. For system (3.87), the associated characteristic equation has the form

$$\mathbf{det}\left[(z-1)I + \sum_{i=1}^{n_d} A_i z^{-d_i}\right] \tag{3.88}$$

and the equilibrium set can be expressed as

$$X_e = \{x \in \Re^n : \mathbb{L}\, x = 0\}.$$

When $\mathbb{L}$ is singular, X_e is a continuum of equilibrium points.

Now, assume that the interconnection topology of the system is described by a connected undirected graph or a digraph containing a globally reachable node. It follows from the properties of the Laplacian that the Laplacian matrix $\mathbb{L}$ has a simple eigenvalue 0, that is, $\mathbf{det}(\mathbb{L}) = 0$ and $\mathbf{rank}(\mathbb{L}) = n - 1$. By the definition of $\mathbb{L}$ we also have $\mathbb{L}\ \mathbf{1}_n = 0$. Hence, all the elements in X_e can be represented as $c\ \mathbf{1}_n = 0$, where c is any constant. System (3.87) achieves a consensus asymptotically if the solution of the system starting from any given initial states $x(-k) \in \Re^n$, $k = 0, 1, \cdots, n_d - 1$, asymptotically converges to an element in X_e.

In view of the foregoing analysis, we can arrive at the following result.

Lemma 3.8. *If the solutions of (3.88) have modulus less than unity except for a root at $z = 1$, then system (3.87) with a connected undirected graph or a digraph containing a globally reachable node achieves a consensus asymptotically.*

It is significant to observe that this lemma implies that under the assumption that the graph is connected or the digraph contains a globally reachable node, the first-order agent system with diverse input delays and communication delays achieves a consensus asymptotically if the closed-loop system is steady semistable with $z = 1$ as a simple pole.

3.6 Notes

This chapter initially examined the dynamic consensus of a linear or linearized multiagent system whose communication topology has a directed spanning tree.

We have studied the asymptotic properties of the consensus value in distributed consensus algorithms over switching, directed random graphs. While different aspects of consensus algorithms over random switching networks, such as conditions for convergence and the speed of convergence, have been widely studied, a characterization of the distribution of the asymptotic consensus for general *asymmetric* random consensus algorithms remains an open problem.

We have derived closed-form expressions for the expectation of the asymptotic consensus value as a function of the set of initial conditions, $\{x_u(0)\}_{u \in V}$, and the set of nodes properties, $\{(p_u, d_u)\}_{u \in V}$. We have also studied the variance of the asymptotic consensus value in terms of several elements that influence it, namely,

(*i*) the initial conditions,

(*ii*) nodes properties, and

(*iii*) the network topology.

We have derived an upper bound for the variance of the asymptotic consensus value that explicitly describes the influence of each one of these elements. We also provide an interpretation of the influence of the network topology on the variance in terms of the eigenvalues of the expected matrix $\mathbb{E}W_k$. From our analysis we conclude that, in most cases, the variance of x^* is primarily governed by how close the subdominant eigenvalues of $\mathbb{E}W_k$ are from 1. We have checked the validity of our predictions with several numerical simulations. This work presented an $\mathbb{H}_2$ performance analysis of the consensus protocol with both process and measurement noises present. In order to perform such an analysis, the consensus model was transformed into an edge representation leading to a minimal realization of the system. While the traditional consensus problem is centered around analysis of the graph Laplacian, the edge variant relies on the edge Laplacian. An advantage of this transformation is the ability to perform analysis of the system orthogonal to the agreement subspace while preserving the strong algebraic properties of the graph via the edge Laplacian.

The $\mathbb{H}_2$ performance of the consensus protocol was shown to be dependent on the number of edges in the graph. Consequently, this suggests that cycles play an important role in terms of the system performance by way of noise propagation. This is in stark contrast to traditional analysis of consensus problems which focuses on the convergence rate of the system. It is well understood that an increase in the number of edges in the graph corresponds to an increase in the second smallest eigenvalue of the graph Laplacian, resulting in faster convergence. However, in this context, the addition of edges produces an adverse effect in performance.

The analysis results showed that certain classes of graphs – spanning trees and k-regular graphs – result in a graph-theoretic description of the system norm. Certainly, it seems that there should be other classes of graphs that allow for similar simplifications, and this is currently under investigation. The observations on the performance ratio relating to the cycle graph suggest that some graphs scale better than others in terms of performance and the number of nodes.

These results were then applied to an optimization problem aimed at selecting sensors for a fixed topology consensus problem. Although the complete statement of the problem is numerically challenging, the relaxation provides a reasonable approach to this problem. This formulation can naturally be pushed further to include, for example, topology design with both favorable convergence rates, good $\mathbb{H}_2$ performance, and low sensor costs. More subtly, it highlights a connection between combinatorial optimization problems and synthesis methods for networked dynamic systems. Consequently, the development of numerically tractable algorithms for the design of these systems is an essential component for the maturation of this field.

Finally, we investigated the consensus problem of second-order discrete-time multiagent systems with nonuniform time delays and dynamically changing communication topologies. A linear consensus protocol was first introduced, and then by model transformations and applying the properties of nonnegative matrices, sufficient conditions were presented to make all agents reach consensus with arbitrary bounded nonuniform time delays and dynamically changing directed graphs that may not have spanning trees.

References

[1] J. Shamma, Cooperative Control of Distributed Multi-agent Systems, John Wiley and Sons, 2008.

[2] Z.K. Li, Z.S. Duan, G.R. Chen, L. Huang, Consensus of multi-agent systems and synchronization of complex networks: a unified viewpoint, IEEE Trans. Circuits Syst. I, Regul. Pap. 57 (1) (2010) 213–224.

[3] M.S. Mahmoud, Y. Xia, Applied Control Systems Design: State-Space Methods, Springer-Verlag, UK, May 2012.

[4] R.A. Horn, C. Johnson, Topics in Matrix Analysis, Cambridge University Press, 1991.

[5] N. Chopra, M.W. Spong, Passivity-based control of multiagent systems, in: S. Kawamura, M. Svinin (Eds.), Advances in Robot Control: From Everyday Physics to Human-like Movements, Springer-Verlag, Berlin, 2006, pp. 107–134.

[6] Y. Hatano, M. Mesbahi, Agreement over random networks, IEEE Trans. Autom. Control 11 (50) (2005) 1867–1872.

[7] D.P. Bertsekas, J.N. Tsitsiklis, Parallel and Distributed Computation, Prentice Hall, 1989.

[8] M. Mesbahi, State-dependent graphs, in: IEEE Conf. Decision and Control, December 2003.

[9] R. Olfati-Saber, J. Fax, R. Murray, Consensus and cooperation in networked multi-agent systems, Proc. IEEE 1 (95) (2007) 1–17.

[10] R. Olfati-Saber, R.M. Murray, Consensus problems in networks of agents with switching topology and time-delays, IEEE Trans. Autom. Control 9 (49) (2004) 1520–1533.

[11] R. Olfati-Saber, J.S. Shamma, Consensus filters for sensor networks and distributed sensor fusion, in: IEEE Conf. Decision and Control, December 2005.

[12] W. Ren, R. Beard, E. Atkins, Information consensus in multivehicle cooperative control, IEEE Control Syst. Mag. 27 (2) (April 2007) 71–82.

[13] W. Ren, Consensus strategies for cooperative control of vehicle formations, IET Control Theory Appl. 2 (1) (2007) 505–512.

[14] D. Spanos, R. Murray, Distributed sensor fusion using dynamic consensus, in: IFAC World Congress, July 2005.

[15] L. Xiao, S. Boyd, S. Kim, Distributed average consensus with least-mean-square deviation, J. Parallel Distrib. Comput. 1 (67) (2007) 33–46.

[16] D. Zelazo, A. Rahmani, M. Mesbahi, Agreement on edges via the edge Laplacian, in: IEEE Conf. Decision and Control, December 2007.

[17] D. Zelazo, M. Mesbahi, Edge agreement: graph-theoretic performance bounds and passivity analysis, IEEE Trans. Autom. Control 56 (3) (March 2011) 544–555.

[18] C. Godsil, G. Royle, Algebraic Graph Theory, Springer, May 2012.

Chapter 4

Energy-based cooperative control

4.1 Dissipative cooperative output synchronization

In the sequel, we present a class of energy-based cooperative control problems that have an explicit connection to convex network optimization problems. The main basis is the passivity and dissipativity concepts. The new notion of maximal equilibrium independent passivity is introduced and it is shown that networks of systems possessing this property asymptotically approach the solutions of a dual pair of network optimization problems, namely, an optimal potential and an optimal flow problem. This connection leads to an interpretation of the dynamic variables, such as system inputs and outputs, to variables in a network optimization framework, such as divergences and potentials, and reveals that several duality relations known in convex network optimization theory translate directly to passivity-based cooperative control problems.

4.1.1 Introduction

Synchronization and cooperation of multiagent systems have attracted considerable attention due to their extensive engineering applications such as in control theory, microgrids, and computer science [1], [2]. In recent years, the cooperation and coordination of multiagent systems have been substantially addressed, which is a basic requirement for studying multiagent systems [3]. Cooperation among agents without centralized controller while each agent can only access local/neighboring information is one of the main challenges in multiagent systems. This problem is one of the most fundamental issues of cooperation control, whose main duty is to implement a distributed protocol via local and neighboring information so that the states or the outputs of agents can reach a prescribed agreement. Analytical and numerical methodologies to solve the synchronization problem have been reported in the literature, which is evident from several monographs and numerous articles in different settings.

However, most attention has been paid to cooperation among homogeneous agents rather than heterogeneous systems. Whereas synchronization methodologies for homogeneous multiagent systems have been thoroughly investigated, synchronization of nonidentical agents still remains a challenging problem. Interesting results have been addressed for the multiagent synchronization problems for heterogeneous dynamics, including descriptor systems.

Discrete Networked Dynamic Systems. https://doi.org/10.1016/B978-0-12-823698-7.00012-6

Consensus protocols for linear and nonlinear multiagent systems are investigated with the help of the Lyapunov method, algebraic graph theory, and matrix theory. Dissipativity as well as its special case passivity has been exploited to analyze the stability of multiagent systems. The finite-time synchronization problem is investigated based on the passivity approach by exploiting an adaptive state feedback with linear and nonlinear couplings [4]. Under switching topologies, a consensus protocol with the help of the dissipativity is proposed to ensure that the agents can reach an agreement [5]. In [6], the distributed suboptimal $\mathbb{H}_\infty$ consensus problem was established based on the dissipativity of uncertain systems.

Most of the synchronization passivity-based methods have considered continuous-time multiagent systems; however, implementation of cooperative discrete-time multiagent systems is needed. The internal model is exploited to convert the synchronization problem into an output stabilization problem. Two types of distributed protocols to solve the synchronization problem for discrete-time dynamical systems are proposed using observer-based methodology [7]. The aforementioned results focus on continuously updated protocols such that every agent needs to continuously employ its state and neighbors' states. On the other hand, to avoid this problem, researchers have investigated the triggering algorithms to reduce communication congestion over networks. Since continuous communication between agents usually leads to the waste of energy and resources, event triggering mechanisms have been tailored to fit synchronization over networks [8].

This section focuses on fully distributed discrete-time protocols using a dissipativity-based approach. In addition, unmodeled dynamics and imperfection of networks such as time delays must be taken into consideration. It further concentrates on dissipativity-based synchronization problems for discrete-time multiagent systems with delays.

4.1.2 Problem preliminaries and formulation

Here we recall the basic graph structure. Let $\mathbb{G} = \{\mathbb{V}, \mathbb{E}, \mathbb{A}\}$ denote an undirected graph of N order, where $\mathbb{V} = \{v_i, i = 1, \dots, N\}$, $\mathbb{E} \subseteq \mathbb{V} \times \mathbb{V}$, and $\mathbb{A} = [a_{ij}] \in \Re^{n\times n}$ are the set of vertices, edges, and the adjacency matrix of the graph $\mathbb{G}$, respectively. It is assumed that each node represents an agent and the information exchange of neighbors of the agent v_i is denoted by $\mathcal{N}_i = \{j : (v_i, v_j) \in \mathbb{E}\}$. If there is an information exchange between agent i and agent j, that is, $(v_i, v_j) \in \mathbb{E}$, then $a_{ij} = a_{ji} = 1$; otherwise $a_{ij} = a_{ji} = 0$. Moreover, $d_i = \sum_{j\in N_i} a_{ij}$ defines the degree of an agent v_i. A sequence of different consecutive nodes starting with v_i and ending with v_j represents a path in the graph $\mathbb{G}$ from v_i to v_j. If there exists a path between any two vertices, then the graph $\mathbb{G}$ is considered to be connected.

Consider a group of N heterogeneous nth-order agents represented by the following descriptor state-space representation with uncertainty for each agent:

$$\begin{aligned} E_i x_i(k+1) &= (A_i + \Delta A_i)x_i(k) + (A_{di} + \Delta A_{di})x_i(k-\tau(k)) + B_i u_i(k), \\ z_i(k) &= C_i x(k) + C_{di} x_i(k-\tau(k)), \\ x_i(k) &= \phi_{i0}(k), \quad k \in [-\bar{\tau}, 0], \end{aligned} \tag{4.1}$$

where $x_i(k) \in \Re^n$ is the agent state, $y_i(k) \in \Re^p$ represents the agent's output, and $u_i(k)$ is the control law of the agent i which can only use the local information of neighbors' outputs; $\phi_0(k)$ is a compatible initial condition. The nominal constant matrices $E_i \in \Re^{n\times n}$, $A_i \in \Re^{n\times n}$, $A_{di} \in \Re^{n\times n}$, $B_i \in \Re^{n\times m}$, $C_i \in \Re^{n\times p}$, $C_{id} \in \Re^{n\times p}$ might be not identical for different agents. The real matrix $E_i \in \Re^{n\times n}$ is assumed to be singular, that is, $\mathbf{rank}(E_i) = r < n$. Here, $\tau(k)$ is a positive integer that represents the time delay and assumed to be time-varying for all agents such that

$$\underline{\tau} \le \tau(k) \le \bar{\tau},$$

where $\underline{\tau}$ and $\bar{\tau}$ are positive integers representing the upper and lower bounds of the time delay; ΔA_i and ΔA_{di} are real unknown time-varying matrix functions representing the parametric uncertainties which are the result of unmodeled dynamics or/and model linearization. Without loss of generality, the following assumption is introduced for technical convenience:

$$\Delta A_i = M_i F_i(k) N_i, \quad \Delta A_{di} = M_i F_i(k) N_{di}, \tag{4.2}$$

where $F_i(k)$ represents unknown matrix functions where its elements are Lebesgue-measurable functions satisfying $F_i^T(k)F_i(k) \le I$. Moreover, the matrices M_i, N_i, and N_{di} are real constant with appropriate dimensions which represent how the uncertain parameters in $F_i(k)$ enter into the nominal matrices.

Definition 4.1. The heterogeneous multiagent dynamical system is considered to achieve output synchronization if for every initial state $x_i(k_0) \in \Re^n$, the following holds:

$$\lim_{k\to\infty} \|z_i(k) - z_j(k)\| = 0$$

for all $i, j \in \mathbb{V}$.

Adopting a time-triggered protocol, we let each agent use its own state to guarantee the dissipativity with respect to the new distributed control law. The following control law is considered:

$$u_i(k) = K_i x_i(k) + \tilde{u}_i(k). \tag{4.3}$$

In (4.3), the constant gain matrix K_i should be designed such that each agent becomes dissipative with respect to the input $\tilde{u}_i(k)$. Then the discrete-time consensus protocol (4.4) for the heterogeneous system (4.1) will be analyzed for both time-triggered and event-triggered mechanisms based on dissipativity analysis. We have

$$\tilde{u}_i(k) = -c_i \sum_{j \in \mathcal{N}_i} (z_i(k) - z_j(k)). \tag{4.4}$$

A general event-triggered control framework for each agent is illustrated in Fig. 4.1. In classical consensus protocol settings, it is assumed that agents can access its continuous measurements and neighboring information. Such an assumption requires enormous computing power and ideal communication settings. This makes the discrete-time protocols more realistic and inevitable. Note that the discrete-time protocol (4.4) must be updated at each time instant $k = 1, 2, \ldots$. However, this kind of consensus protocols might lead to a congestion in the communication network particularly due to the performance requirements. Based on the previous results, we focus on developing an event-triggered scheme such that the multiagent system (4.8) is synchronized. The event triggering error between the current and the last updated state for agent i can be represented as

$$e_i(k) = z_i(k_l^i) - z_i(k),$$

where $z_i(k_l^i)$ is the system output at time k_l^i. The event-triggered policy for the descriptor system (4.1) is governed by

$$\tilde{u}_i(k) = -c_i \sum_{j \in \mathcal{N}_i} (z_i(k_l^i) - z_j(k_l^j)), \quad k \in [k_l^i, k_{l+1}^i). \tag{4.5}$$

The distributed event triggering policy for agent i is designed as follows:

$$k_{l+1}^i = \inf\left\{k : k_l^i \middle| \|e_i(k)\| > \gamma_i \left\| \sum_{j \in \mathcal{N}_i} z_i(k_l^i) - z_j(k_l^j) \right\| \right\}, \tag{4.6}$$

where $\{k_{l+1}^i\}$ represents the event-triggered time sequence, $\{k_l^i\}$ and $\{k_{l+1}^i\}$ are the adjacent sampling instants, and γ_i is a positive scalar to be designed.

Before proceeding, the following definitions and lemmas are introduced to be used in deriving our main results.

Definition 4.2. [9] The following are some essential properties of descriptor systems:

1. Pair (E_i, A_i) is said to be regular if the determinant of the matrix $(zE_i - A_i)$ is not identically zero.
2. If the pair (E_i, A_i) is regular and deg det$(zE_i - A_i) = \text{rank}(E_i)$, then it is said to be causal.

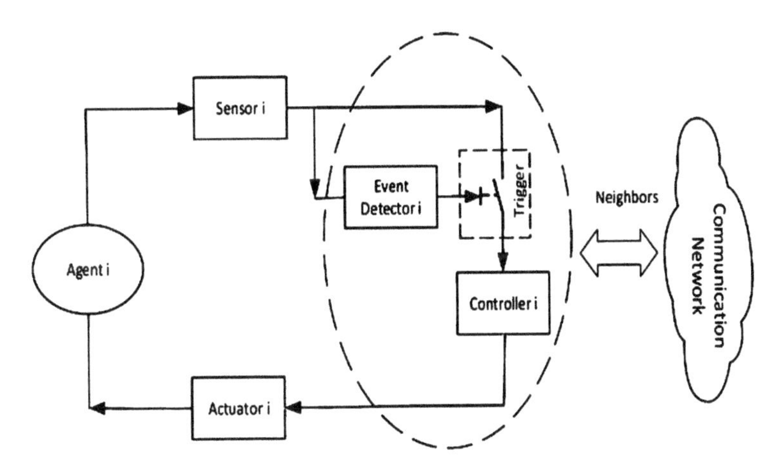

FIGURE 4.1 Schematic diagram of event-triggered control for agent i.

3. For given positive scalars $\underline{\tau}$ and $\bar{\tau}$, the discrete-time descriptor system (4.1) is said to be regular and causal for any time delay $\tau(k)$ satisfying $\underline{\tau} \leq \tau(k) \leq \bar{\tau}$, if the pair (E_i, A_i) is regular and causal. Finally, the dynamical system (4.1) is admissible if it is causal, regular, and stable.

Definition 4.3. The dynamical system of an agent i is called $\mathcal{QSR}$ dissipative if for any integer N, there exist symmetric matrices $\mathcal{Q}_i$, $\mathcal{S}_i$, and $\mathcal{R}_i$ such that the following inequality holds:

$$\sum_{k=0}^{N} z_i^t(k)\mathcal{Q}_i z_i(k) + 2z_i^t(k)\mathcal{S}_i u_i(k) + u_i^t(k)\mathcal{R}_i u_i(k) \geq 0.$$

Remark 4.1. The notion of passivity index $(\mathcal{Q}, \mathcal{S}, \mathcal{R})$ comprises the $\mathcal{H}_\infty$ performance, strict passivity, and shortage of passivity as special cases by selecting different values of the index parameters. Obviously, if $\mathcal{Q}_i = -I$, $\mathcal{R}_i = \gamma^2 I$, and $\mathcal{S}_i = 0$, then dissipative inequality reduces to an $\mathcal{H}_\infty$ performance index. If $\mathcal{Q}_i = 0$, $\mathcal{R}_i < 0$, and $\mathcal{S}_i = I$, then the system is called input excess passive. Moreover, if $\mathcal{Q}_i = 0$, $\mathcal{R}_i > 0$, and $\mathcal{S}_i = I$, then the system is called input shortage passive. In this chapter, we consider the index with a tradeoff between $\mathcal{H}_\infty$ performance and the shortage passivity index.

Lemma 4.1. *Let $x(k) \in \Re^n$ be a vector function. Then, for any positive definite matrix $M > 0$ and integers $m > n$, the following inequality holds:*

$$-(m-n+1)\sum_{r=n}^{m} x^T(r)Mx(r) \leq -\sum_{r=n}^{m} x^T(r)M\sum_{r=n}^{m} x(r).$$

Lemma 4.2. *Let N be the number of agents and a negative scalar $\beta < 0$. Then the following inequality holds:*

$$\beta \sum_{i=1}^{N} \|z_i(k)\|^2 \leq \beta \frac{1}{4N} \sum_{i=1}^{N} \sum_{j \in \mathcal{N}_i} \left\| z_i(k) - z_j(k) \right\|^2 .$$

Proof. It can be easily shown that the sum can be rewritten in the following form:

$$\begin{aligned}
\sum_{i=1}^{N} \|z_i(k)\|^2 &= \frac{1}{2N} \sum_{i=1}^{N} \sum_{j=1}^{N} \left[\|z_i(k)\|^2 + \|z_j(k)\|^2 \right] \\
&= \frac{1}{2N} \sum_{i=1}^{N} \sum_{j=1}^{N} \left[\frac{\|z_i(k) + z_j(k)\|^2}{2} + \frac{\|z_i(k) - z_j(k)\|^2}{2} \right] \\
&\geq \frac{1}{4N} \sum_{i=1}^{N} \sum_{j=1}^{N} \left\| z_i(k) - z_i(k) \right\|^2 .
\end{aligned} \tag{4.7}$$

Since the number of agents connected to the ith agent is less than or equal to N and the right side has positive terms, we obtain

$$\sum_{i=1}^{N} \|z_i(k)\|^2 \geq \frac{1}{4N} \sum_{i=1}^{N} \sum_{j \in \mathcal{N}_i} \left\| z_i(k) - z_j(k) \right\|^2 .$$

Multiplying both sides by the negative scalar β completes the proof. □

Fact 4.1. Given any constant matrices M, N, and Q with appropriate dimensions, where Q a is symmetric matrix, the following holds:

$$Q + MF(k)N + N^T F^T(k) M^T < 0,$$

for every matrix function $F(k)$ satisfying $\|F(k)\| \leq 1$, if and only if there exists a positive scalar ϵ such that

$$Q + \epsilon M M^T + \frac{1}{\epsilon} N^T N < 0.$$

The following inequality will be used in the sequel:

$$\begin{aligned}
z_i^T(k)\tilde{u}_i(k) &= -\frac{1}{2\delta_i}(\tilde{u}_i(k) - \delta_i z_i(k))^T (\tilde{u}_i(k) - \delta_i z_i(k)) \\
&\quad + \frac{1}{2\delta_i}\tilde{u}_i^T(k)\tilde{u}_i(k) + \frac{\delta_i}{2} z_i^T(k) z_i(k) \\
&\leq \frac{1}{2\delta_i}\tilde{u}_i^T(k)\tilde{u}_i(k) + \frac{\delta_i}{2} z_i^T(k) z_i(k).
\end{aligned} \tag{4.8}$$

4.1.3 Design results

In this section, we consider the analysis and synthesis of consensus controllers for descriptor discrete-time multiagent systems with time delay. Now, let us introduce the following theorem about time-triggered consensus protocol.

Theorem 4.1. *Consider the multiagent system (4.1) with the time-triggered consensus protocol (4.4). For given positive integers $\bar{\tau}$ and $\underline{\tau}$ and positive scalars c_i, the agents reach consensus robustly and every agent is dissipative with $(\beta, 1, \alpha_i)$-index if there exist positive scalars ϵ_i, δ_i and matrices $\mathcal{P}_i > 0$, $\mathcal{Q}_1^i > 0$, $\mathcal{Q}_2^i > 0$, $\mathcal{Q}_3^i > 0$, $\mathcal{Z}_1^i > 0$, $\mathcal{Z}_2^i > 0$, $\mathcal{Z}_3^i > 0$, $\mathcal{R}_i$, $L_j^i > 0$, $i = 1, ..., N$, $j = 1, 2, ..., 6$, such that the following inequalities hold:*

$$\Theta = \begin{bmatrix} \Gamma^i & U_1^i & \epsilon_i U_2^i \\ \bullet & -\epsilon_i I & 0 \\ \bullet & \bullet & -\epsilon_i I \end{bmatrix} < 0, \tag{4.9}$$

$$\begin{bmatrix} \mathcal{Z}_3^i & \mathcal{R}_i \\ \mathcal{R}_i & \mathcal{Z}_3^i \end{bmatrix} \geq 0, \tag{4.10}$$

$$(\alpha_i + \frac{1}{2\delta_i})c_i^2 d_i + (\beta + \frac{\delta_i}{2})\frac{1}{4N} < 0, \tag{4.11}$$

where

$$\Gamma^i = \begin{bmatrix} \Gamma_{11}^i & \Gamma_{12}^i & \Gamma_{13}^i & \Gamma_{14} & \Gamma_{15}^i & \Gamma_{16}^i \\ \bullet & \Gamma_{22}^i & \Gamma_{23}^i & E_i^T R_i E_i & -L_2^i & L_2^i B_i \\ \bullet & \bullet & \Gamma_{33}^i & \Gamma_{34}^i & \Gamma_{35}^i & \Gamma_{36}^i \\ \bullet & \bullet & \bullet & \Gamma_{44}^i & -L_4^i & L_4^i B_i \\ \bullet & \bullet & \bullet & \bullet & \Gamma_{55}^i & \Gamma_{56}^i \\ \bullet & \bullet & \bullet & \bullet & \bullet & \Gamma_{66}^i \end{bmatrix}, \tag{4.12}$$

$$\begin{aligned}
\Gamma_{11}^i &= L_1^i(A_i - E_i) + (A_i - E_i)^T L_1^{iT} + (\tau^\star + 1)\mathcal{Q}_3^i \\
&\quad + \mathcal{Q}_1^i + \mathcal{Q}_2^i - E_i^t \mathcal{Z}_1^i E_i - E_i^t \mathcal{Z}_2^i E_i - \beta C_i^t C_i, \\
\Gamma_{12}^i &= E_i^t \mathcal{Z}_2^i E + (A_i - E_i)^t L_2^{it}, \\
\Gamma_{22}^i &= \mathcal{Q}_1^i - E_i^t \mathcal{Z}_2 E_i - E_i^t \mathcal{Z}_3 E_i, \\
\Gamma_{13}^i &= L_1^i A_{di} + (A_i - E_i)^t L_3^{it} - \beta C_i^t C_{di}, \\
\Gamma_{23}^i &= -E_i^t R_i^t E_i + E_i^t \mathcal{Z}_3^{it} E_i + L_2^i A_{di},
\end{aligned}$$

$$
\begin{aligned}
\Gamma_{33}^i &= -\mathcal{Q}_3^i + L_3^i A_{di} + A_{di}^T L_3^{iT} \\
&\quad - 2E_i^T \mathcal{Z}_3^i E_i + E_i^t R_i E_i + E_i^t R_i^t E_i - \beta C_{di}^t C_{di}, \\
\Gamma_{14}^i &= E_i^t \mathcal{Z}_1^i E_i + (A_i - E_i)^t L_4^{it}, \\
\Gamma_{34}^i &= -E_i^t R_i E_i + E_i^t \mathcal{Z}_3^i E_i + A_{di}^t L_4^{it}, \\
\Gamma_{44}^i &= -\mathcal{Q}_2^i - E_i^t \mathcal{Z}_1^i E_i - E_i^t \mathcal{Z}_3^i E_i, \\
\Gamma_{35}^i &= L_3^i + A_{di}^T L_5^{iT}, \\
\Gamma_{15}^i &= E_i^t \mathcal{P}_i - L_1^{iT} + (A - E)^t L_5^{it} - L_2^i, \\
\Gamma_{55}^i &= \bar{\tau}^2 \mathcal{Z}_1^i + \underline{\tau}^2 \mathcal{Z}_2^i + \tau^\star \mathcal{Z}_3^i - L_5^i - L_5^{it} + \mathcal{P}_i, \\
\Gamma_{16}^i &= -C_i^t + L_1^i B_i + (A_i - E_i)^t L_6^i, \\
\Gamma_{36}^i &= -C_{di}^t + A_{di}^t L_6^{it} + L_3^i B_i, \\
\Gamma_{56}^i &= L_5^i B_i + L_6^{it}, \\
\Gamma_{66}^i &= -\alpha_i I + L_5^i B_i + B_i^t L_5^{it}, \\
U_1 &= \begin{bmatrix} M_i^T L_1^{iT} & M_i^t L_2^{it} & \dots & M_i^t L_6^{it} \end{bmatrix}^t, \\
U_2 &= \begin{bmatrix} N_i & 0 & N_{di} & 0 & 0 & 0 \end{bmatrix},
\end{aligned}
$$

and $\tau^\star = \bar{\tau} - \underline{\tau}$.

Proof. Consider the following Lyapunov functionals:

$$
\begin{aligned}
V(k) &= \sum_{i=1}^{N} \sum_{i=1}^{4} V_j^i(k), \\
V_1^i(k) &= x_i^t(k) E_i^t \mathcal{P}_i E_i x_i(k), \\
V_2^i(k) &= \sum_{l=k-\underline{\tau}}^{k-1} x_i^t(l) \mathcal{Q}_1^i x_i(l) + \sum_{l=k-\bar{\tau}}^{k-1} x_i^t(l) \mathcal{Q}_2^i x_i(l), \\
V_3^i(k) &= \sum_{\theta=-\bar{\tau}}^{-\underline{\tau}} \sum_{l=k+\theta}^{k-1} x_i^t(l) \mathcal{Q}_3^i x_i(l), \\
V_4^i(k) &= \bar{\tau} \sum_{\theta=-\bar{\tau}}^{-1} \sum_{l=k+\theta}^{k-1} \eta_i^t(l) E_i^t \mathcal{Z}_1^i E_i \eta_i(l) \\
&\quad + \underline{\tau} \sum_{\theta=-\underline{\tau}}^{-1} \sum_{l=k+\theta}^{k-1} \eta_i^t(l) E_i^T \mathcal{Z}_2^i E_i \eta_i(l) \\
&\quad + \tau^\star \sum_{\theta=-\bar{\tau}}^{\underline{\tau}-1} \sum_{l=k+\theta}^{k-1} \eta_i^T(l) E_i^t \mathcal{Z}_3^i E_i \eta_i(l),
\end{aligned} \tag{4.13}
$$

and $\tau^\star = \bar{\tau} - \underline{\tau}$ and $\eta_i(k) = x_i(k+1) - x_i(k)$. The difference of the Lyapunov functional $V(k)$ can be derived as follows:

$$\begin{aligned}
\Delta V_1^i(k) &= x_i^t(k+1)E_i^t\mathcal{P}_iE_ix_i(k+1) - x_i^t(k)E_i^t\mathcal{P}_iE_ix_i(k) \\
&= \eta_i^t(k)E_i^t\mathcal{P}_iE\eta_i(k) + 2x_i^t(k)E_i^t\mathcal{P}_iE\eta_i(k), \\
\Delta V_2^i(k) &= x_i^t(k)[\mathcal{Q}_1^i + \mathcal{Q}_2^i]x_i(k) - x_i^t(k-\underline{\tau})\mathcal{Q}_1^ix_i(k-\underline{\tau}) \\
&\quad - x_i^t(k-\bar{\tau})\mathcal{Q}_2^ix_i(k-\bar{\tau}), \\
\Delta V_3^i(k) &= (\tau^\star + 1)x_i^t(k)\mathcal{Q}_3^ix_i(k) - \sum_{\ell=k-\bar{\tau}}^{k-\underline{\tau}} x_i^t(\ell)\mathcal{Q}_3^ix_i(\ell) \\
&\le (\tau^\star + 1)x_i^t(k)\mathcal{Q}_3^ix_i(k) - x_i^t(k-\tau(k))\mathcal{Q}_3^ix_i(k-\tau(k)), \\
\Delta V_4^i(k) &= \eta_i^t(k)E_i^t\big[\bar{\tau}^2\mathcal{Z}_1^i + \underline{\tau}^2\mathcal{Z}_2^i + \tau^\star\mathcal{Z}_3^i\big]E_i\eta_i(k) \\
&\quad - \bar{\tau}\sum_{l=k-\bar{\tau}}^{k-1} \eta_i^T(l)E_i^t\mathcal{Z}_1^iE_i\eta_i(l) - \underline{\tau}\sum_{\ell=k-\underline{\tau}}^{k-1} \eta_i^t(\ell)E_i^t\mathcal{Z}_2^iE_i\eta_i(l\ell) \\
&\quad - \tau^\star\sum_{\ell=k-\bar{\tau}}^{k-\underline{\tau}-1} \eta_i^t(\ell)E_i^t\mathcal{Z}_3^iE_i\eta_i(\ell).
\end{aligned} \tag{4.14}$$

According to Lemma 4.1, we obtain

$$\begin{aligned}
&- \bar{\tau}\sum_{\ell=k-\bar{\tau}}^{k-1} \eta_i^t(\ell)E_i^t\mathcal{Z}_1^iE_i\eta_i(\ell) \\
&\le -[x_i(k) - x_i(k-\bar{\tau})]^t E_i^t\mathcal{Z}_1^iE_i[x_i(k) - x_i(k-\bar{\tau})] \\
&- \underline{\tau}\sum_{\ell=k-\underline{\tau}}^{k-1} \eta_i^t(\ell)E_i^T\mathcal{Z}_2^iE_i\eta_i(\ell) \\
&\le -[x_i(k) - x_i(k-\underline{\tau})]^t E_i^t\mathcal{Z}_2^iE_i[x_i(k) - x_i(k-\underline{\tau})].
\end{aligned} \tag{4.15}$$

Using Lemmas 4.1 and 4.2, the following bound holds:

$$\begin{aligned}
&- \tau^\star\sum_{\ell=k-\bar{\tau}}^{k-\underline{\tau}-1} \eta_i^t(\ell)E_i^t\mathcal{Z}_3^iE_i\eta_i(\ell) \\
&= -\tau^\star\sum_{\ell=k-\bar{\tau}}^{k-\tau(k)-1} \eta_i^t(\ell)E_i^t\mathcal{Z}_3^iE_i\eta_i(\ell) - \tau^\star\sum_{\ell=k-\tau(k)}^{k-\underline{\tau}-1} \eta_i^t(\ell)E_i^t\mathcal{Z}_3^iE_i\eta_i(\ell) \\
&\le -\frac{1}{\alpha_1}\Psi_1^{it}(k)E_i^t\mathcal{Z}_3^iE_i\Psi_1^i(k) - \frac{1}{\alpha_1}\Psi_2^{it}(k)E_i^t\mathcal{Z}_3^iE_i\Psi_2^i(k) \\
&\le -\begin{bmatrix} \Psi_1^i(k) \\ \Psi_2^i(k) \end{bmatrix}^t \begin{bmatrix} E_i^t\mathcal{Z}_3^iE_i & E_i^t\mathcal{R}_iE_i \\ E_i^t\mathcal{R}_iE_i & E_i^t\mathcal{Z}_3^iE_i \end{bmatrix} \begin{bmatrix} \Psi_1^i(k) \\ \Psi_2^i(k) \end{bmatrix}
\end{aligned} \tag{4.16}$$

such that

$$\begin{bmatrix} E_i^t \mathcal{Z}_3^i E_i & E_i^t \mathcal{R}^i E_i \\ E_i^t \mathcal{R}^i E_i & E_i^t \mathcal{Z}_3^i E_i \end{bmatrix} \geq 0,$$

where

$$\alpha_1 = \frac{\bar{\tau} - \tau(k)}{\bar{\tau} - \underline{\tau}}, \quad \Psi_1^i(k) = x_i(k - \tau(k)) - x_i(k - \bar{\tau}),$$
$$\alpha_2 = \frac{\tau(k) - \underline{\tau}}{\bar{\tau} - \underline{\tau}}, \quad \Psi_2^i(k) = x_i(k - \underline{\tau}) - x_i(k - \tau(k)).$$

Let

$$\xi_i(k) = \begin{bmatrix} x_i(k) \\ x_i(k - \underline{\tau}) \\ x_i(k - \tau(k)) \\ x_i(k - \bar{\tau}) \\ \eta_i(k) E_i \\ \tilde{u}_i(k) \end{bmatrix}, \quad L_i = \begin{bmatrix} L_1^i \\ L_2^i \\ L_3^i \\ L_4^i \\ L_5^i \\ L_6^i \end{bmatrix},$$

$$2\xi_i^t(k) L_i^t \Big[(A_i - E_i) x_i(k) + A_{di} x_i(k - \tau(k)) + B_i \tilde{u}_i(k) - E_i \eta_i(k) \Big] = 0. \tag{4.17}$$

From the dynamics (4.1) and (4.17), it is straightforward to see that the following equation holds for any matrices L_1^i, L_2^i, L_3^i, L_4^i, L_5^i, L_6^i with appropriate dimensions:

$$\Delta V^i - z_i^T(k)\tilde{u}_i(k) - \alpha_i \tilde{u}_i^T(k)\tilde{u}_i(k) - \beta z_i^T(k) z_i(k) \leq \xi_i^T(k) \Theta \xi_i(k), \tag{4.18}$$

where Θ is given in (4.9). Using Lemma 4.2, the Cauchy–Schwartz inequality, inequalities (4.8), (4.18), and (4.9), and the consensus protocol (4.4), we obtain the following upper bound of the proposed Lyapunov functional:

$$\begin{aligned} \Delta V(k) &= \sum_{i=1}^{N} \Delta V^i(k) \\ &\leq \sum_{i=1}^{N} \Big[z_i^T(k)\tilde{u}_i(k) + \alpha_i \tilde{u}_i^T(k)\tilde{u}_i(k) + \beta z_i^T(k) z_i(k) \Big] \\ &\leq \sum_{i=1}^{N} \Big[(\alpha_i + \frac{1}{2\delta_i}) \|\tilde{u}_i(k)\|^2 + (\beta + \frac{\delta_i}{2}) \|z_i(k)\|^2 \Big] \end{aligned}$$

$$\leq \sum_{i=1}^{N} \Bigg[(\alpha_i + \frac{1}{2\delta_i}) \Bigg\| \sum_{j \in \mathcal{N}_i} c_i (z_i(k) - z_j(k)) \Bigg\|^2$$

$$+ (\beta + \frac{\delta_i}{2}) \frac{1}{4N} \sum_{j = \mathcal{N}_i} \Bigg\| z_i(k) - z_i(k) \Bigg\|^2 \Bigg]$$

$$\leq \sum_{i=1}^{N} \sum_{j \in \mathcal{N}_i} \Bigg[(\alpha_i + \frac{1}{2\delta_i}) c_i^2 d_i$$

$$+ (\beta + \frac{\delta_i}{2}) \frac{1}{4N} \Bigg] \Bigg\| z_i(k) - z_i(k) \Bigg\|^2. \tag{4.19}$$

Thus, choosing the coupling strength c_i such that

$$(\alpha_i + \frac{1}{2\delta_i}) c_i^2 d_i + (\beta + \frac{\delta_i}{2}) \frac{1}{4N} < 0, \tag{4.20}$$

$\|z_i(k) - z_i(k)\|$ converges to zero for all $i, j = 1, 2, ..., N$, which implies that the nominal multiagent system achieves the output synchronization. In the light of the same philosophy, it can be concluded that the multiagent system (4.8) is robustly admissible and reaches consensus for all norm-bounded uncertainties. By considering the uncertainty of the multiagent system, we can establish that

$$\Gamma_i + U_1^{iT} F_i(k) U_2^i + U_2^{iT} F_i(k) U_1^i \leq 0.$$

Then, according to Lemma 4.2 and Fact 4.1, applying the Schur complements completes the proof. □

Remark 4.2. The coupling strength c_i can be designed for each agent independently of the other agents. Moreover, the protocol can be fully implemented in a distributed manner since the coupling strength only depends on its dissipativity parameters and the number of agents. Since β is negative, inequality (4.11) admits a solution of c_i if the linear matrix inequalities (LMIs) in Theorem 4.1 are feasible. The agents reach consensus robustly and every agent is dissipative with $(\beta, 1, \alpha_i)$-index.

4.1.4 Event-triggered protocol

In the sequel, we investigate event-triggered protocols under the limited communication constraint. The following theorem summarizes the main result.

Theorem 4.2. *Given positive integers $\bar{\tau}$ and $\underline{\tau}$ and positive scalars c_i, if there exist positive scalars ϵ_i and matrices $\mathcal{P}_i > 0$, $\mathcal{Q}_1^i > 0$, $\mathcal{Q}_2^i > 0$, $\mathcal{Q}_3^i > 0$, $\mathcal{Z}_1^i > 0$, $\mathcal{Z}_2^i > 0$, $\mathcal{Z}_3^i > 0$, $\mathcal{R}_i$, $L_j^i > 0$, $i = 1, ..., N$, $j = 1, 2, ..., 6$, then the multiagent* *system (4.8) reaches consensus via the event-triggered protocol (4.5) with the triggering condition (4.6) if the inequalities (4.9), (4.10), and*

$$(\alpha_i + \frac{1}{2\delta_i} + \frac{1}{2r} + \gamma_i \frac{1+s+rs/2}{sc_i^2})c_i^2 d_i + (\beta + s + \frac{\delta_i}{2})\frac{1}{4N} < 0 \tag{4.21}$$

hold for positive scalars γ_i, δ_i, s, and r.

Proof. Consider the same Lyapunov functional in Theorem 4.1. Using the LMIs (4.9), (4.18), and (4.10) and the protocol (4.5), we get

$$\begin{aligned}
\Delta V(k) &= \sum_{i=1}^{N} \Delta V^i(k) \\
&\leq \sum_{i=1}^{N} \left[z_i^T(k)\tilde{u}_i(k) + \alpha_i \tilde{u}_i^T(k)\tilde{u}_i(k) + \beta z_i^T(k) z_i(k) \right] \\
&= \sum_{i=1}^{N} \Big[\left[z_i(k_k^i) - e_i(k)\right]^T \tilde{u}_i(k) + \alpha_i \tilde{u}_i^T(k)\tilde{u}_i(k) \\
&\quad + \beta\left[z_i(k_k^i) - e_i(k)\right]^T \left[z_i(k_k^i) - e_i(k)\right] \Big].
\end{aligned} \tag{4.22}$$

Using Fact 4.1, the following inequalities hold:

$$e_i^T(k)u_i(k) \leq \frac{r}{2} e_i^T(k)e_i(k) + \frac{1}{2r} u_i^T(k)u_i(k), \tag{4.23}$$

$$2z_i^T(k_l^i)e_i(k) \leq s z_i^T(k_l^i) z_i(k_l^i) + \frac{1}{s} e_i^T(k)e_i(k) \tag{4.24}$$

for any scalars $s, r > 0$. With the Cauchy–Schwartz inequality, we obtain

$$\left\| \sum_{j \in \mathcal{N}_i} c_i(z_i(k) - z_j(k)) \right\|^2 \leq d_i c_i^2 \sum_{j \in \mathcal{N}_i} \left\| (z_i(k) - z_j(k)) \right\|^2 \tag{4.25}$$

and

$$\left\| \sum_{j \in \mathcal{N}_i} (z_i(k) - z_j(k)) \right\|^2 \leq d_i \sum_{j \in \mathcal{N}_i} \left\| (z_i(k) - z_j(k)) \right\|^2. \tag{4.26}$$

From (4.22)–(4.26), we have

$$\begin{aligned}
\Delta V(k) \leq \sum_{i=1}^{N} \Big[& z_i^T(k_k^i)\tilde{u}_i(k) + (\alpha_i + \frac{1}{2r})\tilde{u}_i^T(k)\tilde{u}_i(k) \\
& + (\beta + s) z_i^T(k_k^i) z_i(k_k^i) + \frac{1+s+rs/2}{s} e_i^T(k)e_i(k) \Big]
\end{aligned}$$

$$\leq \sum_{i=1}^{N} \Bigg[(\alpha_i + \frac{1}{2\delta_i} + \frac{1}{2r}) \Bigg\| \sum_{j\in\mathcal{N}_i} c_i(z_i(k) - z_j(k)) \Bigg\|^2$$

$$+ (\beta + s + \frac{\delta_i}{2})\frac{1}{2N} \sum_{j=\mathcal{N}_i} \frac{\|z_i(k) - z_i(k)\|^2}{2} \Bigg]$$

$$+ \gamma_i \frac{1+s+rs/2}{s} \Bigg\| \sum_{j\in\mathcal{N}_i} (z_i(k_l^i) - z_j(k_l^j)) \Bigg\|$$

$$\leq \sum_{i=1}^{N} \sum_{j\in\mathcal{N}_i} \Bigg[[(\alpha_i + \frac{1}{2\delta_i} + \frac{1}{2r} + \gamma_i \frac{1+s+rs/2}{sc_i^2})c_i^2 d_i$$

$$+ (\beta + s + \frac{\delta_i}{2})\frac{1}{4N}]\|z_i(k) - z_j(k)\|^2 \Bigg]. \tag{4.27}$$

Thus, choosing the coupling strength c_i such that (4.21) is guaranteed forces $\Delta V(k)$ to be negative whenever $\|z_i(k) - z_j(k)\|$ is positive. Then $\|z_i(k) - z_j(k)\|$ converges to zero for all $i, j = 1, 2, ..., N$, which implies that the multiagent system achieves the output synchronization. □

Proposition 4.1. *Given full column matrices $\Omega_i \in \Re^{n\times n-r_i}$ satisfying $E_i^T\Omega_i = 0$ for all $i = 1, 2, ..., N$, if the conditions of Theorem 4.1 are satisfied with Γ_{15}^i being replaced by*

$$\Gamma_{15}^i = S_i\Omega_i^t + E_i^t\mathcal{P}_i - L_1^{it} + (A_i - E_i)^t L_5^{it} - L_2^i, \tag{4.28}$$

then the pair (E_i, A_i) is regular and causal for all $i = 1, 2, \ldots, N$, and the consensus is guaranteed.

Proof. First of all, it is clear that $2x^T(k)S_i\Omega_i^T E_i\eta_i(k) = 0$, which can be added to the difference of the Lyapunov function in equation (4.19) such that Γ_{15}^i in equation (4.28) is modified without affecting the results of Theorem 4.1. Since $rank(E_i) = r_i < n$, there always exist nonsingular matrices H_i and $F_i \in \Re^{n\times n}$ such that

$$\bar{E}_i = H_i E_i F_i = \begin{bmatrix} I_{r_i} & 0 \\ 0 & 0 \end{bmatrix}. \tag{4.29}$$

Then for any nonsingular matrix Φ_i, we have $\Omega_i = H_i \begin{bmatrix} 0 \\ \Phi_i \end{bmatrix}$, which satisfies $E_i^T\Omega_i = 0$. We define the following partitioned matrices:

$$\mathbb{A}_i = H_i A_i F_i = \begin{bmatrix} \mathbb{A}_{11}^i & \mathbb{A}_{12}^i \\ \mathbb{A}_{21}^i & \mathbb{A}_{22}^i \end{bmatrix}, \tag{4.30}$$

$$\mathbb{A}_{di} = H_i A_{di} F_i = \begin{bmatrix} \mathbb{A}^i_{d11} & \mathbb{A}^i_{d12} \\ \mathbb{A}^i_{d21} & \mathbb{A}^i_{d22} \end{bmatrix}, \tag{4.31}$$

$$\mathbb{S}_i = F_i^T S_i = \begin{bmatrix} \mathbb{S}^i_{11} \\ \mathbb{S}^i_{21} \end{bmatrix}. \tag{4.32}$$

It follows from inequality (4.9) with the modified Γ^i_{15} that

$$\Gamma^i = \begin{bmatrix} \Gamma^i_{11} & \Gamma^i_{15} \\ \bullet & \Gamma^i_{55} \end{bmatrix} < 0,$$

which implies the following inequality:

$$\Psi^i = \begin{bmatrix} \Psi^i_{11} & \Psi^i_{15} \\ \bullet & \Psi^i_{55} \end{bmatrix} < 0, \tag{4.33}$$

where

$$\Psi^i_{11} = L^i_1(\bar{A}_i - E_i) + (\bar{A}_i - E_i)^T L^{iT}_1 - E_i^T \mathcal{Z}^i_1 E_i - E_i^T \mathcal{Z}^i_2 E_i,$$
$$\Psi^i_{12} = S_i \Omega_i^T + E_i^T \mathcal{P}_i - L^{iT}_1 + (\bar{A}_i - E)^T L^{iT}_5 - L^i_2,$$
$$\Psi^i_{22} = -L^i_5 - L^{iT}_5.$$

Multiplying both sides of (4.33) by $\begin{bmatrix} I & A_i^t \end{bmatrix}$ and its transpose, respectively, yields

$$E_i^T(\mathcal{P}_i - L^{iT}_3 - L^{iT}_1)A_i + A_i^T(\mathcal{P}_i^T - L^i_3 - L^i_1)E_i^T - L^{iT}_1 E_i$$
$$- E_i^t L^i_1 + S_i \Omega_i^t A + A_i^T \Omega_i S_i^t - E_i^t \mathcal{Z}_1 E_i - E_i^t \mathcal{Z}_1 E_i < 0. \tag{4.34}$$

Multiplying both sides of (4.34) by F_i^t and F_i, respectively, and then using equations (4.29), (4.30), and (4.32), we obtain

$$\begin{bmatrix} ? & ? \\ \bullet & S^i_{21} \Phi_i^t \mathbb{A}^i_{21} + \mathbb{A}^{it}_{21} \Phi_i S^{it}_{21} \end{bmatrix} < 0. \tag{4.35}$$

Let v be a nonzero vector such that $\mathbb{A}^i_{21} v = 0$, i.e., $\mathbb{A}^i_{21}$ is singular. Then, multiplying both sides of (4.35) by $\begin{bmatrix} 0 & v^T \end{bmatrix}$ and its transpose yields a zero value, which is a contradiction. This implies that $\mathbb{A}^i_{21}$ is nonsingular and the pair $(E_i,\ A_i)$ is regular and causal for all $i = 1, 2, ..., N$. This completes the proof. □

Theorem 4.3. *Given positive integers $\bar{\tau}$ and $\underline{\tau}$ and positive scalars c_i, if there exist positive scalars ϵ_i and matrices $\mathcal{P}_i > 0$, $\mathcal{Q}_1^i > 0$, $\mathcal{Q}_2^i > 0$, $\mathcal{Q}_3^i > 0$, $\mathcal{Z}_1^i > 0$, $\mathcal{Z}_2^i > 0$, $\mathcal{Z}_3^i > 0$, $\mathcal{R}_i$, $L_j^i > 0$, $i = 1, ..., N$, $j = 1, 2, ..., 6$, then the multiagent system (4.8) reaches consensus via the event-triggered protocol (4.5) with the triggering condition (4.6) if the inequalities*

$$\begin{bmatrix} \Gamma^i & U_1^i & \epsilon_i U_2^i \\ \bullet & -\epsilon_i I & 0 \\ \bullet & \bullet & -\epsilon_i I \end{bmatrix} < 0, \tag{4.36}$$

$$\begin{bmatrix} \mathcal{Z}_3^i & \mathcal{R}_i \\ \mathcal{R}_i & \mathcal{Z}_3^i \end{bmatrix} \geq 0, \tag{4.37}$$

and

$$(\alpha_i + \frac{1}{2\delta_i} + \frac{1}{2r} + \gamma_i \frac{1 + s + rs/2}{sc_i^2})c_i^2 d_i + (\beta + s + \frac{\delta_i}{2})\frac{1}{4N} < 0 \tag{4.38}$$

hold for positive scalars γ_i, δ_i, s, and r, where

$$\Gamma^i = \begin{bmatrix} \Gamma_{11}^i & \Gamma_{12}^i & \Gamma_{13}^i & \Gamma_{14} & \Gamma_{15}^i & \Gamma_{16}^i \\ \bullet & \Gamma_{22}^i & \Gamma_{23}^i & E_i^T R_i E_i & -L_2^i & L_2^i B_i \\ \bullet & \bullet & \Gamma_{33}^i & \Gamma_{34}^i & \Gamma_{35}^i & \Gamma_{36}^i \\ \bullet & \bullet & \bullet & \Gamma_{44}^i & -L_4^i & L_4^i B_i \\ \bullet & \bullet & \bullet & \bullet & \Gamma_{55}^i & \Gamma_{56}^i \\ \bullet & \bullet & \bullet & \bullet & \bullet & \Gamma_{66}^i \end{bmatrix}, \tag{4.39}$$

$$\begin{aligned}
U_1^i &= \begin{bmatrix} M_i^1 & M_i^2 & \dots & M_i^6 \end{bmatrix}^t, \\
U_2^i &= \begin{bmatrix} N_i & 0 & N_{di} & 0 & 0 & 0 \end{bmatrix}, \\
\Gamma_{11}^i &= A_i^1 + A_i^{1t} + (\tau^\star + 1)\mathcal{Q}_3^i + \mathcal{Q}_1^i + \mathcal{Q}_2^i - E_i^t \mathcal{Z}_1^i E_i \\
&\quad - E_i^t \mathcal{Z}_2^i E_i - \beta \tilde{C}_i^t \tilde{C}_i, \\
\Gamma_{12}^i &= E_i^t \mathcal{Z}_2^i E + A_i^2, \\
\Gamma_{22}^i &= \mathcal{Q}_1^i - E_i^t \mathcal{Z}_2 E_i - E_i^t \mathcal{Z}_3 E_i, \\
\Gamma_{13}^i &= A_{di}^1 + A_i^3 - \beta C_i^t C_{di}, \\
\Gamma_{23}^i &= -E_i^t R_i^t E_i + E_i^t \mathcal{Z}_3^{it} E_i + A_{di}^2, \\
\Gamma_{33}^i &= -\mathcal{Q}_3^i + A_{di}^3 + A_{di}^{3t} - 2E_i^t \mathcal{Z}_3^i E_i + E_i^t R_i E_i \\
&\quad + E_i^T R_i^T E_i - \beta C_{di}^t C_{di},
\end{aligned}$$

$$
\begin{aligned}
&\Gamma_{14}^i = E_i^t \mathcal{Z}_1^i E_i + A_i^{4T}, \\
&\Gamma_{34}^i = -E_i^T R_i E_i + E_i^t \mathcal{Z}_3^i E_i + A_{di}^{4t}, \\
&\Gamma_{44}^i = -\mathcal{Q}_2^i - E_i^t \mathcal{Z}_1^i E_i - E_i^t \mathcal{Z}_3^i E_i, \\
&\Gamma_{35}^i = L_3^i + A_{di}^{5t}, \\
&\Gamma_{15}^i = E_i^T \mathcal{P}_i - L_1^{it} + A_i^{5t} - L_2^i, \\
&\Gamma_{55}^i = \bar{\tau}^2 \mathcal{Z}_1^i + \underline{\tau}^2 \mathcal{Z}_2^i + \tau^\star \mathcal{Z}_3^i - L_5^i - L_5^{it} + \mathcal{P}_i, \\
&\Gamma_{16}^i = -\tilde{C}_i^t + B_i^1 + A_i^{6t}, \\
&\Gamma_{36}^i = -C_{di}^t + A_{di}^{6t} + B_i^3, \\
&\Gamma_{56}^i = B_i^5 + L_6^{it}, \ \Gamma_{66}^i = -\alpha_i I + B_i^5 + B_i^{5t}, \\
&A_i^j = \begin{bmatrix} L_{j1}^i (A_i - E_i) + \lambda_j \mathbb{I} Y_i & L_{j1}^i B_i - \lambda_j \mathbb{I} F_i \\ L_{j2}^i (A_i - E_i) + \lambda_j Y_i & L_{j1}^i B_i - \lambda_j F_i \end{bmatrix}, \\
&A_{di}^j = \begin{bmatrix} L_{j1}^i A_{di} & 0 \\ L_{j2}^i A_{di} & 0 \end{bmatrix}, \\
&\mathbb{I} = \begin{bmatrix} I_m \\ \mathbf{0}_{(n-m)\times m} \end{bmatrix}, \ B_i^j = \begin{bmatrix} L_{j1}^i B_i \\ L_{j2}^i B_i \end{bmatrix}, \ M_i^j = \begin{bmatrix} L_{j1}^i M_i \\ L_{j2}^i M_i \end{bmatrix},
\end{aligned}
$$

$i = 1, 2, \ldots, N$, $j = 1, 2, \ldots, 6$, *and* $\tau^\star = \bar{\tau} - \underline{\tau}$. *Furthermore,* $K_i = F_i^{-1} Y_i$.

Proof. If we view the agents' input $u_i(k)$ as a state component, the multiagent dynamics (4.1) can be represented as follows:

$$
\begin{aligned}
\tilde{E}_i \xi_i(k+1) &= (\tilde{A}_i + \Delta\tilde{A}_i)\xi_i(k) + (\tilde{A}_{di} + \Delta\tilde{A}_{di})\xi_i(k - \tau(k)) + \tilde{B}_i \tilde{u}_i(k), \\
\tilde{z}_i(k) &= \tilde{C}_i \xi(k) + \tilde{C}_{di} \xi_i(k - \tau(k)), \\
\xi_i(k) &= \phi_{i0}(k), \quad k \in [-\bar{\tau}, 0],
\end{aligned}
\tag{4.40}
$$

where $\xi_i(t) = \begin{bmatrix} x_i^T(t) & u_i^T(t) \end{bmatrix}^T$, $\Delta\tilde{A}_i = \tilde{M}_i F(k) \tilde{N}_i$, $\Delta\tilde{A}_{id} = \tilde{M}_i F(k) \tilde{N}_{id}$, $\tilde{C}_i = \begin{bmatrix} C_i & 0 \end{bmatrix}$, $\tilde{C}_i = \begin{bmatrix} C_{di} & 0 \end{bmatrix}$, $\tilde{M}_i = \begin{bmatrix} M_i^T & 0 \end{bmatrix}^T$, $\tilde{N}_i = \begin{bmatrix} N_i & 0 \end{bmatrix}$,

$$
\begin{aligned}
\tilde{E}_i &= \begin{bmatrix} E_i & 0 \\ 0 & 0 \end{bmatrix}, \quad \tilde{A}_i = \begin{bmatrix} A_i & B_i \\ K_i & -I \end{bmatrix}, \\
\tilde{A}_{di} &= \begin{bmatrix} A_{di} & 0 \\ 0 & 0 \end{bmatrix}, \quad \tilde{B}_i = \begin{bmatrix} 0 \\ B_i \end{bmatrix}.
\end{aligned}
$$

We apply Theorem 4.2 to system (4.40) with the following particular configuration of real matrices L_j^i, $(j = 1, 2, 3, 4, 5, 6)$, $i = 1, 2, ..., N$:

$$L_j^i = \begin{bmatrix} L_{j1}^i & \lambda_j \mathbb{I} F \\ L_{j2}^i & \lambda_j F \end{bmatrix}, \quad \mathbb{I} = \begin{bmatrix} I_m \\ 0_{(n-m)\times m} \end{bmatrix}.$$

Condition (4.9) holds by replacing $Y_i = F_i K_i$. It is clear from the inequalities in Theorem 4.2 that a feasible solution fulfills the inequality $\Gamma_{55}^i < 0$ for all $i = 1, \ldots, N$. This implies that L_5^i are nonsingular matrices for all $i = 1, \ldots, N$ and thus F_i is invertible. This completes the proof. □

4.1.5 Simulation example 4.1

In the sequel, a simulation example of a heterogeneous multiagent system is given to show the effectiveness of the considered time-triggered and event-triggered methodologies. The connected graph of all agents is illustrated in Fig. 4.2. We assume that only the edge output information is available and the state information cannot be obtained. We consider the descriptor multiagent system (4.1) which is governed by the following matrices:

$$A_1 = \begin{bmatrix} 0.89 & 1 \\ 0 & 0.5 \end{bmatrix}, \quad A_2 = \begin{bmatrix} 0.79 & 0.5 \\ -0.15 & -0.25 \end{bmatrix},$$

$$A_3 = \begin{bmatrix} 1.39 & 0.5 \\ 0.75 & -0.25 \end{bmatrix}, \quad A_4 = \begin{bmatrix} -0.36 & 1.35 \\ -1.875 & 1.025 \end{bmatrix},$$

$$A_{di} = \begin{bmatrix} 0.1 & 0 \\ 0.1 & 0 \end{bmatrix}, \quad E_i = \begin{bmatrix} 1 & 0 \\ 0 & 0 \end{bmatrix},$$

$$B_i = \begin{bmatrix} 1 \\ 1.5 \end{bmatrix}, \quad C_i = \begin{bmatrix} 1 & 1.6 \end{bmatrix}, \quad C_{di} = \begin{bmatrix} 0.1 & 0.1 \end{bmatrix},$$

$$M_i = \begin{bmatrix} 1 & 0 \\ 1 & 0 \end{bmatrix}, \quad N_i = \begin{bmatrix} 1 & 1 \\ 0 & 1 \end{bmatrix}, \quad N_{di} = \begin{bmatrix} 0 & 1 \\ 2 & 1 \end{bmatrix}.$$

Let the initial state vectors $x_{01} = \begin{bmatrix} -3 & 1 \end{bmatrix}^T$, $x_{02} = \begin{bmatrix} -1.9 & -1 \end{bmatrix}^T$, $x_{03} = \begin{bmatrix} 1.5 & 1 \end{bmatrix}^T$, $x_{04} = \begin{bmatrix} 1 & 1 \end{bmatrix}^T$. The number of agents $N = 4$ and the degrees of the agents are given as $d_1 = 1$, $d_2 = d_3 = d_4 = 2$. First, we simulate the synchronization protocol under the time-triggered methodology. The coupling gains are set to $c_1 = 0.417$ and $c_2 = c_3 = c_4 = 0.59$, and the triggering parameters are given by $\gamma_i = 1$. Using the LMI conditions (4.9), (4.10), and (4.11), we obtain $\alpha = 2.0355$ and $\beta = -1.3838$. Using the same LMIs, one can

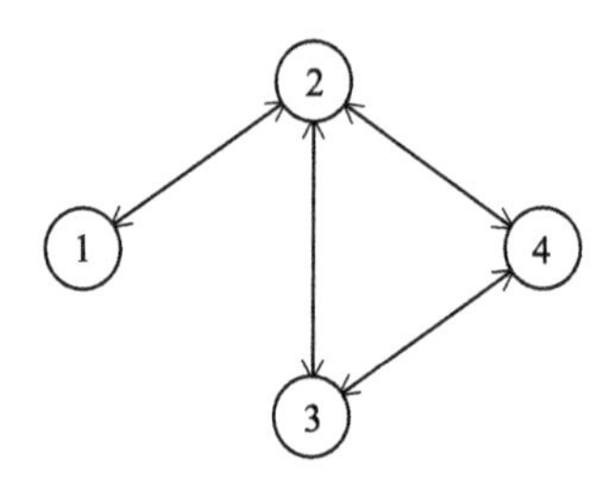

FIGURE 4.2 Connected topology of the multiagent system.

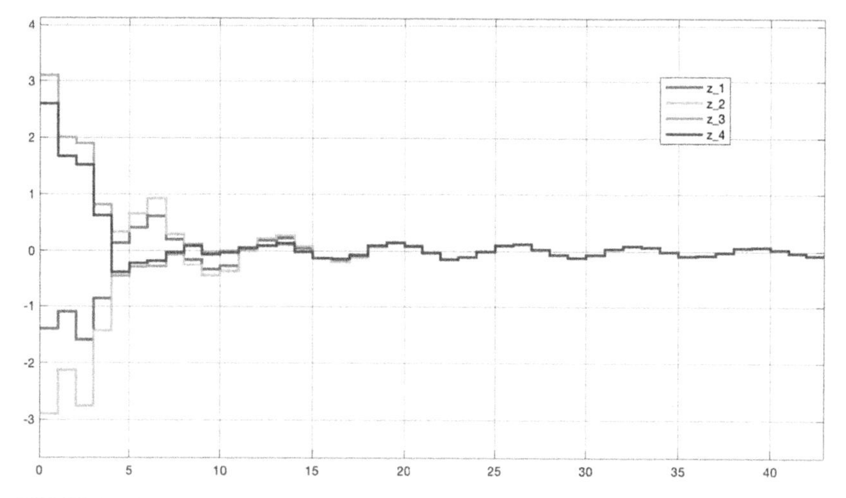

FIGURE 4.3 Synchronized agents' outputs by Theorem 4.1.

obtain the following gains:

$$K_1 = \left[\begin{array}{cc} 0 & 0 \end{array} \right], \qquad K_2 = \left[\begin{array}{cc} 0.1 & 0.5 \end{array} \right],$$
$$K_3 = \left[\begin{array}{cc} -0.5 & 0.5 \end{array} \right], \; K_4 = \left[\begin{array}{cc} 1.25 & -0.35 \end{array} \right].$$

Fig. 4.3 shows the output trajectories of the multiagent system under a time-triggered strategy. One can clearly decide that all agent's outputs are synchronized, which clearly demonstrates the effectiveness of the considered methodology.

Moreover, using Theorem 4.3, we can obtain the same controller gains to use them for the event triggering policy since they satisfy inequalities (4.36) and (4.37). The same parameters $\alpha = 2.0355$, $\beta = -1.3838$ and inequality (4.38) can be used to obtain the event triggering parameters and coupling gains as follows: $\gamma_i = 0.1$, $c_1 = 0.024$, and $c_2 = c_3 = c_4 = 0.04$. Fig. 4.4 depicts the output trajectories of the multiagent system under an event-triggered strategy. One can see that all agents' outputs are synchronized under the proposed event-triggered

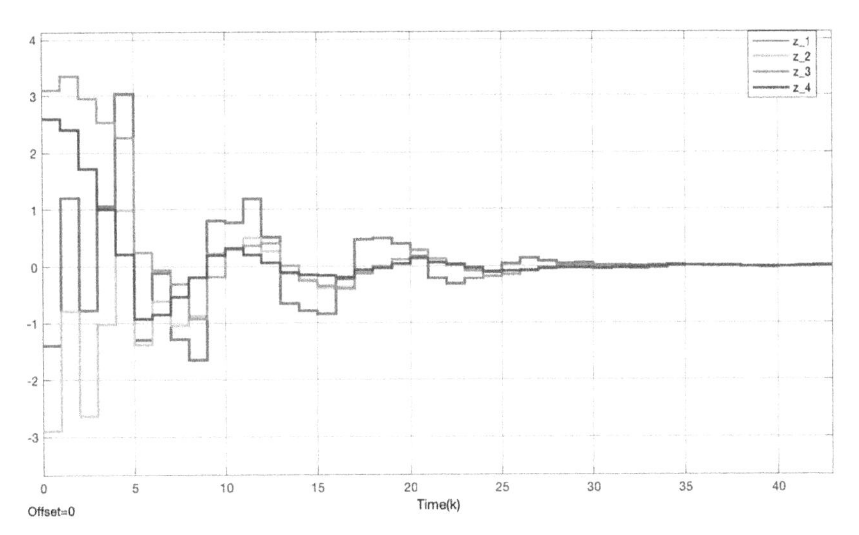

FIGURE 4.4 Synchronized agents' outputs by Theorems 4.1–4.3.

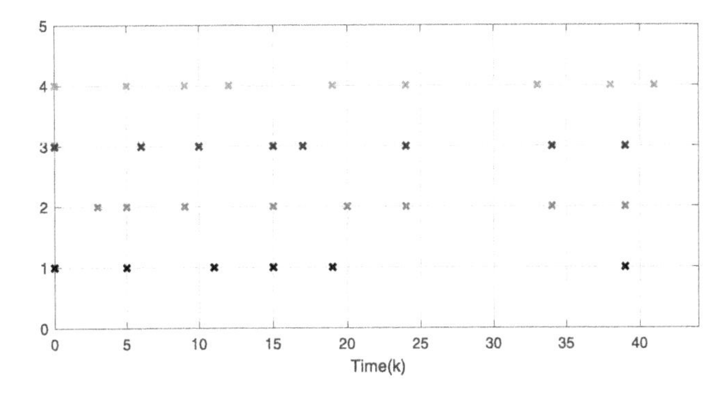

FIGURE 4.5 Triggering instants of each agent by Theorem 4.2.

policy. Fig. 4.5 shows the transmission instants of every agent on the interval $[0, 40\text{ s})$, where the average release time for every agent is 8, 6, 6, and 5 samples, respectively. It can be observed from the simulation results in Figs. 4.3 and 4.4 that although Theorem 4.3 does not require communication with neighboring agents periodically, the synchronization is still achieved as in the case with Theorem 4.1.

4.2 Passivity analysis of time delay systems

In recent years, neural networks have received many scholars' extensive attention due to their wide application in various areas such as data mining, signal processing, telecommunications, and combinatorial optimization problems [10], [11], [12], [13], [14], [15]. Meanwhile, time delays often occur in many engineering and physical systems. The existence of time delay is a main source of poor performance and instability of neural networks. Therefore, more and more effort has been expended on stability analysis of time delay systems, and many excellent results have been reported in the literature. In addition, when we investigate the dynamical characteristic of neural networks in a digital way, the discrete-time neural networks become more important than their continuous-time counterparts in engineering applications. Therefore, it is important to study the stability of discrete-time delay systems.

An important work of stability analysis for the discrete-time delay systems is to find the maximal admissible delay upper bound (MADUB) such that the considered systems remain stable. The Lyapunov–Krasovskii method is the most popular method used in the stability analysis of time delay systems, because its analytical procedure can fully utilize the information on time delay systems.

4.2.1 Introduction

In order to find the MADUB, there are two effective ways. One way is to choose a suitable Lyapunov–Krasovskii functional (LKF), and the other is to obtain the tighter bounds for the forward difference of LKF. Methods for constructing a delicate LKF include delay partitioning approaches [16], triple summation terms [17], [18], and augmented vectors [19]. Techniques for obtaining tighter bounds of the forward differences include the summation inequality approach, the auxiliary function-based inequality method [20], [21], and the free-weighting matrix method [22]. The discrete Jensen inequality is a popular summation inequality employed to estimate the summation term $\sum_{s=k-b}^{k-a-1} \Delta x^T(s)R\Delta x(s)$, which often occurs in the forward difference of the LKF. A Wirtinger-based discrete inequality, which covers the Jensen inequality as a special case, was proposed in [23], [24]. A new free matrix-based summation inequality was also developed in [25], which is less conservative than the Wirtinger-based inequality. Recently, an improved integral inequality was proposed in [26] by introducing some free matrices to the Wirtinger inequality, which further extended the free matrix-based integral inequality in [10] and the Wirtinger-based integral inequality in [27]. Based on this integral inequality and modified augmented LKF, some new delay-dependent stability criteria are presented. These stability criteria are less conservative than the existing results. Inspired by the idea of [26], we aim to establish a new general free matrix-based summation inequality to extend the free matrix-based summation inequality in [25]. Furthermore, the new summation inequality is applied to the stability analysis for the discrete-time systems with time-varying delay and a less conservative stability criterion will be given.

The problem of passivity for the discrete-time systems with time-varying delay has also attracted increasing attention, due to its theoretical importance [28], [29], [30], [31], [32], [33], [34], [35]. The passivity for discrete-time switched systems with various activation functions and mixed time delays is investigated in [31], and a sufficient condition for exponential stability of the considered neural networks with passivity is obtained. In [32], the problem of passivity for the discrete-time recurrent neural networks with parameter uncertainties was studied. To the best of our knowledge, there are few results on the passivity of discrete-time linear systems.

Motivated by the above discussion, in this chapter, we will consider the problem of stability and passivity for a class of discrete-time linear systems. The main contributions of this chapter are summarized as follows:

(1) The adaptation of integral inequalities into the discrete-time framework does not seem to be an easy task. Inspired by the idea of [26], a new general free matrix-based summation inequality is developed, which can be regarded as a discrete-time version of the integral inequality in [26]. By employing the novel summation inequality, a less conservative stability condition is derived in this paper.

(2) Different from some existing results in [31], [32], [33], a new LKF term $V_4(k)$ is introduced. Based on the new summation inequality, the advance difference of the LKF term $V_4(k)$ is estimated. A less conservative stability criterion and a passivity criterion for discrete-time linear systems are obtained.

4.2.2 System formulation

Consider the following discrete-time linear system with time-varying delay:

$$\begin{aligned} x(k+1) &= Ax(k) + Bx(k-\tau(k)), \\ x(k) &= \phi(k), \quad k = -\tau_2, -\tau_2+1, \ldots, 0, \end{aligned} \tag{4.41}$$

where $x(k) \in \mathbb{R}^n$ is the system state vector, $\phi(k)$ is the initial condition, A and B are constant matrices, $\tau(k)$ is the time-varying delay, which is a positive integer, and $1 \le \tau_1 \le \tau(k) \le \tau_2$.

When an external disturbance appears in system (4.41), the discrete-time system can be written as follows:

$$\begin{aligned} x(k+1) &= Ax(k) + Bx(k-\tau(k)) + C\omega(k), \\ y(k) &= Dx(k) + E\omega(k), \\ x(k) &= \phi(k), \quad k = -\tau_2, -\tau_2+1, \ldots, 0, \end{aligned} \tag{4.42}$$

where $\omega(k)$ is the disturbance input and $y(k)$ is the output of the system.

The following definition and lemmas are useful in the proof of the design results.

Definition 4.4. [31] System (4.41) is said to be passive. If $\forall \alpha > 0$, the following inequality holds:

$$2\sum_{i=0}^{k} y^T(i)\omega(i) \ge -\alpha \sum_{i=0}^{k} \omega^T(i)\omega(i), \quad \forall k \ge 0. \tag{4.43}$$

Lemma 4.3 (The discrete Wirtinger-based inequality). *[23] For a given matrix $R > 0$, three given nonnegative integers a, b, k satisfying $a < b \leq k$, and a vector function $x(\cdot) \in \mathbb{R}^n$, the following inequality holds:*

$$\sum_{s=k-b}^{k-a-1} \Delta x^T(s) R \Delta x(s) \geq \frac{1}{b-a} \Omega_0^T R \Omega_0 + \frac{3}{b-a} \Omega_1^T R \Omega_1,$$

where $\Delta x(s) = x(s+1) - x(s)$ and $\Omega_0 = x(k-a) - x(k-b)$, $\Omega_1 = x(k-a) + x(k-b) - \frac{2}{b-a+1} \sum_{s=k-b}^{k-a} x(s)$.

The following initial result is established.

Lemma 4.4. *For a given matrix $R \in \Re^{n \times n} > 0$, three given nonnegative integers a, b, k satisfying $a < b \leq k$, and a vector function $x(\cdot) \in \Re^n$, taking matrices $\Theta_i \in \Re^{n \times m}$ $(i = 1, 2)$ and a vector $\zeta \in \Re^m$ such that*

$$x(k-a) - x(k-b) = \Theta_1 \zeta,$$

$$x(k-a) + x(k-b) - \frac{2}{b-a+1} \sum_{s=k-b}^{k-a} x(s) = \Theta_2 \zeta,$$

for any matrices $N_i \in \Re^{n \times m}$ $(i = 1, 2)$, the following inequality holds:

$$-\sum_{s=k-b}^{k-a-1} \Delta x^T(s) R \Delta x(s) \leq \zeta^T \left[\Theta^T N + N^T \Theta + (b-a) N^T \bar{R}^{-1} N \right] \zeta,$$

where $\Theta = \begin{bmatrix} \Theta_1 \\ \Theta_2 \end{bmatrix}$, $N = \begin{bmatrix} N_1 \\ N_2 \end{bmatrix}$, $\bar{R} = \begin{bmatrix} R & 0 \\ 0 & 3R \end{bmatrix}$.

Proof. From Lemma 4.3, we have

$$\begin{aligned} &(b-a) \sum_{s=k-b}^{k-a-1} \Delta x^t(s) R \Delta x(s) \\ &\geq \zeta^t \begin{bmatrix} \Theta_1 \\ \Theta_2 \end{bmatrix}^t \bar{R} \begin{bmatrix} \Theta_1 \\ \Theta_2 \end{bmatrix} \zeta \\ &= \zeta^t \Theta^t \bar{R} \Theta \zeta. \end{aligned}$$

For any constant matrix N, the following inequality holds:

$$\zeta^t [\bar{R}\Theta + (b-a)N]^t \bar{R}^{-1} [\bar{R}\Theta + (b-a)N] \zeta \geq 0.$$

Hence

$$\zeta^t \Theta^t \bar{R} \Theta \zeta \geq -(b-a) \zeta^t [\Theta^t N + N^t \Theta + (b-a) N^t \bar{R}^{-1} N] \zeta.$$

This completes the proof. □

Remark 4.3. The summation inequality in Lemma 4.4 covers the Wirtinger-based summation inequality in [23] as a special case. In fact, let $\zeta = [x^t(k-a), x^t(k-b), \frac{1}{b-a+1}\sum_{s=k-b}^{k-a} x^t(s)]^t$, $\Theta_1 = [I, -I, 0]$, $\Theta_2 = [I, I, -2I]$, $N_1 = -\frac{1}{b-a}R\Theta_1$, and $N_2 = -\frac{3}{b-a}R\Theta_2$. It is easy to calculate

$$\begin{aligned}
&\zeta^t[\Theta^t N + N^t\Theta + (b-a)N^t\bar{R}^{-1}N]\zeta \\
&= -\frac{1}{b-a}\zeta^t\Theta^t\bar{R}\Theta\zeta \\
&= -\frac{1}{b-a}\zeta^t[\Theta_1^t R\Theta_1 + 3\Theta_2^t R\Theta_2]\zeta \\
&= -\frac{1}{b-a}\Omega_0^t R\Omega_0 - \frac{3}{b-a}\Omega_1^t R\Omega_1,
\end{aligned}$$

where Ω_0, Ω_1 are defined as in Lemma 4.3. So the inequality in Lemma 4.4 implies the inequality in Lemma 4.3.

Remark 4.4. Let $\zeta = [x^t(k-a), x^t(k-b), \chi^t(k,a,b)]^t$, $\Theta_1 = [I, -I, 0]$, $\Theta_2 = [I, I, -I]$, $N_1 = -\frac{1}{b-a}R\Theta_1$, $N_2 = -\frac{3}{b-a}R\Theta_2$, where

$$\chi(k,a,b) = -\frac{1}{b-a}[2\sum_{s=k-b}^{k-a-1} x(s) + x(k-a) - x(k-b)].$$

By employing the similar calculation as in Remark 4.3, it is easy to prove that the discrete Wirtinger-based inequality in [24] is a special case of the inequality in Lemma 4.4.

Remark 4.5. The summation inequalities obtained in [23], [24], [25] are used to estimate the upper bound of

$$-\sum_{s=k-b}^{k-a-1} \Delta x^T(s)R\Delta x(s).$$

It is not easy however to use them to estimate the upper bound of $-\sum_{s=k-b}^{k-a-1} x^t(s)Rx(s)$.

However, Lemma 4.4 can be used to estimate the upper bound of

$$-\sum_{s=k-b}^{k-a-1} x^t(s)Rx(s)$$

by selecting the appropriate matrices N_i and Θ_i.

For instance, let $z(s) = \sum_{i=k-b}^{k+s-1} x(i)$. Then $\Delta z(s) = z(s+1) - z(s) = x(k+s)$. Let $\zeta = [\sum_{s=k-b}^{k-a} x^t(s), x^t(k-a), \frac{1}{b-a+1}\sum_{s=-b}^{-a}\sum_{u=k+s}^{k-a} x^t(u)]^t$,

$\Theta_1 = [I, -I, 0]$, $\Theta_2 = [-I, -I, 2I]$, $N_1 = -\frac{1}{b-a}R\Theta_1$, $N_2 = -\frac{3}{b-a}R\Theta_2$, $\Theta = [\Theta_1^t, \Theta_2^t]^t$. Then

$$z(-a) - z(-b) = \Theta_1\zeta,$$
$$z(-a) + z(-b) - \frac{2}{b-a+1}\sum_{s=-b}^{-a} z(s) = \Theta_2\zeta.$$

In fact,

$$\begin{aligned}
&z(-a) + z(-b) - \frac{2}{b-a+1}\sum_{s=-b}^{-a} z(s) \\
&= \sum_{i=k-b}^{k-a-1} x(i) + 0 - \frac{2}{b-a+1}\sum_{s=-b}^{-a}\sum_{i=k-b}^{k+s-1} x(i) \\
&= \sum_{i=k-b}^{k-a-1} x(i) - \frac{2}{b-a+1}[\sum_{s=-b}^{-a}\sum_{i=k-b}^{k-a} x(i) - \sum_{s=-b}^{-a}\sum_{i=k+s}^{k-a} x(i)] \\
&= \sum_{i=k-b}^{k-a-1} x(i) - \frac{2}{b-a+1}[(b-a+1)\sum_{i=k-b}^{k-a} x(i) - \sum_{s=-b}^{-a}\sum_{i=k+s}^{k-a} x(i)] \\
&= \sum_{i=k-b}^{k-a-1} x(i) - 2\sum_{i=k-b}^{k-a} x(i) + \frac{2}{b-a+1}\sum_{s=-b}^{-a}\sum_{i=k+s}^{k-a} x(i) \\
&= -x(k-a) - \sum_{i=k-b}^{k-a} x(i) + \frac{2}{b-a+1}\sum_{s=-b}^{-a}\sum_{i=k+s}^{k-a} x(i) \\
&= \Theta_2\zeta.
\end{aligned}$$

Using Lemma 4.4 gives

$$\begin{aligned}
&-\sum_{s=k-b}^{k-a-1} x^T(s)Rx(s) \\
&= -\sum_{s=-b}^{-a-1} x^T(k+s)Rx(k+s) \\
&= -\sum_{s=-b}^{-a-1} \Delta z^T(s)R\Delta z(s) \\
&\le \zeta^T[\Theta^T N + N^T\Theta + (b-a)N^T\bar{R}^{-1}N]\zeta \\
&= -\frac{1}{b-a}\zeta^T[\Theta_1^T R\Theta_1 + 3\Theta_2^T R\Theta_2]\zeta.
\end{aligned}$$

That is,

$$
\begin{aligned}
&-\sum_{s=k-b}^{k-a-1} x^T(s)Rx(s) \\
&\leq -\frac{1}{b-a}[(\sum_{i=k-b}^{k-a} x^T(i) - x^T(k-a))R \\
&\times (\sum_{i=k-b}^{k-a} x(i) - x(k-a)) \\
&+3(\sum_{i=k-b}^{k-a} x^T(i) + x^T(k-a) \\
&-\frac{2}{b-a+1}\sum_{s=-b}^{-a}\sum_{i=k+s}^{k-a} x^T(i))R \\
&\times (\sum_{i=k-b}^{k-a} x(i) + x(k-a) - \frac{2}{b-a+1}\sum_{s=-b}^{-a}\sum_{i=k+s}^{k-a} x(i))].
\end{aligned}
$$

Therefore, from Lemma 4.4, we can derive the results of [21].

4.2.3 Stability results

In this section, we will develop a new stability criterion for system (4.41). In order to simplify the representation of vectors and matrices, the following notations are used:

$$
\begin{aligned}
\Delta x(k) &= x(k+1) - x(k), \quad \tau_{12} = \tau_2 - \tau_1, \\
\xi(k) &= [x^T(k), x^T(k-\tau_1), x^T(k-\tau(k)), \\
&\quad x^T(k-\tau_2), \frac{1}{\tau_1+1}\sum_{s=k-\tau_1}^{k} x^T(s), \\
&\quad \frac{1}{\tau(k)-\tau_1+1}\sum_{s=k-\tau(k)}^{k-\tau_1} x^T(s), \\
&\quad \frac{1}{\tau_2-\tau(k)+1}\sum_{s=k-\tau_2}^{k-\tau(k)} x^T(s)]^T, \\
\eta(k) &= [x^T(k), \sum_{s=k-\tau_1}^{k-1} x^T(s), \sum_{s=k-\tau_2}^{k-\tau_1-1} x^T(s)]^T,
\end{aligned}
$$

$$\begin{aligned}
\Pi_1 &= [e_1 + e_s, (\tau_1 + 1)e_5 - e2, (\tau(k) - \tau_1 + 1)e_6 \\
&\quad + (\tau_2 - \tau(k) + 1)e_7 - e_3 - e_4], \\
\Pi_2 &= [e_1, (\tau_1 + 1)e_5 - e1, (\tau(k) - \tau_1 + 1)e_6 \\
&\quad + (\tau_2 - \tau(k) + 1)e_7 - e_3 - e_2], \\
\Xi_1 &= \frac{1}{2} Sym\{(\Pi_1 + \Pi_2)P(\Pi_1 - \Pi_2)^T\} + e_1 Q_1 e_1^T \\
&\quad - e_2 Q_1 e_2^T + e_2 Q_2 e_2^T - e_4 Q_2 e_4^T \\
&\quad + \tau_1 e_s R_1 e_s^T + \tau_{12} e_s R_2 e_s^T, \\
\Xi_2 &= Sym\{\sum_{i=1}^{3} \Upsilon_i N_i^T\}, \\
\Xi_3 &= \tau_1 N_1 \bar{R}_1^{-1} N_1^T + (\tau(k) - \tau_1) N_2 \bar{R}_2^{-1} N_2^T \\
&\quad + (\tau_2 - \tau(k)) N_3 \bar{R}_2^{-1} N_3^T, \\
\Upsilon_1 &= [e_1 - e_2, e_1 + e_2 - 2e_5], \\
\Upsilon_2 &= [e_2 - e_3, e_2 + e_3 - 2e_6], \\
\Upsilon_3 &= [e_3 - e_4, e_3 + e_4 - 2e_7], \\
\bar{R}_i &= \begin{bmatrix} R_i & 0 \\ 0 & 3R_i \end{bmatrix}, \quad i = 1, 2, \\
e_s &= (A - I)e_1 + Be_3, \\
e_i &= [0_{n \times (i-1)n}, I_{n \times n}, 0_{n \times (7-i)n}]^T, \quad i = 1, 2, \ldots, 7.
\end{aligned}$$

Theorem 4.4. *For given positive integers $\tau_1 < \tau_2$, system (4.41) is asymptotically stable if there exist matrices $P \in \Re^{3n \times 3n} > 0$, $Q_i \in \Re^{n \times n} > 0$, $R_i \Re^{n \times n} > 0$ $(i = 1, 2)$, and $N_i \Re^{7n \times 2n}$ $(i = 1, 2, 3)$ such that the following LMIs hold:*

$$\begin{bmatrix} (\Xi_1 + \Xi_2)_{[\tau(k)=\tau_1]} & \tau_1 N_1 & \tau_{12} N_3 \\ * & -\tau_1 \bar{R}_1 & 0 \\ * & * & -\tau_{12} \bar{R}_2 \end{bmatrix} < 0, \tag{4.44}$$

$$\begin{bmatrix} (\Xi_1 + \Xi_2)_{[\tau(k)=\tau_2]} & \tau_1 N_1 & \tau_{12} N_2 \\ * & -\tau_1 \bar{R}_1 & 0 \\ * & * & -\tau_{12} \bar{R}_2 \end{bmatrix} < 0. \tag{4.45}$$

Proof. Consider the following LKF candidate taken from [23], [25], [36]:

$$V(k) = \sum_{i=1}^{3} V_i(k), \tag{4.46}$$

where

$$
\begin{aligned}
V_1(k) &= \eta^T(k)P\eta(k), \\
V_2(k) &= \sum_{s=k-\tau_1}^{k-1} x^T(s)Q_1x(s) + \sum_{s=k-\tau_2}^{k-\tau_1-1} x^T(s)Q_2x(s), \\
V_3(k) &= \sum_{s=-\tau_1}^{-1}\sum_{u=k+s}^{k-1} \Delta x^T(u)R_1\Delta x(u) \\
&\quad + \sum_{s=-\tau_2}^{-1-\tau_1}\sum_{u=k+s}^{k-1} \Delta x^T(u)R_2\Delta x(u).
\end{aligned}
$$

Next, we will calculate the forward differences of $V_i(k)$, $i = 1, 2, 3$. We have

$$
\begin{aligned}
\Delta V_1(k) &= \begin{bmatrix} x(k+1) \\ \sum_{s=k-\tau_1}^{k} x(s) - x(k-\tau_1) \\ \sum_{s=k-\tau_2}^{k-\tau_1} x(s) - x(k-\tau_2) \end{bmatrix}^T P \\
&\quad \times \begin{bmatrix} x(k+1) \\ \sum_{s=k-\tau_1}^{k} x(s) - x(k-\tau_1) \\ \sum_{s=k-\tau_2}^{k-\tau_1} x(s) - x(k-\tau_2) \end{bmatrix} \\
&\quad - \begin{bmatrix} x(k) \\ \sum_{s=k-\tau_1}^{k} x(s) - x(k) \\ \sum_{s=k-\tau_2}^{k-\tau_1} x(s) - x(k-\tau_1) \end{bmatrix}^T P \\
&\quad \times \begin{bmatrix} x(k) \\ \sum_{s=k-\tau_1}^{k} x(s) - x(k) \\ \sum_{s=k-\tau_2}^{k-\tau_1} x(s) - x(k-\tau_1) \end{bmatrix}, \qquad (4.47)
\end{aligned}
$$

$$
\begin{aligned}
\Delta V_2(k) &= x^T(k)Q_1x(k) - x^T(k-\tau_1)Q_1x(k-\tau_1) \\
&\quad + x^T(k-\tau_1)Q_2x(k-\tau_1) \\
&\quad - x^T(k-\tau_2)Q_2x(k-\tau_2), \qquad (4.48)
\end{aligned}
$$

$$
\begin{aligned}
\Delta V_3(k) &= \tau_1\Delta x^T(k)R_1\Delta x(k) - \sum_{s=k-\tau_1}^{k-1} \Delta x^T(s)R_1\Delta x(s) \\
&\quad + \tau_{12}\Delta x^T(k)R_2\Delta x(k)
\end{aligned}
$$

$$
\begin{aligned}
& - \sum_{s=k-\tau(k)}^{k-1-\tau_1} \Delta x^T(s) R_2 \Delta x(s) \\
& - \sum_{s=k-\tau_2}^{k-1-\tau(k)} \Delta x^T(s) R_2 \Delta x(s).
\end{aligned} \tag{4.49}
$$

Utilizing Lemma 4.4 to estimate the second, fourth, and fifth summation terms in (4.49) yields

$$
- \sum_{s=k-\tau_1}^{k-1} \Delta x^t(s) R_1 \Delta x(s) \tag{4.50}
$$

$$
\leq \xi^t(k)[\Upsilon_1 N_1^t + N_1 \Upsilon_1^t + \tau_1 N_1 \bar{R}_1^{-1} N_1^t]\xi(k),
$$

$$
- \sum_{s=k-\tau(k)}^{k-1-\tau_1} \Delta x^t(s) R_2 \Delta x(s) \tag{4.51}
$$

$$
\leq \xi^t(k)[\Upsilon_2 N_2^t + N_2 \Upsilon_2^t + (\tau(k) - \tau_1) N_2 \bar{R}_2^{-1} N_2^t]\xi(k),
$$

and

$$
- \sum_{s=k-\tau_2}^{k-1-\tau(k)} \Delta x^t(s) R_2 \Delta x(s) \tag{4.52}
$$

$$
\leq \xi^T(k)[\Upsilon_3 N_3^t + N_3 \Upsilon_3^t + (\tau_2 - \tau(k)) N_3 \bar{R}_2^{-1} N_3^t]\xi(k).
$$

From (4.47)–(4.52), it follows that

$$
\Delta V(k) \leq \xi^T(k)(\Xi_1 + \Xi_2 + \Xi_3)\xi(k). \tag{4.53}
$$

Noting that $\Xi_1 + \Xi_2 + \Xi_3$ is linear with respect to $\tau(k)$ and $\tau_1 \leq \tau(k) \leq \tau_2$, the Schur complement implies that $\Xi_1 + \Xi_2 + \Xi_3 < 0$ if and only if (4.44) and (4.45) hold. This completes the proof. □

Remark 4.6. Since $\Xi_1 + \Xi_2 + \Xi_3$ is linear with respect to $\tau(k)$ and $\tau_1 \leq \tau(k) \leq \tau_2$, we have $\Xi_1 + \Xi_2 + \Xi_3 < 0$ if and only if $(\Xi_1 + \Xi_2 + \Xi_3)[\tau(k) = \tau_1] < 0$ and $(\Xi_1 + \Xi_2 + \Xi_3)[\tau(k) = \tau_2] < 0$; $(\Xi_1 + \Xi_2 + \Xi_3)[\tau(k) = \tau_1] < 0$ is equivalent to $\Xi_1 + \Xi_2 + \tau_1 N_1 \bar{R}_1^{-1} N_1^T + (\tau_2 - \tau_1) N_3 \bar{R}_2^{-1} N_3^T < 0$. By using the Schur complement, $\Xi_1 + \Xi_2 + \tau_1 N_1 \bar{R}_1^{-1} N_1^T + (\tau_2 - \tau_1) N_3 \bar{R}_2^{-1} N_3^T < 0$ is equivalent to the LMI (4.44). So $(\Xi_1 + \Xi_2 + \Xi_3)[\tau(k) = \tau_1] < 0$ if and only if the LMI (4.44) holds. Similarly, $(\Xi_1 + \Xi_2 + \Xi_3)[\tau(k) = \tau_2] < 0$ if and only if the LMI (4.45) holds. Therefore, $\Xi_1 + \Xi_2 + \Xi_3 < 0$ if and only if LMIs (4.44) and (4.45) hold.

Remark 4.7. Upon introducing the free matrices N_i, Θ_i $(i = 1, 2)$, the new summation inequality in Lemma 4.4 can provide more freedom in bounding the quadratic summation. This summation inequality is with less conservativeness. As mentioned in Remarks 4.3 and 4.4, this new summation inequality includes the discrete Wirtinger-based summation inequalities in [23], [24] as special cases. Theorem 4.4 is derived by this summation inequality. Therefore, Theorem 4.4 in this chapter has the potential to provide less conservatism than those stability criteria in [23], [24].

In order to reduce the stability criterion conservatism in Theorem 4.4, an LKF term $V_4(k) = \sum_{s=-\tau_2}^{-\tau_1-1} \sum_{u=k+s}^{k-1} x^T(u) R_3 x(u)$ will be introduced. The asymptotic stability of system (4.41) will be reconsidered. For the sake of simplicity, some notations are defined as

$$
\begin{aligned}
\check{\xi}(k) = [&x^T(k), x^T(k-\tau_1), x^T(k-\tau(k)), x^T(k-\tau_2), \\
&\frac{1}{\tau_1+1}\sum_{s=k-\tau_1}^{k} x^T(s), \frac{1}{\tau(k)-\tau_1+1}\sum_{s=k-\tau(k)}^{k-\tau_1} x^T(s), \\
&\frac{1}{\tau_2-\tau(k)+1}\sum_{s=k-\tau_2}^{k-\tau(k)} x^T(s), \frac{1}{\tau(k)-\tau_1+1}\sum_{s=-\tau(k)}^{-\tau_1}\sum_{u=k+s}^{k-\tau_1} x^T(u), \\
&\frac{1}{\tau_2-\tau(k)+1}\sum_{s=-\tau_2}^{-\tau(k)}\sum_{u=k+s}^{k-\tau(k)} x^T(u)]^T, \\
\check{\Pi}_1 = [&\check{e}_1 + \check{e}_s, (\tau_1+1)\check{e}_5 - \check{e}_2, (\tau(k)-\tau_1+1)\check{e}_6 \\
&+ (\tau_2-\tau(k)+1)\check{e}_7 - \check{e}_3 - \check{e}_4], \\
\check{\Pi}_2 = [&\check{e}_1, (\tau_1+1)\check{e}_5 - \check{e}_1, (\tau(k)-\tau_1+1)\check{e}_6 \\
&+ (\tau_2-\tau(k)+1)\check{e}_7 - \check{e}_3 - \check{e}_2], \\
\check{\Xi}_1 = &\frac{1}{2}Sym\{(\check{\Pi}_1+\check{\Pi}_2)P(\check{\Pi}_1-\check{\Pi}_2)^T\} + \check{e}_1 Q_1 \check{e}_1^T \\
&- \check{e}_2 Q_1 \check{e}_2^T + \check{e}_2 Q_2 \check{e}_2^T - \check{e}_4 Q_2 \check{e}_4^T \\
&+ \tau_1 \check{e}_s R_1 \check{e}_s^T + \tau_{12} \check{e}_s R_2 \check{e}_s^T + \tau_{12} \check{e}_1 R_3 \check{e}_1^T, \\
\check{\Xi}_2 = &Sym\{\sum_{i=1}^{5} \check{\Upsilon}_i N_i^T\}, \\
\check{\Xi}_3 = &\tau_1 N_1 \bar{R}_1^{-1} N_1^T + (\tau(k)-\tau_1) N_2 \bar{R}_2^{-1} N_2^T \\
&+ (\tau_2-\tau(k)) N_3 \bar{R}_2^{-1} N_3^T + (\tau(k)-\tau_1) N_4 \bar{R}_3^{-1} N_4^T \\
&+ (\tau_2-\tau(k)) N_5 \bar{R}_3^{-1} N_5^T, \\
\check{\Upsilon}_1 = [&\check{e}_1 - \check{e}_2, \check{e}_1 + \check{e}_2 - 2\check{e}_5], \\
\check{\Upsilon}_2 = [&\check{e}_2 - \check{e}_3, \check{e}_2 + \check{e}_3 - 2\check{e}_6], \\
\check{\Upsilon}_3 = [&\check{e}_3 - \check{e}_4, \check{e}_3 + \check{e}_4 - 2\check{e}_7],
\end{aligned}
$$

$$\check{\Upsilon}_4 = [(\tau(k) - \tau_1 + 1)\check{e}_6 - \check{e}_2, (\tau(k) - \tau_1 + 1)\check{e}_6 + \check{e}_2 - 2\check{e}_8],$$
$$\check{\Upsilon}_5 = [(\tau_2 - \tau(k) + 1)\check{e}_7 - \check{e}_3, (\tau_2 - \tau(k) + 1)\check{e}_7 + \check{e}_3 - 2\check{e}_9],$$
$$\bar{R}_i = \begin{bmatrix} R_i & 0 \\ 0 & 3R_i \end{bmatrix}, \quad i = 1, 2, 3,$$
$$\check{e}_s = (A - I)\check{e}_1 + B\check{e}_3,$$
$$\check{e}_i = [0_{n\times(i-1)n}, I_{n\times n}, 0_{n\times(9-i)n}]^T, \quad i = 1, 2, \ldots, 9.$$

Theorem 4.5. *For given positive integers τ_1, τ_2 satisfying $\tau_1 < \tau_2$, system (4.41) is asymptotically stable if there exist a $3n \times 3n$-matrix $P > 0$, $n \times n$-matrices $Q_1 > 0$, $Q_2 > 0$, $R_i > 0$ $(i = 1, 2, 3)$, and $9n \times 2n$-matrices N_i $(i = 1, 2, 3, 4, 5)$ such that the following LMIs hold:*

$$\begin{bmatrix} (\check{\Xi}_1 + \check{\Xi}_2)_{[\tau(k)=\tau_1]} & \tau_1 N_1 & \tau_{12} N_3 & \tau_{12} N_5 \\ * & -\tau_1 \bar{R}_1 & 0 & 0 \\ * & * & -\tau_{12}\bar{R}_2 & 0 \\ * & * & * & -\tau_{12}\bar{R}_3 \end{bmatrix} < 0, \tag{4.54}$$

$$\begin{bmatrix} (\check{\Xi}_1 + \check{\Xi}_2)_{[\tau(k)=\tau_2]} & \tau_1 N_1 & \tau_{12} N_2 & \tau_{12} N_4 \\ * & -\tau_1 \bar{R}_1 & 0 & 0 \\ * & * & -\tau_{12}\bar{R}_2 & 0 \\ * & * & * & -\tau_{12}\bar{R}_3 \end{bmatrix} < 0. \tag{4.55}$$

Proof. Consider the following LKF candidate:

$$V(k) = \sum_{i=1}^{4} V_i(k), \tag{4.56}$$

where $V_i(k)$ $(i = 1, 2, 3)$ are defined as in (4.46), and

$$V_4(k) = \sum_{s=-\tau_2}^{-\tau_1-1} \sum_{u=k+s}^{k-1} x^T(u) R_3 x(u).$$

Calculating the forward difference of $V_4(k)$ yields

$$\Delta V_4(k) = \tau_{12} x^T(k) R_3 x(k) - \sum_{s=k-\tau(k)}^{k-1-\tau_1} x^T(s) R_3 x(s) - \sum_{s=k-\tau_2}^{k-1-\tau(k)} x^T(s) R_3 x(s). \tag{4.57}$$

Utilizing the last inequality in Remark 4.5 to estimate the second and third summation terms in (4.57) yields, respectively,

$$-\sum_{s=k-\tau(k)}^{k-1-\tau_1} x^t(s)R_3x(s) \le \breve{\xi}^t(k)[\Upsilon_4 N_4^t + N_4\Upsilon_4^t + (\tau(k)-\tau_1)N_4\bar{R}_3^{-1}N_4^t]\breve{\xi}(k) \tag{4.58}$$

and

$$-\sum_{s=k-\tau_2}^{k-1-\tau(k)} x^t(s)R_3x(s) \le \breve{\xi}^t(k)[\Upsilon_5 N_5^t + N_5\Upsilon_5^t + (\tau_2-\tau(k))N_5\bar{R}_3^{-1}N_5^t]\breve{\xi}(k). \tag{4.59}$$

Following the similar proof of Theorem 4.4, we have

$$\Delta V(k) \le \breve{\xi}^t(k)(\breve{\Xi}_1 + \breve{\Xi}_2 + \breve{\Xi}_3)\breve{\xi}(k). \tag{4.60}$$

It is easy to prove that $\breve{\Xi}_1 + \breve{\Xi}_2 + \breve{\Xi}_3 < 0$ if and only if (4.54) and (4.55) hold. This completes the proof. □

Remark 4.8. An LKF term $V_4(k) = \sum_{s=-\tau_2}^{-\tau_1-1}\sum_{u=k+s}^{k-1} x^T(u)R_3x(u)$ is introduced in this chapter. However, this LKF term is not taken into account in [23], [25], [36]. Since the inequality with double summation term in Remark 4.5 is used to estimate the forward difference of $V_4(k)$, it may bring less conservatism than Theorem 4.4. Example 4.2 also shows that Theorem 4.5 in this chapter is with less conservatism than Theorem 4.4.

Next, based on the foregoing stability criterion, the passivity analysis for system (4.42) will be carried out. For simplicity, some notations are defined as

$$\begin{aligned}
\tilde{\xi}(k) = [&x^t(k), x^t(k-\tau_1), x^t(k-\tau(k)), x^t(k-\tau_2), \\
&\frac{1}{\tau_1+1}\sum_{s=k-\tau_1}^{k} x^t(s), \frac{1}{\tau(k)-\tau_1+1}\sum_{s=k-\tau(k)}^{k-\tau_1} x^t(s), \\
&\frac{1}{\tau_2-\tau(k)+1}\sum_{s=k-\tau_2}^{k-\tau(k)} x^T(s), \frac{1}{\tau(k)-\tau_1+1}\sum_{s=-\tau(k)}^{-\tau_1}\sum_{u=k+s}^{k-\tau_1} x^t(u), \\
&\frac{1}{\tau_2-\tau(k)+1}\sum_{s=-\tau_2}^{-\tau(k)}\sum_{u=k+s}^{k-\tau(k)} x^t(u), \omega^T(k)]^t, \\
\tilde{\Pi}_1 = [&\tilde{e}_1+\tilde{e}_s, (\tau_1+1)\tilde{e}_5-\tilde{e}_2, (\tau(k)-\tau_1+1)\tilde{e}_6 \\
&+(\tau_2-\tau(k)+1)\tilde{e}_7-\tilde{e}_3-\tilde{e}_4],
\end{aligned}$$

$$\begin{aligned}
\tilde{\Pi}_2 &= [\tilde{e}_1, (\tau_1+1)\tilde{e}_5 - \tilde{e}_1, (\tau(k)-\tau_1+1)\tilde{e}_6 \\
&\quad + (\tau_2 - \tau(k)+1)\tilde{e}_7 - \tilde{e}_3 - \tilde{e}_2], \\
\tilde{\Xi}_1 &= \frac{1}{2} Sym\{(\tilde{\Pi}_1 + \tilde{\Pi}_2)P(\tilde{\Pi}_1 - \tilde{\Pi}_2)T\} + \tilde{e}_1 Q_1 \tilde{e}_1^t \\
&\quad - \tilde{e}_2 Q_1 \tilde{e}_2^T + \tilde{e}_2 Q_2 \tilde{e}_2^T - \tilde{e}_4 Q_2 \tilde{e}_4^T \\
&\quad + \tau_1 \tilde{e}_s R_1 \tilde{e}_s^t + \tau_{12} \tilde{e}_s R_2 \tilde{e}_s^t + \tau_{12} \tilde{e}_1 R_3 \tilde{e}_1 \\
&\quad - \{\tilde{e}_1 D^t \tilde{e}_{10}^t + \tilde{e}_{10} E^t \tilde{e}_{10}^t\} - \alpha \tilde{e}_{10} \tilde{e}_{10}^t, \\
\tilde{\Xi}_2 &= Sym\{\sum_{i=1}^{5} \tilde{\Upsilon}_i N_i^t\}, \\
\tilde{\Xi}_3 &= \tau_1 N_1 \bar{R}_1^{-1} N_1^t + (\tau(k)-\tau_1) N_2 \bar{R}_2^{-1} N_2^t \\
&\quad + (\tau_2 - \tau(k)) N_3 \bar{R}_2^{-1} N_3^t + (\tau(k)-\tau_1) N_4 \bar{R}_3^{-1} N_4^t \\
&\quad + (\tau_2 - \tau(k)) N_5 \bar{R}_3^{-1} N_5^t, \\
\tilde{\Upsilon}_1 &= [\tilde{e}_1 - \tilde{e}_2, \tilde{e}_1 + \tilde{e}_2 - 2\tilde{e}_5], \\
\tilde{\Upsilon}_2 &= [\tilde{e}_2 - \tilde{e}_3, \tilde{e}_2 + \tilde{e}_3 - 2\tilde{e}_6], \\
\tilde{\Upsilon}_3 &= [\tilde{e}_3 - \tilde{e}_4, \tilde{e}_3 + \tilde{e}_4 - 2\tilde{e}_7], \\
\tilde{\Upsilon}_4 &= [(\tau(k)-\tau_1+1)\tilde{e}_6 - \tilde{e}_2, (\tau(k)-\tau_1+1)\tilde{e}_6 + \tilde{e}_2 - 2\tilde{e}_8], \\
\tilde{\Upsilon}_5 &= [(\tau_2-\tau(k)+1)\tilde{e}_7 - \tilde{e}_3, (\tau_2-\tau(k)+1)\tilde{e}_7 + \tilde{e}_3 - 2\tilde{e}_9], \\
\tilde{e}_s &= (A-I)\tilde{e}_1 + B\tilde{e}_3 + C\tilde{e}_{10}, \\
\tilde{e}_i &= [0_{n\times(i-1)n}, I_{n\times n}, 0_{n\times(10-i)n}]^T, \quad i = 1, 2, \ldots, 10.
\end{aligned}$$

Theorem 4.6. *For given positive integers $\tau_1 < \tau_2$, system (4.42) is passive if there exist a $3n \times 3n$-matrix $P > 0$, $n \times n$-matrices $Q_1 > 0$, $Q_2 > 0$, $R_i > 0$ $(i = 1, 2, 3)$, any $10n \times 2n$-matrices N_i $(i = 1, 2, 3, 4, 5)$, and a scalar $\alpha > 0$ such that the following LMIs hold:*

$$\begin{bmatrix} (\tilde{\Xi}_1 + \tilde{\Xi}_2)_{[\tau(k)=\tau_1]} & \tau_1 N_1 & \tau_{12} N_3 & \tau_{12} N_5 \\ * & -\tau_1 \bar{R}_1 & 0 & 0 \\ * & * & -\tau_{12} \bar{R}_2 & 0 \\ * & * & * & -\tau_{12} \bar{R}_3 \end{bmatrix} < 0, \tag{4.61}$$

$$\begin{bmatrix} (\tilde{\Xi}_1 + \tilde{\Xi}_2)_{[\tau(k)=\tau_2]} & \tau_1 N_1 & \tau_{12} N_2 & \tau_{12} N_4 \\ * & -\tau_1 \bar{R}_1 & 0 & 0 \\ * & * & -\tau_{12} \bar{R}_2 & 0 \\ * & * & * & -\tau_{12} \bar{R}_3 \end{bmatrix} < 0. \tag{4.62}$$

Proof. To analyze the passivity of system (4.42), we consider the same LKF as in (4.56), and then it is easy to deduce that

$$\Delta V(k) - 2y^T(k)\omega(k) - \alpha\omega^T(k)\omega(k) \le \tilde{\xi}^T(k)(\tilde{\Xi}_1 + \tilde{\Xi}_2 + \tilde{\Xi}_3)\tilde{\xi}(k). \tag{4.63}$$

By utilizing the Schur complement, $\tilde{\Xi}_1 + \tilde{\Xi}_2 + \tilde{\Xi}_3 < 0$ if and only if (4.61) and (4.62) hold. Hence, we obtain

$$\Delta V(k) - 2y^T(k)\omega(k) - \alpha\omega^T(k)\omega(k) \le 0. \tag{4.64}$$

From the zero initial condition, we obtain

$$\begin{aligned} 2\sum_{i=0}^{k} y^T(i)\omega(i) &\ge \sum_{i=0}^{k} \Delta V(i) - \alpha \sum_{i=0}^{k} \omega^T(i)\omega(i) \\ &= V(k+1) - V(0) - \alpha \sum_{i=0}^{k} \omega^T(i)\omega(i) \\ &= V(k+1) - \alpha \sum_{i=0}^{k} \omega^T(i)\omega(i) \\ &\ge -\alpha \sum_{i=0}^{k} \omega^T(i)\omega(i), \ \forall k \ge 0. \end{aligned} \tag{4.65}$$

According to Definition 4.4, system (4.42) is passive, which completes the proof. □

In the sequel, several numerical examples are given to show the effectiveness and less conservatism of our main results.

4.2.4 Simulation example 4.2

Consider the discrete-time linear system (4.41) with

$$A = \begin{bmatrix} 0.8 & 0 \\ 0.05 & 0.9 \end{bmatrix}, \quad B = \begin{bmatrix} -0.1 & 0 \\ -0.2 & -0.1 \end{bmatrix}.$$

For different values τ_1, the maximal admissible delay upper bounds (MADUBs) obtained by Theorem 4.4 and other stability criteria are given in Tables 4.1 and 4.2. From Tables 4.1 and 4.2, it can be easily seen that the MADUBs obtained by Theorem 4.4 are larger than, or at least as large as, those obtained by the stability criteria in [21], [22], [23], [24], [25], [37], [38]. Although the triple-summation terms are employed in [19], the MADUBs obtained by our method are still greater than or equal to those obtained in [19]. Therefore, the stability criterion obtained in this chapter is more effective. Consider the time-varying delay $\tau(k) = int[\frac{25}{2} - \frac{19}{2}\sin(\frac{k\pi}{4})]$ and $x(0) = [0.3, -0.4]^T$. The state responses of system (4.41) are shown in Fig. 4.6, which shows the system in Example 4.2 is stable.

TABLE 4.1 The MADUBs τ_2 for different τ_1 (Example 4.3).

τ_1	1	3	5	7	11	13	20
[37]	12	13	14	15	17	19	24
[22]	17	17	17	18	20	22	25
[38]	17	17	18	18	20	23	27
[19]	22	22	22	22	23	24	28
[23]	20	21	21	22	23	24	29
[25]	21	21	21	22	23	24	29
Theorem 4.4	22	22	22	22	24	24	29

TABLE 4.2 The MADUBs τ_2 for different τ_1 (Example 4.3).

τ_1	2	4	6	10	15	20	25
[21] (Remark 4.3)	20	21	21	23	25	29	32
[21] (Theorem 4.4)	20	21	21	23	25	29	32
[24] (Remark 4.4)	20	21	21	23	25	29	32
Theorem 4.4	22	22	22	23	25	29	32

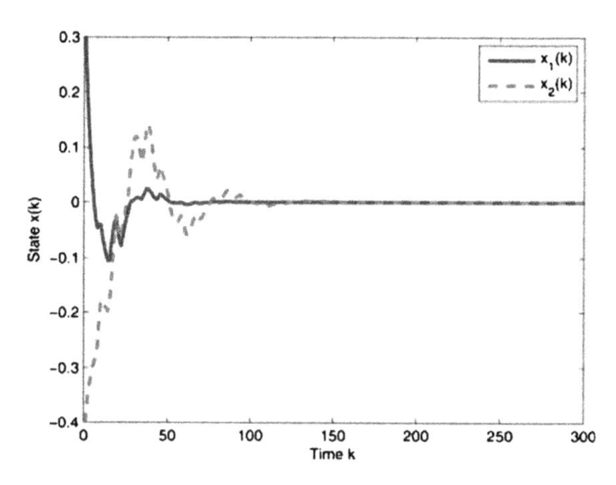

FIGURE 4.6 State responses of the system considered in Example 4.2.

4.2.5 Simulation example 4.3

Consider the discrete-time linear system (4.41) with

$$A = \begin{bmatrix} 0.648 & 0.04 \\ 0.12 & 0.654 \end{bmatrix}, \quad B = \begin{bmatrix} -0.1512 & -0.0518 \\ 0.0259 & -0.1091 \end{bmatrix}.$$

This example is considered in [25], [36]. The MADUBs obtained by Theorems 4.4 and 4.5, together with the results reported in several other publications, are listed in Table 4.3. From Table 4.3, compared with the stability criteria obtained through the Wirtinger-based inequality [23], the auxiliary function-based inequality [21], the free matrix-based summation inequality [25], and the relaxed summation inequality [36], Theorems 4.4 and 4.5 can provide less conservativeness. Obviously, Theorem 4.5 is less conservative than Theorem 4.4 due to the introduction of the LKF term $V_4(k)$. With the time-varying delay $\tau(k) = int[\frac{27}{2} - \frac{17}{2}\sin(\frac{k\pi}{4})]$ and initial state $x(0) = [0.15, -0.2]^T$, the state responses of system (4.41) for initial state $x(0) = [0.15, -0.2]^T$ are given in Fig. 4.7, which shows that the considered discrete-time linear system is stable.

TABLE 4.3 The MADUBs τ_2 for different τ_1 (Example 4.4).

τ_1	5	7	11	13	20
[23]	20	22	25	27	34
[21]	20	22	26	28	34
[25]	21	22	26	27	34
[36]	21	22	26	28	35
Theorem 4.4	22	23	27	28	35
Theorem 4.5	22	24	28	30	37

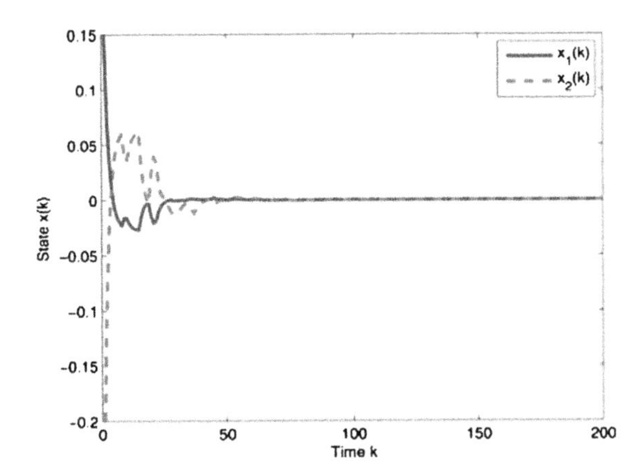

FIGURE 4.7 State responses of the system considered in Example 4.3.

4.2.6 Simulation example 4.4

Consider the discrete-time system (4.42) with the following matrices:

$$A = \begin{bmatrix} 0.8 & 0 \\ 0 & 0.4 \end{bmatrix}, \quad B = \begin{bmatrix} -0.2 & 0 \\ -0.2 & -0.1 \end{bmatrix},$$

$$C=\begin{bmatrix}1 & 0\\ 0 & 1\end{bmatrix},$$

$$D=\begin{bmatrix}0.1 & 0.1\\ 0 & 0.1\end{bmatrix},\quad E=\begin{bmatrix}-0.2 & 0.1\\ 0 & 0.2\end{bmatrix}.$$

When $\tau_1 = 2, 6, 10$, by utilizing Theorem 4.5, the MADUBs that keep the passivity of system (4.42) are $\tau_2 = 10, 13, 14$, respectively. Assume the time-varying delay $\tau(k) = int[6 - 4\sin(\frac{k\pi}{4})]$ and initial state $x(0) = [0.15, -0.2]^t$. Fig. 4.8 gives the state responses of system (4.42) with disturbance input $\omega(k) = [10 * (0.5)^k, -10 * (0.4)^k]t$, which shows that the considered discrete-time linear system is stable.

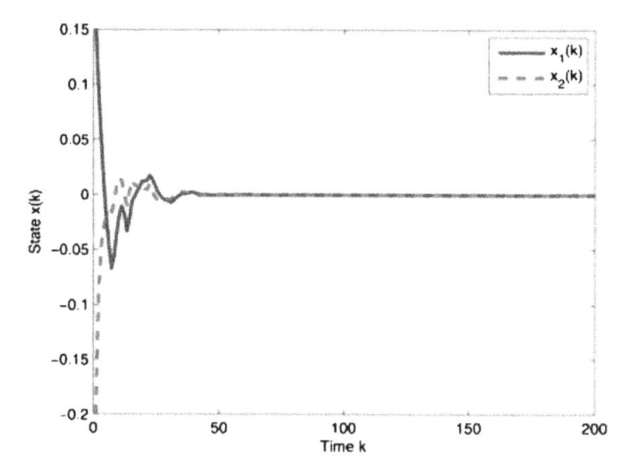

FIGURE 4.8 State responses of the system considered in Example 4.4.

4.3 Consensus tracking of saturated systems

We learned throughout the foregoing chapters that cooperative control of networked systems in which multiple agents work together to accomplish a task has been an interesting topic for decades [39–45]. The distributed interaction among agents induces some merits, such as high computational efficiency and low energy consumption, and research investigations were performed focusing on consensus [39,42,45], synchronization [46,47], and controllability [48,49], to mention a few.

4.3.1 Introduction

One of the common engineering constraints is the input saturation, which means the control input is asked to locate in a bounded region [50]. Earlier results on the coordination of networked systems subject to the input saturation are reported in [51–55], [57], [58] where semiglobal coordinated control of general linear multiagent systems with input saturation has been addressed using low-gain feedback and each agent is made asymptotically null controllable with bounded control. Observe that the developed results mean that the initial states of all agents should be chosen from an (arbitrarily large) bounded region, and the role of the low-gain feedback technique is to tune the control input to be small enough to avoid the saturation. Note that references [55–57] took the disturbance into consideration.

It is quite evident that one should seek to get rid of the restriction on the initial states of the agents, thereby paving the way to global coordinated control of multiagent systems [59–66]. Therefore, we are going to deal with more general neutrally stable systems and extend the systems with double integrator dynamics to those with high-order integrator-type dynamics.

In what follows, we consider the global consensus tracking problem of a saturated system with N agents. We use a graph $\mathbb{G} = \{\mathbb{V}, \mathbb{E}, \mathbb{W}\}$ to describe the communication relationship among the N agents, in which the vertex set $\mathbb{V} = \{v^1, v^2, \ldots, v^N\}$ and the edge set $\mathbb{E} = \{(v^i, v^j) \mid \text{if there exist an information channel between agent } v^i \text{ and agent } v^j\}$ denote the agents in the network and the neighboring interaction, respectively; $\mathbb{W} = (w^{ij}) \in \Re^{N \times N}$ is the adjacency matrix, where

$$w^{ij} = \begin{cases} 1, & \text{if } (v^i, v^j) \in \mathbb{E}, \\ 0, & \text{otherwise}\,. \end{cases}$$

Let $\mathbb{L} = \mathbb{D} - \mathbb{W}$ be the Laplacian matrix of $\mathbb{G}$, where $\mathbb{D}$ is a diagonal matrix with the ith diagonal element being $\sum_{j=1}^{N} w^{ij}$. Introduce $\overline{\mathbb{G}}$, a graph generated by $\mathbb{G}$ and an added vertex labeled by v^0. In $\overline{\mathbb{G}}$, v^0 can affect the agents in $\mathbb{G}$ but not vice versa.

We introduce the diagonal matrix $\mathbb{H} = \operatorname{diag}\{h^1, h^2, \ldots, h^N\}$ to represent the communication relationship between agents in $\mathbb{G}$ and v^0, and then

$$h^i = \begin{cases} 1, & \text{if the } i\text{th agent in } \mathbb{G} \text{ can receive the information from } v^0, \\ 0, & \text{otherwise}, \end{cases}$$

and $\mathbb{L} + \mathbb{H}$ is named the generated Laplacian matrix of $\overline{\mathbb{G}}$. As shown in [42], $\mathbb{L} + \mathbb{H} \succ 0$ if $\mathbb{G}$ is connected and there is at least one agent in $\mathbb{G}$ informed by v^0.

4.3.2 Problem statement

We aim at investigating the global consensus of discrete-time networked systems subject to input saturation. Consider a networked system consisting of N agents, in which each agent moves in n-dimensional Euclidean space and regulates itself according to the following dynamics:

$$x^i(k+1) = Ax^i(k) + Bu^i(k), \tag{4.66}$$

where $x^i(k) \in \Re^{n\times 1}$ is the state of the ith agent at time step k, $u^i(k) \in \Re^{m\times 1}$ is the control input, $u^i(k) = \begin{pmatrix} u_1^i(k) & u_2^i(k) & \cdots & u_m^i(k)\end{pmatrix}^t$ is the input control, and it is asked to meet

$$-\Delta \le u_q^i(k) \le \Delta \;\; \text{for all } q = 1, 2, \dots, m,$$

where $\Delta > 0$ is the saturation level; A and B are the system matrices with

$$A = \begin{pmatrix} 1 & 1 & 0 & \cdots & 0 & 0 \\ 0 & 1 & 1 & \cdots & 0 & 0 \\ 0 & 0 & 1 & \cdots & 0 & 0 \\ \vdots & \vdots & \vdots & \ddots & \vdots & \vdots \\ 0 & 0 & 0 & \cdots & 1 & 1 \\ 0 & 0 & 0 & \cdots & 0 & 1 \end{pmatrix} \in \Re^{n\times n}, \quad B = \begin{pmatrix} 0 \\ 0 \\ 0 \\ \vdots \\ 0 \\ 1 \end{pmatrix} \in \Re^{n\times m}. \tag{4.67}$$

The objective now is to guide system (4.66) to track a virtual leader $x^0(k)$ in global sense, in which the dynamics of the virtual leader are

$$x^0(k+1) = Ax^0(k), \tag{4.68}$$

with $x^0(k) \in \Re^{n\times 1}$ being the state of the virtual leader at time step k. The global consensus tracking is defined in Definition 4.5 below.

Definition 4.5. The system (4.66) can achieve global consensus if for $x^i(0) \in \Re^{n\times 1}$ $(i = 1, \dots, N)$ there exists

$$\lim_{k\to\infty} \left\| x^i(k) - x^0(k) \right\| = 0, \quad i = 1, 2, \dots, N.$$

To proceed further, define

$$\tilde{x}^i(k) = x^i(k) - x^0(k), \quad \tilde{x}(k) = \begin{pmatrix} \tilde{x}^1(k) \\ \tilde{x}^2(k) \\ \vdots \\ \tilde{x}^N(k) \end{pmatrix}.$$

Then the error system of (4.66) and (4.68) is

$$\tilde{x}^i(k) = A\tilde{x}^i(k) + Bu^i(k), \tag{4.69}$$

and the global consensus tracking problem of system (4.66) is equivalent to the global stabilization of the error system (4.69).

For a family of real numbers $\Theta = \{\theta_m\}_{m=1,2,\ldots,n}$, we define matrices

$$A_\Theta = \begin{pmatrix} 1 & \theta_2 & \theta_3 & \cdots & \theta_{n-1} & \theta_n \\ 0 & 1 & \theta_3 & \cdots & \theta_{n-1} & \theta_n \\ \vdots & \vdots & \vdots & \ddots & \vdots & \vdots \\ 0 & 0 & 0 & \cdots & 1 & \theta_n \\ 0 & 0 & 0 & \cdots & 0 & 1 \end{pmatrix}, \quad B_\Theta = \begin{pmatrix} 1 \\ 1 \\ \vdots \\ 1 \end{pmatrix}. \tag{4.70}$$

Based on these two matrices, we further define the system

$$y^i(k+1) = A_\Theta y^i(k) + B_\Theta \operatorname{sat}_\Delta\left(u^i(k)\right), \tag{4.71}$$

where $\operatorname{sat}_\Delta\left(u^i(k)\right) = \left(\operatorname{sat}_\Delta\left(u_1^i(k)\right) \quad \operatorname{sat}_\Delta\left(u_2^i(k)\right) \quad \cdots \quad \operatorname{sat}_\Delta\left(u_m^i(k)\right)\right)^T$ is a saturation function with saturation level Δ. For convenience, we use the pairs (A, B) and (A_Θ, B_Θ) to represent systems (4.66) and (4.71), respectively, throughout the chapter. Lemma 4.5 discloses the relationship between (A, B) and (A_Θ, B_Θ).

Lemma 4.5. *[66] For any family of real numbers $\Theta = \{\theta^m\}_{m=1,2,\ldots,n}$, let (A_Θ, B_Θ) be defined by (4.70). Then, for any controllable pair (A_Θ, B_Θ), there exists a coordinate change $y^i(k) = P_\Theta \tilde{x}^i(k)$ such that system (4.69) becomes system (4.71).*

In view of Lemma 4.5, the global consensus of system (4.66) can be transformed to that of system (4.71), and therefore, in the sequel, we focus on system (4.71).

4.3.3 Control design

Initially, the following result is established.

Theorem 4.7. *Consider a connected networked systems consisting of N agents in which each agent is steered by dynamics (4.66). If there exists a family of real numbers $\Theta = \{\theta_m\}_{m=1,2,\ldots,n}$ satisfying $0 < \sum_{m=1}^{k-1} \theta_m < \theta_k < 1$ for $k \in \{2, 3, \ldots, n\}$, (A_Θ, B_Θ) is controllable, and $A_\Theta - \lambda^i B_\Theta \theta$ is Schur stable for $i = 1, 2, \ldots, N$, then the control laws*

$$u^i(k) = -\sigma \sum_{m=1}^{n} \theta_m \mathbf{sat}_{M_m}\left[\frac{1}{\sigma} \sum_{j \in N(i)} w^{ij}\left(y_m^i(k) - y_m^j(k) - \frac{1}{\sigma} h^i y_m^i(k)\right)\right] \tag{4.72}$$

can guide system (4.66) to achieve global consensus, where

$$\theta = \begin{pmatrix} \theta_1 & \theta_2 & \cdots & \theta_n \end{pmatrix}$$

along with

$$\begin{aligned} &M_n = 1, \\ &M_m = 1 + \alpha_m \tfrac{\theta_{m+1}}{\theta_m} \left[M_{m+1} - \left| \mathbf{sat}_{M_{m+1}} \left(\tfrac{y^i_{m+1}(k)}{\sigma} \right) \right| \right], \\ &m = 1, 2, \ldots, n-1, \end{aligned} \tag{4.73}$$

with $y^i = P_\Theta x^i$, $\sigma = \frac{\Delta}{\sum_{m=1}^{m} \theta_m}$, *and* $\alpha_m \in [0,\ 1]$ *(*$m = 1, 2, \ldots, n$*).*

Proof. We introduce

$$\begin{aligned} z^i(k) &= \frac{1}{\sigma} y^i(k), \\ \Xi^i_m(k) &= \sum_{j \in N(i)} w^{ij} \left(z^i_m(k) - z^j_m(k) \right) - h^i z^i_m(k), \\ \Xi^i(k) &= \sum_{j \in N(i)} w^{ij} \left(z^i(k) - z^j(k) \right) - h^i z(k), \\ z(k) &= \begin{pmatrix} z^1(k) \\ z^2(k) \\ \vdots \\ z^N(k) \end{pmatrix}. \end{aligned}$$

Then

$$z^i(k+1) = A_\theta z^i(k) + B_\theta W^i(k)$$

and

$$\begin{aligned} W^i(k) &= - \sum_{m=1}^{n} \theta_m \, \mathrm{sat}_{M_m} \left(\sum_{j \in N(i)} w^{ij} \left(z^i_m(k) - z^j_m(k) - h^i z^i_m(k) \right) \right) \\ &= - \sum_{m=1}^{n} \theta_m \, \mathrm{sat}_{M_m} \left(\Xi^i_m(k) \right) \in \Re. \end{aligned} \tag{4.74}$$

Now, we construct the Lyapunov functional

$$
\begin{aligned}
V(k) &= \sum_{i=1}^{N} \left(\mathrm{s}^i(k)\right)^T \left(\mathrm{s}^i(k)\right) = \sum_{i=1}^{N}\sum_{m=1}^{n} \left(\mathrm{z}_m^i(k)\right)^2 \\
&:= \sum_{m=1}^{n} V_m(k)
\end{aligned}
\tag{4.75}
$$

with

$$
V_m(k) = \sum_{i=1}^{N} \left(\Xi_m^i(k)\right)^2, \quad \forall m = 1, 2, \ldots, n. \tag{4.76}
$$

The consensus of system (4.76) will be proved by taking the difference of $V_m(k)$ $(m = 1, 2, \ldots, n)$ according to (4.76).

According to (4.76), the difference of $V_n(k)$ yields

$$
\begin{aligned}
V_n(k+1) - V_n(k) &= \sum_{i=1}^{N} \left(\Xi_n^i(k+1)\right)^2 - \sum_{i=1}^{N} \left(\Xi_n^i(k)\right)^2 \\
&= \sum_{i=1}^{N} \left(\Xi_n^i(k) + W^i(k)\right)^2 - \sum_{i=1}^{N} \left(\Xi_n^i(k)\right)^2 \\
&= \sum_{i=1}^{N} \left[\left(W^i(k)\right)^2 - 2\Xi_n^i(k) \sum_{m=1}^{n} \theta_m \, \mathrm{sat}_{M_m}\left(\Xi_m^i(k)\right)\right].
\end{aligned}
$$

Since $0 < \sum_{m=1}^{k-1} \theta_m < \theta_k < 1$ for $k = 2, 3, \ldots, n$, we have

$$
\begin{aligned}
V_n(k+1) - V_n(k) &= \sum_{i=1}^{N} \left[\left(W^i(k)\right)^2 - 2\left|\Xi_n^i(k) W^i(k)\right|\right] \\
&= \sum_{i=1}^{N} \left|W^i(k)\right| \left[\left|W_i(k)\right| - 2\left|\Xi_n^i(k)\right|\right]
\end{aligned}
$$

and

$$
\begin{aligned}
\left|W^i(k)\right| &= \left| - \sum_{m=1}^{n} \theta_m \mathbf{sat}_{M_m}\left(\Xi_m^i(k)\right)\right| \\
&\geq \theta_n \left|\mathbf{sat}_{M_n}\left(\mathrm{E}_n^i(k)\right)\right| - \sum_{m=1}^{n-1} \theta_m \left|\mathbf{sat}_{M_m}\left(\Xi_m^i(k)\right)\right|.
\end{aligned}
$$

If $\Xi_n^i(k) \notin [-1, 1]$, then $\left|\mathbf{sat}_{M_n}\left(\Xi_n^i(k)\right)\right| = 1$ and further, we get

$$
\begin{aligned}
\left|W^i(k)\right| &\geq \theta_n - \sum_{m=1}^{n-1} \theta_m \left|\mathbf{sat}_{M_m}\left(\Xi_m^i(k)\right)\right| \\
&\geq \theta_n - \theta_1 M_1 - \sum_{m=2}^{n-1} \theta_m \left|\mathbf{sat}_{M_m}\left(\Xi_m^i(k)\right)\right| \\
&= \theta_n - \theta_1 \left(1 + \alpha_1 \frac{\theta_2}{\theta_1}\left(M_2 - \left|\mathbf{sat}_{M_2}\left(\Xi_2^i(k)\right)\right|\right)\right) \\
&\quad - \sum_{m=2}^{n-1} \theta_m \left|\mathbf{sat}_{M_m}\left(\Xi_m^i(k)\right)\right| \\
&= \theta_n - \theta_1 - \theta_2 M_2 + \theta_2 \left[M_2 - \left|\mathbf{sat}_{M_2}\left(\Xi_2^i(k)\right)\right|\right] \\
&\quad - \alpha_1 \theta_2 \left(M_2 - \left|\mathbf{sat}_{M_2}\left(\Xi_2^i(k)\right)\right|\right) \\
&\quad - \sum_{m=3}^{n-1} \theta_m \left|\mathbf{sat}_{M_m}\left(\Xi_m^i(k)\right)\right| \\
&= \theta_n - \theta_1 - \theta_2 M_2 + \theta_2 (1 - \alpha_1)\left(M_2 - \left|\mathbf{sat}_{M_2}\left(\Xi_2^i(k)\right)\right|\right) \\
&\quad - \sum_{m=3}^{n-1} \theta_m \left|\mathbf{sat}_{M_m}\left(\Xi_m^i(k)\right)\right| \\
&\geq \theta_n - \theta_1 - \theta_2 M_2 - \sum_{m=3}^{n-1} \theta_m \left|\mathbf{sat}_{M_m}\left(\Xi_m^i(k)\right)\right|.
\end{aligned}
$$

On substituting M_2 from (4.73) into the above inequality, it follows that

$$
\left|W^i(k)\right| \geq \theta_n - \theta_1 - \theta_2 - \theta_3 M_3 - \sum_{m=4}^{n-1} \theta_m \left|\mathbf{sat}_{M_m}\left(\Xi_m^i(k)\right)\right|.
$$

Repeating the same operation, it turns out that

$$
\left|W^i(k)\right| \geq \theta_n - \sum_{m=1}^{n-1} \theta_m > 0. \tag{4.77}
$$

Further manipulation yields

$$
\begin{aligned}
\left|W^i(k)\right| &\leq \sum_{m=1}^{n} \theta_m = \theta_n + \sum_{m=1}^{n-1} \theta_m \\
&\leq 2 - \theta_n + \sum_{m=1}^{n-1} \theta_m.
\end{aligned} \tag{4.78}
$$

In view of (4.78) and the fact that $\Xi_i^n(k) \notin [-1, 1]$, one obtains

$$\begin{aligned}\left|W^i(k)\right| - 2\left|\Xi_i^n(k)\right| &\le \left|W^i(k)\right| - 2 \\ &\le -\theta_n + \sum_{m=1}^{n-1} \theta_m \\ &= -\left(\theta_n - \sum_{m=1}^{n-1} \theta_m\right) < 0. \end{aligned} \tag{4.79}$$

This leads by (4.77) and (4.79) to

$$\begin{aligned} V_n(k+1) - V_n(k) &\le -\left(\theta_n - \sum_{m=1}^{n-1} \theta_m\right)^2 \\ &< 0, \end{aligned}$$

which implies that $\Xi_n^i(k)$ would decrease and enter into $[-1,\ 1]$ as time progresses, and then remains in $[-1,\ 1]$.

Taking the difference of $V_{n-1}(k)$ according to (4.74) results in

$$\begin{aligned} V_{n-1}(k+1) - V_{n-1}(k) &= \sum_{i=1}^{N} \left(\Xi_{n-1}^i(k+1)\right)^2 - \sum_{i=1}^{N} \left(\Xi_{n-1}^i(k)\right)^2 \\ &= \sum_{i=1}^{N} \left(\Xi_{n-1}^i(k) + \theta_n \Xi_n^i(k) + W^i(k)\right)^2 - \sum_{i=1}^{N} \left(\Xi_{n-1}^i(k)\right)^2 \\ &= \sum_{i=1}^{N} \left\{\left[\Xi_{n-1}^i(k) - \sum_{m=1}^{n-1} \theta_m \mathbf{sat}_{M_m}\left(\Xi_m^i(k)\right)\right]^2 - \left(\mathbf{B}_{n-1}^i(k)\right)^2\right\} \\ &= \sum_{i=1}^{N} \left|\sum_{m=1}^{n-1} \theta_m \mathbf{sat}_{M_m}\left(\Xi_m^i(k)\right)\right| \left(\left|\sum_{m=1}^{n-1} \theta_m \mathbf{sat}_{M_m}\left(\Xi_m^i(k)\right)\right| - 2\left|\Xi_{n-1}^i(k)\right|\right). \end{aligned}$$

By similarity to the analysis for $V_n(k+1) - V_n(k)$, if $\Xi_{n-1}^i(k) \notin [-1, 1]$,

$$\begin{aligned} V_{n-1}(k+1) - V_{n-1}(k) &= -\left(\theta_{n-1} - \sum_{m=1}^{n-2} \theta_m\right)^2 \\ &< 0, \end{aligned}$$

which further implies that $\Xi_{n-1}^i(k)$ will enter into and remain in $[-1,\ 1]$.

Proceeding further, we consider $\Xi^i_{n-1}(k)$ for $i = 1, 2, \ldots, n-2$, and iterating, it turns out that

$$\Xi^i(k) \in \underbrace{[-1,\ 1] \times [-1,\ 1] \times \cdots \times [-1, 1]}_{n}$$

and the saturation on $u^i(k)$ in (4.72) can be bypassed. Then, the updating dynamics of $z_i(k)$ are

$$\begin{aligned} & z^i(k+1) = \\ & A_\Theta z^i(k) + B_\Theta \left\{ -\sum_{m=1}^{n} \theta_m \sum_{j \in N(i)} w^{ij} \left(z^i_m(k) - z^j_m(k) \right) - \sum_{m=1}^{n} \theta_m h^i z^i_m(k) \right\} \\ & = A_\Theta z^i(k) - B_\Theta \theta \, \Xi^i(k). \end{aligned} \tag{4.80}$$

Additionally,

$$Z(k+1) = [I_N \otimes A_\Theta - (\mathbb{L} + \mathbb{H}) \otimes B_\Theta \theta] \, Z(k). \tag{4.81}$$

The symmetry of $\mathbb{L}$ leads to the fact that there exists an orthogonal matrix $\mathbb{S}$, such that

$$\begin{aligned} \mathbb{L} + \mathbb{H} = \mathbb{S}^t & \begin{pmatrix} \lambda^1 & 0 & \cdots & 0 \\ 0 & \lambda^2 & \cdots & 0 \\ \vdots & \vdots & \ddots & \vdots \\ 0 & 0 & \cdots & \lambda^N \end{pmatrix} \mathbb{S} \\ & := \mathbb{S}^t \, \Sigma \, \mathbb{S}. \end{aligned}$$

Considering $V(k)$ in (4.75) and letting

$$\begin{aligned} \bar{z}(k) &= (\mathbb{S} \otimes I_n) \, z(k), \\ \bar{A_\Theta} &= (I_N \otimes A_\Theta - (\mathbb{L} + \mathbb{H}) \otimes B_\Theta \theta), \end{aligned}$$

we arrive at

$$\begin{aligned} V(k+1) &= z^t(k+1) \left((\mathbb{L} + \mathbb{H})^2 \otimes I_n \right) z(k+1) \\ &= z^t(k) \left[\bar{A_\Theta}^t \left((\mathbb{L} + \mathbb{H})^2 \otimes I_n \right) \bar{A_\Theta} \right] \\ &= \bar{z}^t(k) \Big[(I_N \otimes A_\Theta - \Sigma \otimes B_\Theta \theta)^t \left(\Sigma^2 \otimes I_n \right) \\ & \times (I_N \otimes A_\Theta - \Sigma \otimes B_\Theta \theta) \Big] \bar{z}(k) \\ &= \sum_{i=1}^{N} \left(\lambda^i \right)^2 \left(\bar{z}^i(k) \right)^t \left[\left(A_\Theta - \lambda^i B_\Theta \theta \right)^t \left(A_\Theta - \lambda^i B_\Theta \theta \right) \right] \bar{z}^i(k). \end{aligned}$$

Since $A_\Theta - \lambda^i B_\Theta \theta$ is Schur stable for $i = 1, \ldots, N$, we have

$$(A_\Theta - \lambda^i B_\Theta \theta)^t (A_\Theta - \lambda^i B_\Theta \theta) - I_n \prec 0,$$

and it follows that

$$\begin{aligned} V(k+1) - V(k) &\leq \sum_{i=1}^{N} \left(\lambda^i\right)^2 \left(\bar{z}^i(k)\right)^t \bar{z}^i(k) \\ &\quad - z^t(k)\left[(\mathbb{L}+\mathbb{H})^2 \otimes I_n\right] z(k) \\ &= z^t(k)\left[(\mathbb{L}+\mathbb{H})^2 \otimes I_n\right] z(k) \\ &\quad - z^T(k)\left[(\mathbb{L}+\mathbb{H})^2 \otimes I_n\right] z(k) \\ &\leq 0, \end{aligned} \tag{4.82}$$

where $V(k+1) - V(k) = 0$ if and only if $z(k) = \mathbf{0}$. Moreover, the relationship among $\tilde{x}^i(k)$, $y^i(k)$ and $z^i(k)$ induces that

$$\lim_{k \to \infty} \tilde{x}^i(k) = \mathbf{0}, \quad i = 1, 2, \ldots, N,$$

which further guarantees that the global consensus tracking of system (4.66) with the virtual leader (4.68) can be achieved. This completes the proof. □

4.3.4 Simulation example 4.5

In this section, we provide numerical simulations to support our findings. We select

$$A = \begin{bmatrix} 1 & 1 \\ 0 & 1 \end{bmatrix}, \quad B = \begin{bmatrix} 0 \\ 1 \end{bmatrix}.$$

The saturation level is $\Delta = 2$. To meet the premise of Theorem 4.7, we choose

$$\theta = [\theta_1 \quad \theta_2] = \begin{bmatrix} \frac{1}{6} & \frac{1}{4} \end{bmatrix}.$$

Obviously, one has

$$A_\Theta = \begin{bmatrix} 1 & \frac{1}{4} \\ 0 & 1 \end{bmatrix}, \quad B_\Theta = \begin{bmatrix} 1 \\ 1 \end{bmatrix},$$

and

$$P_\Theta = \begin{bmatrix} 1 & 1 \\ 0 & -4 \end{bmatrix}.$$

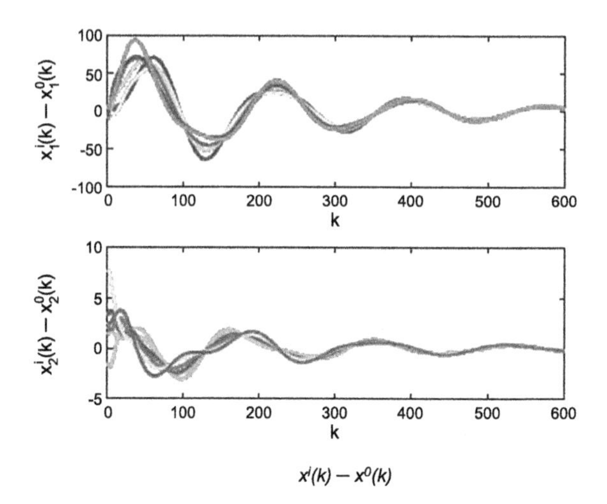

FIGURE 4.9 Global consensus tracking results: States.

We choose a networked system consisting of $N = 6$ agents and one virtual leader, where the interaction among all agents and the virtual leader is described by matrices W and H as follows:

$$W = \frac{1}{10} \times \begin{bmatrix} 0 & 1 & 0 & 0 & 0 & 0 \\ 1 & 0 & 1 & 0 & 0 & 0 \\ 0 & 1 & 0 & 1 & 0 & 0 \\ 0 & 0 & 1 & 0 & 1 & 0 \\ 0 & 0 & 0 & 1 & 0 & 1 \\ 0 & 0 & 0 & 0 & 1 & 0 \end{bmatrix}, \quad H = \frac{1}{10} \times \begin{bmatrix} 0 & 0 & 0 & 0 & 0 & 0 \\ 0 & 1 & 0 & 0 & 0 & 0 \\ 0 & 0 & 0 & 0 & 0 & 0 \\ 0 & 0 & 0 & 0 & 0 & 0 \\ 0 & 0 & 0 & 0 & 1 & 0 \\ 0 & 0 & 0 & 0 & 0 & 0 \end{bmatrix}.$$

In addition, we choose $\alpha_1 = \frac{1}{2} \in [0, 1]$. The state of each agent, including that of the virtual leader, is randomly selected from $[-10, 10] \times [-10, 10]$.

Fig. 4.9 shows the convergence of the error states of all agents. It is shown that the error state between each agent and the virtual leader would converge to zero as time goes by. As displayed in Fig. 4.10, $y^i(k) - y^0(t)$ also converges to zero. The control input u^i in Fig. 4.11 is always located in $[-\Delta, \ \Delta] = [-2, \ 2]$ and approaches zero. All simulations verify the effectiveness of the theoretical results.

4.4 Notes

In this chapter, a new distributed event-triggered protocol based on the dissipativity approach has been investigated to solve the synchronization problem for discrete-time uncertain linear multiagent systems. The measurement output of each agent would be transmitted to the neighboring agents only when its corre-

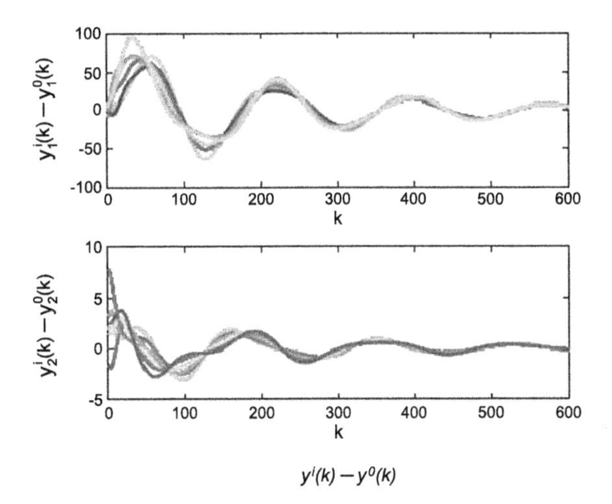

FIGURE 4.10 Global consensus tracking results: Outputs.

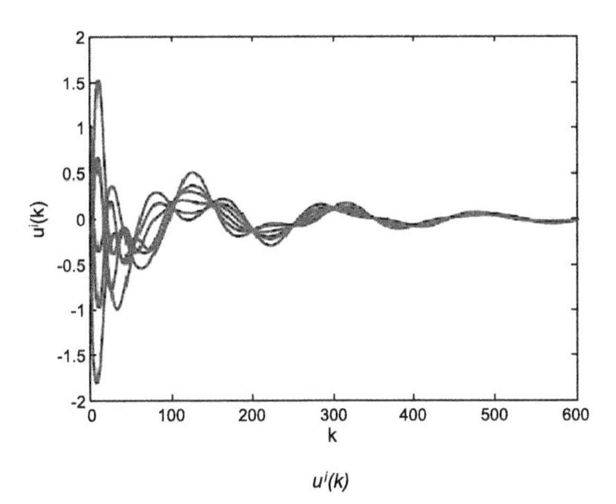

FIGURE 4.11 Global consensus tracking results: Control inputs.

sponding triggering condition is violated. Compared with the previous related results, the main contribution is that we have proposed both time-triggered and event-triggered consensus protocols that do not require continuous monitoring of measurement nor continuous communication among agents. The proposed triggering mechanisms do not reveal the Zeno behavior due to its discrete nature. Furthermore, the event/time triggering controller gains are solved by obtaining a feasible solution for a set of LMIs. Lastly, the effectiveness of the synchronization protocols has been demonstrated by a numerical simulation example. Additional research directions should include the extension of the proposed results to synthesize distributed event-based protocols for nonlinear discrete-time systems using dissipation theory, or considering different triggering mechanisms for different agents.

Next, the problem of stability and passivity has been investigated for the discrete delayed linear system. A new general free matrix-based summation inequality is proved, which can be seen as a discrete-time version of the integral inequality in [26]. By utilizing this extended summation inequality, two improved delay-dependent sufficient conditions on the stability of system (4.41) have been obtained. Finally, the numerical examples demonstrate the effectiveness of the proposed method.

Finally, we have taken the networked systems with input saturation into consideration and focused on the global consensus tracking of this kind of systems. For saturated networked systems with discrete-time high-order integrator dynamics, nonlinear feedback laws have been constructed to directly avoid the saturation function, and then sufficient conditions have been provided to ensure the global consensus tracking of the systems. Numerical simulations verified the theoretical results we have obtained. In the near future, we will devote ourselves to the global coordinated control of networked systems with input saturation and input additive disturbance as well as dead zone.

References

[1] R. Olfati-Saber, J.A. Fax, R.M. Murray, Consensus and cooperation in networked multi-agent systems, Proc. IEEE 95 (1) (2007) 215–233.

[2] Z. Zuo, Q. Han, B. Ning, X. Ge, X. Zhang, An overview of recent advances in fixed-time cooperative control of multi-agent systems, IEEE Trans. Ind. Inform. 14 (6) (2018) 2322–2334.

[3] Y. Feng, Z. Duan, W. Ren, G. Chen, Consensus of multi-agent systems with fixed inner connections, Int. J. Robust Nonlinear Control 28 (1) (2018) 154–173.

[4] S.Y. Ren, W. Wang, W. Jin-Liang, Passivity-based finite-time synchronization of nonlinear multi-agent systems, IEEE Trans. Netw. Sci. Eng. (2020).

[5] P. Shi, Y. Jiafeng, Dissipativity-based consensus for fuzzy multi-agent systems under switching directed topologies, IEEE Trans. Fuzzy Syst. (2020), https://doi.org/10.1109/TFUZZ.2020.2969391.

[6] V. Ugrinovskii, C. Langbort, Distributed $\mathcal{H}_\infty$ consensus-based estimation of uncertain systems via dissipativity theory, IET Control Theory Appl. 5 (12) (2011) 1458–1469.

[7] X. Xu, S. Chen, W. Huang, L. Gao, Leader-following consensus of discrete-time multi-agent systems with observer-based protocols, Neurocomputing 118 (2013) 334–341.

[8] Y. Liu, Y. Jia, Event-based consensus control of multi-agent systems by $\mathcal{L}_\infty$ theory, Int. J. Control. Autom. Syst. 16 (3) (2018) 1254–1262.

[9] L. Dai, Singular Control Systems, Lecture Notes in Control and Information Sciences, vol. 118, Springer, New York, 1989.

[10] H.B. Zeng, Y. He, M. Wu, J. She, Free-matrix-based integral inequality for stability analysis of systems with time-varying delay, IEEE Trans. Autom. Control 60 (10) (2015) 2768–2772.

[11] Y. Wei, Z. Lin, Stabilization of exponentially unstable discrete-time linear systems by truncated predictor feedback, Syst. Control Lett. 97 (2016) 27–35.

[12] K. Mathiyalagan, H. Su, P. Shi, R. Sakthivel, Exponential H_∞ filtering for discrete-time switched neural networks with random delays, IEEE Trans. Cybern. 45 (2015) 676–687.

[13] K. Mathiyalagan, R. Sakthivel, S.M. Anthoni, Exponential stability result for discrete-time stochastic fuzzy uncertain neural networks, Phys. Lett. A 376 (2012) 901–912.

[14] C.K. Zhang, Y. He, L. Jiang, M. Wu, H.B. Zeng, Delay-variation-dependent stability of delayed discrete-time systems, IEEE Trans. Autom. Control 61 (9) (2016) 2663–2669.

[15] X.M. Zhang, Q.L. Han, Abel lemma-based finite-sum inequality and its application to stability analysis for linear discrete time-delay systems, Automatica 57 (2015) 199–202.
[16] X. Meng, J. Lam, B. Du, H.J. Gao, A delay-partitioning approach to the stability analysis of discrete-time systems, Automatica 46 (3) (2010) 610–614.
[17] O.M. Kwon, M.J. Park, J.H. Park, S.M. Lee, E.J. Cha, Improved robust stability criteria for uncertain discrete-time systems with interval time-varying delays via new zero equalities, IET Control Theory Appl. 6 (16) (2012) 2567–2575.
[18] F.X. Wang, X.G. Liu, M.L. Tang, Y.J. Shu, Stability analysis of discrete-time systems with variable delays via some new summation inequalities, Adv. Differ. Equ. 95 (2016) 1–20.
[19] O.M. Kwon, M.J. Park, J.H. Park, Stability and stabilization for discrete-time systems with time-varying delays via augmented Lyapunov–Krasovskii functional, J. Franklin Inst. 350 (3) (2013) 521–540.
[20] X.G. Liu, F.X. Wang, M.L. Tang, Auxiliary function-based summation inequalities and their applications to discrete-time system, Automatica 78 (2017) 211–215.
[21] P.T. Nam, H. Trinh, P.N. Pathirana, Discrete inequalities based on multiple auxiliary functions and their applications to stability analysis of time-delay systems, J. Franklin Inst. 352 (2015) 5810–5831.
[22] Y. He, M. Wu, G.P. Liu, Output feedback stabilization for a discrete-time system with a time-varying delay, IEEE Trans. Autom. Control 53 (10) (2008) 2372–2377.
[23] A. Seuret, F. Gouaisbaut, E. Fridman, Stability of discrete-time systems with time-varying delays via a novel summation inequality, IEEE Trans. Autom. Control 60 (10) (2015) 2740–2745.
[24] P.T. Nam, P.N. Pathirana, H. Trinh, Discrete Wirtinger-based inequality and its application, J. Franklin Inst. 352 (5) (2015) 1893–1905.
[25] J. Chen, J. Lu, S. Xu, Summation inequality and its application to stability analysis for time-delay systems, IET Control Theory Appl. 10 (4) (2016) 391–395.
[26] Y. Liu, H.P. Ju, B. Guo, Results on stability of linear systems with time varying delay, IET Control Theory Appl. 11 (1) (2016) 129–134.
[27] A. Seuret, F. Gouaisbaut, Wirtinger-based integral inequality: application to time-delay systems, Automatica 49 (8) (2013) 2860–2866.
[28] S. Ramasamy, G. Nagamani, Q. Zhu, Robust dissipativity and passivity analysis for discrete-time stochastic T-S fuzzy Cohen–Grossberg Markovian jump neural networks with mixed time delays, Nonlinear Dyn. 85 (4) (2016) 2777–2799.
[29] G. Nagamani, T. Radhika, P. Balasubramaniam, A delay decomposition approach for robust dissipativity and passivity analysis of neutral-type neural networks with leakage time-varying delay, Complexity 21 (2016) 248–264.
[30] G. Nagamani, S. Ramasamy, P. Balasubramaniam, Robust dissipativity and passivity analysis for discrete-time stochastic neural networks with time-varying delay, Complexity 21 (3) (2016) 47–58.
[31] D. Zhang, L. Yu, Passivity analysis for discrete-time switched neural networks with various activation functions and mixed time delays, Nonlinear Dyn. 67 (2012) 403–411.
[32] Y. Shu, X. Liu, Y. Liu, Stability and passivity analysis for uncertain discrete-time neural networks with time-varying delay, Neurocomputing 173 (2016) 1706–1714.
[33] K. Yu, C. Lien, J. Chen, L. Chung, Passivity analysis and passive control for uncertain discrete switched time-delay systems via a simple switching signal design, Adv. Differ. Equ. 104 (2016) 1–24.
[34] M.S. Mahmoud, L. Xie, Passivity analysis and synthesis for uncertain time-delay systems, Math. Probl. Eng. 7 (2001) 455–484.
[35] J. Zhu, Q. Zhang, Z. Yuan, Delay-dependent passivity criterion for discrete-time delayed standard neural network model, Neurocomputing 73 (2010) 1384–1393.
[36] C.K. Zhang, Y. He, L. Jiang, M. Wu, An improved summation inequality to discrete-time systems with time-varying delay, Automatica 74 (2016) 10–15.
[37] B. Zhang, S. Xu, Y. Zou, Improved stability criterion and its applications in delayed controller design for discrete-time systems, Automatica 44 (11) (2008) 2963–2967.

[38] H. Shao, Q.L. Han, New stability criteria for linear discrete-time systems with interval-like time-varying delays, IEEE Trans. Autom. Control 56 (3) (2011) 619–625.
[39] X.L. Wang, X.F. Wang, Semi-global consensus of multi-agent systems with intermittent communications and low-gain feedback, IET Control Theory Appl. 9 (5) (2015) 766–774.
[40] W. Ren, R.W. Beard, Consensus seeking in multiagent systems under dynamically changing interaction topologies, IEEE Trans. Autom. Control 50 (5) (2005) 655–661.
[41] H. Su, Z. Li, M.Z.Q. Chen, Distributed estimation and control for two-targets tracking mobile sensor networks, J. Franklin Inst. 354 (7) (2017) 2994–3007.
[42] H. Su, G. Chen, X. Wang, Z. Lin, Adaptive second-order consensus of networked mobile agents with nonlinear dynamics, Automatica 47 (2) (2011) 368–375.
[43] Q. Liang, Z. She, L. Wang, H. Su, General Lyapunov functions for consensus of nonlinear multi-agent systems, IEEE Trans. Circuits Syst. II, Express Briefs 64 (10) (2017) 1232–1236.
[44] C. Li, X. Yu, W. Yu, T. Huang, Z. Liu, Distributed event-triggered scheme for economic dispatch in smart grids, IEEE Trans. Ind. Inform. 12 (5) (2016) 1775–1785.
[45] X.L. Wang, H. Su, X. Wang, B. Liu, Second-order consensus of multi-agent systems via periodically intermittent pinning control, Circuits Syst. Signal Process. 35 (7) (2016) 2413–2431.
[46] H. Su, X. Wang, Z. Lin, Synchronization of coupled harmonic oscillators in a dynamic proximity network, Automatica 45 (2009) 2286–2291.
[47] L. Wang, M.Z.Q. Chen, Q. Wang, Bounded synchronization of a heterogeneous complex switched network, Automatica 56 (2015) 19–24.
[48] B. Liu, Y. Han, F. Jiang, H. Su, J. Zou, Group controllability of discrete-time multi-agent systems, J. Franklin Inst. 353 (2016) 3524–3559.
[49] Y. Bai, L. Wang, M.Z.Q. Chen, N. Huang, Controllability emerging from conditional path reachability in complex networks, Int. J. Robust Nonlinear Control 27 (2017) 4919–4930.
[50] Z. Lin, Low Gain Feedback, Springer, 1999.
[51] H. Su, M.Z.Q. Chen, J. Lam, Semi-global leader-following consensus of linear multi-agent systems with input saturation via low gain feedback, IEEE Trans. Circuits Syst. I, Regul. Pap. 60 (7) (2013) 1881–1889.
[52] H. Su, G. Jia, M.Z.Q. Chen, Semi-global containment control of multi-agent systems with intermittent input saturation, J. Franklin Inst. 352 (9) (2015) 3504–3525.
[53] X.L. Wang, X.F. Wang, Semi-global consensus of multi-agent systems with intermittent communications and low-gain feedback, IET Control Theory Appl. 9 (5) (2015) 766–774.
[54] X.L. Wang, H. Su, X.F. Wang, G. Chen, Fully distributed event-triggered semi-global consensus of multiagent systems with input saturation, IEEE Trans. Ind. Electron. 64 (6) (2017) 5055–5064.
[55] X.L. Wang, H. Su, M.Z.Q. Chen, X.F. Wang, Observer-based robust coordinated control of multi-agent systems with input saturation, IEEE Trans. Neural Netw. Learn. Syst. 29 (5) (2017) 1933–1946.
[56] X.L. Wang, H. Su, X.F. Wang, G. Chen, Robust semi-global swarm tracking of coupled harmonic oscillators with input saturation and external disturbance, Int. J. Robust Nonlinear Control 28 (2018) 1566–1582.
[57] H. Su, Y. Qiu, L. Wang, Semi-global output consensus of discrete-time multi-agent systems with input saturation and external disturbances, ISA Trans. 67 (2017) 131–139.
[58] M.Z.Q. Chen, L. Zhang, H. Su, G. Chen, Stabilizing solution and parameter dependence of modified algebraic Riccati equation with application to discrete-time network synchronization, IEEE Trans. Autom. Control 61 (1) (2016) 228–233.
[59] Z. Meng, Z. Zhao, Z. Lin, On global leader-following consensus of identical linear dynamic systems subject to actuator saturation, Syst. Control Lett. 62 (2) (2013) 132–142.
[60] T. Yang, Z. Meng, D.V. Dimarogonas, K.H. Johansson, Global consensus for discrete-time multi-agent systems with input saturation constraints, Automatica 50 (2) (2014) 499–506.
[61] L. Zhang, M.Z.Q. Chen, Event-based global stabilization of linear systems via a saturated linear controller, Int. J. Robust Nonlinear Control 26 (5) (2016) 1073–1091.

[62] H. Su, M.Z.Q. Chen, X. Wang, Global coordinated tracking of multi-agent systems with disturbance uncertainties via bounded control inputs, Nonlinear Dyn. 82 (4) (2015) 2059–2068.
[63] Z. Zhao, Z. Lin, Discrete-time global leader-following consensus of a group of general linear systems using bounded controls, Int. J. Robust Nonlinear Control 27 (2017) 3433–3465.
[64] Z. Jin, R.M. Murray, Multi-hop relay protocols for fast consensus seeking, in: Proceedings of the 45th IEEE Conference on Decision and Control, San Diego, USA, 2006, pp. 1001–1006.
[65] W. Yang, X. Wang, H. Shi, Fast consensus seeking in multi-agent systems with time delay, Syst. Control Lett. 62 (2013) 269–276.
[66] N. Marchand, A. Hably, A. Chemori, Global stabilization with low computational cost of the discrete-time chain integrators by means of bounded controls, IEEE Trans. Autom. Control 52 (5) (2007) 948–952.

Chapter 5

Performance of consensus algorithms

5.1 Introduction

Various natural and man-made systems have been modeled under the network theory framework. Different network models with distinct design principles have been proposed to better understand these real-world networks. The eigenvalue spectrum of these networks not only contains information about structural characteristics of underlying networks but also provides insight into dynamical behavior and stability of corresponding complex systems. Depending on the structural characteristics of underlying model networks, the spectra of these networks exhibit specific features. All these ascertain that the spectra of networks can be used as a practical tool for classifying and understanding different real-world systems represented as networks. In this chapter, we first discuss the features which the different regions of spectra furnish, in case of the model networks. Further, we go on to discuss spectral characteristics of real-world networks, with particular emphasis on their extent of similarities and differences with the spectra of model networks.

Networks present a simple framework to model complex systems comprised of interacting elements. The two basic ingredients of a network are its nodes and connections. Mathematically, a network or a graph is defined as a set of N nodes and N_c connections which can be represented in terms of an adjacency matrix, $\mathbb{A} = [A_{ij}]$, as

$$A_{ij} = \begin{cases} 1 & i \to j, \\ 0 & \text{otherwise.} \end{cases} \tag{5.1}$$

It is useful to recall here the degree of a node, the largest degree, and the average degree of a network (see also Chapter 2).

Degree d_i of a node i refers to the number of nodes i is connected to and can be calculated from (5.1) as

$$d_i = \sum_{j=1}^{N} A_{ij}. \tag{5.2}$$

Discrete Networked Dynamic Systems. https://doi.org/10.1016/B978-0-12-823698-7.00013-8

Similarly the largest degree (d_M) is $\max(d_i)$, where $1 \leq i \leq N$, and the average degree d_{av} is

$$d_{av} = \frac{\sum_{i=1}^{N} d_i}{N}. \tag{5.3}$$

In the 2000s and 2010s, network science has shown tremendous success in modeling and understanding of complex phenomena in various model and real-world systems and in network science.

Networks present a simple framework to model complex systems comprised of interacting elements. The two basic ingredients of a network are its nodes and connections. Mathematically, a network or a graph is defined as a set of N nodes and N_c connections which can be represented in terms of an adjacency matrix, $\mathbf{A}$, as

$$A_{ij} = \begin{cases} 1 & \text{if } i \sim j, \\ 0 & \text{otherwise.} \end{cases} \tag{5.4}$$

5.2 The agreement algorithm

The "agreement algorithm" is basically an iterative procedure for the solution of the distributed consensus problem. In this section, we describe and analyze the agreement algorithm [1]. We consider a set $\mathbf{N} = \{1, 2, \cdots, n\}$ of agents embedded, at each time k, in a directed graph $\mathbb{G}(k) = (\mathbf{N}, \mathbb{E}(k))$, where k lies in some discrete set of times which we will take, for simplicity, to be the nonnegative integers. Each agent i starts with a scalar value $x_i(0)$: The vector with the values of all agents at time k will be denoted by $x(k) = (x_1(k), \cdots, x_n(k))$. The agreement algorithm updates $x(k)$ according to the equation $x(k+1) = \mathbf{A}(k)x(k)$, or

$$x_i(k+1) = \sum_{n}^{j=1} \mathbf{a}_{ij}(k)\, x_i(k),$$

where $\mathbf{A}(k) := [\mathbf{a}_{ij}(k)]$ is a nonnegative stochastic matrix whose row-sums are equal to 1. Observe that $x_i(k+1)$ is a weighted average of the values $x_i(k)$ held by the agents at time k. We next state some conditions under which the agreement algorithm is guaranteed to converge.

Assumption 5.1. There exists a positive constant α such that:

(A) $\mathbf{a}_{ij}(k) > \alpha, \forall i, k;$

(B) $\mathbf{a}_{ij}(k) \in \{0\} \cup [\alpha, 1], \forall i, j, k;$

(C) $\sum_{n}^{j=1} \mathbf{a}_{ij}(k) = 1, \forall i, k.$

Intuitively, whenever $\mathbf{a}_{ij}(k) > 0$, agent j communicates its current value $x_j(k)$ to agent i. Each agent i updates its own value, by forming a weighted average of its own value and the values it has just received from other agents.

In terms of the directed graph $\mathbb{G}(k) = (\mathbf{N}, \mathbb{E}(k))$, we introduce an arc $(j, i) \in \mathbb{E}(k)$ if and only if $\mathbf{a}_{ij}(k) > 0$. Note that $(i, i) \in \mathbb{E}(k)$, $\forall k$. A minimal assumption, which is necessary for consensus to be reached and for each agent to have an effect on the final value, requires that following an arbitrary time k, and for any i, j, there is a sequence of communications through which agent i will influence (directly or indirectly) the value held by agent j.

Assumption 5.2 (Connectivity). The graph $(\mathbf{N}, \cup_{s \geq k} \mathbb{E}(s))$ is strongly connected for all $k > 0$.

Various special cases of possible interest are noted:

Time-invariant model There is a fixed matrix $\mathbf{A}$, with entries $\mathbf{a}_{ij}$, such that, for each k, we have $\mathbf{a}_{ij}(k) = \mathbf{a}_{ij}$.

Symmetric model If $(i, j) \in \mathbb{E}(k)$, then $(j, i) \in \mathbb{E}(k)$. That is, whenever i communicates to j, there is a simultaneous communication from j to i.

Equal neighbor model Here,

$$\mathbf{a}_{ij} = \begin{cases} 1/d_i(k) & \text{if } j \in \mathbf{N}_i(k), \\ 0 & \text{if } j \notin \mathbf{N}_i(k), \end{cases}$$

where $\mathbf{N}_i(k) = \{j | (j, i) \in \mathbb{E}(k)\}$ is the set of agents j whose value is taken into account by i at time k, and $d_i(k)$ is its cardinality. Note that here the constant α of Assumption 5.1 can be taken to be $1/n$.

Assumption 5.3 (Bounded intercommunication times). There is some B such that

$$(\mathbf{N}, \mathbb{E}(kB)) \cup \mathbb{E}(kB + 1) \cup \cdots \cup \mathbb{E}((k + 1)B - 1)$$

is strongly connected for all integer k.

Theorem 5.1. *[1] Under Assumptions 5.1, 5.2, and 5.3, the agreement algorithm guarantees asymptotic consensus.*

Theorem 5.2. *[1] Under Assumptions 5.1 and 5.2, for the symmetric model the agreement algorithm guarantees asymptotic consensus.*

It is interesting to note that Theorems 5.1–5.2 can be reformulated as results on the convergence of products of stochastic matrices (refer to Chapter 1).

Corollary 5.1. *Consider an infinite sequence of stochastic matrices*

$$\mathbf{A}(0), \ \mathbf{A}(1), \ \mathbf{A}(2), \ \cdots$$

that satisfies Assumptions 5.1 and 5.2. If either Assumption 5.3 is satisfied or we have a symmetric model, then there exists a nonnegative vector $\mathbf{d}$ *such that*

$$\lim_{k \to \infty} \mathbf{A}(k), \ \mathbf{A}(k - 1), \ \cdots, \ \mathbf{A}(1), \ \mathbf{A}(0) = \mathbf{1}\mathbf{d}^t.$$

Here, **1** is a column vector whose elements are all equal to one.

According to [2] convergence occurs whenever the matrices are all taken from a finite set of ergodic matrices, and the finite set is such that any finite product of matrices in that set is again ergodic. To this end, Corollary 5.1 extends [2] by not requiring the matrices $A(k)$ to be ergodic, though it is limited to matrices with positive diagonal entries.

The presence of long matrix products suggests that convergence to consensus in the linear iteration

$$x(k+1) = \mathbf{A}(k)\, x(k),$$

with $\mathbf{A}(k)$ being stochastic, might be characterized in terms of a joint spectral radius. The joint spectral radius $\varrho(M)$ of a set of matrices M is a scalar that measures the maximal asymptotic growth rate that can be obtained by forming long products of matrices taken from the set M:

$$\varrho(M) = \lim_{k\to\infty} \sup_{M_{i1}, M_{i2}, \cdots, M_{ik} \in M} ||M_{i1}, M_{i2}, \cdots, M_{ik}||^{1/k}.$$

This quantity does not depend on the norm used. Moreover, for any $q > \varrho(M)$ there exists a C for which

$$||M_{ik}, M_{i2}, \cdots, M_{i1}\mathbf{y}||^{1/k} \geq C\, q^k ||\mathbf{y}||, \ \forall \mathbf{y}, \ M_{ij} \in M.$$

Stochastic matrices satisfy $||\mathbf{A}\, x||_\infty \leq ||x||_\infty$ and $\mathbf{A1} = \mathbf{1}$ and so they have a spectral radius equal to one. The product of two stochastic matrices is again stochastic and so the joint spectral radius of any set of stochastic matrices is equal to one.

5.3 Performance and robustness of averaging algorithms

We lay special emphasis on a subcase of the distributed consensus problem, the distributed averaging problem. While a consensus algorithm combines the measurements of the individual nodes into a global value, an averaging algorithm further guarantees that the limit will be the exact average of the individual values.

One of the simplest examples of coordinated control is the so-called *rendezvous problem*. Assume that agents have dynamics of type

$$x_s(k+1) = x_s(k) + u_s(k), \ \ k \in \mathbb{Z}_{\geq 0}, \tag{5.5}$$

with $x_s \in \Re^n$ and $u_s \in \Re^n$, $\forall s \in \mathcal{S}$. The control goal is to make all agents converge their state to the same point. These moving agents with the state representing position define the *rendezvous problem*. Indeed, as we learned before, there are many variants of this problem.

The question now is, *What are exactly the issues we want to analyze?*

Here is a brief list:

(A) Given a graph $\mathbb{G}$, find out whether there exists a control scheme $u_s = g_s(x)$ adapted to $\mathbb{G}$ such that the state evolutions governed by the equations

$$x_s(k+1) = x_s(k) + g_s(x)$$

all converge to the same point, namely, for all initial conditions $x(0)$, there exists $x^* \in \Re^n$ such that

$$\lim_{k\to\infty} x_s(k) = x^*, \quad \forall s \in \mathcal{S}. \tag{5.6}$$

(B) In case when (A) has a positive answer, we would like to find effective ways for producing the control scheme. Indeed, in general, there will be many possible control schemes and the choice can be dictated to optimize certain performance indices:
(B1) the velocity of convergence to the rendezvous point;
(B2) the displacement of x^* from the initial condition.

Remark 5.1. It is significant to note that, without further assumptions, the problem as stated in (A) above is always solvable and with no communication among agents. Simply, it is sufficient to put $u_s = -x_s$, and we will have $x_s(k) = 0$, $\forall s \in \mathcal{S}$, $\forall k \geq 1$, and this is known as a "deadbeat control" in the standard theory of control. This is not a feasible solution since it implicitly requires that agents have already agreed to make O their rendezvous point, and in other terms, they have already coordinated off-line.

Remark 5.2. The prior coordination addressed in Remark 5.1 is not practical since the origin may be far off from their initial condition and thus an unreasonable choice (in general not optimizing (B2)). Henceforth, we make the following extra assumption on the rendezvous point x^* which automatically excludes the deadbeat control scheme above.

We require that, translating all initial conditions $x_s(0) \to x_s(0) + a$ with the same vector, also the rendezvous point translates the same way $x^* \to x^* + a$. We will refer to this as to the translation invariance requirement.

Seeking a linear solution to the foregoing problem, we consider controllers of the type

$$u_s(k) = \sum_{m\in\mathcal{S}} K_{sm}\, x_m(k), \quad K_{sm} \in \Re^{\mathcal{S}\times\mathcal{S}}. \tag{5.7}$$

Incorporating controller (5.7) into the agent dynamics, we readily obtain

$$x_s(k+1) = \sum_{m\in\mathcal{S}} G_{sm}\, x_m(k),$$
$$G_{sm} = K_{sm} + I. \tag{5.8}$$

This type of models (5.8) has applications much broader than just in the rendezvous problem for mobile agents. Instead of a position, the state $x_s(k)$ can as well be interpreted as an estimation or as an opinion on some fact possessed by unit s at time k and the common convergence to the same value is a phenomenon known as consensus.

Note that the dimension of the state does not play any particular role in the dynamics (5.8) as all components of the state vectors $x_s(k)$ evolve separately all with the same dynamics given by the matrix G. For this reason, from now on, we will assume that the state $x_s(k)$ of each agent is one-dimensional, namely, a scalar. In this setting, (5.8) can be rewritten in a more compact form simply as

$$x(k+1) = G\,x(k) \longrightarrow x(k) = G^k\,x(0). \tag{5.9}$$

Observe that the translation invariance, in this context, amounts to requiring that $G^k\mathbf{1} \to \mathbf{1}$ for $k \to \infty$. Since $G^{k+1}\mathbf{1} = GG^k\mathbf{1}$ then converges both to $\mathbf{1}$ and to $G^k\mathbf{1}$, the translation invariance is also equivalent to require $G\mathbf{1} = \mathbf{1}$ (each row of G sums to 1).

Note moreover that the feedback law (5.7) is adapted to $\mathbb{G}$ if K (or equivalently G) is adapted to $\mathbb{G}$. Therefore, to exhibit a solution to the rendezvous problem with translation invariance, it is sufficient to exhibit $G \in \Re^{\mathcal{S}\times\mathcal{S}}$ adapted to $\mathbb{G}$ such that $G\mathbf{1} = \mathbf{1}$ with $G_{sm} \geq 0$, $\forall s, m \in \mathcal{S}$. It turns out that G belongs to the class of *stochastic matrices* which leads to an interesting property that the associated Laplacian matrix is $\mathbb{L}(G) = I - G$. This makes (5.9) expressed as

$$x(k+1) = x(k) - \mathbb{L}(G)x(k), \tag{5.10}$$

or, component-wise,

$$x_s(k+1) = x_s(k) + \sum_m G_{sm}[x_m(k) - x_s(k)]. \tag{5.11}$$

We observe from (5.11) that this only involves the state of s and differences between the states of s and of its neighbors m. Thus, there is no need for the nodes to exchange information in an absolute reference frame, but only relative information suffices.

To this end, the first general observation to be made on stochastic matrices is that the set of stochastic matrices is closed under a number of important operations:

(1) If $P,\ Q \in \Re^{\mathcal{S}\times\mathcal{S}}$ are stochastic, then $\lambda\,P + (1-\lambda)\,Q$ is stochastic for any $\lambda \in (0,\ 1)$.

(2) If $P,\ Q \in \Re^{\mathcal{S}\times\mathcal{S}}$ are stochastic, then $P\,Q$ is stochastic. In particular, P^k is stochastic for any $k \in \mathbb{N}$.

(3) If P_n is a sequence of stochastic matrices such that $P_n \longrightarrow P$ for $n \longrightarrow \infty$, then P is stochastic.

Towards establishing the main behavior of powers of stochastic matrices, we recall the following lemma.

Lemma 5.1 (Contraction principle). *Let $Q \in \Re^{\mathcal{S}\times\mathcal{S}}$ be a stochastic matrix such that there exist $\alpha > 0$ and $m \in \mathcal{S}$ such that $Q_{sm} \geq \alpha$, for all $m \in \mathcal{S}$. Then, for all $x \in \Re^{\mathcal{S}}$, it holds true that $y = Q\,x$ satisfies*

$$\max_{s\in\mathcal{S}} y_s - \min_{s\in\mathcal{S}} y_s \leq (1-\alpha)(\max_{s\in\mathcal{S}} x_s - \min_{s\in\mathcal{S}} x_s). \tag{5.12}$$

Proof. We observe that

$$\begin{aligned} y_s = \sum_{m\in\mathcal{S}} Q_{sm}\,x_m &= \sum_{m\in\mathcal{S}} Q_{sm}[x_m - \min_{u\in\mathcal{S}} x_u] \\ &\quad + \sum_{m\in\mathcal{S}} Q_{sm} \min_{u\in\mathcal{S}} x_u \\ &\geq \alpha[x_m - \min_{u\in\mathcal{S}} x_u] + \min_{u\in\mathcal{S}} x_u \\ &= \alpha\, x_m + (1-\alpha)\min_{u\in\mathcal{S}} x_u. \end{aligned} \tag{5.13}$$

In a similar way,

$$\begin{aligned} y_s = \sum_{m\in\mathcal{S}} Q_{sm}\,x_m &= \sum_{m\in\mathcal{S}} Q_{sm}[x_m - \max_{u\in\mathcal{S}} x_u] \\ &\quad + \sum_{m\in\mathcal{S}} Q_{sm} \max_{u\in\mathcal{S}} x_u \\ &\leq \alpha[x_m - \max_{u\in\mathcal{S}} x_u] + \max_{u\in\mathcal{S}} x_u \\ &= \alpha\, x_m + (1-\alpha)\max_{u\in\mathcal{S}} x_u. \end{aligned} \tag{5.14}$$

Combining inequalities (5.13) and (5.14) yields (5.12) as desired. □

Lemma 5.2 (Convergence to consensus). *Let $P \in \Re^{\mathcal{S}\times\mathcal{S}}$ be a stochastic matrix such that $\mathbb{G}(P)$ admits a globally reachable aperiodic vertex. Then, the following two equivalent facts hold true.*

(A) *The dynamics (5.9) are such that, for any initial condition $x(0) = x_0 \in \Re^{\mathcal{S}}$, there exists a scalar such that*

$$x(k) = G^k\,x(0) \longrightarrow \alpha\,\mathbf{1}, \quad k \longrightarrow \infty,$$

which implies that all components $x_s(k)$ converge to the same value α.

(B) *There exists a vector $\pi \in \Re^{\mathcal{S}}$ such that $\pi_s \geq 0$ for all s, $\sum_{s\in\mathcal{S}} \pi_s = 1$, and*

$$\lim_{k\longrightarrow\infty} G^k = \mathbf{1}\,\pi^*. \tag{5.15}$$

In other terms, G^k converges to a matrix having all rows equal to the row vector π^*. Furthermore, $\alpha = \pi^* x(0)$.

Proof. We let $s \in \mathcal{S}$ be the aperiodic vertex which is reachable from all others. This means that there exists $k^* \in \mathbb{N}$ such that $Q := P^{k^*}$ is such that $Q_{ms} > 0$ for all $m \in \mathcal{S}$. Let $\alpha = \min\{Q_{ms} : m \in \mathcal{S}\} > 0$. By letting $y^0 \in \Re^{\mathcal{S}}$ and $y^1 = Q\, y^0$, Lemma 5.1 implies that

$$\max_{s\in\mathcal{S}} y_s^1 - \min_{s\in\mathcal{S}} y_s^1 \leq (1-\alpha)(\max_{s\in\mathcal{S}} y_s^0 - \min_{s\in\mathcal{S}} y_s^0). \tag{5.16}$$

Now fix $x(0) = x^0 \in \Re^{\mathcal{S}}$ arbitrarily and consider (5.9). Define

$$M_k = \max_{s\in\mathcal{S}} x_s(k), \quad m_k = \min_{s\in\mathcal{S}} x_s(k)$$

and take into consideration that the components of $x(k)$ are convex combinations of those of $x(k-1)$; the sequences M and m are therefore bounded and, respectively, nonincreasing and nondecreasing (and hence convergent).

Next, the sequence $\Delta_k = M_k - m_k$ converges. Moreover, it holds that $\Delta_{nk^*} \leq (1-\alpha)^n\, \Delta_0$, which implies that $\Delta_{nk^*} \longrightarrow 0$ for $n \longrightarrow \infty$. Hence, all components of $x(k)$ will converge to the same limit, thus proving the first claim (A). On applying this result choosing $x(0) = e_m$, the mth element of the canonical basis of $\Re^{\mathcal{S}}$, we obtain that all elements of the mth column of P^k will converge to the same limit. This clearly yields the second claim (B). □

5.3.1 Simulation example 5.1

Consider the stochastic matrix

$$P = \begin{bmatrix} 1/2 & l/2 & 0 \\ l/3 & l/3 & l/3 \\ 1/3 & 0 & 2/3 \end{bmatrix}.$$

It is evident that P is irreducible and aperiodic. Let us compute the invariant probability π. From $\pi^* P = \pi^*$, we get

$$\begin{cases} -1/2\pi_1 + 1/3\pi_2 + 1/3\pi_3 = 0, \\ 1/2\pi_1 - 2/3\pi_2 = 0, \\ 1/3\pi_2 - 1/3\pi_3 = 0. \end{cases}$$

This leads to $\pi_2 = \pi_3$ and $\pi_1 = 4/3\pi_2$. On using the normalization condition $1/\pi_1 + \pi_2 + \pi_3 = 1$, we get finally $\pi = [2/5,\ 3/10,\ 3/10]^*$. Fig. 5.1 depicts the convergence behavior. The left diagram displays the entries of the third row of P^k, which converge to π^*. The right diagram plots the associated averaging dynamics (5.9) from random initial conditions within $(0,\ l)$.

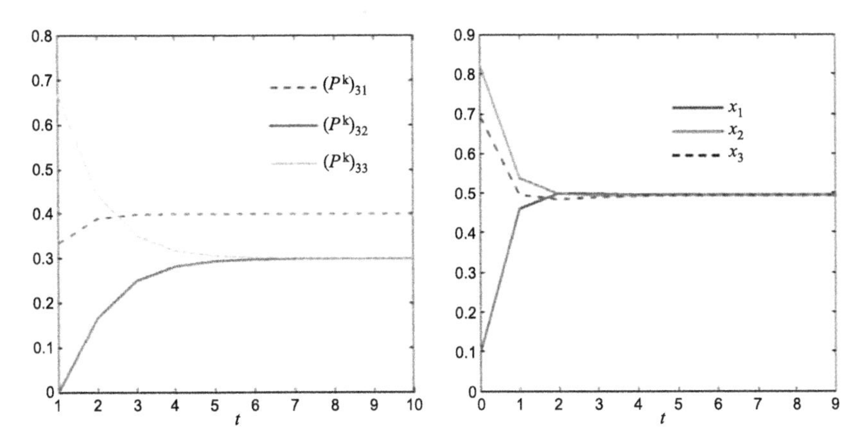

FIGURE 5.1 Illustration of convergence.

5.3.2 Convergence rate and eigenvalues

This section deals more precisely with convergence speed. The speed of convergence of (5.15) is dictated by the magnitude of the eigenvalues of P. We start by recalling the following result.

Lemma 5.3. *Let $M \subset \Re^{\mathcal{S}\times\mathcal{S}}$ be any matrix and let λ_i be its eigenvalues. Let $\varrho = \max|\lambda_i|$ be the spectral norm of M. Then for every $\varepsilon > 0$, there exists a constant C_ε such that*

$$||M^k x_o||_2 \leq C_\varepsilon(\varrho+\varepsilon)^k ||x_0||_2, \quad \forall\, k. \tag{5.17}$$

A simple application of this lemma allows us to obtain the following results.

Lemma 5.4. *Let $P \in \Re^{\mathcal{S}\times\mathcal{S}}$ be a stochastic matrix such that $\mathbb{G}(P)$ admits a globally reachable aperiodic node. Consider all its eigenvalues μ_i but one and define $\varrho_2 = \max\{|\mu_i| < 1\}$. Then, for every $\varepsilon > 0$, there exists a constant C_ε such that*

$$||(P^k - \mathbf{1}\pi^*)x_o||_2 \leq C_\varepsilon(\varrho_2+\varepsilon)^k ||x_0||_2, \quad \forall\, k. \tag{5.18}$$

Note in Lemma 5.4 that letting $R = P - \mathbf{1}\pi^*$, it follows that $R^k = P^k - \mathbf{1}\pi^*$, and $R\mathbf{1} = 0$. The parameter ϱ_2, introduced in the statement of Lemma 5.4, is also called the *second eigenvalue of P*, and the difference $l - \varrho_2$ is called the *spectral gap of P*.

The foregoing result essentially discloses that convergence to rendezvous happens exponentially fast as ϱ_2^k. This is only approximately true because of the arbitrarily small ε we have to fix. When P is symmetric, things are much simpler and we can indeed prove the following result.

Lemma 5.5 (Convergence rate for symmetric P). *Let $P \in \Re^{\mathcal{S}\times\mathcal{S}}$ be a symmetric stochastic matrix such that $\mathbb{G}(P)$ is strongly connected and aperiodic. Then*

$$||(P^k - N^{-1}\mathbf{11}^*)x_o||_2 \leq \varrho_2^k \,||x_0||_2. \tag{5.19}$$

5.3.3 Linear quadratic control

We consider the usual time-invariant consensus dynamics

$$x(k+1) = P\,x(k), \tag{5.20}$$

where the matrix P is adapted to a strongly connected aperiodic graph $\mathbb{G} = (\mathbb{V}, \mathbb{E})$ of order N. For simplicity, we assume that the matrix P is symmetric, although this assumption can be relaxed to some extent. The eigenvalues of P are denoted as μ_i for $i \in \{1, \cdots, N\}$, $\mu_1 = l$. We recall the second eigenvalue $\varrho_2 = \max\{|\mu_i|,\ i = 2, \cdots, N\}$. When convenient, we will also make suitable assumptions on the statistics of the initial condition.

Earlier, the speed of convergence to the consensus value of dynamics (5.21) was estimated in terms of the second eigenvalue ϱ_2 of the matrix P. From Lemma 5.5, we can recall that the second eigenvalue determines the convergence rate according to the estimate

$$N^{-1}||(P^k - N^{-1}\mathbf{11}^*)x_o||^2 \leq \varrho_2^{2k}\, N^{-1}\,||x_0||_2. \tag{5.21}$$

Note that different from (5.19), we have multiplied both sides of the inequality in (5.21) by N^{-1}. Indeed, in the limit $N \to \infty$, it makes sense to consider the normalized version of the squared norm $N^{-1}||N_n||^2$, because $|x_s(0)| \leq \varepsilon$ for every $s \in \mathcal{S}$ yields $N^{-1}||N_n||^2 \leq \varepsilon^2$.

Considering the initial conditions $x_s(0)$ are assumed independent random variables with mean m and variance σ^2, by taking the mean value of (5.21) we obtain

$$\begin{aligned}
&\frac{1}{N}\mathbb{E}[||P^k x_o - N^{-1}\mathbf{11}^* x_o||^2] \\
&= \frac{1}{N}\mathbb{E}[||(P^k - N^{-1}\mathbf{11}^*)x_o||^2] \\
&= \frac{1}{N}\mathbb{E}[\mathbf{trace}((P^k - N^{-1}\mathbf{11}^*)(x_o x_o^*)(P^k - N^{-1}\mathbf{11}^*))] \\
&= \frac{1}{N}[\mathbf{trace}((P^k - N^{-1}\mathbf{11}^*)\mathbb{E}(x_o x_o^*)(P^k - N^{-1}\mathbf{11}^*))] \\
&= \frac{\sigma^2}{N}[\mathbf{trace}(P^{2k} - N^{-1}\mathbf{11}^*)],
\end{aligned} \tag{5.22}$$

which can be further expressed in terms of the Frobenius norm (see Chapter 1) as

$$\frac{1}{N}\mathbb{E}[||P^k x_o - N^{-1}\mathbf{11}^* x_o||^2] = \frac{\sigma^2}{N}||P^k - N^{-1}\mathbf{11}^*||_F$$
$$= \frac{\sigma^2}{N}\sum_{i=2}^{N} |\mu_i|^{2k}. \tag{5.23}$$

Remark 5.3. It is significant to observe that the estimate (5.21) can be readily derived from estimate (5.23) by simply using the bounding relation $||P^k - N^{-1}\mathbf{11}^*||_F \leq N\, \varrho_2^k$. In the case of complete graphs, $P = N^{-1}\mathbf{11}^*$ and therefore $||P^k - N^{-1}\mathbf{11}^*||_F = 0$ and $\varrho_2 = 0$. Thus, the two estimates coincide in this case.

We proceed further and recap the consensus dynamics in the context of the rendezvous application by interpreting it as a closed-loop feedback control as

$$x(k+1) = x(k) + u(k), \quad u(k) = [P - I]x(k). \tag{5.24}$$

Given a scalar $\varphi > 0$, we measure the performance of system (5.24) by the standard quadratic cost functionals

$$J \triangleq J_x + \varphi\, J_u,$$
$$J_x = N^{-1}\sum_{m=0}^{\infty} \mathbb{E}[||x(m) - N^{-1}\mathbf{11}^* x_o||^2],$$
$$J_u = N^{-1}\sum_{m=0}^{\infty} \mathbb{E}||u(m)||^2. \tag{5.25}$$

The term J_x measures the speed of convergence to consensus, whereas J_u measures the control effort needed to achieve it. The two functionals J_x and J_u can be expressed in terms of the eigenvalues $\{\mu_i\}$. The following result is established.

Lemma 5.6 (Linear quadratic cost). *If the stochastic matrix $P \in \Re^{S \times S}$ is irreducible aperiodic and symmetric, then*

$$J_x = \frac{\sigma^2}{N}\sum_{m=0}^{\infty} ||P^m - N^{-1}\mathbf{11}^*||_F^2 = \frac{\sigma^2}{N}\sum_{j=2}^{\infty}\frac{1}{1-\mu_j^2}, \tag{5.26}$$

$$J_u = \frac{\sigma^2}{N}\sum_{m=0}^{\infty} ||P^{m+1} - P^m||_F^2 = \frac{\sigma^2}{N}\sum_{j=2}^{\infty}\frac{1-\mu_j}{1+\mu_j}. \tag{5.27}$$

Proof. It is a straightforward task to show that expression (5.26) is a consequence of (5.23). Regarding J_u, the first equality comes from a computation analogous to (5.22). The second one instead follows from the observation that the eigenvalues of $(P^{m+1} - P^m)^2$ are given by $(\mu_j^{m+1} - \mu_j^m)^2$, which equals $\{\mu_j^{2m}(1-\mu_j)^2\}$. This completes the proof. □

Remark 5.4. In general, it is immediate to see that

$$\frac{N-1}{N} \leq \frac{J_x}{\sigma^2} \leq \frac{1}{1-\varrho_2^2},$$

for which both bounds might be tight. On the other hand, using the Gershgorin theorem (see Chapter 1) and letting $P = [P_{sm}]$, $s, m \in \mathcal{S}$ such that $P_{ss} > 0$, $\forall s \in \mathcal{S}$ and $\beta = \min_s P_{ss}$, it follows that

$$J_u \leq \frac{1-\beta}{\beta}.$$

5.3.4 Performance with noise

We continue further and treat noisy agent dynamics and analyze the behavior and performance of consensus algorithms under the presence of noise in the dynamics. As we will see, cost functionals similar to those introduced above naturally qualify to this case.

We start by considering the case when noise enters additively in the update equation:

$$x(k+1) = P\,x(k) + \omega(k). \tag{5.28}$$

We recall the standing assumption that P is a symmetric stochastic matrix. We assume $\omega_s(k)$ to be independent random variables with zero mean and variance σ^2. Initially, we note that the mean value is governed by $\mathbb{E}[x(k+1)] = P\,\mathbb{E}[x(k)]$ so that, if P is irreducible and aperiodic, we have convergence to the average consensus: $\mathbb{E}[x(k)] \longrightarrow N^{-1}\mathbf{1}\mathbf{1}^* x(0)$ (here, $x(0)$ is seen as deterministic).

We introduce the mean terms

$$m(k) = N^{-1}\sum_s x_s(k), \quad \mu(k) = N^{-1}\sum_s \omega_s(k).$$

Then (5.28) implies that

$$m(k+1) = m(k) + \mu(k) \tag{5.29}$$

and consequently

$$m(k) = m(0) + \sum_s^{k-1} \mu(s). \tag{5.30}$$

Observe that each $\mu(k)$ is a random variable with zero mean and variance σ^2/N. Therefore, we can conclude that $m(k)$ is a random process with

$$\mathbb{E}[m(k)] = m(0), \quad \mathbf{Var}[m(k)] = \sigma^2/N.$$

This shows how noise accumulates into the linear dynamics (essentially because of its marginally stable structure) and creates such unbounded effects on the average dynamics. A similar phenomenon takes place if we measure the distance of the process from a consensus point. Consider the following functional:

$$J_\omega = \frac{1}{N} \lim_{k\to\infty} \mathbb{E}||x(k) - N^{-1}\mathbf{11}^* x(k)||^2. \tag{5.31}$$

It is significant to observe that J_ω coincides with the functional J_x introduced to describe the linear-quadratic cost functional.

Lemma 5.7. *If the stochastic matrix $P \in \Re^{\mathcal{S}\times\mathcal{S}}$ is irreducible aperiodic and symmetric, then*

$$J_\omega = J_x = \frac{\sigma^2}{N} \sum_{m=0}^{\infty} ||P^{2m} - N^{-1}\mathbf{11}^*||_F^2. \tag{5.32}$$

Proof. It follows from (5.28) that, for every time k, we have

$$x(k) = P^k m(0) + \sum_{s=0}^{k-1} P^s\, \omega(k-s-1),$$

which, in turn, yields

$$\begin{aligned}
&\mathbb{E}||x(k) - N^{-1}\mathbf{11}^* x(k)||^2 \\
&= \mathbb{E}||(P^k - N^{-1}\mathbf{11}^*)x(0)||^2 \\
&+ \sum_{s=0}^{k-1}\sum_{j=0}^{k-1} \mathbb{E}[(P^k - N^{-1}\mathbf{11}^*)\omega(k-s-1)^*][(P^j - N^{-1}\mathbf{11}^*)\omega(k-j-1)^*] \\
&+ 2\sum_{s=0}^{k-1} \mathbb{E}[(P^k - N^{-1}\mathbf{11}^*)x(0)^*(P^j - N^{-1}\mathbf{11}^*)\omega(k-j-1)^*)].
\end{aligned} \tag{5.33}$$

The first term in (5.33) converges to 0 when $k \to \infty$, due to the assumptions made on P. The third term equals zero because all noises are zero mean. Finally, the second term can be rewritten as

$$\sigma^2 \sum_{s=0}^{k-1} \mathbf{trace}[(P^{2s} - N^{-1}\mathbf{11}^*)] = \sigma^2 \sum_{s=0}^{k-1} [P^s - N^{-1}\mathbf{11}^*]_F^s,$$

where the independence assumption on the noises has been used. By taking the limit as $k \to \infty$, we obtain the desired result. □

5.4 Leader following consensus

In recent years, much attention has been paid to distributed coordinated control of multiagent systems due to their potential applications in various areas, such as smart grids [6], aircraft control [23], formation control [5], [9], rendezvous [13], robotics [22], sensor networks [20], and cooperative unmanned air vehicles [3]. Although each agent has limited processing ability, a certain number of agents connected by a network as a whole can perform complex missions in a coordinated way. An important and fundamental issue in coordinated control of multiagent systems is the consensus problem, which aims to develop a distributed protocol for each agent based on the limited information about itself and its neighbors to make all agents reach an agreement on certain quantities of interest.

5.4.1 Review

Consensus problems have a long history in the fields of computer science and distributed computing [7], [10], [11], [14–16], [19], [29], [31], [32]. An important embranchment of consensus problems is the leader following consensus problem, and a number of researchers have made great achievements in the literature (see for instance [8], [12], [26], [27], [30]). It should be noted that time delays cannot be avoided in the research of networked control systems [17], [18]. Ding et al. [8] proposed a network-based control protocol and a delay-dependent stability criterion to ensure the leader following consensus. Tang et al. studied the leader following consensus for a class of nonlinear stochastic multiagent systems with bounded time-varying delays in [27] and some conditions were proposed for global exponential leader following consensus.

On the other hand, the study of discrete-time multiagent systems with time delay has attracted more and more research attention during the 2010s. In [24], a new concept, preleader–follower decomposition, was introduced, and a necessary and sufficient condition of consensus was provided. Several criteria were provided in [28] for consensus of nonlinear discrete-time multiagent systems with time-varying delay. In [25], Wang et al. studied the consensus problem of a first-order discrete-time multiagent system with both constant and time-varying delays, and provided the maximum delay bound for each case.

It is however worth noting that all aforementioned papers studied the consensus problem with small delay sequences (SDSs), that is, for the time delays less than an upper bound, consensus of the corresponding multiagent system can always be guaranteed. However, due to the presence of the communication network in a multiagent system, the consecutive data dropouts are unavoidable. If the zero-order hold technique is adopted, this may lead to the large delay sequence (LDS) phenomenon, that is, the time delays are larger than that in the above-mentioned papers. In such a situation, the existing methods in the literature are not applicable anymore. This motivated us to propose a new consensus analysis framework for such kinds of multiagent systems.

Large delay theory (LDT) was first presented in [21], which was used to analyze the stability of time delay systems with LDS. Such a method can provide a larger upper bound of time delays, and thus is less conservative. Based on the LDT, the leader following consensus problem is investigated for a multiagent system with LDS.

5.4.2 Problem formulation

The multiagent system under consideration consists of N following agents and a leader agent v_0, whose dynamics are described by

$$x_0(k+1) = Ax_0(k), \tag{5.34}$$

where $x_0 \in \Re^n$ is the state of the leader v_0 and $A \in \Re^{n\times n}$ is a known constant matrix. The dynamics of the following agents are given by

$$x_i(k+1) = Ax_i(k) + Bu_i(k), \tag{5.35}$$

where $x_i \in \Re^n$ is the state of agent v_i, $u_i \in \Re^m$ is the control input of agent v_i, $i \in \mathcal{N}$, and $B \in \Re^{n\times m}$ is a known constant matrix. Protocol u_i is said to solve the leader following consensus problem if for any initial conditions the states of agents satisfy

$$\lim_{k\to+\infty} x_i(k) - x_0(k) = 0, \quad i \in \mathcal{I}. \tag{5.36}$$

We recall from previous chapters that to solve the leader following consensus problem, the following control protocol is quite effective:

$$u_i(k) = -K\left[\sum_{v_j\in\mathcal{N}_i} \alpha_{ij}(k)(x_j(k) - x_i(k)) + m_i(x_0(k) - x_i(k))\right]. \tag{5.37}$$

By taking the communication time delay into account, we modify (5.37) and apply the following control protocol:

$$\begin{aligned} u_i(k) = -K\big[&\sum_{v_j\in\mathcal{N}_i} \alpha_{ij}(k)(x_j(k-\tau(k)) - x_i(k-\tau(k))) \\ &+ m_i(x_0(k-\tau(k)) - x_i(k-\tau(k)))\big], \end{aligned} \tag{5.38}$$

where $K \in \Re^{m\times n}$ is a gain matrix to be determined and $\tau(k)$ is a time-varying integer delay satisfying $0 \le d_1 \le \tau(k) \le d_3$. With protocol (5.38), the closed-loop form of system (5.35) can be rewritten as

$$x_i(k+1) = Ax_i(k) - BK[\sum_{v_j \in \mathcal{N}_i} \alpha_{ij}(x_j(k-\tau(k)) - xi(k-\tau(k)))$$
$$+ m_i(x_0(k-\tau(k)) - x_i(k-\tau(k)))]. \tag{5.39}$$

To render the solvability of the leader following consensus tractable, we introduce the following assumptions and definitions.

Assumption 5.4. The matrix pair $(A,\ B)$ is controllable.

Assumption 5.5. Every vertex $v_i, i = 1, \ldots, \mathbb{V}$, is reachable from v_0 in $\mathbb{G}$.

Assumption 5.6. The leader following consensus problem of (5.39) can be solved if $0 \le d_1 \le \tau(k) \le d_2$ based on the existing literature, but it cannot be solved if $d_2 < \tau(k) \le d_3$ for $\forall k \in \mathbb{N}$.

Definition 5.1. Given a time sequence $\mathbb{S} = \{i, i+1, i+2, \ldots, i+j\}, i, j \in \mathbb{N}$, $\mathbb{S}$ is called an LDS if $d_2 < \tau(k) \le d_3$ for $\forall k \in \mathbb{S}$.

Definition 5.2. Given a time sequence $\mathbb{S} = \{i, i+1, i+2, \ldots, i+j\}, i, j \in \mathbb{N}$, $\mathbb{S}$ is called an SDS if $d_1 \le \tau(k) \le d_2$ for $\forall k \in \mathbb{S}$.

Not only do we investigate the general communication delay bound, but we also consider the occasional appearance of large time delay, which exceeds the general delay bound.

Remark 5.5. For any $k_1, k_2 \in \mathbb{N},\ k_1 < k_2$, the number of LDSs in the sequence $\mathbb{S} = \{k_1, k_1+1, \ldots, k_2-1\}$ is denoted as $N_l(k_1,\ k_2)$, and the number of switches in sequence $\mathbb{S}$ is denoted as $N_l(k_1,\ k_2)$. It is obvious that $N_\sigma(k_1,\ k_2) \le 2N_l(k_1,\ k_2)$.

In the sequel, we focus on the analysis of the consensus condition of systems (5.34) and (5.39) with large time delays. In addition, it will be investigated that the lengths and frequencies of LDSs in the interval $[k_0,\ k]$ can ensure consensus.

5.4.3 Design result

In this section, we will analyze the leader following consensus problem. Before presenting the main result, we first propose a novel time delay system and then provide two basic lemmas that are essential to the development of the main result of this section.

In practical applications of multiagent systems, due to the fact that agents are connected through networks, communication failures or consecutive packet dropouts may happen randomly, which will lead to the appearance of LDSs. Thus, LDSs and SDSs may appear alternatively. In order to illustrate such a case clearly, we introduce the following time delay system with switched time delays:

$$x_i(k+1) = Ax_i - BK\Big[\sum_{v_j \in \mathcal{N}_i} \alpha_{ij} x_j(k - \tau_{\sigma(k)}(k)) - x_i(k - \tau_{\sigma(k)}(k)) \\ + m_i(x_0(k - \tau_{\sigma(k)}(k)) - x_i(k - \tau_{\sigma(k)}(k))\Big], \tag{5.40}$$

where

$$\sigma(k) = \begin{cases} 1, & d_1 \le \tau(k) \le d_2, \\ 2, & d_2 < \tau(k) \le d_3 \end{cases}$$

is a piece-wise constant function. It implies that system (5.40) operates in SDSs when $\sigma(k) = 1$ and system (5.40) runs in LDSs when $\sigma(k) = 2$.

Let $\delta_i(k) = x_i(k) - x_0(k)$, $i \in \mathcal{I}$. Then, we obtain the closed-loop system of the multiagent system (5.34)–(5.35)

$$\begin{aligned} \delta(k+1) &= (I_N \otimes A)\delta(k) + (L \otimes BK)\delta(k - \tau_{\sigma(k)}(k)) \\ &\quad + (\Lambda \otimes BK)\delta(k - \tau_{\sigma(k)}(k)) \\ &= (I_N \otimes A)\delta(k) + (H \otimes BK)\delta(k - \tau_{\sigma(k)}(k)), \end{aligned} \tag{5.41}$$

where $\delta(k) = \left[\delta_1^T(k), \ldots, \delta_N^T(k)\right]^T$, $x(k) = \left[x_1^T(k), \ldots, x_N^T(k)\right]^T$, and $H = L + \Lambda$ with $\Lambda = \text{diag}(m_1, m_2, \ldots, m_N)$.

Remark 5.6. Obviously, if system (5.41) is asymptotically stable, then the leader following consensus for the multiagent system (5.34)–(5.35) is reached by protocol (5.38). Based on Assumption 5.6, when $d_2 < \tau(k) \le d_3$, the leader following consensus cannot be guaranteed. Therefore, the proposed methods in the existing literature are not able to solve such a problem, and hence a new methodology should be developed to solve this problem.

Next, we analyze the stability of system (5.41). First, the case of $\sigma(k) = 1$ is considered:

$$\begin{aligned} \delta(k+1) &= (I_N \otimes A)\delta(k) + (H \otimes BK)\delta(k - \tau(k)), \\ \delta(k) &= \theta(k), \quad k = -d_3, -d_3 + 1, \ldots, 0, \end{aligned} \tag{5.42}$$

where $d_1 \le \tau(k) \le d_2$ and $\theta(k)$ is a function defined in the interval $[-d_3, 0]$. We choose the following Lyapunov functional candidate:

$$V_1(k) = \sum_{i=1}^{5} V_{1i}(k), \tag{5.43}$$

where

$$
\begin{aligned}
V_{11}(k) &= \delta^T(k)P_1\delta(k), \\
V_{12}(k) &= \sum_{i=k-d_2}^{k-1} e^{\gamma_1(i-k+1)}\delta^T(k)Q_1\delta(k), \\
V_{13}(k) &= \sum_{i=k-d_3}^{k-d_2-1} e^{\gamma_1(i-k+1)}\delta^T(k)Q_2\delta(k), \\
V_{14}(k) &= \sum_{j=-d_2}^{-1}\sum_{i=k+j}^{k-1} e^{\gamma_1(i-k+1)}\mathfrak{R}^T(i)(E_1+E_2)\mathfrak{R}(i), \\
V_{15}(k) &= \sum_{j=-d_3}^{-d_2-1}\sum_{i=k+j}^{k-1} e^{\gamma_1(i-k+1)}\mathfrak{R}^T(i)E_3\mathfrak{R}(i),
\end{aligned}
$$

with $\mathfrak{R}(k) = \delta(k+1) - \delta(k)$, and P_1, Q_1, Q_2, and E_v $(v = 1, 2, 3)$ are positive definite matrices to be determined.

Lemma 5.8. *Assume that Assumptions 5.4–5.6 hold. For a given constant $\gamma_1 > 0$, if there exist matrices $P_1 > 0$, $Q_i > 0$ $(i = 1, 2)$, $E_j > 0$ $(j = 1, 2, 3)$ and matrices M_i $(i = 1, 2, 3, 4)$ with appropriate dimensions such that the following linear matrix inequality (LMI) holds,*

$$
\begin{bmatrix} \Phi_1 + \Phi_2 + \Phi_2^T & \Phi_3 \\ & \Phi_4 \end{bmatrix} < 0, \tag{5.44}
$$

where

$$
\Phi_1 = \begin{bmatrix} \phi_{11} & \phi_{12} & 0 & 0 \\ & \phi_{22} & 0 & 0 \\ & * & e^{-d_2\gamma_1}Q_2 - e^{-d_2\gamma_1}Q_1 & 0 \\ & * & * & -e^{-d_3\gamma_1}Q_2 \end{bmatrix},
$$

$$
\begin{aligned}
\phi_{11} &= (I_N \otimes A^T)P_1(I_N \otimes A) - e^{-\gamma_1}P_1 + Q_1 \\
&\quad + d_2(I_N \otimes (A-I)^T)(E_1+E_2)(I_N \otimes (A-I)) \\
&\quad + (d_3 - d_2)(I_N \otimes (A-I)^T)E_3(I_N \otimes (A-I)), \\
\phi_{12} &= (I_N \otimes A^T)P_1(H \otimes BK) \\
&\quad + d_2(I_N \otimes (A-I)^T)(E_1+E_2)(H \otimes BK) \\
&\quad + (d_3 - d_2)(I_N \otimes (A-I)^T)E_3(H \otimes BK), \\
\phi_{22} &= (H^T \otimes K^T B^T)P_1(H \otimes BK) \\
&\quad + d_2(H^T \otimes K^T B^T)(E_1+E_2)(H \otimes BK) \\
&\quad + (d_3 - d_2)(H^T \otimes K^T B^T)E_3(H \otimes BK),
\end{aligned}
$$

$$\Phi_2 = [M_1 + M_3 \quad M_2 - M_1 \quad M_4 - M_2 - M_3 \quad -M_4],$$
$$\Phi_3 = [\sqrt{\rho_1} M_1 \quad \sqrt{\rho_2} M_2 \quad \sqrt{\rho_1} M_3 \quad \sqrt{\rho_3} M_4],$$
$$\Phi_4 = diag\{-E_1, -E_1, -E_2, -E_3\},$$
$$\rho_1 = \frac{e^{(d_2+1)\gamma_1} - e^{\gamma_1}}{e^{\gamma_1} - 1}, \quad \rho_2 = \frac{e^{(d_2+1)\gamma_1} - e^{\gamma_1(d_1+1)}}{e^{\gamma_1} - 1},$$
$$\rho_3 = \frac{e^{(d_3+1)\gamma_1} - e^{(d_2+1)\gamma_1}}{e^{\gamma_1} - 1},$$

then along the trajectory of system (5.42), we have $V_1(k+1) \leq e^{-\gamma_1} V_1(k)$.

Proof. Define $V_1 \triangleq \sum_{i=1}^{5} \Delta V_{1i} = \sum_{i=1}^{5}(V_{1i}(k+1) - e^{-\gamma_1} V_{1i}(k))$ and $\Gamma(k) \triangleq [\delta^T(k) \ \delta^T(k-\tau(k)) \ \delta^T(k-d_2) \ \delta^T(k-d_3)]^T$. Along the trajectory of system (5.42), we have

$$\begin{aligned}
\Delta V_1 &= \delta^t(k)(I_N \otimes A^t) P_1 (I_N \otimes A)\delta(k) \\
&+ \delta^t(k)(I_N \otimes A^t) P_1 (H \otimes BK)\delta(k-\tau(k)) \\
&+ \delta^t(k-\tau(k))(H^t \otimes K^t B^t) P_1 (I_N \otimes A)\delta(k) - e^{-\gamma_1}\delta^t(k) P_1 \delta(k) \\
&+ \delta^t(k-\tau(k))(H^t \otimes K^t B^t) P_1 (H \otimes BK)\delta(k-\tau(k)) + \delta^t(k) Q_1 \delta(k) \\
&- e^{-d_2\gamma_1}\delta^t(k-d_2) Q_1 \delta(k-d_2) + e^{-d_2\gamma_1}\delta^T(k-d_2) Q_2 \delta(k-d_2) \\
&- e^{-d_3\gamma_1}\delta^t(k-d_3) Q_2 \delta(k-d_3) + d_2 \mathfrak{R}^t(k)(E_1 + E_2)\mathfrak{R}(k) \\
&- \sum_{i=k-\tau(k)}^{k-1} e^{\gamma_1(i-k)} \mathfrak{R}^t(i) E_1 \mathfrak{R}(i) - \sum_{i=k-d_2}^{k-\tau(k)-1} e^{\gamma_1(i-k)} \mathfrak{R}^t(i) E_1 \mathfrak{R}(i) \\
&- \sum_{i=k-d_2}^{k-1} e^{\gamma_1(i-k)} \mathfrak{R}^t(i) E_2 \mathfrak{R}(i) + (d_3 - d_2)\mathfrak{R}^t(k) E_3 \mathfrak{R}(k) \\
&- \sum_{i=k-d_3}^{-k-d_2-1} e^{\gamma_1(i-k)} \mathfrak{R}^T(i) E_3 \mathfrak{R}(i) \\
&+ 2\Gamma^t(k) M_1 \left[\delta(k) - \delta(k-\tau(k)) - \sum_{i=k-\tau(k)}^{k-1} \mathfrak{R}(i) \right] \\
&+ 2\Gamma^t(k) M_2 \left[\delta(k-\tau(k)) - \delta(k-d_2) - \sum_{i=k-d_2}^{k-\tau(k)-1} \mathfrak{R}(i) \right] \\
&+ 2\Gamma^t(k) M_3 \left[\delta(k) - \delta(k-d_2) - \sum_{i=k-d_2}^{k-1} \mathfrak{R}(i) \right] \\
&+ 2\Gamma^t(k) M_4 \left[\delta(k-d_2) - \delta(k-d_3) \sum_{i=k-d_3}^{-k-d_2-1} \mathfrak{R}(i) \right]
\end{aligned}$$

$$
\begin{aligned}
&\leq \delta^t(k)(I_N \otimes A^t)P_1(I_N \otimes A)\delta(k) - e^{-\gamma_1}\delta^t(k)P_1\delta(k) \\
&+ \delta^t(k)(I_N \otimes A^t)P_1(H \otimes BK)\delta(k-\tau(k)) \\
&+ \delta^t(k-\tau(k))(H^t \otimes K^tB^t)P_1(I_N \otimes A)\delta(k) \\
&+ \delta^t(k-\tau(k))(H^t \otimes K^tB^t)P_1(H \otimes BK)\delta(k-\tau(k)) \\
&+ \delta^t(k)Q_1\delta(k) - e^{-d_2\gamma_1}\delta^T(k-d_2)Q_1\delta(k-d_2) \\
&+ e^{-d_2\gamma_1}\delta^T(k-d_2)Q_2\delta(k-d_2) - e^{-d_3\gamma_1}\delta^t(k-d_3)Q_2\delta(k-d_3) \\
&+ d_2\delta^t(k)(I_N \otimes (A-I)^t)(E_1+E_2)(I_N \otimes (A-I))\delta(k) \\
&+ d_2\delta^t(k)(I_N \otimes (A-I)^t)(E_1+E_2)(H \otimes BK)\delta(k-\tau(k)) \\
&+ d_2\delta^t(k-\tau(k))(H^t \otimes K^tB^t)(E_1+E_2)(I_N \otimes (A-I))\delta(k) \\
&+ d_2\delta^t(k-\tau(k))(H^t \otimes K^tB^t)(E_1+E_2)(H \otimes BK)\delta(k-\tau(k)) \\
&+ (d_3-d_2)\delta^t(k)(I_N \otimes (A-I)^t)E_3(I_N \otimes (A-I))\delta(k) \\
&+ (d_3-d_2)\delta^t(k)(I_N \otimes (A-I)^t)E_3(H \otimes BK)\delta(k-\tau(k)) \\
&+ (d_3-d_2)\delta^t(k-\tau(k))(H^t \otimes K^tB^t)E_3(I_N \otimes (A-I))\delta(k) \\
&+ (d_3-d_2)\delta^t(k-\tau(k))(H^t \otimes K^tB^t)E_3(H \otimes BK)\delta(k-\tau(k)) \\
&+ \Gamma^t(k)(\Phi_2+\Phi_2^t)\Gamma(k) + \rho_1\Gamma^t(k)M_1E_1^{-1}M_1^t\Gamma(k) \\
&+ \rho_2\Gamma^t(k)M_2E_1^{-1}M_2^t\Gamma(k) + \rho_1\Gamma^t(k)M_3E_2^{-1}M_3^t\Gamma(k) \\
&+ \rho_4\Gamma^t(k)M_4E_3^{-1}M_4^t\Gamma(k) \\
&\leq \Gamma^T(k)[\Phi_1+\Phi_2+\Phi_2^T+\rho_1M_1E_1^{-1}M_1^t+\rho_2M_2E_1^{-1}M_2^T \\
&+ \rho_1M_3E_2^{-1}M_3^t+\rho_4M_4E_3^{-1}M_4^t]\Gamma(k).
\end{aligned} \tag{5.45}
$$

In the first inequality above, the following inequality is used:

$$
\begin{aligned}
&-2\Gamma^t(k)M_1\sum_{i=k-\tau(k)}^{k-1}\Re(i) \\
&\leq \sum_{i=k-\tau(k)}^{k-1} e^{\gamma_1(i-k)}\Re^T(i)E_1\Re(i) + \sum_{i=k-\tau(k)}^{k-1} e^{-\gamma_1(i-k)}\Gamma^t(k)M_1E_1^{-1}M_1^t\Gamma(k) \\
&\leq \sum_{i=k-\tau(k)}^{k-1} e^{\gamma_1(i-k)}\Re^t(i)E_1\Re(i) + \sum_{i=k-d_2}^{k-1} e^{-\gamma_1(i-k)}\Gamma^t(k)M_1E_1^{-1}M_1^t\Gamma(k) \\
&= \sum_{i=k-\tau(k)}^{k-1} e^{\gamma_1(i-k)}\Re^t(i)E_1\Re(i) + \rho_1\Gamma^t(k)M_1E_1^{-1}M_1^T\Gamma(k).
\end{aligned}
$$

Thus, if LMI (5.44) holds, by using the Schur complement [4], we have $\Phi_1+\Phi_2+\Phi_2^T+\rho_1M_1E_1^{-1}M_1^T+\rho_2M_2E_1^{-1}M_2^T+\rho_1M_3E_2^{-1}M_3^T+\rho_4M_4E_3^{-1}M_4^T<0$. We can obtain $V_1(k)\leq 0$, namely, $V_1(k+1)\leq e^{-\gamma_1}V_1(k)$. The proof is completed. □

Now, we consider the case of $\sigma(k) = 2$:

$$\begin{aligned}\delta(k+1) &= (I_N \otimes A)\delta(k) + (H \otimes BK)\delta(k - \tau(k)),\\ \delta(k) &= \theta(k), \quad k = -d_3, -d_3+1, \ldots, 0,\end{aligned} \tag{5.46}$$

where $d_2 < \tau(k) \le d_3$ and $\theta(k)$ is a function defined in the interval $[-d_3, 0]$.

We choose the Lyapunov functional candidate in the following form:

$$V_2(k) = \sum_{i=1}^{5} V_{2i}(k), \tag{5.47}$$

where

$$\begin{aligned}
V_{21}(k) &= \delta^t(k) P_2 \delta(k),\\
V_{22}(k) &= \sum_{i=k-d_2}^{k-1} e^{\gamma_2(k-1-i)} \delta^t(i) Q_3 \delta(i),\\
V_{23}(k) &= \sum_{i=k-d_3}^{k-d_2-1} e^{\gamma_2(k-1-i)} \delta^t(i) Q_4 \delta(i),\\
V_{24}(k) &= \sum_{j=-d_2}^{-1} \sum_{i=k+j}^{k-1} e^{\gamma_2(k-1-i)} \mathfrak{R}^t(i)(E_4 + E_5)\mathfrak{R}(i),\\
V_{25}(k) &= \sum_{j=-d_3}^{-d_2-1} \sum_{i=k+j}^{k-1} e^{\gamma_2(k-1-i)} \mathfrak{R}^t(i) E_6 \mathfrak{R}(i),
\end{aligned}$$

and P_2, Q_3, Q_4, E_i $(i = 4, 5, 6)$ are positive definite matrices to be determined.

Lemma 5.9. *Assume that Assumptions 5.4–5.6 hold. For a given constant $\gamma_2 > 0$, if there exist matrices $P_2 > 0$, $Q_i > 0$ $(i = 3, 4)$, and $E_j > 0$ $(j = 4, 5, 6)$ and matrices S_l $(l = 1, 2, 3)$ with appropriate dimensions such that the following LMI holds,*

$$\begin{bmatrix} \Psi_1 + \Psi_2 + \Psi_2^T & \Psi_3 \\ & \Psi_4 \end{bmatrix} < 0, \tag{5.48}$$

where

$$\Psi_1 = \begin{bmatrix} \psi_{11} & \psi_{12} & 0 & 0 \\ & \psi_{22} 0 & 0 & \\ & * & e^{d_2\gamma_2} Q_4 - e^{d_2\gamma_2} Q_3 & 0 \\ & * & 0 & -e^{d_3\gamma_2} Q_4 \end{bmatrix},$$

$$
\begin{aligned}
\psi_{11} &= (I_N \otimes A^t)P_2(I_N \otimes A) - e^{\gamma_2}P_2 \\
&\quad + (d_3 - d_2)(I_N \otimes (A-I)^t)E_6(I_N \otimes (A-I)) \\
&\quad + Q_3 + d_2(I_N \otimes (A-I)^T)(E_4 + E_5)(I_N \otimes (A-I)), \\
\psi_{12} &= (I_N \otimes A^T)P_2(H \otimes BK) \\
&\quad + d_2(I_N \otimes (A-I)^T)(E_4 + E_5)(H \otimes BK) \\
&\quad + (d_3 - d_2)(I_N \otimes (A-I)^T)E_6(H \otimes BK), \\
\psi_{22} &= (H^T \otimes K^T B^T)P_2(H \otimes BK) \\
&\quad + d_2(H^T \otimes K^T B^T)(E_4 + E_5)(H \otimes BK) \\
&\quad + (d_3 - d_2)(H^T \otimes K^T B^T)E_6(H \otimes BK), \\
\Psi_2 &= \begin{bmatrix} S_1 & -S_2 + S_3 & S_2 - S_1 & -S_3 \end{bmatrix}, \\
\Psi_3 &= \begin{bmatrix} \sqrt{\rho_4}S_1 & \sqrt{\rho_5}S_2 & \sqrt{\rho_5}S_3 \end{bmatrix}, \\
\Psi_4 &= diag\{-E_4 - E_5, -E_6, -E_6\}, \\
\rho_4 &= \frac{1 - e^{-d_2\gamma_2}}{e^{\gamma_2} - 1}, \quad \rho_5 = \frac{e^{-d_2\gamma_2} - e^{-d_3\gamma_2}}{e^{\gamma_2} - 1},
\end{aligned}
$$

then the following inequality holds:

$$
V_2(k+1) \le e^{\gamma_2} V_2(k). \tag{5.49}
$$

Proof. Define $\Delta V_2(k) = V_2(k+1) - e^{\gamma_2}V_2(k)$. Along the trajectory of system (5.46), we have

$$
\begin{aligned}
\Delta V_2 &= \delta^t(k)(I_N \otimes A^t)P_2(I_N \otimes A)\delta(k) - e^{\gamma_2}\delta^t(k)P_2\delta(k) \\
&\quad + \delta^t(k)(I_N \otimes A^t)P_2(H \otimes BK)\delta(k-\tau(k)) \\
&\quad + \delta^t(k-\tau(k))(H^T \otimes K^T B^T)P_2(I_N \otimes A)\delta(k) \\
&\quad + \delta^t(k)Q_3\delta(k) - e^{d_2\gamma_2}\delta^T(k-d_2)Q_3\delta(k-d_2) \\
&\quad + \delta^t(k-\tau(k))(H^t \otimes K^t B^t)P_2(H \otimes BK)\delta(k-\tau(k)) \\
&\quad + e^{d_2\gamma_2}\delta^t(k-d2)Q_4\delta(k-d_2) - e^{d_3\gamma_2}\delta^t(k-d_3)Q_4\delta(k-d_3) \\
&\quad + d_2\mathfrak{R}^t(k)(E_4+E_5)\mathfrak{R}(k) + (d_3-d_2)\mathfrak{R}^t(k)E_6\mathfrak{R}(k) \\
&\quad - \sum_{i=k-d_2}^{-k-1} e^{\gamma_2(k-i)}\mathfrak{R}^T(i)(E_4+E_5)\mathfrak{R}(i) \\
&\quad - \sum_{i=k-d_3}^{k-\tau(k)-1} e^{\gamma_2(k-i)}\mathfrak{R}^t(i)E_6\mathfrak{R}(i) \\
&\quad - \sum_{i=k-\tau(k)}^{-k-d_2-1} e^{\gamma_2(k-i)}\mathfrak{R}^t(i)E_6\mathfrak{R}(i)
\end{aligned}
$$

$$+2\Gamma^t(k)S_1\left[\delta(k)-\delta(k-d_2)-\sum_{i=k-d_2}^{k-1}\mathfrak{R}(i)\right]$$
$$+2\Gamma^t(k)S_2\left[\delta(k-d_2)-\delta(k-\tau(k))-\sum_{i=k-\tau(k)}^{k-d_2-1}\mathfrak{R}(i)\right]$$
$$+2\Gamma^t(k)S_3\left[\delta(k-\tau(k))-\delta(k-d_3)-\sum_{i=k-d_3}^{k-\tau(k)-1}\mathfrak{R}(i)\right]$$
$$\leq \Gamma^t(k)[\Psi_1+\Psi_2+\Psi_2^T+\rho_4 S_1(E_4+E_5)^{-1}S_1^T$$
$$+\rho_5 S_2 E_6^{-1}S_2^T+\rho_5 S_3 E_6^{-1}S_3^T]\Gamma(k). \tag{5.50}$$

By using the Schur complement, we have $\Psi_1+\Psi_2+\Psi_2^T+\rho_4 S_1(E_4+E_5)^{-1}S_1^T+\rho_5 S_2 E_6^{-1}S_2^T+\rho_5 S_3 E_6^{-1}S_3^T<0$ if LMI (5.48) holds. Therefore $V_2(k)\leq 0$, that is, (5.49) holds. The proof is completed. □

We are now in a position to consider the stability of the switched time-varying delay system (5.41). As mentioned above, if system (5.41) is asymptotically stable, the leader following consensus problem of the multiagent system (5.34)–(5.35) is solved.

Theorem 5.3. *Assume that Assumptions 5.4–5.6 hold. For given constants $\gamma_1>0$, $\gamma_2>0$, if there exist matrices $P_i>0\ (i=1,2)$, $Q_j>0\ (j=1,2,3,4)$, $E_l>0\ (l=1,2,\ldots,6)$ and matrices $M_i\ (i=1,2,3,4)$, $S_j\ (j=1,2,3)$ with appropriate dimensions, such that LMIs (5.44) and (5.48) hold, and the following conditions are satisfied:*

- **(C1)** $\frac{\mathbb{T}^+(k_0,k)}{\mathbb{T}^-(k_0,k)}\leq\frac{\gamma_1-\gamma^*}{\gamma^*+\gamma_2}$, *where $\gamma^*\in(0,\gamma_1)$ and $\mathbb{T}^+(k_0,k)$ and $\mathbb{T}^-(k_0,k)$ denote the total time lengths of LDSs and SDSs in time interval $[k_0,k]$, respectively;*
- **(C2)** $\zeta>1$ *satisfies* $P_1\leq\zeta P_2$, $P_2\leq\zeta P_1$, $Q_1\leq\zeta Q_3$, $Q_3\leq\zeta Q_1$, $Q_2\leq\zeta Q_4$, $Q_4\leq\zeta Q_2$, $E_i\leq\zeta E_j$, $\forall\{i,j\}$ *or* $\{j,i\}\in\{\{1,4\},\{2,5\},\{3,6\}\}$;
- **(C3)** $F_l(k_0,k)\leq\frac{\gamma}{\ln(\zeta^2\zeta_1)}$, $\gamma\in(0,\gamma^*)$, $\zeta_1=e^{d_3(\gamma_1+\gamma_2)}$, $F_l(k_0,k)=\frac{N_l(k_0,k)}{k-k_0}$,

then system (5.41) is asymptotically stable.

Proof. Without loss of generality, for system (5.41), assume that $k_1<k_2<\cdots$ are the switching instants, $\mathbb{S}_1=\{k_{2j},k_{2j}+1,\ldots,k_{2j+1}-1\}$ denotes the SDSs, and $\mathbb{S}_2=\{k_{2j+1},k_{2j+1}+1,\ldots,k_{2j+2}-1\}$ denotes the LDS. Choose a piecewise Lyapunov functional as follows:

$$V(k)=V_\sigma(k)=\begin{cases}V_1(k), & k\in\mathbb{S}_1,\\ V_2(k), & k\in\mathbb{S}_2,\end{cases} \tag{5.51}$$

where $V_1(k)$ and $V_2(k)$ are the same as in (5.43) and (5.47), respectively. From C2, we can see that

$$V_1(k) \leq \zeta V_2(k),\ V_2(k) \leq \zeta \zeta_1 V_1(k). \tag{5.52}$$

Based on Lemmas 5.8 and 5.9, we have

$$V(k) \leq \begin{cases} e^{-\gamma_1 (k-k_{2j})} V_1(k_{2j}), & k \in \mathbb{S}_1, \\ e^{\gamma_2 (k-k_{2j+1})} V_2(k_{2j+1}), & k \in \mathbb{S}_2. \end{cases} \tag{5.53}$$

Then, along the trajectory of system (5.41), we have

$$\begin{aligned} V(k) &\leq e^{\gamma_2 (k-k_{2j+1})} V_2(k_{2j+1}) \\ &\leq \zeta \zeta_1 e^{\gamma_2 (k-k_{2j+1})} V_1(k_{2j+1}) \\ &\leq \zeta \zeta 1 e^{-\gamma_1 (k_{2j+1}-k_{2j})} e^{\gamma_2 (k-k_{2j+1})} V_1(k_{2j}) \\ &= \zeta \zeta_1 e^{-\gamma_1 \mathbb{T}^-(k_{2j},k)} e^{\gamma_2 \mathbb{T}^+(k_{2j},k)} V_1(k_{2j}) \\ &\leq \cdots \\ &\leq \zeta^{N_\sigma(k_0,k)} \zeta_1^{N_f(k_0,k)} e^{-\gamma_1 \mathbb{T}^-(k_0,k) + \gamma_2 \mathbb{T}^+(k_0,k)} V_1(k_0). \end{aligned}$$

From C1, we have

$$-(\gamma_1 - \gamma^*)\mathbb{T}^-(k_0, k) + (\gamma_2 + \gamma^*)\mathbb{T}^+(k_0, k) \leq 0. \tag{5.54}$$

Then the length bound of LDSs in $[k_0, k]$ can be calculated as

$$\mathbb{T}^+(k_0, k) \leq \frac{\gamma_1 - \gamma^*}{\gamma_1 + \gamma_2} \mathbb{T}^-(k_0, k) < \frac{\gamma_1 - \gamma^*}{\gamma_1 + \gamma_2}(k - k_0). \tag{5.55}$$

From (5.54) we obtain

$$\begin{aligned} &-\gamma_1 \mathbb{T}^-(k_0, k) + \gamma_2 \mathbb{T}^+(k_0, k) = \\ &\quad -\gamma^*(k - k_0) - (\gamma_1 - \gamma^*)\mathbb{T}^-(k_0, k) + (\gamma_2 + \gamma^*)\mathbb{T}^+(k_0, k). \end{aligned}$$

Therefore, it is obvious that

$$-\gamma_1 \mathbb{T}^-(k_0, k) + \gamma_2 \mathbb{T}^+(k_0, k) \leq -\gamma^*(k - k_0). \tag{5.56}$$

Then,

$$V(k) \leq \zeta^{N_\sigma(k_0,k)} \zeta_1^{N_l(k_0,k)} e^{-\gamma^*(k-k_0)} V_1(k_0). \tag{5.57}$$

By applying C3, we have

$$\begin{aligned} N_l(k_0, k) &\leq \frac{\gamma(k - k_0)}{\ln(\zeta^2 \zeta_1)}, \\ N_\sigma(k_0, k) &\leq 2N_l(k_0, k) = \frac{2\gamma(k - k_0)}{\ln(\zeta^2 \zeta_1)}. \end{aligned} \tag{5.58}$$

Based on (5.57) and (5.58), we have

$$\begin{aligned}
\zeta^{N_\sigma(k_0,k)}\zeta_1^{N_l(k_0,k)} &= e^{N_\sigma(k_0,k)\ln\zeta + N_l(k_0,k)\ln\zeta_1} \\
&\le e^{2N_l(k_0,k)\ln\zeta + N_l(k_0,k)\ln\zeta_1} \\
&= e^{N_l(k_0,k)\ln(\zeta^2\zeta_1)} \\
&\le e^{\gamma(k-k_0)}, \\
V(k) &\le e^{\gamma(k-k_0)}e^{-\gamma^*(k-k_0)}V_1(k_0) \\
&= e^{(\gamma-\gamma^*)(k-k_0)}V_1(k_0).
\end{aligned} \tag{5.59}$$

Let $\lambda = \min\{\lambda_{\min}(P_1), \lambda_{\min}(P_2)\}$. Then we see that

$$V(k) \ge \lambda||\delta(k)||^2,$$

which implies

$$\begin{aligned}
\lambda||\delta(k)||^2 &\le e^{(\gamma-\gamma^*)(k-k_0)}V_1(k_0), \\
||\delta(k)|| &\le \sqrt{\frac{V_1(k_0)}{\lambda}}e^{\frac{1}{2}(\gamma-\gamma^*)(k-k_0)}.
\end{aligned}$$

Thus, system (5.41) is asymptotically stable and therefore the proof is completed. □

Remark 5.7. Theorem 5.3 shows that by imposing certain restrictions on LDSs, the leader following consensus of multiagent systems with LDS can be achieved. The quantitative requirements on the LDSs are explicitly presented, where C1 is a constraint on the length of the LDSs and C3 is a constraint on the frequency of the LDSs in time interval $[k_0, k]$.

5.4.4 Simulation example 5.2

In this section, an example is provided to illustrate the theoretical result proposed in the previous section. In this example, the multiagent system consists of a leader and three followers. The state-space matrices in (5.34) and (5.35) and matrix H are given as follows:

$$A = \begin{bmatrix} 1 & 0.5 \\ 0 & 1 \end{bmatrix}, \quad B = \begin{bmatrix} 2 \\ -1 \end{bmatrix}, \quad H = \begin{bmatrix} 1 & 0 & 0 \\ -1 & 1 & 0 \\ -1 & 0 & 1 \end{bmatrix}.$$

The gain matrix K is obtained by the pole placement method. Here, let $K = [0.0250, 0.3414]$. The other parameters are given as $d_1 = 2$, $d_2 = 3$, $d_3 = 16$, $\gamma_1 = 0.1015$, $\gamma_2 = 0.778$, $\gamma^* = 0.0705$, $\gamma = 0.07049$, $\zeta = 1.001$, $k_0 = 0$, $k = 300$, $x_0(0) = [5, 4]^T$, $x_1(0) = [2, 1]^T$, $x_2(0) = [2, 1.5]^T$, $x_3(0) = [10, 5]^T$.

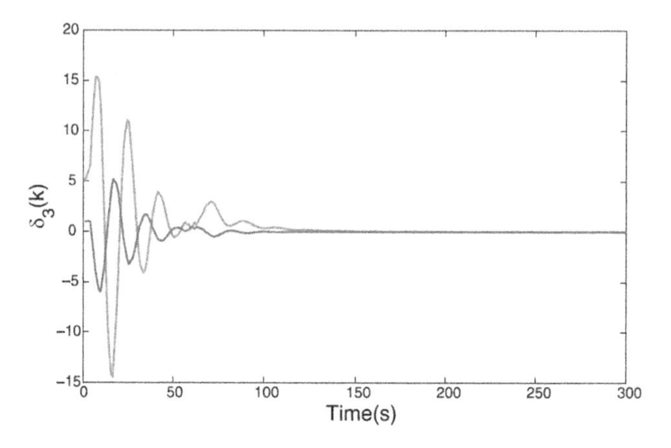

FIGURE 5.2 The error trajectories between the leader and followers.

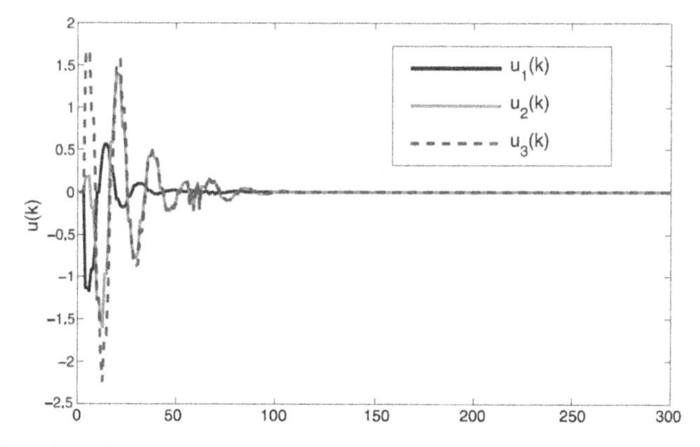

FIGURE 5.3 Control input signal.

The simulation results are shown in Figs. 5.2–5.4: Fig. 5.2 shows the simulation results under the proposed scheme (transforming a leader following consensus problem to a stability problem of a switching system with two constraint conditions on LDSs); Fig. 5.3 shows the control input signals of the three following agents; and Fig. 5.4 shows the simulation results that the multiagent system always runs in the delay range $d_1 \leq \tau(k) \leq d_3$. From Fig. 5.2, it is observed that the errors between the leader agent and follower agents tend to zero asymptotically, which implies that the protocol (5.38) solves the leader following consensus problem with two constraints on the LDSs: length and frequency. From Fig. 5.4, we can see that if the multiagent system always runs in the delay range $d_1 \leq \tau(k) \leq d_3$, that is, there are no constraints on the LDSs, then the leader following consensus cannot be achieved. For the case of $1 \leq \tau_1(k) \leq 3$,

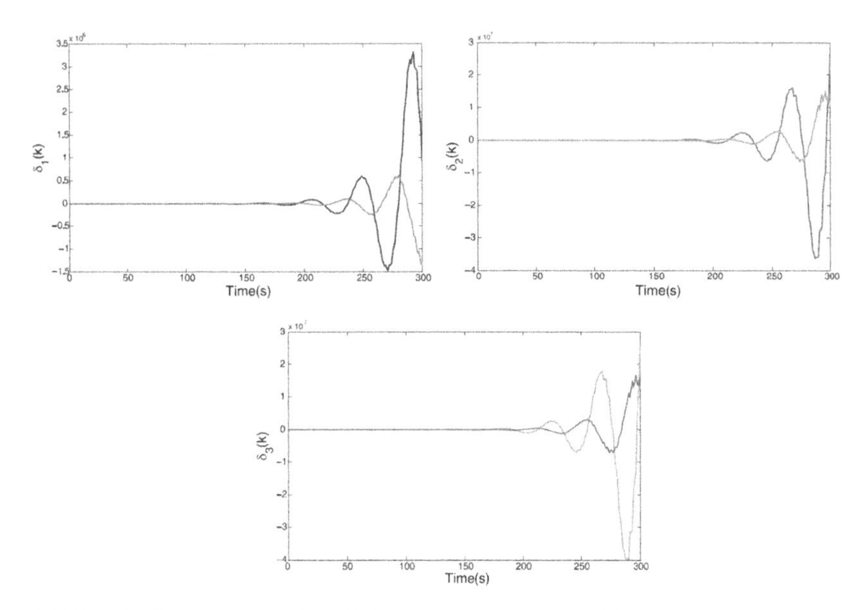

FIGURE 5.4 The error trajectories with the delay range $\tau(k) \in [d_1, d_3]$.

TABLE 5.1 The bound of $\frac{\mathbb{T}^+(k_0,k)}{\mathbb{T}^-(k_0,k)}$, $F_l(k_0,k)$, and d_3 for $\tau_1(k) \in [1,3]$ for different $\tau_2(k)$.

d_2	(3, 10]	(3, 12]	(3, 16]
$\frac{\mathbb{T}^+(k_0,k)}{\mathbb{T}^-(k_0,k)}$	0.0476	0.0476	0.0476
$F_l(k_0,k)$	0.0058	0.0048	0.0036
d_3	10	12	16

the bounds of $\frac{\mathbb{T}^+(k_0,k)}{\mathbb{T}^-(k_0,k)}$, $F_l(k_0,k)$, and d_3 can be seen in Table 5.1 under different $\tau_2(k)$.

5.5 Stochastic approximation algorithms

Consider n agents distributed according to a directed graph (or digraph) $\mathbb{G} = (\mathbf{N}, \mathbb{E})$ consisting of a set of nodes $\mathbf{N} = \{1, 2, \cdots, n\}$ and a set of edges $\mathbb{E} \subset \mathbf{N} \times \mathbf{N}$. In the digraph, an edge from node i to node j is denoted as an ordered pair (i, j), where $i \neq j$ (so there is no edge between a node and itself). A path (from i_1 to i_ℓ) consists of a sequence of nodes $i_1, i_2, \cdots, i_\ell$, $\ell \geq 2$, such that $(i_k, i_{k+1}) \in \mathbb{E}$ for $k = 1, \cdots, \ell - 1$.

We say node i is connected to node $j(\neq i)$ if there exists a path from i to j. The graph $\mathbb{G}$ is said to be strongly connected if each node i is connected to any other node j by a path.

For convenience of exposition, the two names, agent and node, will be used alternatively. The agent A_k (respectively, node k) is a neighbor of A_j (respectively, node j) if $k,\ j \in \mathbb{E}$, where $k \neq j$. Denote the neighbors of node i by $\mathbf{N}_i = \{k|(k,\ i)\,\mathbb{E}\}$.

In practical applications, the information exchange between different agents may involve the usage of sensors, quantization, and wireless fading channels, which makes it unlikely to have noise-free data delivery. In such models with noisy measurements, the traditional algorithms involving a constant (or nonvanishing) step size in general cannot ensure convergence. This calls for randomized or stochastic approximation-type algorithms for consensus seeking where the data transmitted from other agents are corrupted by noises (see Fig. 5.5). In

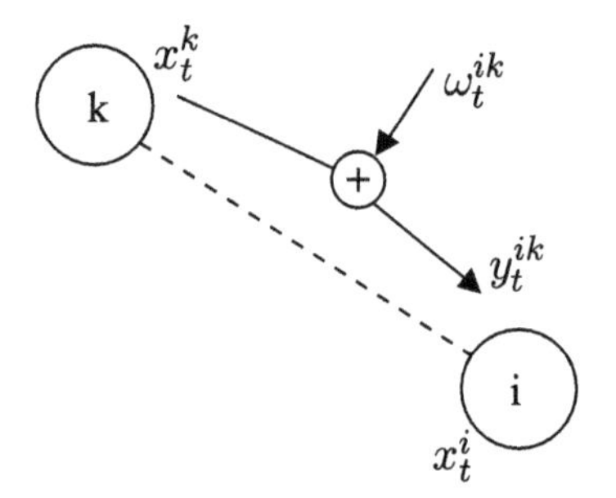

FIGURE 5.5 Control input signal.

developing the averaging scheme it is crucial to maintain a tradeoff in attenuating the noise and ensuring a suitable stabilizing capability to drive the individual states toward each other. To achieve this objective, the step size can be decreased neither too slowly, nor too quickly. In particular, almost sure convergence results have been obtained in directed graph models satisfying a circulant invariance property [33], and mean square convergence has been established for connected undirected graphs by stochastic Lyapunov analysis [34].

5.5.1 Problem formulation

The state of each agent is updated by the rule

$$x_{k+1}^i = (1 - \alpha_k \beta_{ii})\, x_k^i + \alpha_k \sum_{k \in \mathbf{N}_i} \beta_{ik}\, y_k^{ik}, \quad i \in \mathbf{N}_i,\ \ k \geq 0, \tag{5.60}$$

where the step size at $\alpha_k \geq 0$, $\beta_{ik} > 0$ for $k \in \mathbf{N}_i$, and $\beta_{ii} = \sum_{k \in \mathbf{N}_i} \beta_{ik}$. We call β_{ik}, $k \in \mathbf{N}_i$, the relative weight that A_i assigns to its neighbor A_k. We restrict that $\alpha_k \beta^* \in [0,\ 1]$, where

$$\beta^* = \max_{k \in \mathbf{N}} \beta_{ii}.$$

Thus the right hand side of (5.60) is a convex combination of the agent's state and its $|\mathbf{N}_i|$ observations. In the sequel, we use $|\mathbf{S}|$ to denote the cardinality of a set $\mathbf{S}$. The objective of the consensus problem is to select $\{\alpha_k,\ k \geq 0\}$ so that the individual states converge to a common limit in a certain sense.

5.5.2 The measurement model

For agent A_i, we denote its state at time k by $x_k^i \in \Re$, where $k \in \mathbb{Z}_+ = \{0,\ 1,\ 2,\ \cdots\}$. For each $i \in \mathbf{N}$, agent A_i receives noisy measurements of the states of its neighbors. We denote the resulting measurement by agent A_i of agent A_j's state by

$$y_k^{ij} = x_k^j + \omega_k^{ij},\ \ k \in \mathbb{Z}_+,\ j \in \mathbf{N}_i, \tag{5.61}$$

where $\omega_k^{ij} \in \Re$ is the additive noise, as illustrated in Fig. 5.5. The underlying probability space is denoted by Ω, $\mathbb{F}$, P. We call y_k^{ij} the observation of the state of A_j obtained by A_i, and we assume each A_i knows its own state x_k^i exactly.

A natural interpretation for the additive noise is that x_k^i is corrupted by noise during interagent communication. To proceed further, we introduce the followings assumptions:

(A1) The graph $\mathbb{G} = (\mathbf{N},\ \mathbb{E})$ is strongly connected.

(A2) The noises ω_k^{ij}, $k \in \mathbb{Z}_+$, $i \in \mathbf{N}$, $j \in \mathbf{N}_i$, are independent with respect to the indices $i,\ j,\ k$ and also independent of the initial states x_0^i, $i \in \mathbf{N}$, and each ω_k^{ij} has zero mean and variance $Q_k^{i,j} \geq 0$. In addition,

$$\sup i \in \mathbf{N}\mathbb{E}|x_0^i|^2 < \infty,\ \ \sup k \geq 0, i \in \mathbf{N} \sup j \in \mathbf{N}\ Q_k^{i,j} < \infty.$$

(A3) Since the additive noise is contained in the term $\sigma_k^i = \sum_{j \in \mathbf{N}_i} \beta_{ij}\ y_k^{ij} \geq 0$, each state x_k^i will have long-term fluctuations if the step size α_k is selected as a constant. With the aim of getting a stable behavior for the agents, a vanishing sequence $\{\alpha_k,\ k \geq 0\}$ must be used. Specifically, the sequence $\alpha_k,\ k \geq 0$, satisfies
i) $\alpha_k \in [0,\ (\beta^*)^{-1}]$; and
ii) there exists $T_0 \geq 1$ such that

$$\frac{\alpha}{k^\gamma} \leq \frac{\beta}{k^\gamma} \tag{5.62}$$

for all $k \geq T_0$, where $\gamma \in (0.5,\ 1]$ and $0 < \alpha \leq \beta < \infty$.

Observe that Condition A2 means that the noises are all independent random variables with respect to both space (as indexed by different pairs of neighboring nodes) and time.

5.5.3 The algorithm

For each i, we further define

$$\beta_{ij} = 0, \quad j \notin \mathbf{N}_i \cup \{i\},$$
$$\mathbf{B} = \begin{bmatrix} -\beta_{11} & \beta_{12} & \cdots & \beta_{1n} \\ \beta_{21} & -\beta_{22} & \cdots & \beta_{2n} \\ \vdots & \vdots & \vdots & \vdots \\ \beta_{n1} & \beta_{n2} & \cdots & -\beta_{nn} \end{bmatrix}. \tag{5.63}$$

Let $\tilde{\omega}_k^i = \sum_{j \in \mathbf{N}_i} \beta_{ij} \omega_k^{ij}$ and introduce

$$x_k = [x_k^1, \cdots, x_k^n]^t, \quad \tilde{\omega}_k = [\tilde{\omega}_k^1, \cdots, \tilde{\omega}_k^n]^t. \tag{5.64}$$

We now rewrite the updating procedure (5.60) in either of the following compact forms:

$$x_{k+1} = x_k + \alpha_k \, \mathbf{B} \alpha_k \, \tilde{\omega}_k, \tag{5.65}$$
$$x_{k+1}^i = x_k^i + \alpha_k \, [\sigma_k^i - \beta_{ii} \, x_k^i], \tag{5.66}$$
$$\sigma_k^i = \sum_{j \in \mathbf{N}_i} \beta_{ij} \, y_k^{ij}.$$

Observe that $[\sigma_k^i - \beta_{ii} \, x_k^i]$ provides a correction term controlled by the step size α_k.

Note that under A1, $\beta^* > 0$. In further analysis, the parameters T_0, α, β, γ are treated as fixed constants associated with $\{\alpha_k, \; k \geq 0\}$. Note that A3 implies

$$\sum_{k=0}^{\infty} \alpha_k = 0, \quad \sum_{k=0}^{\infty} \alpha_k^2 < \infty, \tag{5.67}$$

which is a typical property for step size sequences used in classical stochastic approximation theory. We can see that when $\alpha_k \to 0$ (5.60), the signal x_k^j (contained in y_k^{ij}), as the state of A_k, is attenuated together with the noise. Hence, α_k cannot decrease too fast since otherwise, the agents may prematurely converge to different individual limits.

Further stochastic consensus algorithms with measurement noise in strongly connected digraph models can be found in [35,36]. Basically, two different approaches, namely, Lyapunov analysis and double array analysis, are developed, leading to mean square and almost sure convergence results, respectively.

5.5.4 Simulation example 5.3

For the purpose of illustration, we consider a digraph with 5 nodes, as shown in Fig. 5.6. In simulation, the variance of the independent and identically distributed Gaussian measurement noises is $\sigma^2 = 0.01$. The initial state vector is $x_k|_{k=0} = [4, 3, 1, 6, 1]^t$. In Fig. 5.7, the simulation of the standard averaging rule with equal weights for an agent's neighbors and itself is depicted, for instance,

$$(x_{k+1}^1 = [x_k^1 + y_k^{12} + y_k^{15}]/3, \ k \geq 0),$$

where no convergence is achieved. Fig. 5.8 shows mean square and strong consensus as achieved by algorithm (5.65) with $\beta_{ij} = |\mathbf{N}_i|^{-1}$, and the step size sequence $\{\alpha_k = (k+5)^{-0.85}, \ k \geq 0\}$, where the note trajectories all merge toward a constant.

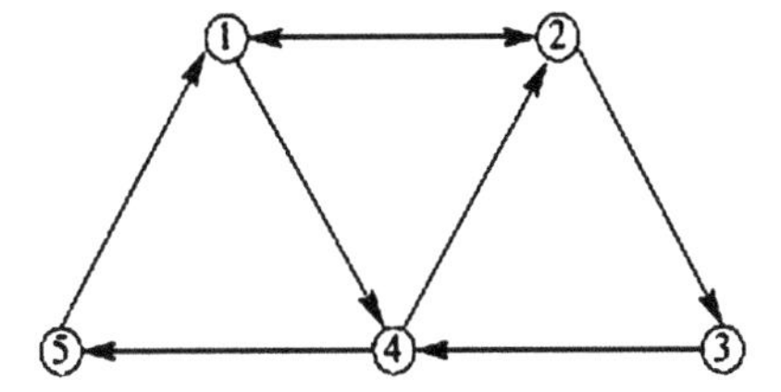

FIGURE 5.6 A digraph with five nodes.

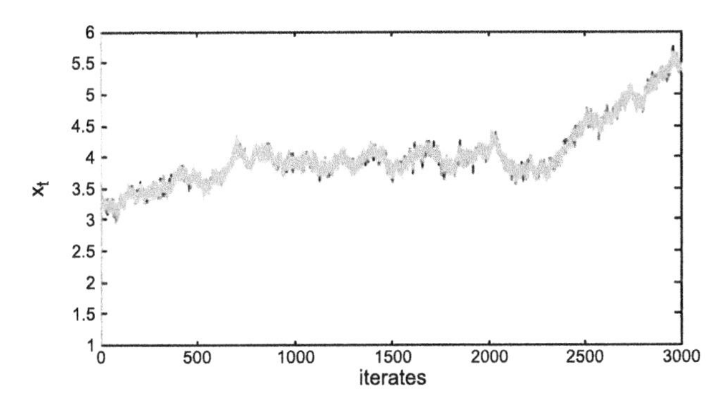

FIGURE 5.7 The trajectories using fixed weights.

5.6 Notes

In this chapter, the leader following consensus problem of discrete-time multiagent systems with LDSs has been investigated. By imposing some constraints on the length and frequency of the LDSs, a sufficient condition guaranteeing consensus of the considered multiagent system has been provided. The sufficient condition is given in the form of LMIs. For the permissible range of LDSs, it is shown that the result we obtain in this chapter is less conservative.

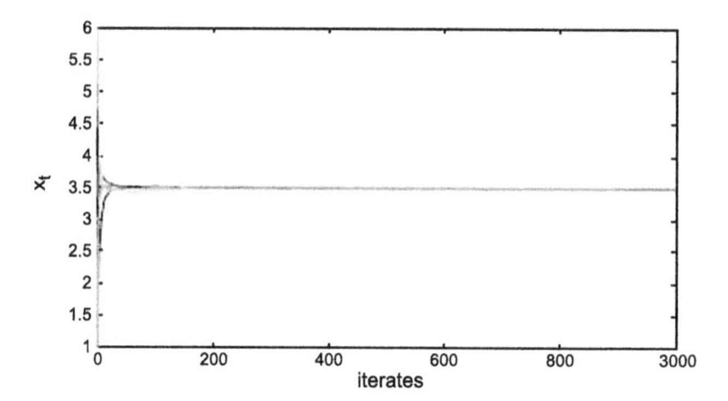

FIGURE 5.8 The trajectories using the same constant level with a decreasing step size.

References

[1] J.N. Tsitsiklis, D.P. Bertsekas, M. Athans, Distributed asynchronous deterministic and stochastic gradient optimization algorithms, IEEE Trans. Autom. Control 31 (9) (1986).

[2] J. Wolfowitz, Products of indecomposable, aperiodic, stochastic matrices, Proc. Am. Math. Soc. 15 (1963) 733–737.

[3] R.W. Beard, T.W. McLain, M.A. Goodrich, E.P. Anderson, Coordinated target assignment and intercept for unmanned air vehicles, IEEE Trans. Robot. Autom. 18 (6) (2002) 911–922.

[4] S. Boyd, L.E. Ghaoui, E. Feron, V. Balakrishnan, Linear Matrix Inequalities in System and Control Theory, Studies in Applied Mathematics, vol. 15, Society for Industrial and Applied Mathematics, 1994.

[5] H. Chung, S.S. Sastry, Autonomous helicopter formation using model predictive control, in: AIAA Guidance, Navigation, and Control Conference and Exhibit, 2006, pp. 21–24.

[6] D. Chunxia, Y. Dong, J.M. Guerrero, X. Xiangpeng, Multiagent system-based distributed coordinated control for radial DC microgrid considering transmission time delays, IEEE Trans. Smart Grid 99 (2017) 1–12.

[7] M.H. DeGroot, Reaching a consensus, J. Am. Stat. Assoc. 69 (345) (1974) 118–121.

[8] L. Ding, Q. Han, G. Guo, Network-based leader-following consensus for distributed multiagent systems, Automatica 49 (7) (2013) 2281–2286.

[9] X. Dong, Q. Li, R. Wang, Z. Ren, Time-varying formation control for second-order swarm systems with switching directed topologies, Inf. Sci. 369 (2016) 1–13.

[10] S.L. Du, W. Xia, X.M. Sun, W. Wang, Sampled-data-based consensus and L_2-gain analysis for heterogeneous multiagent systems, IEEE Trans. Cybern. 47 (6) (2017) 1523–1531.

[11] A. Jadbabaie, J. Lin, et al., Coordination of groups of mobile autonomous agents using nearest neighbor rules, IEEE Trans. Autom. Control 48 (6) (2003) 988–1001.

[12] H. Li, X. Liao, T. Huang, Y. Wang, Q. Han, T. Dong, Algebraic criteria for second-order global consensus in multi-agent networks with intrinsic nonlinear dynamics and directed topologies, Inf. Sci. 259 (1) (2014) 25–35.

[13] J. Lin, A.S. Morse, B.D. Anderson, The multi-agent rendezvous problem, in: 42nd IEEE Conference on Decision and Control, vol. 2, 2003, pp. 1508–1513.

[14] N.A. Lynch, Distributed Algorithms, Morgan Kaufmann, 1996.

[15] R. Olfati-Saber, J. Fax, R.M. Murray, Consensus and cooperation in networked multi-agent systems, Proc. IEEE 95 (1) (2007) 215–233.

[16] R. Olfati-Saber, R.M. Murray, Consensus problems in networks of agents with switching topology and time-delays, IEEE Trans. Autom. Control 49 (9) (2004) 1520–1533.

[17] Z.H. Pang, G.P. Liu, D. Zhou, D. Sun, Data-driven control with input design-based data dropout compensation for networked nonlinear systems, IEEE Trans. Control Syst. Technol. 25 (2) (2017) 628–636.

[18] Z.H. Pang, G.P. Liu, D. Zhou, D. Sun, Input design-based compensation control for networked nonlinear systems with random delays and packet dropouts, IEEE Trans. Circuits Syst. II 64 (3) (2017) 299–303.
[19] X. Qing, H.R. Karimi, Y. Niu, X. Wang, Decentralized unscented Kalman filter based on a consensus algorithm for multi-area dynamic state estimation in power systems, Int. J. Electr. Power Energy Syst. 65 (2015) 26–33.
[20] J. Stankovic, T.F. Abdelzaher, C. Lu, L. Sha, J.C. Hou, et al., Real-time communication and coordination in embedded sensor networks, Proc. IEEE 91 (7) (2003) 1002–1022.
[21] X. Sun, G. Liu, D. Rees, W. Wang, Delay-dependent stability for discrete systems with large delay sequence based on switching techniques, Automatica 44 (11) (2008) 2902–2908.
[22] Y. Tang, X. Xing, H. Karimi, L. Kocarev, Tracking control of networked multi-agent systems under new characterizations of impulses and its applications in robotic systems, IEEE Trans. Ind. Electron. 417 (1) (2015) 1–12.
[23] C. Tomlin, G. Pappas, S. Sastry, Conflict resolution for air traffic management: a study in multiagent hybrid systems, IEEE Trans. Autom. Control 43 (4) (1998) 509–521.
[24] L. Wang, F. Xiao, A new approach to consensus problems in discrete-time multiagent systems with time-delays, Sci. China, Ser. F 50 (4) (2007) 625–635.
[25] Z. Wang, J. Xu, H. Zhang, Consensus seeking for discrete-time multi-agent systems with communication delay, IEEE/CAA J. Autom. Sin. 2 (2) (2015) 151–157.
[26] W. Xing, Y. Zhao, H.R. Karimi, Convergence analysis on multi-AUV systems with leader-follower architecture, IEEE Access 5 (2017) 853–868.
[27] T. Yang, H. Gao, W. Zhang, J. Kurths, Leader-following consensus of a class of stochastic delayed multi-agent systems with partial mixed impulses, Automatica 53 (2015) 346–354.
[28] C. Yao, L. Jinhu, L. Zongli, Consensus of discrete-time multi-agent systems with transmission nonlinearity, Automatica 49 (6) (2013) 1768–1775.
[29] H. Zhang, D. Yue, X. Yin, S. Hu, C.x. Dou, Finite-time distributed event-triggered consensus control for multi-agent systems, Inf. Sci. 339 (2016) 132–142.
[30] X. Zhang, L. Liu, G. Feng, Leader follower consensus of time-varying nonlinear multi-agent systems, Automatica 52 (2015) 8–14.
[31] X. Zhao, C. Ma, X. Xing, X. Zheng, A stochastic sampling consensus protocol of networked Euler-Lagrange systems with application to two-link manipulator, IEEE Trans. Ind. Inform. 11 (4) (2015) 907–914.
[32] Y. Zhao, B. Li, J. Qin, H. Gao, H.R. Karimi, $\mathbb{H}_\infty$ consensus and synchronization of nonlinear systems based on a novel fuzzy model, IEEE Trans. Cybern. 43 (6) (2013) 2157–2169.
[33] M. Huang, J.H. Manton, Stochastic double array analysis and convergence of consensus algorithms with noisy measurements, in: Proc. American Control Conference, New York, July 2007, pp. 705–710.
[34] M. Huang, J.H. Manton, Stochastic Lyapunov analysis for consensus algorithms with noisy measurements, in: Proc. American Control Conference, New York, July 2007, pp. 1419–1424.
[35] L. Xiao, S. Boyd, Fast linear iterations for distributed averaging, Syst. Control Lett. 53 (2004) 65–78.
[36] L. Xiao, S. Boyd, S.-J. Kim, Distributed average consensus with least-mean-square deviation, J. Parallel Distrib. Comput. 67 (2007) 33–46.

Chapter 6

Event-based coordination control

6.1 Event-based tracking control

Recent years have witnessed the increasing interest in the study of the coordination problem for multiagent systems. The consensus of multiagent systems is a central problem, which has had many applications in the fields of biology, computer science, and control engineering. During the past decades, the study of the consensus problem has attracted much attention for systems of single-integrator kinematics, double-integrator dynamics, and high-order dynamics.

6.1.1 Introduction

Decentralized consensus control for multiagent systems is currently facilitated by recent technological advances on computing and communication resources. Each agent can be equipped with a small embedded microprocessor, which will be responsible for collecting the information from neighboring nodes and actuating the controller updates according to some ruling. However, the embedded processors are usually resource-limited. One of the most important aspects in the implementation of decentralized consensus algorithms is the communication and control actuation schemes.

6.1.2 Problem statement

Suppose that agent i takes the following dynamics:

$$x_i(k+1) = x_i(k) + Tu_i(k),\ i = 1, 2, \cdots, n, \tag{6.1}$$

where the integer k is the discrete-time index, T is the sampling period which is assumed to be given a priori, and $x_i(k) \in R$ and $u_i(k) \in R$, respectively, represent the agent's state and the control input of the ith agent at time $t = kT$. In this chapter, the consensus tracking problem will be considered, where the time-varying reference state is denoted by $x^r(k)$. If the ith agent can access the leader's state information, then $a_{(i,n+1)} > 0$; otherwise, $a_{(i,n+1)} = 0$.

Throughout this section, for simplicity, we use (k) to represent (kT), for example, $x_i(k) = x_i(kT)$, $x^r(k) = x^r(kT)$, etc.

Discrete Networked Dynamic Systems. https://doi.org/10.1016/B978-0-12-823698-7.00014-X

Definition 6.1. The solution of a dynamic system is said to be uniformly ultimately bounded (UUB) if for a compact set $\mathbb{U}$ of $\Re^n$ and for all $x(k_0) = x_0 \in \mathbb{U}$, there exists an $\varepsilon > 0$ and a number $T(\varepsilon, x_0)$ such that $||x(k)|| < \varepsilon$ for all $k > k_0 + T$.

In the sequel, we assume that the communication graph between the agents and the leader is fixed and directed and that no data loss and transmission delay occur in the network communication.

The following consensus algorithm was shown to be able to guarantee the tracking of states of agents (6.1) with a time-varying reference state $x^r(k)$:

$$\begin{aligned} u_i(k) = & \frac{1}{T\sum_{j=1}^{n+1} a_{(i,j)}} \sum_{j=1}^{n} a_{(i,j)}[x_j(k) - x_j(k-1) - T\gamma\{x_i(k) - x_j(k)\}] \\ & + \frac{a_{(i,n+1)}}{T\sum_{j=1}^{n+1} a_{(i,j)}}[x^r(k) - x^r(k-1) - T\gamma\{x_i(k) - x_r(k)\}], \end{aligned} \tag{6.2}$$

where γ is a positive gain.

The process of implementing an algorithm with the tracking controller (6.2) under an event-triggered scheme can be illustrated as follows. First, all agents compute their trigger functions based on the past and current sampled signals; if the trigger condition is fulfilled for one agent, the agent will broadcast its actual measurement value to its neighbors. The time when the agent sends the measurement value of the state out to its neighbors is called the release time. Each agent's controller $u_i(k)$ is updated by evaluation using the latest information from its neighbors. In order to model the event triggers for agents, here we denote the release times for the ith agent and its neighbors by s_m^i and s_m^j $(m = 0, 1, 2, \cdots, j \in N_i)$ and for the leader by s_m^r $(m = 0, 1, 2, \cdots)$. The broadcast states for agents and the leader can be described by the following piece-wise constant functions:

$$\tilde{x}_i(k) = x_i(s_m^i),\ k \in (s_m^i, s_{m+1}^i), \tag{6.3}$$

$$\tilde{x}_r(k) = x_r(s_m^r),\ k \in (s_m^r, s_{m+1}^r), \tag{6.4}$$

where $s_m^i - s_{m+1}^i$ is an integer and $s_m^i T - s_{m+1}^i T$ is a multiple of T.

Define the trigger functions for agent i as $f_i(\cdot)$, $i = 1, 2, \cdots, n$, and the trigger function for the leader as $g(\cdot)$. For agent i, the states $x_j(k)$, $j \in N_i$, are unknown; only $x_i(k)$ and $\tilde{x}_i(k)$ and the broadcast values $\tilde{x}_j(k)$ of the neighbors $j \in N_i$ are available. An event for agent i about the state is triggered only when the trigger condition

$$f_i(k, x_i(k), \tilde{x}_i(k)) < 0,\ \ i = 1, 2, \cdots, n, \tag{6.5}$$

is violated while the state of the leader is triggered as soon as

$$g(k, x^r(k), \tilde{x}^r(k)) < 0 \tag{6.6}$$

is violated.

Then, an appropriate consensus algorithm with the tracking controller can be described as

$$\begin{aligned}
u_i(k) &= \frac{1}{T\sum_{j=1}^{n+1} a_{(i,j)}} \sum_{j=1}^{n} a_{(i,j)}(\tilde{x}_j(k) - \tilde{x}_j[k-1] - T\gamma\{\tilde{x}_i(k) - \tilde{x}_j(k)\}) \\
&+ \frac{a_{(i,n+1)}}{T\sum_{j=1}^{n+1} a_{(i,j)}} (\tilde{x}^r(k) - \tilde{x}^r[k-1] - T\gamma\{\tilde{x}_i(k) - \tilde{x}^r(k)\}) \\
&= \frac{1}{T\sum_{j=1}^{n+1} a_{(i,j)}} \sum_{j=1}^{n} a_{(i,j)}(x_j[s^j_{m_j}(k)] - xj[s^j_{m_j}(k-1)] \\
&- T\gamma xi[s^i_m] - xj[s^j_{m_j}(k)]) \\
&+ \frac{a_{(i,n+1)}}{T\sum_{j=1}^{n+1} a_{(i,j)}} (x^r[s^r_{n(k)}] - x^r[s^r_{n(k-1)}] \\
&- T\gamma\{x_i[s^i_m] - x^r[s^r_n(k)]\}), \ \ k \in [s^i_m, s^i_{m+1}),
\end{aligned} \tag{6.7}$$

where

$$\begin{aligned}
m_j(k) &\triangleq arg \min_{p \in N: s^j_p \le k} \{k - s^j_p\}, \\
n(k) &\triangleq arg \min_{q \in N: s^r_q \le k} \{k - s^r_q\}, \quad k \in [s^i_m, s^i_{m+1}),
\end{aligned} \tag{6.8}$$

and $N = \{1, 2, 3, \cdots\}$, $\{s^j_p, p \in \mathbb{Z}^+\}$ represents the set of release times for the jth agent before time k and $\{s^r_q, q \in \mathbb{Z}^+\}$ represents the set of release times for the leader before time k.

Remark 6.1. It is observed that $s^i_{m+1} - s^i_m$ in (6.8) may be larger than T. Moreover, in the time interval $[s^i_m, s^i_{m+1})$, the events triggered by the neighbors of the ith agent and the leader may occur, which means that $s^j_{m_j(k)}$ and $s^r_{n(k)}$ may change for $k \in [s^i_m, s^i_{m+1})$. Therefore, $u_i(k)$ in (6.8) may be time-varying for $k \in [s^i_m, s^i_{m+1})$ depending on the variations of $s^j_{m_j(k)}$ and $s^r_{n(k)}$.

6.1.3 Design results

In this section, the tracking problem for system (6.1) under the event-triggered consensus algorithm (6.8) will be investigated.

To proceed further, we define the state error between the current sampled time and the last release time of the ith agent and the leader, respectively, as

$$\begin{aligned}
e_i(k) &= x_i[s^i_m] - x_i(k), k \in [s^i_m, s^i_{m+1}), \\
e_j(k) &= x_j[s^j_m] - x_j(k), k \in [s^j_m, s^j_{m+1}),
\end{aligned} \tag{6.9}$$

along with

$$e^r(k) = x^r[s_m^r] - x^r(k), k \in [s_m^r, s_{m+1}^r), \tag{6.10}$$

where $m = 0, 1, \cdots$. Hence, $u_i(k)$ in (6.8) can be rewritten as

$$\begin{aligned}
u_i(k) = & \frac{1}{T\sum_{j=1}^{n+1} a_{(i,j)}} \sum_{j=1}^{n} a_{(i,j)}(x_j(k) + e_j(k) - x_j[k-1] - e_j[k-1] \\
& - T\gamma x_i(k) + e_i(k) - x_j(k) - e_j(k)) \\
& + \frac{a_{(i,n+1)}}{T\sum_{j=1}^{n+1} a_{(i,j)}}(x^r(k) + e^r(k) - x^r[k-1] - e^r[k-1] \\
& - T\gamma\{x_i(k) + e_i(k) - x_r(k) - e^r(k)\}).
\end{aligned} \tag{6.11}$$

Now, introducing the tracking error between the ith agent and the leader as $\delta_i(k) \triangleq x_i(k) - x^r(k)$, we have

$$x_i[k+1] - x_i(k) = \delta_i[k+1] + x^r[k+1] - \delta_i(k) - x^r(k), \tag{6.12}$$

together with

$$\begin{aligned}
u_i(k) = & \frac{1}{T\sum_{j=1}^{n+1} a_{(i,j)}} \sum_{j=1}^{n} a_{(i,j)}(\delta_j(k) + x^r(k) + e_j(k) \\
& - \delta_j[k-1] - x^r[k-1] - e_j[k-1] \\
& - \gamma\{\delta_i(k) + x^r(k) + e_i(k) - \delta_j(k) - x^r(k) - e_j(k)\}) \\
& + \frac{a_{(i,n+1)}}{T\sum_{j=1}^{n+1} a_{(i,j)}}(x^r(k) + e^r(k) - x^r[k-1] - e^r[k-1] \\
& - T\gamma\{\delta_i(k) + e_i(k) - e^r(k)\}).
\end{aligned} \tag{6.13}$$

By combining (6.12) and (6.13) into (6.1) and making algebraic manipulations, we obtain, for $k \in [s_m^i, s_{m+1}^i)$,

$$\begin{aligned}
\delta_i[k+1] = & 2x^r(k) - x^r[k+1] - x^r[k-1] \\
& + \frac{a_{(i,n+1)}}{\sum_{j=1}^{n+1} a_{(i,j)}}\{e^r(k) + T\gamma e^r(k) - e^r[k-1]\} \\
& + (1 - T\gamma)\delta_i(k) - T\gamma e_i(k) \\
& + \frac{1}{\sum_{j=1}^{n+1} a_{(i,j)}} \sum_{j=1}^{n} a_{(i,j)}[(1 + T\gamma)(\delta_j(k) + e_i(k)) \\
& - \delta_j[k-1] - e_j[k-1]].
\end{aligned} \tag{6.14}$$

We introduce the following matrix and augmented variable definitions:

$$
\begin{aligned}
B &= \mathbf{diag}\{a_{(1,n+1)}, \cdots, a_{(n,n+1)}\}, \\
D &= \mathbf{diag}\{\sum_{j=1}^{n+1} a_{(1,j)}, \cdots, \sum_{j=1}^{n+1} a_{(n,j)}\}, \\
\Delta[k+1] &= (\delta_1[k+1], \cdots, \delta_n[k+1])^{\top}, \\
e(k) &= (e_1(k), \cdots, e_n(k))^{\top}, \\
\theta^r(k) &= 2x^r(k) - x^r[k+1] - x^r[k-1], \\
\beta^r(k) &= e^r(k) + T\gamma e^r(k) - e^r[k-1].
\end{aligned}
$$

Therefore, (6.14) can be rewritten as

$$
\begin{aligned}
\Delta(k+1) = {} & \theta^r(k)1_n + D^{-1}B\beta^r(k)1n + (1-T\gamma)I_n\Delta(k) - T\gamma e(k) \\
& + D^{-1}A(1+T\gamma)\Delta(k) + D^{-1}A(1+T\gamma)e(k) \\
& - D^{-1}A\Delta[k-1] - D^{-1}Ae[k-1],
\end{aligned} \tag{6.15}
$$

and then

$$
\begin{bmatrix} \Delta[k+1] \\ \Delta(k) \end{bmatrix} = \tilde{A}\begin{bmatrix} \Delta(k) \\ \Delta[k-1] \end{bmatrix} + \tilde{B}\begin{bmatrix} e(k) \\ e[k-1] \end{bmatrix} + \tilde{C}\begin{bmatrix} \theta^r(k)1_n \\ \beta^r(k)1_n \end{bmatrix}, \tag{6.16}
$$

where

$$
\begin{aligned}
\tilde{A} &= \begin{bmatrix} (1-T\gamma)I_n + (1+T\gamma)D^{-1}A & -D^{-1}A \\ I_n & 0 \end{bmatrix}, \\
\tilde{B} &= \begin{bmatrix} (1+T\gamma)D^{-1}A - T\gamma I_n & -D^{-1}A \\ 0 & 0 \end{bmatrix}, \\
\tilde{C} &= \begin{bmatrix} I_n & D^{-1}B \\ 0 & 0 \end{bmatrix}.
\end{aligned} \tag{6.17}
$$

We define the following augmented variables as

$$
Y(k) = \begin{bmatrix} \Delta(k) \\ \Delta[k-1] \end{bmatrix}, \ \omega(k) = \begin{bmatrix} e(k) \\ e[k-1] \end{bmatrix}, \ z(k) = \begin{bmatrix} \beta^r(k)1_n \\ \theta^r(k)1_n] \end{bmatrix},
$$

so that (6.16) and (6.17) become

$$
Y[k+1] = \tilde{A}Y(k) + \tilde{B}\omega(k) + \tilde{C}z(k). \tag{6.18}
$$

By successive iterations, the solution of (6.18) can be cast into

$$Y(k) = \tilde{A}^k Y(0) + \sum_{s=1}^{k} \tilde{A}^{k-s}\{\tilde{B}\omega(s-1) + \tilde{C}z(s-1)\}. \tag{6.19}$$

In the following, the convergence analysis of (6.18) will be carried out based on (6.19). For this purpose, the following lemma is needed.

Lemma 6.1. *Assume that the leader has a directed path to all agents from 1 to n and let λ_i be the eigenvalue of $D^{-1}A$. Then $\tau_i > 0$ holds, where*

$$\tau_i \triangleq 2|1-\lambda_i|2\{2[1-Re(\lambda_i)] \\ -|1-\lambda_i|2\}/(|1-\lambda_i|^4 + 4[Im(\lambda_i)]^2),$$

where $Re(\cdot)$ and $Im(\cdot)$ denote, respectively, the real and imaginary parts of a number. If positive scalars T and γ satisfy

$$T\gamma < \min\{1, \min_{i=1,\dots,n} \tau_i\}, \tag{6.20}$$

then $\tilde{A}$, defined in (6.17), has all eigenvalues within the unit circle.

In the following, the bound for the tracking error is estimated under the control (6.8) with the event-based scheme.

Theorem 6.1. *Assume that the leader has a directed path to all agents from 1 to n and its states $x^r(k)$ satisfy $|(x^r(k) - x^r[k-1])/T| \le \bar{\xi}$ (i.e., the changing rate of $x^r(k)$ is bounded), and*

$$|e_i[s]| \le \alpha_1 + c_1 e^{-\beta_1 sT}, \ i = 1, 2, \cdots, n, \tag{6.21}$$

$$|e^r[s]| \le \alpha_2 + c_2 e^{-\beta_2 sT}, \tag{6.22}$$

where $\alpha_j \ge 0$, $\beta_j \ge 0$, $c_j \ge 0$, $j = 1, 2$, are some constants, $s \in Z^+$. If positive scalars γ and T satisfy (6.20), under the control algorithm (6.8), the infinite norm of the solution of (6.18) is UUB by

$$||Y(k)||_\infty \le [2\alpha_1(1+T\gamma) + 2c_1(1+T\gamma)e^{-\beta_1 T} + b_1]||(I_{2n} - \tilde{A})^{-1}||_\infty, \\ (k \to \infty), \tag{6.23}$$

where $b_1 = \max\{2T\bar{\xi}, (2+T\gamma)(\alpha_2 + c2e^{-\beta_2 T})\}$.

Proof. From the definition of $||\cdot||_\infty$, it can be seen from the condition (6.22) that

$$||\omega[s]||_\infty \le \alpha_1 + c_1 e^{-\beta_1}(s-1)T. \tag{6.24}$$

Furthermore, it is easy to see from the definitions of $\tilde{B}$, $\tilde{C}$, and $z[s]$ that

$$||\tilde{B}||_\infty \le 2(1+T\gamma),$$

and

$$||\tilde{C}z[s]||_\infty \le \max\{2T\bar{\xi}, (2+T\gamma)(\alpha_2 + c_2 e^{-\beta_2 T})\} \triangleq b_1. \tag{6.25}$$

Under the assumptions of this theorem, it follows from Lemma 6.1 that $\tilde{A}$ has all eigenvalues within the unit circle. According to [1], there exists a matrix norm $||\cdot||$ such that $||\tilde{A}|| < 1$. Therefore, the following relation can be deduced:

$$\lim_{k\to\infty} ||\sum_{s=0}^{k-1} \tilde{A}^s||_\infty \le ||(I_{2n} - \tilde{A})^{-1}||_\infty. \tag{6.26}$$

Then, combining (6.19), (6.22), and (6.24)–(6.26), it can be concluded that

$$\begin{aligned}
||Y(k)|_\infty &\le ||\tilde{A}^k Y[0]||_\infty + ||\sum_{s=1}^{k} \tilde{A}^{k-s}\tilde{B}\omega[s-1]||_\infty \\
&\quad + ||\sum_{s=1}^{k} \tilde{A}^{k-s}\tilde{C}z[s-1]||_\infty \\
&\le ||\tilde{A}^k Y[0]||_\infty + (\alpha_1 + c1e^{-\beta_1 T})||\tilde{B}||_\infty ||\sum_{s=1}^{k} \tilde{A}^{k-s}||_\infty \\
&\quad + b_1 ||\sum_{s=1}^{k} \tilde{A}^{k-s}||_\infty \\
&\le ||\tilde{A}^k||_\infty ||Y[0]||_\infty + 2(1+T\gamma)(\alpha_1 + c1e^{-\beta_1 T})||\sum_{s=1}^{k} \tilde{A}^{k-s}||_\infty \\
&\quad + b_1 ||\sum_{s=1}^{k} \tilde{A}^{k-s}||_\infty \\
&\le [2\alpha_1(1+T\gamma) + 2c_1(1+T\gamma)e^{-\beta_1 T} + b_1]||(I_{2n} - \tilde{A})^{-1}||_\infty,
\end{aligned}$$

which ends the proof. □

Remark 6.2. From the definition of $||\cdot||_\infty$ and inequality (6.23), we have

$$|\delta_i(k)| \le [2\alpha_1(1+T\gamma) + 2c_1(1+T\gamma)e^{-\beta_1 T} + b_1]||(I_{2n} - \tilde{A})^{-1}||_\infty$$

as $k \to \infty$, $i = 1, 2, \cdots, n$, which means the tracking error between the ith agent and the leader is ultimately bounded.

Remark 6.3. For all agents and the leader, we define the event-triggered functions, respectively, as

$$f_i(s, |e_i[s]|) \triangleq |e_i[s]| - (\alpha_1 + c1e^{-\beta_1 sT}), \; i = 1, 2, \cdots, n, \tag{6.27}$$

$$g(s, |e^r[s]|) \triangleq |e^r[s]| - (\alpha_2 + c_2 e^{-\beta_2 sT}). \tag{6.28}$$

The events are triggered for the ith agent and the leader when $f_i(s, |e_i[s]|) > 0$ and $g(s, |e^r[s]|) > 0$, respectively. Under the above event-triggered schemes, the conditions in (6.21)–(6.22) can be guaranteed for all $s \in Z^+$, which can be concluded from the following analysis: For agent i, $\forall s \in Z^+$, there exists one interval $[s_l^i, s_{l+1}^i)$, such that $s \in [s_l^i, s_{l+1}^i)$, and in the time interval (s_l^i, s_{l+1}^i) no event has occurred.

Case 1: We have $s = s_p^i$, i.e., an event is triggered for agent i at time s. According to equation (6.9), it can be obtained that $e_i[s] = 0$, so the condition in (6.21) is guaranteed.

Case 2: We have $s \in (s_l^i, s_{l+1}^i)$. Because no event has occurred in this time interval, it can be concluded that the trigger function in (6.27) satisfies $f_i(s, e_i[s]) < 0$. So the condition in (6.21) can be guaranteed. Also, (6.22) can be guaranteed by using a similar analysis method as above. From the structure of event triggers (6.27) and (6.28), it can also be seen that the triggered mechanism used in each agent is decentralized and therefore realizable.

Remark 6.4. The discrete exponentially decreasing threshold $\alpha + ce^{-\beta kT}$ provides a very flexible event-triggered control strategy for multiagent systems. The parameter α can be used to adjust the state errors convergence region. The parameter c can be tuned in such a way that the events are not too dense for small times kT. The parameter β can be used to determine the speed of convergence. For small times kT, the event times depend dominantly on c, so the density of events does not increase with decreasing α. For larger times kT, the density does not increase with decreasing c either.

In Theorem 6.1, the parameters β_1 and β_2 in (6.22) and (6.23), respectively, are assumed to be constant. In the following, revised versions of (6.22) and (6.23) are proposed by setting β_1 and β_2 as time-varying functions which can lead to a smaller upper bound for $||Y(k)||_\infty$ compared with that in Theorem 6.1.

To show this fact, the following lemma is needed.

Lemma 6.2. *If function $f(k) = kC_k^m a^{k-m}$, where $0 < a < 1$, and m is a finite positive integer, then $\lim_{k \to +\infty} f(k) = 0$.*

Proof. It can be easily concluded based on the L'Hopital rule in [2]. □

Theorem 6.2. *Assume that the leader has a directed path to all agents from 1 to n and its state $x^r(k)$ satisfies $|(x^r(k) - x^r[k-1])/T| \leq \bar{\xi}$ (i.e., the changing* *rate of $x^r(k)$ is bounded), and*

$$||e_i[s]|| \leq \alpha_1 + c_1 e^{-\beta_1(s)sT}, \ i = 1, 2, \cdots, n, \tag{6.29}$$

$$||e^r[s]|| \leq \alpha_2 + c_2 e^{-\beta_2(s)sT}, \tag{6.30}$$

where

$$\beta_1(s) = -\frac{ln\rho(\tilde{A})s}{sT}, \beta_2(s) = -\frac{ln\rho(\tilde{A})s}{sT}, \tag{6.31}$$

and $\alpha_j \geq 0$, $c_j \geq 0$, $j = 1, 2$, are some constants, $s \in Z^+$. If positive scalars γ and T satisfy (6.20), under the control algorithm (6.8), the infinite norm of the solution of (12) is UUB by

$$||Y(k)||_\infty \leq [2\alpha_1(1 + T\gamma) + b_2]||(I_{2n} - \tilde{A})^{-1}||_\infty, \ (k \to \infty), \tag{6.32}$$

where $b_2 = \max\{2T\bar{\xi}, \alpha_2(2 + T\gamma)\}$.

Proof. It follows from (6.19) that

$$||Y(k)||_\infty \leq ||\tilde{A}^k Y[0]||_\infty + ||\sum_{s=1}^{k} \tilde{A}^{k-s}\tilde{B}\omega[s-1]||_\infty$$
$$+ ||\sum_{s=1}^{k} \tilde{A}^{k-s}\tilde{C}z[s-1]||_\infty.$$

Using the conditions in (6.29)–(6.31), we can easily show that

$$||\omega[s-1]||_\infty \leq \alpha_1 + c_1\rho(\tilde{A})^{s-1}$$

and

$$||\tilde{C}z[s-1]||_\infty \leq \max\{2T\bar{\xi}, (2 + T\gamma)(\alpha_2 + c_2\rho(\tilde{A})^{s-1})\} \triangleq b(s). \tag{6.33}$$

Then, from [3] it can be concluded that

$$||Y(k)||_\infty \leq ||\tilde{A}^k||_\infty||Y[0]||_\infty + \alpha_1||\tilde{B}||_\infty||\sum_{s=1}^{k} \tilde{A}^{k-s}||_\infty$$
$$+ c_1||\tilde{B}||_\infty \sum_{s=1}^{k} ||\tilde{A}^{k-s}||_\infty \rho(\tilde{A})^{s-1} + ||\sum_{s=1}^{k} \tilde{A}^{k-s} b(s)||_\infty. \tag{6.34}$$

As shown in [4], there exists an invertible matrix P such that $\tilde{A}$ is similar to a Jordan canonical matrix J, i.e., $P^{-1}\tilde{A}P = J = diag\{J_1, J_2, \cdots, J_l\}$, where $J_s, s = 1, 2, \cdots, l$, are upper triangular Jordan blocks, whose principal diagonal elements are the eigenvalues of $\tilde{A}$. Then, for the third term on the right hand side of inequality (6.34), we have

$$
\begin{aligned}
& c_1||\tilde{B}||_\infty \sum_{s=1}^{k} ||\tilde{A}^{k-s}||_\infty \rho(\tilde{A})^{s-1} \\
\leq & 2(1+T\gamma)\sum_{s=1}^{k} ||\tilde{A}^{k-s}||_\infty \rho(\tilde{A})^{s-1} \\
= & 2(1+T\gamma)\sum_{s=1}^{k} ||(PJP^{-1})^{k-s}||_\infty \rho(\tilde{A})^{s-1} \\
\leq & 2(1+T\gamma)\cdot c\sum_{s=1}^{k} ||J||_\infty^{k-s} \rho(\tilde{A})^{s-1} \\
\leq & 2c(1+T\gamma)\sum_{s=1}^{k}[\rho(\tilde{A})^{k-s} + C_{k-s}^{1}\rho(\tilde{A})^{k-s-1} + \cdots \\
& + C_{k-s}^{m}\rho(\tilde{A})^{k-s-m+1}]\rho(\tilde{A})^{s-1} \\
\leq & 2c(1+T\gamma)[k\rho(\tilde{A})^{k-1} + kC_k^1\rho(\tilde{A})^{k-2} + \cdots + kC_k^m\rho(\tilde{A})^{k-m}],
\end{aligned}
$$

where $c \triangleq ||P||_\infty \cdot ||P-1||_\infty$ and m is the maximum order of J_s, $s = 1, 2, \cdots, l$. According to Lemma 6.2 and the above analysis, it can be easily obtained that

$$
c_1||\tilde{B}||_\infty \sum_{s=1}^{k} ||\tilde{A}_\infty^{k-s} \rho(\tilde{A})^{s-1} \to 0, \ (k \to \infty). \tag{6.35}
$$

By a similar proof process for (6.35) and using (6.26), the following relation can be deduced:

$$
||\sum_{s=1}^{k} \tilde{A}^{k-s} b(s)||_\infty \leq b_2||(I_{2n} - \tilde{A})^{-1}||_\infty, \ (k \to \infty), \tag{6.36}
$$

where $b_2 = \max\{2T\bar{\xi}, \alpha_2(2+T\gamma)\}$.

Since all the eigenvalues of $\tilde{A}$ are within the unit circle as stated in Lemma 6.1, we can obtain that $\lim_{k\to\infty} \tilde{A}^k = 0_{2n\times 2n}$. Combining (6.34)–(6.36), we can obtain

$$
||Y(k)||_\infty \leq [2(1+T\gamma)\alpha_1 + b_2]||(I_{2n} - \tilde{A})^{-1}||_\infty, \ (k \to \infty). \tag{6.37}
$$

This ends the proof. □

Remark 6.5. Define the trigger functions for each agent i and the leader, respectively, as follows:

$$f_i(s, |e_i[s]|) \triangleq |e_i[s]| - (\alpha_1 + c_1 e^{-\beta_1(s)sT}),\ i = 1, 2, \cdots, n, \tag{6.38}$$

$$g(s, |e^r[s]|) \triangleq |e^r[s]| - (\alpha_2 + c_2 e^{-\beta_2(s)sT}). \tag{6.39}$$

Similarly to the statements in Remark 6.5, it is known that, under the event-triggered schemes with trigger functions above, the conditions in (6.29) and (6.30) can be guaranteed for all $s \in Z^+$.

Remark 6.6. Similarly to Remark 6.2 and from inequality (6.32), it follows that

$$|\delta_i(k)| \le 2[(1 + T\gamma)\alpha_1 + b_2]\|(I_{2n} - \hat{A})^{-1}\|_\infty$$

as $k \to \infty$, $i = 1, 2, \cdots, n$. Obviously, under the new event-triggered scheme, the resulting upper bound for the tracking error $\delta_i(k)$ is smaller than that under the event-triggered scheme in Remark 6.4. However, since the introduction of the time-varying parameters β_1 and β_2, the relatively heavier computation is needed when using the event-triggered scheme in Remark 6.6.

Remark 6.7. If we choose $\alpha_i = 0$, $c_i = 0$, $(i = 1, 2)$ in Theorems 6.1 and 6.2, then the event-triggered communication scheme proposed in this chapter reduces to the time-triggered scheme which has been studied in [3], and the upper bound of tracking errors between agents and the leader is $2T\bar{\xi}\|(I_{2n} - \tilde{A})^{-1}\|_\infty$.

6.1.4 Simulation example 6.1

In this section, a numerical example with four agents and a time-varying reference satisfying the same communication graph as in [3] is employed to validate the main results of this section (see Fig. 6.1). If agent j is a neighbor of agent i, we let $a_{(i,j)} = 1$; otherwise $a_{(i,j)} = 0$. Then, for this example, the corresponding adjacency matrix is

$$A = \begin{bmatrix} 0 & 1 & 0 & 0 \\ 1 & 0 & 1 & 0 \\ 1 & 0 & 0 & 0 \\ 0 & 1 & 1 & 0 \end{bmatrix}.$$

Since agent 3 can have access to the leader, $a_{(3,5)} = 1$, $a_{(1,5)} = 0$, $a_{(2,5)} = 0$, and $a_{(4,5)} = 0$. The reference state is chosen as $x^r(k) = \sin(kT) + kT$.

This example has been studied in [3] by using the synchronous communication scheme. In the following, we will study this example by using the asynchronous communication scheme, that is, by using the event-triggered scheme proposed in this section.

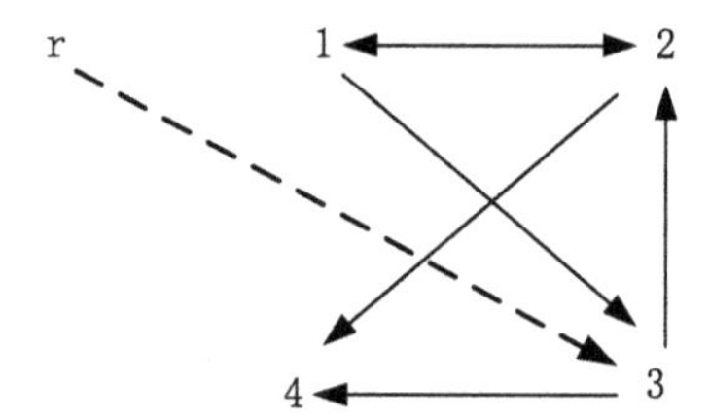

FIGURE 6.1 Directed graph for four agents with a leader.

The initial states of the four agents are chosen as

$$[x_1[0], x_2[0], x_3[0], x_4[0]] = [2, 1, -1, -3].$$

Without loss of generality, we suppose that

$$[x_1[-1], x_2[-1], x_3[-1], x_4[-1]] = [0, 0, 0, 0].$$

Let $T = 0.1$ and $\gamma = 3$. By simple computation, it can be seen that the condition (6.20) holds.

In the following simulation, two cases for the selection of α_j, c_j, and β_j $(j = 1, 2)$ will be considered, which correspond to different types of triggered thresholds.

Case 1: We have $\alpha_1 = 0.01$, $\alpha_2 = 0.04$, $c_1 = 0.1$, $c_2 = 0.5$, $\beta_1 = 0.5$, $\beta_2 = 0.5$. Under the tracking control (6.8) with the event-triggered schemes in Remark 6.5, the dynamic responses for the states $x_i(k)$ and the tracking errors $\delta_i(k) = x_i(k) - x^r(k)$, $i = 1, \cdots, 4$, are shown in Figs. 6.2 and 6.3, respectively. By simple computation according to Table 6.1, we obtain that only 80.68% of sampled states of all agents and the leader are needed to be sent out to their neighbors.

TABLE 6.1 Sample and release times for Case 1.

	Leader	Agent 1	Agent 2	Agent 3	Agent 4
Sample times	500	500	500	500	500
Release times	354	411	416	427	409

Case 2: Parameters α_j and c_j $(j = 1, 2)$ are the same as in Case 1, but the parameters β_1 and β_2 are chosen as $\beta_1(s) = -ln\varrho(\tilde{A})s/0.1s$, $\beta_2(s) = -ln\rho(\tilde{A})s/0.1s$, $s \in Z^+$ and $\rho(\tilde{A}) = 0.9405$. Under the tracking control (6.8) with the event-triggered schemes in Remark 6.7, the dynamic responses for the states $x_i(k)$ and the tracking errors $\delta_i(k) = x_i(k) - x^r(k)$, $i = 1, \cdots, 4$, are shown in Figs. 6.4 and 6.5, respectively. By simple computation according to Table 6.2, we obtain that only 79.84% of sampled states of all the agents and the leader need to be sent to their neighbors.

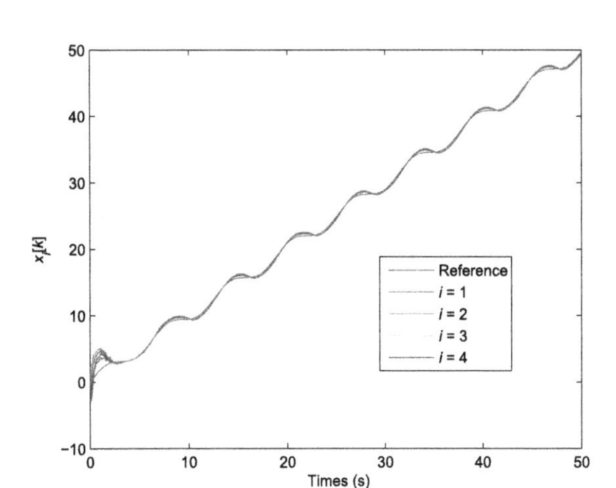

FIGURE 6.2 State responses of system (6.1) in Case 1.

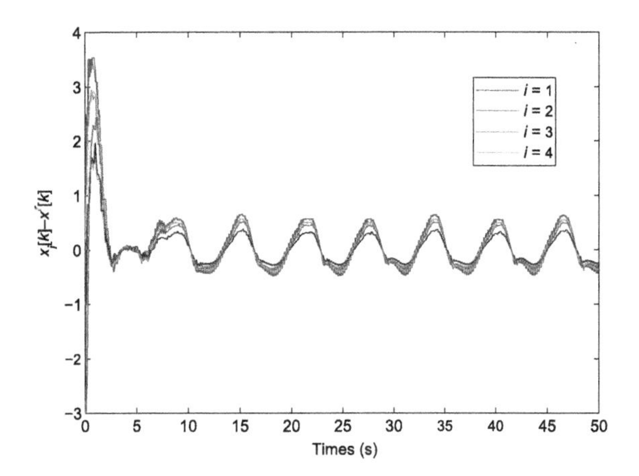

FIGURE 6.3 Tracking errors in Case 1.

TABLE 6.2 Sample and release times for Case 2.

	Leader	Agent 1	Agent 2	Agent 3	Agent 4
Sample times	500	500	500	500	500
Release times	356	412	409	420	399

Case 3: We have $\alpha_j = 0$ and $c_j = 0$ $(j = 1, 2)$. Under the tracking control (6.2) with the time-triggered scheme, the dynamic responses for the states $x_i(k)$ and the tracking errors $\delta_i(k) = x_i(k) - x^r(k)$, $i = 1, \cdots, 4$, are shown in Figs. 6.6 and 6.7, respectively. As obtained above, only around 80% of sampled states of all agents and the leader need to be transmitted through the network for Cases 1 and 2 in order to guarantee the tracking performance. Therefore, around

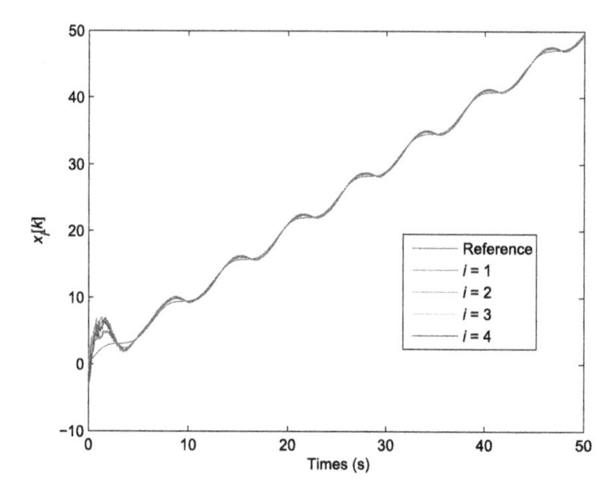

FIGURE 6.4 State responses of system (6.1) in Case 2.

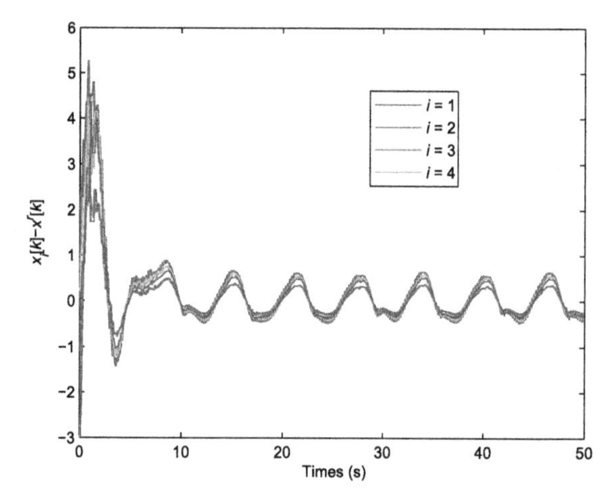

FIGURE 6.5 Tracking errors in Case 2.

20% of communication resource is saved. According to the simulation, when the time is longer than 30 s, it can be computed that the bound of the tracking error for Case 1 is 0.6497 and for Case 2 it is 0.6396. It is clear that $0.6396 < 0.6497$, which is consistent with the analysis in Remark 6.7 and validates the foregoing theoretical results.

6.2 Discrete two-timescale systems

Several dynamical processes evolving on different timescales and influencing each other often characterize physical systems. When several orders of magnitude differentiate the various timescales, the analysis of the overall systems

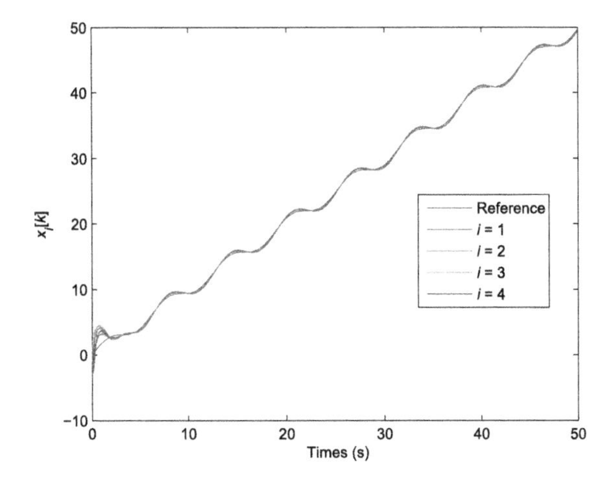

FIGURE 6.6 State responses of system (6.1) in Case 3.

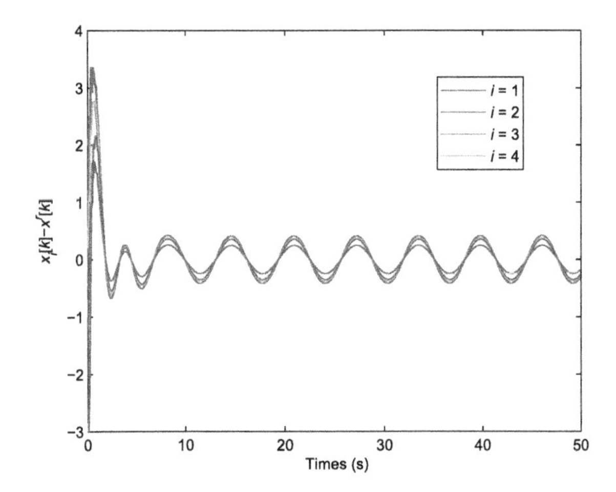

FIGURE 6.7 Tracking errors in Case 3.

becomes more difficult. In this case, standard control techniques lead to ill-conditioned problems. To overcome this, singular perturbation theory proposes to approximate the dynamics by decoupling the slow dynamical process from the faster ones. Consequently, the control design is also decoupled with respect to each timescale and then it is proven that the joint actions perform well when applied to the overall system.

On another research avenue, it was recently shown that aperiodic sampling provides key advantages over periodic sampling in terms of network utilization and computational load. There are two types of such sampling or triggering schemes, namely, event triggering (ET) and self-triggering (ST). In the former, the plant's state is checked continuously or periodically at the sensor node and is transmitted to the controller only when some predefined condition is violated. On the other hand, ST does not require continuous monitoring of the plant state and rather predicts the next triggering instant based on plant dynamics and the latest available state information.

Event-based control of two-timescale systems has recently received some attention. Specifically, Refs. [11], [12] presented a scheme whereby separate ET conditions are designed for slow and fast subsystems of the two-timescale systems and [13] provided input-to-state stability (ISS)-Lyapunov stability results for such systems.

6.2.1 Preliminaries

We start by introducing the following notations. A function $f : \Re^n \to \Re^m$ is said to be locally Lipschitz if for every compact set $\mathbf{S} \subset \Re^n$ there exists a constant $L \in \Re^+$ such that $|f(x) - f(y)| \leq L|x - y|$, $\forall x, y \in \mathbf{S}$. For a function $f : \Re^n \to \Re^n$ we denote by $f_j : \Re^n \to \Re$ the function whose image is the projection of f on its jth coordinate, that is, $(f_j)(x) = \Pi(f(x))$. Given a Lipschitz continuous function f, we also denote by L_{f_j} the Lipschitz constant of f_j. A function $\gamma : [0, a) \to \Re_0^+$ is of class $\mathcal{K}$ if it is continuous and strictly increasing and $\gamma(0) = 0$; if, furthermore, $a \to \infty$ and $\gamma(s) \to \infty$ as $s \to \infty$, then γ is said to be of class $\mathcal{K}_\infty$. A continuous function $\beta : \Re_0^+ \times \Re_0^+ \to \Re_0^+$ is of class $\mathcal{KL}$ if $\beta(., \tau)$ of class $\mathcal{K}$ for each fixed $\tau \geq 0$ and for each fixed $s \geq 0$ $\beta(s, \tau)$ is decreasing with respect to τ and $\beta(s, \tau) \to 0$ for $\tau \to \infty$. Given an essentially bounded function $\delta : \Re_0^+ \to \Re^m$ we denote by the $\mathcal{L}_\infty$-norm $||\delta||_\infty = ess\ \sup_{\{t \in \Re_0^+\}}\{||\delta||\} < \infty$.

A vector or matrix $\pi(\sigma)$ of a positive scalar σ is said to be $\mathbf{O}(\sigma)$ if there exist positive constants d and σ^* such that $||\pi(\sigma)|| \leq d\sigma^*$ for all $\sigma \leq \sigma^*$. We let $\Re^+$ be the closed right half complex plane. The matrices, if their dimensions are not explicitly stated, are assumed to be compatible for algebraic operations.

Definition 6.2. A continuous function $0 \leq V : \Re^n \to \mathbb{R}$ is an ISS-Lyapunov function for the system $x(k+1) = f(x(k), \phi(x(k) + e(k)))$ if there exist $\mathcal{K}_\infty$-functions θ_1, θ_2, such that

$$\theta_1(||x||) \leq V(x) \leq \theta_2(||x||) \tag{6.40}$$

and for some class $\mathcal{K}_\infty$ function ϱ and some class $\mathcal{K}$ function γ, $V(x)$ also satisfies

$$V(f(x(k), \phi(x(k) + e(k)))) - V(x(k)) \leq -\varrho(||x(k)||) + \gamma(||e(k)||). \tag{6.41}$$

It follows from [5] that an ISS-stable system admits an ISS-Lyapunov function.

6.2.2 Global analytic results

In the sequel, we consider a class of discrete networked dynamic systems described by the linear model

$$x(k+1) = \mathcal{A}x(k) + \mathcal{B}u(k), \tag{6.42}$$

where the vectors $x(k) \in \Re^n$, $u(k) \in \Re^m$, $x(k+1) \in \Re^n$ are the state, control input, and successor state vectors, respectively, for each time instant $k \in \mathbb{Z}_+$. Let system (6.42) be stabilizable in the sense that there exists a matrix $\mathcal{K}$ such that $|\lambda(\mathcal{A} + \mathcal{B}\mathcal{K})| < 1$. This means that system (6.42) is also ISS-stabilizable with respect to the measurement error $e(k)$, which implies that there exists a stabilizing feedback control law $u(k) = \mathcal{K}(x(k) + e(k))$, where $e(k)$ is the measurement error seen as a new input and $\mathcal{K}$ is the feedback gain matrix. The resulting closed-loop system is

$$x(k+1) = (\mathcal{A} + \mathcal{B}\mathcal{K})x(k) + \mathcal{B}\mathcal{K}e(k), \tag{6.43}$$

where

$$\mathcal{A} = \begin{bmatrix} (I_1 + \varepsilon A_1) & \varepsilon A_2 \\ A_3 & A_4 \end{bmatrix}, \quad \mathcal{B} = \begin{bmatrix} \varepsilon B_1 \\ B_2 \end{bmatrix}, \tag{6.44}$$
$$x(k) = col[x_1(k) \ \ x_2(k)],$$

where the vectors $x_1(k) \in \Re^{n_1}$, $x_2(k) \in \Re^{n_2}$, $u(k) \in \Re^m$ are the state components and control input, respectively, and $\varepsilon > 0$ is a small parameter.

It follows that system (6.43) admits a quadratic ISS-Lyapunov function of the form

$$0 < \mathbf{V}(x) = x^t(k)\mathbf{P}x(k), \ \ \mathbf{P} > 0. \tag{6.45}$$

Function $\mathbf{V}$ is considered to be radially unbounded and satisfies property (6.40) with $\theta_1(||x||) = \lambda_m(\mathbf{P})||x||^2$ and $\theta_2(||x||) = \lambda_M(\mathbf{P})||x||^2$. Given $\mathbf{Q} > 0$, let $\mathbf{P}$ be the unique positive definite solution of

$$(\mathcal{A} + \mathcal{B}\mathcal{K})^t\mathbf{P}(\mathcal{A} + \mathcal{B}\mathcal{K}) - \mathbf{P} = -\mathbf{Q}. \tag{6.46}$$

The difference of the ISS-Lyapunov function is

$$\begin{aligned} &\mathbf{V}(x(k+1)) - \mathbf{V}(x(k)) \\ &\quad = -x^t(k)\mathbf{Q}x(k) + 2x^t(k)(\mathcal{A} + \mathcal{B}\mathcal{K})^t\mathbf{P}\mathcal{B}\mathcal{K}e(k) + e^t(k)\mathcal{K}^t\mathcal{B}^t\mathbf{P}\mathcal{B}\mathcal{K}e(k). \end{aligned} \tag{6.47}$$

Hence property (6.41) holds with

$$\varrho(||x||) = \frac{1}{2}\lambda_m(\mathcal{Q})||x||^2,$$
$$\gamma(||e||) = \left(\frac{2||(\mathcal{A}+\mathcal{BK})^t\mathbf{P}\mathcal{BK}||^2}{\lambda_m(\mathbf{Q})} + ||\mathcal{K}^t\mathcal{B}^t\mathbf{P}\mathcal{BK}||\right)||e||^2. \tag{6.48}$$

6.2.3 Nonsingular transformation

Consider a class of linear shift-invariant systems with mode-separation property described by (6.42) with (6.44). We use the nonsingular transformation [6–9] in (6.95) with $u(k) \equiv 0$,

$$z(k) = \begin{bmatrix} z_1(k) \\ z_2(k) \end{bmatrix} = T\begin{bmatrix} x_1(k) \\ x_2(k) \end{bmatrix} = Tx(k), \tag{6.49}$$
$$T = \begin{bmatrix} I_1 - \varepsilon L_1 L_2 & -\varepsilon L_1 \\ L_2 & I_2 \end{bmatrix}, \quad T^{-1} = \begin{bmatrix} I_1 & \varepsilon L_1 \\ -L_2 & I_2 - \varepsilon L_2 L_1 \end{bmatrix}$$

and choose $L_1 \in \Re^{n_1 \times n_2}$, $L_2 \in \Re^{n_2 \times n_1}$ to satisfy

$$0 = A_3 + L_2 - A_4 L_2 + \varepsilon L_2[A_1 - A_2 L_2],$$
$$0 = A_2 + L_1 - L_1 A_4 + \varepsilon[A_1 - A_2 L_2]L_1 - \varepsilon L_1 L_2 A_2. \tag{6.50}$$

The existence of matrices L_1 and L_2 and the convergence of (6.50) can be shown by verifying that, for sufficiently small ε, the norm conditions of [7] are satisfied. Deploying the implicit function theorem under the nonsingularity of $(I_2 - A_4)$, that is, $det[I_2 - A_4] \neq 0$, guarantees that $L_1(\varepsilon)$ and $L_2(\varepsilon)$ are analytic at $\varepsilon = 0$ with $L_1(0) = -A_2[I_2 - A_4]^{-1}$ and $L_2(0) = -[I_2 - A_4]^{-1}A_3$. The transformed system is given by

$$z(k+1) = \begin{bmatrix} I_1 + \varepsilon(A_s + \mathbf{O}(\varepsilon)) & 0 \\ 0 & A_{22} + \mathbf{O}(\varepsilon) \end{bmatrix} z(k). \tag{6.51}$$

It has been established that system (6.95) can be decomposed into a slow subsystem of order n_1 represented by

$$x_s(k+1) = [I_1 + \varepsilon A_s]x_s(k) + \varepsilon B_s u_s(k), \tag{6.52}$$
$$A_s = A_1 + A_2(I_2 - A_4)^{-1}A_3,$$
$$B_s = B_1 + A_2(I_2 - A_4)^{-1}B_2 \tag{6.53}$$

and a fast subsystem with the order n_2 given by

$$x_f(k+1) = A_4 x_f(k) + B_2 u_f(k). \tag{6.54}$$

Analogous results can be derived by the discrete quasisteady state [6,9,8].

6.2.4 Event-based control

It has been customary for two-timescale systems to design a composite controller

$$u(k) = K_1 x_1(k) + K_2 x_2(k) \tag{6.55}$$

to ensure satisfactory closed-loop responses for system (6.42) for all $\varepsilon \in [0,\ \varepsilon^*)$, $\varepsilon^* > 0$ and determine the gains K_1, K_2 based on the slow and fast subsystems. In the sequel, we implement this design under event triggering in which the event-driven control is to update the control signal if and only if the state is outside a given set. We let s_k be the kth sampling time and we let $T_k = s_{k+1} - s_k$ represent the kth sampling period. During T_k, the control signal is held constant by the zero-order hold method. Hence, the control law (6.55) becomes

$$u(k) := u(s_k) = K_1 x_1(s_k) + K_2 x_2(s_k), \quad \forall k \in [s_k,\ s_{k+1}), \tag{6.56}$$

where $s_k,\ s_{k+1},\ s_{k+2}, \ldots$ is a sequence of update times at which new measurements are acquired and control input is recomputed. The corresponding state errors are defined by

$$\begin{aligned} e_1(k) &= x_1(s_k) - x_1(k), \\ e_2(k) &= x_2(s_k) - x_2(k), \quad k \in [s_k, s_{k+1}). \end{aligned} \tag{6.57}$$

In terms of (6.56) and (6.57), we rewrite (6.44) as

$$\begin{aligned} x_1(k+1) &= (I_1 + \varepsilon D_1) x_1(k) + \varepsilon D_2 x_2(k) \\ &\quad + \varepsilon B_1 K_1 e_1(k) + \varepsilon B_1 K_2 e_2(k), \\ x_2(k+1) &= D_3 x_1(k) + D_4 x_2(k) \\ &\quad + B_2 K_1 e_1(k) + B_2 K_2 e_2(k), \\ D_1 &= A_1 + B_1 K_1,\ \ D_2 = A_2 + B_1 K_2, \\ D_3 &= A_3 + B_2 K_1,\ \ D_4 = A_4 + B_2 K_2. \end{aligned} \tag{6.58}$$

In the manner of (6.52)–(6.54), the closed-loop slow and fast components of (6.58) are

$$x_s(k+1) = [I_1 + \varepsilon D_s] x_s(k) + \varepsilon B_s \omega(k), \tag{6.59}$$

$$x_f(k+1) = D_4 x_f(k) + B_2 \omega(k), \tag{6.60}$$

$$\begin{aligned} D_s &= D_1 + D_2 (I_2 - D_4)^{-1} D_3, \\ B_s &= B_1 + D_2 (I_2 - D_4)^{-1} B_2. \end{aligned} \tag{6.61}$$

It is noted that (6.59) is asymptotically stable $\forall\, 0 < \varepsilon < \varepsilon^*$ and sufficiently small ε^* if and only if $\lambda(I_1 + \varepsilon D_s) < 1$ or equivalently if and only if $\lambda(D_s) < 0$. Similarly, (6.60) is asymptotically stable $\forall\ 0 < \varepsilon < \varepsilon^*$ and sufficiently small

ε^* if and only if $\lambda(D_4) < 1$. System (6.59)–(6.61) in compact form has the following form:

$$\xi(k+1) = \Xi(\varepsilon)\xi(k) + \Pi K\omega(k), \tag{6.62}$$

$$\Xi(\varepsilon) = \begin{bmatrix} (I_1 + \varepsilon D_s) & 0 \\ 0 & D_4 \end{bmatrix}, \ \xi(k) = \begin{bmatrix} x_s(k) \\ x_f(k) \end{bmatrix},$$
$$\Pi = \begin{bmatrix} \varepsilon B_1 \\ B_2 \end{bmatrix}, \ \omega(k) = \begin{bmatrix} e_1(k) \\ e_2(k) \end{bmatrix},$$
$$\mathcal{K} = \begin{bmatrix} K_1 & K_2 \end{bmatrix}. \tag{6.63}$$

Note that system (6.62) characterizes an event-based control system with $\xi(k)$ and $\omega(k)$ being the state and error vectors, respectively. It further captures the two-timescale property of system (6.42). Recall that the sampled states $x_1(s_k)$ and $x_2(s_k)$ remain intact between two consecutive events. Then in each interval we get

$$\omega(k+1) = -\Xi(\varepsilon)\xi(k) - \varepsilon\Pi\mathcal{K}\omega(k). \tag{6.64}$$

We seek to stabilize system (6.62) irrespective of the measurement error $\omega(k)$. We achieve this by determining the sequence of transmission instants $\{s_k\}$, $k \in \mathbb{Z} \geq 0$, such that the time difference $\{s_{k+1} - s_k\}$ is uniformly lower-bounded by $\gamma > 0$.

6.2.5 Event-based conditions

We now focus on the generation of events. In principle, when $||\omega(k)||$ exceeds a certain prescribed level an event is triggered. To watch this, we introduce the event triggering condition by

$$\omega^t(k)\omega(k) < \beta\xi^t(k)\xi x(k), \ k \in [s_{k+1}, s_k). \tag{6.65}$$

This implies that the $(j+1)$th control task is implemented whenever condition (6.65) is not satisfied. The following result provides LMI-based conditions to find a suitable β such that the closed-loop system (6.59)–(6.61) under event condition (6.65) is asymptotically stable. In view of (6.57), (6.62) can be expressed as

$$\xi(k+1) = \Xi(\varepsilon)\xi(k) + \Pi K\omega(s_k), \tag{6.66}$$

where s_k is the latest actuation update instant. We invoke the results of (6.45)–(6.48) with $\mathcal{P} > 0$ and $\mathcal{Q} > 0$ satisfying

$$\Xi^t(\varepsilon)\mathcal{P}\Xi(\varepsilon) - \mathcal{P} = -\mathcal{Q}. \tag{6.67}$$

We are in a position to present the initial design result.

Theorem 6.3. *Consider system (6.62) and the trigger function (6.65). Select gain matrices K_1 and K_s such that $\lambda(D_s) < 0$ and $\lambda(D_4) < 1$. Then there exists ε^* such that $\forall\, 0 < \varepsilon < \varepsilon^*$ such that the following bounds hold:*

$$||\omega|| \leq \upsilon||\xi||, \tag{6.68}$$

$$\upsilon = (\frac{4||\Xi^t(\varepsilon)\mathcal{P}\Pi\mathcal{K}||^2}{\sigma\lambda_m^2(\mathcal{Q})} + 2\frac{||\mathcal{K}^t\Pi^t\mathcal{P}\Pi\mathcal{K}||}{\sigma\lambda_m(\mathcal{Q})})^{-1}.$$

Proof. In view of (6.57), (6.62) can be expressed as

$$\xi(k+1) = \Xi(\varepsilon)\xi(s_k) + \Pi\mathcal{K}\omega(s_k), \tag{6.69}$$

where s_k is the latest actuation update instant. We invoke the results of (6.45)–(6.40) with $\mathcal{P} > 0$ and $\mathcal{Q} > 0$ satisfying (6.67). The control updates of the two-timescale system (6.66) under the effect of event triggering should be enforced for some ε^*, σ as long as (6.68) is satisfied such that $\forall\, 0 < \varepsilon < \varepsilon^*$ $0 < \sigma < 1$. □

Remark 6.8. Since matrix $\Xi(\varepsilon)$ is block-diagonal, it would be beneficial to consider matrix $\mathcal{P}$ block-diagonal which leads to $\mathcal{P}$ being block-diagonal. This simplifies the computational burden in (6.67) and the subsequent expressions.

Next, we examine the minimum time between two consecutive executions. A sufficient condition for nontrivial interexecution times is provided by the following theorem.

Theorem 6.4. *Consider system (6.62)–(6.63) and the trigger function (6.65). Select gain matrices K_1 and K_s such that $\lambda(D_s) < 0$ and $\lambda(D_4) < 0$. Define $\min k = k^*$ as the minimum instant that violates condition (6.68). Suppose there exists ε^* such that $\forall\, 0 < \varepsilon < \varepsilon^*$, the inequality*

$$k^* = arg\min_{k\in\mathbb{N}}\{||\sum_{j=0}^{k-1}\Xi^{k-j-1}\eta_2||$$
$$\geq \upsilon\,||\Xi^k\omega(s_k) + \sum_{j=0}^{k-1}\Xi^{k-j-1}\eta_1||\} \tag{6.70}$$

has a solution $k^* > 1$, $\forall k$. Then the event-triggered criterion (6.68) holds and it takes at least two steps for the next controller update.

Proof. In the sequel, it is crucial to examine how the state and error propagate with time. We set η_1 a constant given by $\eta_1 = \Pi K\xi(s_k)$ and therefore the solution (6.66) becomes

$$\xi(k) = \Xi^k\omega(s_k) + \sum_{j=0}^{k-1}\Xi^{k-j-1}\eta_1. \tag{6.71}$$

It is straightforward to see that the error at the next discrete-time instant is given by $\omega(k+1) = \xi(s_k) - \xi(k+1)$. Hence manipulating (6.62) yields

$$\omega(k+1) = -\Xi\omega(k) + (I + \Xi - \Pi K)\xi(s_k). \tag{6.72}$$

The solution of (6.72) is given by

$$\omega(k) = \sum_{j=0}^{k-1} \Xi^{k-j-1}\eta_2, \tag{6.73}$$

where η_2 is a constant given by $\eta_2 = (I + \Xi - \Pi K)\xi(s_k)$, $\omega(s_k) = 0$. If condition (6.70) holds for some ε^* satisfying $\forall\, 0 < \varepsilon < \varepsilon^*$ with $k^* > 1$, $\forall k$, then the triggering rule (6.68) is nontrivial, which completes the proof. □

6.2.6 Self-triggered control

The ST formulation provides an alternative approach to finding sampling periods by predicting the next sampling time as opposed to the ET counterpart which relies on continuous monitoring of the plant's state. Motivated by the corresponding self-triggered notion [10], hereafter we are going to establish results for discrete systems with two timescales. Towards our goal, we will consider system (6.62)–(6.63); recall that in (6.62), $\xi(k)$ represents the vector of $x_s(k)$ and $x_f(k)$ related to the error $\omega(k)$ which depends on the discrete system's original states $x_1(k)$ and $x_2(k)$. To proceed, it is therefore required to transform the system back to its original states as follows:

$$\xi(k) = TX(k), \tag{6.74}$$

where

$$X(k) = \begin{bmatrix} x_1(k) \\ x_2(k) \end{bmatrix}.$$

Rewriting (6.62),

$$\begin{aligned} TX(k+1) &= \Xi(\varepsilon)TX(k) + \Pi K\omega(k), \\ X(k+1) &= AX(k) + B\omega(k), \end{aligned} \tag{6.75}$$

where

$$\begin{aligned} A(\varepsilon) &= T^{-1}\Xi(\varepsilon)T, \\ B &= T^{-1}\Pi K. \end{aligned} \tag{6.76}$$

Substituting $\omega(k) = X(k) - X(s_k)$ in (6.75),

$$X(k+1) = (A(\varepsilon) + B)X(k) - BX(s_k). \tag{6.77}$$

The condition for ISS-Lyapunov stability is now formally stated for system (6.76)–(6.77).

Lemma 6.3. *For the two-timescale discrete system* (6.77) *with* ε^* *such that* $\forall\, 0 < \varepsilon < \varepsilon^*$, *following* Definition 6.2 *for ISS-Lyapunov stability, if the Lyapunov function chosen as* $V(X(k)) = X^t(k)PX(k)$, *where* $0 < Q$, $0 < P$, *satisfies*

$$(A(\varepsilon) + B)^t P(A(\varepsilon) + B) - P = -Q, \tag{6.78}$$

the following inequality must hold for class $\mathcal{K}_\infty$ *functions* f, h *and the class* $\mathcal{K}$ *function* $g\ \forall\, k \in [s_k, s_{k+1})$*:*

$$V(k+1) - V(k) \le -f(||X(k)||) + g(||\omega(k)||) - h(||X(s_k)||), \tag{6.79}$$

with

$$g(||\omega(k)||) \le h(||X(s_k)||) \tag{6.80}$$

satisfying the stability condition, where

$$\begin{aligned} f(||X(k)||) &= ||\sqrt{Q}X(k)||_2^2, \\ h(||X(s_k)||) &= ||\sqrt{M}X(s_k)||_2^2, \\ g(||\omega(k)||) &= ||\sqrt{N}\omega(k)||_2^2, \end{aligned} \tag{6.81}$$

where

$$\begin{aligned} M(\varepsilon) &= B^T PB - A(\varepsilon)^T PBB^T PA(\varepsilon) - 2I, \\ N(\varepsilon) &= (A(\varepsilon) + B)^t PBB^T P(A(\varepsilon) + B). \end{aligned} \tag{6.82}$$

Proof. The difference of the Lyapunov function for successive discrete-time instants is given as

$$V(k+1) - V(k) = X^t(k+1)PX(k+1) - X^t(k)PX(k)$$

or

$$\begin{aligned} &V(k+1) - V(k) \\ &= -X^t(k)QX(k) + X^t(s_k)B^T PBX(s_k) \\ &\quad - 2X^t(s_k)B^T P(A(\varepsilon) + B)X(k) \\ &= -X^t(k)QX(k) - X^t(s_k)B^T PBX(s_k) \\ &\quad - 2X^t(s_k)B^T PA(\varepsilon)X(s_k) - 2X^t(s_k)B^T P(A(\varepsilon) + B)\omega(k). \end{aligned}$$

The upper bound on the last two terms is given as

$$\begin{aligned} -2X^t(s_k)B^T PA(\varepsilon)X(s_k) &\le ||X(s_k)||_2^2 + ||B^T PA(\varepsilon)X(s_k)||_2^2, \\ -2X^t(s_k)B^T P(A(\varepsilon) + B)\omega(k) &\le ||X(s_k)||_2^2 + ||B^T P(A(\varepsilon) + B)\omega(k)||_2^2, \end{aligned}$$

which renders

$$\begin{aligned} V(k+1)-V(k) \leq & -X^t(k)QX(k) \\ & - X^t(s_k)[B^T PB - A(\varepsilon)^T PBB^T PA(\varepsilon) - 2I]X(s_k) \\ & + \omega^t(k)[(A(\varepsilon)+B)^t PBB^T P(A(\varepsilon)+B)]\omega(k) \end{aligned}$$

or

$$\begin{aligned} V(k+1)-V(k) \leq & -||\sqrt{Q}X(k)||_2^2 - ||\sqrt{M(\varepsilon)}X(s_k)||_2^2 \\ & + ||\sqrt{N(\varepsilon)}\omega(k)||_2^2, \end{aligned} \tag{6.83}$$

with $M(\varepsilon)$ and $N(\varepsilon)$ defined in (6.82). This completes the proof. □

According to (6.80), the stability condition is fulfilled until

$$||\sqrt{N(\varepsilon)}\omega(k)||_2^2 \leq ||\sqrt{M(\varepsilon)}X(s_k)||_2^2,$$

which hints that when the *upper bound* on the error term,

$$E(k) := \omega^t(k)N(\varepsilon)\omega(k) = ||\sqrt{N(\varepsilon)}\omega(k)||_2^2, \tag{6.84}$$

approaches $||\sqrt{M(\varepsilon)}X(s_k)||_2^2$, the state should be transmitted. Following this line, the forthcoming lemma shows that the error bound indeed exists.

Lemma 6.4. *For the two-timescale discrete system* (6.77) *with* ε^* *such that* $\forall\, 0 < \varepsilon < \varepsilon^*$, *the error defined in* (6.84), *where* $\omega(k) = X(k) - X(s_k)$ *and* $E(s_k) = 0$, *is bounded* $\forall k \in [s_k, s_{k+1})$ *by*

$$E(k) \leq \frac{H(X(s_k),\varepsilon)}{G(\varepsilon)} \left[(1+G(\varepsilon))^{(s_{k+1}-s_k)} - 1 \right], \tag{6.85}$$

where

$$\begin{aligned} G(\varepsilon) &= ||\sqrt{N(\varepsilon)}(A+B-I)\sqrt{N(\varepsilon)}^{-1}||_2^2, \\ H(X(s_k),\varepsilon) &= ||\sqrt{N(\varepsilon)}(A(\varepsilon)-I)X(s_k)||_2^2, \end{aligned} \tag{6.86}$$

and $N(\varepsilon)$ *is given by* (6.82).

Proof. To find an upper bound on $E(k)$, it is noted that for an ET counterpart the state at every sampling instant of a discrete-time system is compared with that of the last transmitted state. Following this idea, we take the difference between successive error terms $\forall k \in [s_k, s_{k+1})$, i.e.,

$$\begin{aligned} \Delta E(k) := E(k+1) - E(k) &\leq ||\sqrt{N(\varepsilon)}\Delta\omega(k)||_2^2 \\ &\leq ||\sqrt{N(\varepsilon)}(\omega(k+1)-\omega(k))||_2^2 \\ &\leq ||\sqrt{N(\varepsilon)}(X(k+1)-X(k))||_2^2. \end{aligned} \tag{6.87}$$

From (6.77),

$$\begin{aligned}\Delta E(k) &\leq ||\sqrt{N(\varepsilon)}[(A(\varepsilon)+B)X(k)-BX(s_k)-X(k)]||_2^2\\&\leq ||\sqrt{N(\varepsilon)}[(A(\varepsilon)+B-I)X(k)-BX(s_k)]||_2^2\\&\leq ||\sqrt{N(\varepsilon)}[(A(\varepsilon)+B-I)(\omega(k)+(A(\varepsilon)-I)X(s_k))]||_2^2\\&\leq G(\varepsilon)||\sqrt{N(\varepsilon)}\omega(k)||_2^2+H(X(s_k),\varepsilon),\end{aligned} \tag{6.88}$$

where $G(\varepsilon)$ and $H(X(s_k),\varepsilon)$ are given in (6.86).

Inequality (6.88) can also be written as

$$E(k+1)-(1+G(\varepsilon))E(k)\leq H(X(s_k),\varepsilon), \tag{6.89}$$

which is a first-order difference equation with the solution

$$\begin{aligned}E(k)\leq\,&(1+G(\varepsilon))^{(s_{k+1}-s_k)}E(s_k)\\&+H(X(s_k),\varepsilon)\left(\frac{(1+G(\varepsilon))^{(s_{k+1}-s_k)}-1}{G(\varepsilon)}\right).\end{aligned}$$

With the initial condition $E(s_k)=0$, one can easily arrive at the error bound (6.85), and this completes the proof. □

With the error bound given by Lemma 6.4, (6.85) is equated to $||\sqrt{M(\varepsilon)}X(s_k)||_2^2$ to find the next triggering time. This is the main result of this section given in the following theorem.

Theorem 6.5. *For the two-timescale discrete system (6.77) with ε^* such that $\forall\, 0<\varepsilon<\varepsilon^*$, the sampling time $T_s\in\Re^+$, and the transformation matrix defined in (6.109), the ISS-Lyapunov condition (6.80)–(6.81) is satisfied with the following choice of the next triggering time for each time instant $k\in\mathbb{Z}_+$ and $0<\eta\in\mathbb{Z}_+$:*

$$s_{k+1}=s_k+\eta T_s,$$

where

$$\eta=\Delta\{\frac{ln\left(1+\frac{G(\varepsilon)}{H(X(s_k),\varepsilon)}||\sqrt{M(\varepsilon)}T^{-1}\xi(s_k)||_2^2\right)}{T_s\times ln(1+G(\varepsilon))}\}, \tag{6.90}$$

with $G(\varepsilon)$, $H(X(s_k),\varepsilon)$, and $M(\varepsilon)$ defined in (6.86) and (6.82), and $\Delta\{.\}$, which represents the floor function, takes $a\in\mathbb{R}$, and gives an integer less than or equal *to a, is used to ensure that the next triggering time is an integer multiple of the sampling time.*

Proof. Plugging the error bound (6.85) into (6.83), we get

$$V(k+1) - V(k) \leq -X^t(k)QX(k) - X^t(s_k)M(\varepsilon)X(s_k) + \frac{H(X(s_k),\varepsilon)}{G(\varepsilon)}\left[(1+G(\varepsilon))^{(s_{k+1}-s_k)} - 1\right], \quad (6.91)$$

and summing up $\forall k \in [s_k, s_{k+1})$,

$$\sum_{k=s_k}^{s_{k+1}} [V(k+1) - V(k)] \leq - \sum_{k=s_k}^{s_{k+1}} [X^t(k)QX(k)] - \sum_{k=s_k}^{s_{k+1}} [X^t(s_k)M(\varepsilon)X(s_k)] + \sum_{k=s_k}^{s_{k+1}} \left[\frac{H(X(s_k),\varepsilon)}{G(\varepsilon)}[(1+G(\varepsilon))^{(s_{k+1}-s_k)} - 1]\right]. \quad (6.92)$$

Therefore, equating the last two terms gives

$$(1+G(\varepsilon))^{(s_{k+1}-s_k)} = 1 + \frac{G(\varepsilon)}{H(X(s_k),\varepsilon)}||\sqrt{M(\varepsilon)}X(s_k)||_2^2,$$

$$s_{k+1} = s_k + \frac{ln\left(1 + \frac{G(\varepsilon)}{H(X(s_k),\varepsilon)}||\sqrt{M(\varepsilon)}X(s_k)||_2^2\right)}{ln(1+G(\varepsilon))}$$

or

$$s_{k+1} = s_k + \Delta\{\frac{ln\left(1 + \frac{G(\varepsilon)}{H(X(s_k),\varepsilon)}||\sqrt{M(\varepsilon)}T^{-1}\xi(s_k)||_2^2\right)}{T_s \times ln(1+G(\varepsilon))}\}T_s.$$

From (6.74), $X(k) = T^{-1}\xi(k)$, which gives (6.90). This completes the proof. □

6.2.7 Simulation example 6.2

The triggering mechanisms described above for two-timescale systems are tested on a numerical example given in [11], which has the following matri-

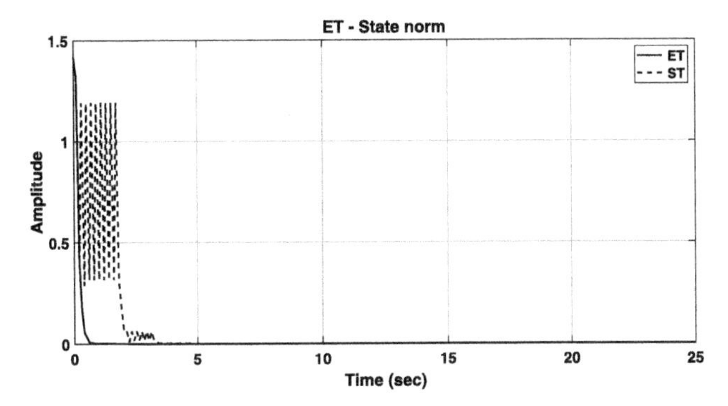

FIGURE 6.8 State norm. Solid line: event-triggered system; dashed line: self-triggered system.

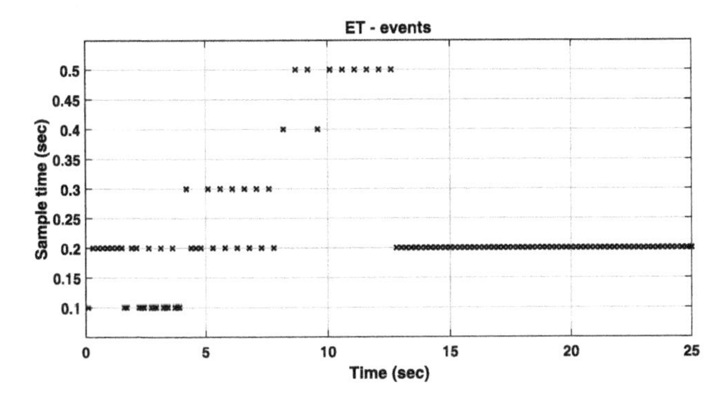

FIGURE 6.9 Sampling times for event-triggered systems.

ces:

$$\mathbb{A} = \begin{bmatrix} 0 & 0.4 & 0 & 0 \\ 0 & 0 & 0.345 & 0 \\ 0 & -0.524 & -0.465 & 0.262 \\ 0 & 0 & 0 & -1 \end{bmatrix}, \quad \mathbb{B} = \begin{bmatrix} 0 \\ 0 \\ 0 \\ 1 \end{bmatrix}, \tag{6.93}$$

and $n_1 = 2$, $n_2 = 2$, $\varepsilon = 0.1$, and initial state $x_0(k) = [1\ \ 0\ \ 1\ \ 0]^t$. Composite control of the form (6.55) is used with

$$K_1 = -[1\ \ 0.86], \quad K_2 = -[0.18\ \ 0.05]. \tag{6.94}$$

It can be seen from Fig. 6.8 that the states of the system for both triggering schemes are regulated in finite time with ST taking a slightly higher time.

This is due to longer sampling intervals resulting from the proposed ST scheme, as shown in Figs. 6.9 and 6.10.

The resulting error $\omega(k)$ and the control input are also shown in Figs. 6.11 and 6.12.

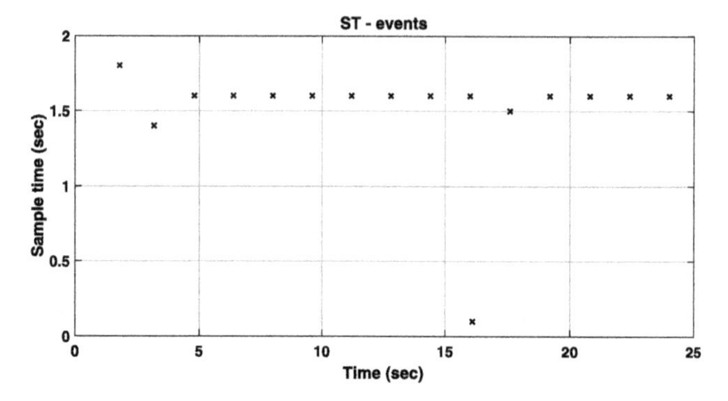

FIGURE 6.10 Sampling times for self-triggered systems.

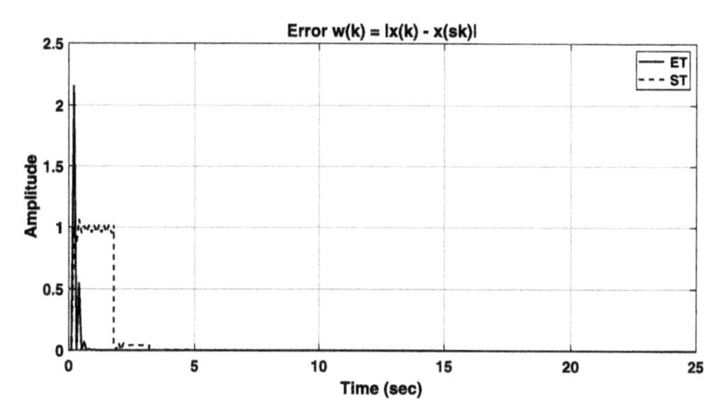

FIGURE 6.11 Control input. Solid line: event-triggered system; dashed line: self-triggered system.

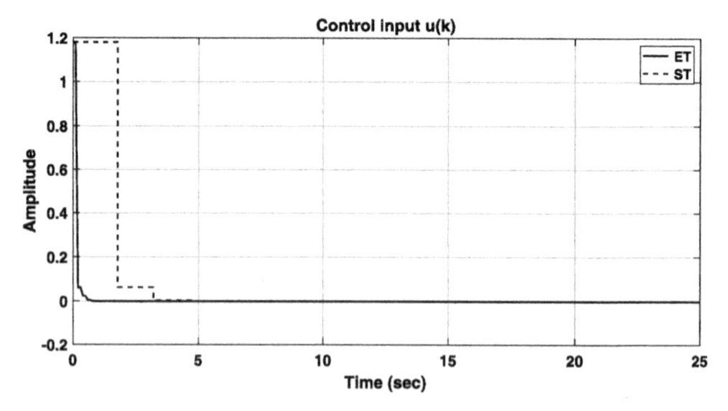

FIGURE 6.12 Control input. Solid line: event-triggered system; dashed line: self-triggered system.

6.3 Networks of two-timescale systems

In what follows, we consider a network of n identical, linear discrete systems with mode separation. For any $j = 1, \cdots, n$, the jth system at time k is characterized by the state $x(k) = col[x_1(k) \;\; x_2(k)] \in \Re^{n_1+n_2}$ and there exists a small $\varepsilon > 0$ such that its dynamics are given by

$$\begin{bmatrix} x_{i1}(k+1) \\ x_{i2}(k+1) \end{bmatrix} = \begin{bmatrix} (I_1 + \varepsilon A_{i1}) & \varepsilon A_{i2} \\ A_{i3} & A_{i4} \end{bmatrix} \begin{bmatrix} x_{i1}(k) \\ x_{i2}(k) \end{bmatrix} + \begin{bmatrix} \varepsilon B_{i1} \\ B_{i2} \end{bmatrix} u_i(k), \tag{6.95}$$

where $u(k) \in \Re^m$ is the control input and $\mathbf{rank}(B_{i1}) = \mathbf{rank}(B_{i2}) = m$. To ensure the well-posedness of (6.95), we invoke the standard assumption that the matrix $I_2 - A_{i4}$ is invertible.

With the network of n systems we associate a graph $\mathbb{G} = (\mathbb{V}; \mathbb{E}; \mathbb{A})$ of order n composed of a *vertex set* $\mathbb{V} = \{v_1, \; v_2, \; ..., \; v_n\}$, a set of ordered pairs of vertices called *edges* $\mathbb{E} = \{e_{ij} = (v_i, \; v_j)\} \subseteq \mathbb{V} \times \mathbb{V}$, and a weighted adjacency matrix $\mathbb{A} = [a_{ij}]$ with nonnegative adjacent elements a_{ij}. For emphasis, we denote by $\mathbb{V}(\mathbb{G})$ and $\mathbb{E}(\mathbb{G})$ the vertex set and edge set of graph $\mathbb{G}$, respectively. The node indices belong to a finite index set $\mathbb{I} = \{1, 2, \ldots, n\}$. Moreover, $a_{ij} > 0$ if $(v_i, \; v_j) \in \mathbb{E}$ and $a_{ij} = 0$ if $(v_i, \; v_j) \notin \mathbb{E}$ for all $i = 1, \; ..., \; n$. Also, $(v_i, \; v_j) \in \mathbb{E}$ if and only if the ith agent can receive information from the jth agent directly.

The (graph) Laplacian of $\mathbb{G}$ is a rank-deficient, symmetric, and positive semidefinite matrix defined by

$$\begin{aligned} \mathbb{L}(\mathbb{G}) &:= \mathbb{E}(\mathbb{G})\mathbb{E}(\mathbb{G})^t = \Delta(\mathbb{G}) - A(\mathbb{G}) \\ &:= [\ell_{ij}], \;\; \ell_{ij} = -a_{ij}, \;\; \ell_{ii} = \sum_{j=1}^{n} a_{ij}. \end{aligned}$$

Definition 6.3. A *path of length p* in the graph $\mathbb{G} = (\mathbb{V}; \mathbb{E}; \mathbb{A})$ is a union of edges

$$\bigcup_{k=1}^{p} (i_k, \; j_k), \;\; \Rightarrow \;\; i_{k+1} = j_k, \;\; \forall \, k \in \{1, \cdots .p-1\}.$$

The node j is *connected* with node i in $\mathbb{G} = (\mathbb{V}; \mathbb{E}; \mathbb{A})$ if there exists at least a path in $\mathbb{G}$ from i to j, that is, ($i_1 = i$ and $j_p = j$). A *connected edge* is such that any of its two distinct elements is connected.

From now onwards, we assume that the undirected graph $\mathbb{G}$ is connected. Invoking the properties of the graph Laplacian $\mathbb{L}$, we further assume that there exists an orthonormal matrix $\mathbb{T} \in \Re^{n \times n}$, that is, ($\mathbb{T}\mathbb{T}^t = \mathbb{T}^t\mathbb{T} = I_n$), such that

$$\mathbb{T}\mathbb{L}\mathbb{T}^t = \mathbb{D} = \mathbf{diag}(\lambda_1, \lambda_2, \cdots, \lambda_n).$$

Definition 6.4. The n two-timescale systems defined by (6.95) achieve asymptotic synchronization using local information if there exists a state feedback controller of the form

$$u_i(k) = K_{i1} \sum_{j=1}^{n} a_{ij}[x_{1i}(k) - x_{1j}(k)] + K_{i2} \sum_{j=1}^{n} a_{ij}[x_{2i}(k) - x_{2j}(k)], \quad K_{i1} \in \Re^{m\times n_1}, \; K_{i2} \in \Re^{m\times n_2} \tag{6.96}$$

such that

$$\lim_{k\to\infty} ||x_{1i}(k) - x_{1j}(k)|| = 0 \quad \text{and} \quad \lim_{k\to\infty} ||x_{2i}(k) - x_{2j}(k)|| = 0.$$

6.3.1 Synchronization using local information

The main goal of this section is the characterization of the feedback controllers that use local information and asymptotically synchronize the singularly perturbed systems defined by (6.95). In order to do that we firstly define the collective dynamics describing the behavior of the overall network of n feedback coupled systems.

In the sequel, we introduce the augmented state and control vectors as follows:

$$x_1(k) = [x_{1i}^t(k), \; \cdots, \; x_{1n}^t(k)]^t \in \Re^{n\times n_1},$$
$$x_2(k) = [x_{2i}^t(k), \; \cdots, \; x_{2n}^t(k)]^t \in \Re^{n\times n_2}.$$

Remark 6.9. It is significant to observe that the asymptotic synchronization is equivalent with

$$\lim_{k\to\infty} (\mathbb{L} \otimes I_{n_1})x_1 = 0 \quad \text{and} \quad \lim_{k\to\infty} (\mathbb{L} \otimes I_{n_2})x_2 = 0.$$

Moreover, due to the fact that $\mathbb{TL} = \mathbb{DT}$, the synchronization can be cast into the form

$$\lim_{k\to\infty} (\mathbb{D} \otimes I_{n_1})(\mathbb{T} \otimes I_{n_1})x_1 = 0,$$
$$\lim_{k\to\infty} (\mathbb{D} \otimes I_{n_2})(\mathbb{T} \otimes I_{n_1})x_2 = 0. \tag{6.97}$$

Substituting (6.96) into (6.95) results in the following collective closed-loop dynamics:

$$\begin{bmatrix} x_1(k+1) \\ x_2(k+1) \end{bmatrix} = \begin{bmatrix} (I_1 + \varepsilon \hat{A}_1) & \varepsilon \hat{A}_2 \\ \hat{A}_3 & \hat{A}_4 \end{bmatrix} \begin{bmatrix} x_{i1}(k) \\ x_{i2}(k) \end{bmatrix}, \tag{6.98}$$

where

$$
\begin{aligned}
\hat{A}_1 &= (I_n \otimes A_1) - (I_n \otimes B_1 K_1)(\mathbb{L} \otimes I_{n_1}), \\
\hat{A}_2 &= (I_n \otimes A_2) - (I_n \otimes B_1 K_2)(\mathbb{L} \otimes I_{n_1}), \\
\hat{A}_3 &= (I_n \otimes A_3) - (I_n \otimes B_2 K_1)(\mathbb{L} \otimes I_{n_1}), \\
\hat{A}_4 &= (I_n \otimes A_4) - (I_n \otimes B_2 K_2)(\mathbb{L} \otimes I_{n_1}).
\end{aligned}
\tag{6.99}
$$

Unlike the invertibility of matrix $I_2 - A_4$, one has no guarantee that the matrix $I_2 - \hat{A}_4$ is invertible. Therefore, the well-posedness of the closed-loop system (6.98) has also to be ensured by the choice of the controller gains.

We now proceed by making a final change of variables,

$$
\begin{aligned}
\widehat{x}_1(k) &= (\mathbb{T} \otimes I_{n_1}) x_1(k), \\
\widehat{x}_2(k) &= (\mathbb{T} \otimes I_{n_2}) x_2(k),
\end{aligned}
\tag{6.100}
$$

which converts the aggregate dynamics (6.98) into the form

$$
\begin{bmatrix} \widehat{x}_1(k+1) \\ \widehat{x}_2(k+1) \end{bmatrix} = \begin{bmatrix} (I_1 + \varepsilon \widehat{A}_1) & \varepsilon \widehat{A}_2 \\ \widehat{A}_3 & \widehat{A}_4 \end{bmatrix} \begin{bmatrix} \widehat{x}_1(k) \\ \widehat{x}_2(k) \end{bmatrix},
\tag{6.101}
$$

where

$$
\begin{aligned}
\widehat{A}_1 &= (I_n \otimes A_1) - (I_n \otimes B_1 K_1)(\mathbb{D} \otimes I_{n_1}), \\
\widehat{A}_2 &= (I_n \otimes A_2) - (I_n \otimes B_1 K_2)(\mathbb{D} \otimes I_{n_1}), \\
\widehat{A}_3 &= (I_n \otimes A_3) - (I_n \otimes B_2 K_1)(\mathbb{D} \otimes I_{n_1}), \\
\widehat{A}_4 &= (I_n \otimes A_4) - (I_n \otimes B_2 K_2)(\mathbb{D} \otimes I_{n_1}).
\end{aligned}
\tag{6.102}
$$

The following important observations stand out:

A) Invoking the properties of Kronecker products from Chapter 1, the closed-loop system (6.102) can be interestingly decoupled in n independent two-timescale systems. Precisely, one uses the fact that for any matrices $\mathcal{M}, \mathcal{N}$ of appropriate dimensions we have

$$
\begin{aligned}
(I_n \otimes \mathcal{M}) &- (I_n (I_n \otimes \mathcal{N})(\mathbb{D} \otimes I_m) \\
&= (I_n \otimes \mathcal{M}) - (\mathbb{D} \otimes \mathcal{N}) \\
&= \mathbf{diag}[\mathcal{M}, \cdots, \mathcal{M}] - \mathbf{diag}[\lambda_1 \mathcal{N}, \cdots, \lambda_n \mathcal{N}] \\
&= \mathbf{diag}[\mathcal{M} - \lambda_1 \mathcal{N}, \cdots, \mathcal{M} - \lambda_n \mathcal{N}],
\end{aligned}
\tag{6.103}
$$

which eventually results in

$$
\begin{aligned}
\widehat{A}_1 &= \mathbf{diag}[A_1 - \lambda_1 B_1 K_1, \cdots, A_1 - \lambda_n B_1 K_1], \\
\widehat{A}_2 &= \mathbf{diag}[A_2 - \lambda_1 B_1 K_2, \cdots, A_1 - \lambda_n B_1 K_2],
\end{aligned}
$$

$$\widehat{A}_3 = \mathbf{diag}[A_3 - \lambda_1 B_2 K_1, \cdots, A_1 - \lambda_n B_1 K_1],$$
$$\widehat{A}_4 = \mathbf{diag}[A_{i4} - \lambda_1 B_2 K_2, \cdots, A_1 - \lambda_n B_2 K_1]. \tag{6.104}$$

This in turn casts the closed-loop system (6.101) into the form

$$\begin{bmatrix} \widehat{x}_{i1}(k+1) \\ \widehat{x}_{i2}(k+1) \end{bmatrix} = \begin{bmatrix} (I_1 + \varepsilon A_{i1} - \lambda_i B_{i1} K_{i1}) & \varepsilon(A_{i2} - \lambda_i B_{i1} K_{i2}) \\ (A_{i3} - \lambda_i B_{i2} K_{i1}) & (A_{i4} - \lambda_i B_{i2} K_{i2}) \end{bmatrix} \times \begin{bmatrix} \widehat{x}_{i1}(k) \\ \widehat{x}_{i2}(k) \end{bmatrix},$$
$$i = 1, \cdots, n. \tag{6.105}$$

B) Essentially, the asymptotic synchronization problem with local information becomes a problem of simultaneous stabilization of systems in (6.105) for $1, \cdots, n$. Hence, (6.97) can be seen as

$$\lim_{k \to \infty} (\mathbb{D} \otimes I_{n_1}) \widehat{x}_1 = 0,$$
$$\lim_{k \to \infty} (\mathbb{D} \otimes I_{n_2})(\mathbb{T} \otimes I_{n_1}) \widehat{x}_2 = 0. \tag{6.106}$$

In view of the fact that $\mathbb{D} = \mathbf{diag}[\lambda_1, \cdots, \lambda_n]$, $\lambda \equiv 0$, it follows that the asymptotic synchronization condition reduces to

$$\lim_{k \to \infty} \widehat{x}_{i1} = 0,$$
$$\lim_{k \to \infty} \widehat{x}_{i2} = 0,$$
$$i = 2, \cdots, n. \tag{6.107}$$

C) It follows from the definition of $\mathbb{T}$ that the change of variables

$$\widehat{x}_1(k) = (\mathbb{T}^t \otimes I_{n_1}) x_1(k),$$
$$\widehat{x}_2(k) = (\mathbb{T}^t \otimes I_{n_2}) x_2(k)$$

also holds. Following the timescale theory, the synchronization manifold depends on the dynamics of $[\widehat{x}_1(k); \widehat{x}_2(k)]$. Effectively, if the system

$$\begin{bmatrix} \widehat{x}_1(k+1) \\ \widehat{x}_2(k+1) \end{bmatrix} = \begin{bmatrix} (I_1 + \varepsilon A_1) & \varepsilon A_2 \\ A_3 & A_4 \end{bmatrix} \begin{bmatrix} \widehat{x}_1(k) \\ \widehat{x}_2(k) \end{bmatrix} \tag{6.108}$$

has a stable equilibrium point $[\widehat{x}_1^*(k); \widehat{x}_2^*(k)]$, then system (6.95) will asymptotically reach a finite consensus. If (6.108) is unstable, then all the systems in (6.95) will synchronize on divergent trajectories.

D) The well-posedness of system (6.98) is equivalent to the one of system (6.101), which in turn is ensured if all systems in (6.105) are well-posed. It must be emphasized that for $i = 1$, the system is well-posed due to the nonsingularity of $I_2 - A_{i4}$. The rest of the systems in (6.101) are well-posed if K_{i2} is chosen such that $(A_{i4} - \lambda_i B_{i2} K_{i2})$ is invertible for $i = 2, \cdots, n$.

6.3.2 Control design

In this section we aim to provide a control design method that allows to completely decouple the slow and fast dynamics that occur in the overall system. Towards our goal, there are two possible methods [6,9]:

(A) The nonsingular transformation method. According this method, system (6.95) with $u(k) \equiv 0$ can be decoupled using the explicitly invertible linear transformation

$$z_i(k) = \begin{bmatrix} z_{i1}(k) \\ z_{i2}(k) \end{bmatrix} = T_i \begin{bmatrix} x_{i1}(k) \\ x_{i2}(k) \end{bmatrix} = T_i x_i(k), \tag{6.109}$$

$$T_i = \begin{bmatrix} I_1 - \varepsilon L_{i1} L_{i2} & -\varepsilon L_{i1} \\ L_{i2} & I_2 \end{bmatrix}, \quad T_i^{-1} = \begin{bmatrix} I_1 & \varepsilon L_{i1} \\ -L_{i2} & I_2 - \varepsilon L_{i2} L_{i1} \end{bmatrix}$$

and by choosing $L_{i1} \in \Re^{n_1 \times n_2}$, $L_{i2} \in \Re^{n_2 \times n_1}$ to satisfy

$$\begin{aligned} 0 &= A_{i3} + L_{i2} - A_{i4} L_{i2} + \varepsilon L_{i2}[A_{i1} - A_{i2} L_{i2}], \\ 0 &= A_{i2} + L_{i1} - L_{i1} A_{i4} + \varepsilon [A_{i1} - A_{i2} L_{i2}] L_{i1} - \varepsilon L_{i1} L_{i2} A_{i2}. \end{aligned} \tag{6.110}$$

The existence of matrices L_{i1} and L_{i2} and the convergence of (6.50) can be shown by verifying that, for sufficiently small ε, the norm conditions of [7] are satisfied. Deploying the implicit function theorem under the nonsingularity of $(I_2 - A_{i4})$, that is, $det[I_2 - A_{i4}] \neq 0$, guarantees that $L_{i1}(\varepsilon)$ and $L_{i2}(\varepsilon)$ are analytic at $\varepsilon = 0$ with $L_{i1}(0) = -A_{i2}[I_2 - A_{i4}]^{-1}$ and $L_{i2}(0) = -[I_2 - A_{i4}]^{-1} A_{i3}$.

(B) The timescale analysis method. Following the discrete quasisteady state concept [6,8], the fast modes corresponding to the small eigenvalues are important only during a short transient period. After that period, they are negligible, and the behavior of the discrete system can be described by its slow modes. Formally, letting $x_{i2}(k+1) = x_{i2}(k) \triangleq x_{i2s}(k)$ in (6.95) is equivalent to neglecting the effect of the fast modes. Under this condition, the discrete quasisteady state is given by

$$x_{i2s}(k) = [I_2 - A_{i4}]^{-1} [A_{i3} x_{i1s}(k) + B_{i2} u_{is}(k)]. \tag{6.111}$$

Adopting either method to the discrete network system under consideration, it follows that the aggregate model (6.101)–(6.102) can be readily decoupled into a *slow subsystem* of order n_1 described by

$$\begin{aligned}
\widehat{x}_{i1s}(k+1) &= [I_1 + \varepsilon A_{io} - \lambda_i B_o K_o]\widehat{x}_{is}(k), \\
\widehat{x}_{i1s}(0) &= \widehat{x}_{i1}(0), \\
\widehat{x}_{i2s}(k) &= [I_2 - A_{i4}]^{-1}[A_{i3} x_{i1s}(k) - \lambda_i B_2 K_o]\widehat{x}_{i1s}(k), \qquad (6.112) \\
A_{io} &= A_{i1} + A_{i2}[I_2 - A_{i4}]^{-1} A_{i3}, \\
B_{io} &= B_{i1} + A_{i2}[I_2 - A_{i4}]^{-1} B_{i2}, \qquad (6.113)
\end{aligned}$$

In addition to a *fast subsystem* with the order n_2 given by

$$\begin{aligned}
\widehat{x}_{if}(k+1) &= [A_{i4} - \lambda_i B_2 K_2]\widehat{x}_{if}(k), \\
\widehat{x}_{if}(0) &= \widehat{x}_{i2}(0) - \widehat{x}_{i2s}(0). \qquad (6.114)
\end{aligned}$$

We are now in a position to establish the following theorem.

Theorem 6.6. *Let the gain matrices K_o and K_2 be designed such that for $i = 2, \cdots, n$ the matrices*

$$[A_{i4} - \lambda_i B_2 K_2], \ [I_1 + \varepsilon A_{io} - \lambda_i B_o K_o]$$

are all Schur stable. Then, there exists $\varepsilon^* > 0$ such that the controllers (6.96) with

$$K_1 = \left(I_m - K_2[I_2 - A_{i4}]^{-1} B_{i2}\right) K_o - K_2[I_2 - A_{i4}]^{-1} A_{i3} \qquad (6.115)$$

asymptotically synchronize system (6.95) with local information.

Proof. Following the results of [8] and selecting the gain matrices K_o and K_2 to stabilize the slow and fast subsystems (6.113) and (6.114), respectively, for $i = 2, \cdots, n$, guarantees that

$$\begin{aligned}
\widehat{x}_{i1}(k) &= \widehat{x}_{i1s}(k) + \mathbf{O}(\varepsilon), \\
\widehat{x}_{i2}(k) &= [I_2 - A_{i4}]^{-1}[A_{i3} x_{i1s}(k) - \lambda_i B_2 K_o]\widehat{x}_{i1s}(k) + \widehat{x}_{if}(k) + \mathbf{O}(\varepsilon)
\end{aligned} \qquad (6.116)$$

hold for all finite $\varepsilon > 0$ and all $k \in [0, \infty)$. Recall that the asymptotic synchronization corresponds to

$$\begin{aligned}
\lim k &\to \infty(\mathbb{L} \otimes I_{n1}) x_1(k) = 0, \\
\lim k &\to \infty(\mathbb{L} \otimes I_{n2}) x_2(k) = 0,
\end{aligned}$$

which holds true in view of

$$\begin{aligned}
(\mathbb{L} \otimes I_{n1}) x_1(k) &= (\mathbb{D} \otimes I_{n1}) \widehat{x}_1(k) \\
&= [0, \ \lambda_2 \widehat{x}_{12}, \ \cdots, \ \lambda_n \widehat{x}_{1n}]^t,
\end{aligned}$$

$$(\mathbb{L} \otimes I_{n2})x_2(k) = (\mathbb{D} \otimes I_{n2})\widehat{x_2}(k)$$
$$= [0,\ \lambda_2\widehat{x}_{22},\ \cdots,\ \lambda_n\widehat{x}_{2n}]^t. \qquad \square$$

Basically, Theorem 6.6 guarantees that in order to asymptotically synchronize system (6.95) we have to separately synchronize the fast and slow dynamics by stabilizing the dynamics of the error between the different systems. Note that when A_4 has a spectral radius less than 1, then a lower-order controller with $K_2 = 0$ will asymptotically synchronize system (6.95) as well.

6.3.3 Simulation example 6.3

We consider the synchronization of three agents described by an undirected graph $\mathbb{G}$. The system is given by (6.95), where

$$A_1 = \begin{bmatrix} 0.3 & -0.6 \\ -0.2 & 0.2 \end{bmatrix},\ A_2 = \begin{bmatrix} 0.2 & 0 \\ 0.3 & -0.2 \end{bmatrix},\ A_3 = \begin{bmatrix} 0.1 & 0.1 \\ -0.2 & 0.2 \end{bmatrix},$$
$$A_4 = \begin{bmatrix} 0.5 & -0.1 \\ -0.1 & 0.3 \end{bmatrix},\ B_1 = \begin{bmatrix} 1 \\ 1 \end{bmatrix},\ B_2 = \begin{bmatrix} 2 \\ 1 \end{bmatrix}.$$

We assign to each agent a vector state having four components characterized by slow and fast dynamics. For any agent $i \in \{1, \cdots, n\}$, let us denote by $x_{i,1s}$, $x_{i,2s}$, and $x_{i,1f}, x_{i,2f}$ its slow and fast state components, respectively. The Laplacian matrix describing the undirected topology of the graph $\mathbb{G}$ is defined by

$$\mathbb{L} = \begin{bmatrix} 2 & -1 & -1 \\ -1 & 2 & -1 \\ -1 & -1 & 2 \end{bmatrix}.$$

In simulation we choose $\varepsilon = 0.002$ and all the components of the initial condition are chosen within $[-4\ 5]$. Since A_4 is Schur stable, we proceed to stabilize the dynamics (6.105) such that $K_1 = K_o$. Figs. 6.13 and 6.14 show that the slow and fast models approximate the decoupled system within $\mathbf{O}(\varepsilon)$ neighborhood. Observe in these figures that the slow states are x_{211}, x_{221} associated with agent 2 and x_{311}, x_{321} associated with agent 3. In a similar way, the fast states are x_{212}, x_{222} associated with agent 2 and x_{312}, x_{322} associated with agent 3. The closeness of trajectories is observed. The state trajectories of all agents are also illustrated in Fig. 6.15.

6.4 Consensus of multiagent delay systems with adversaries

Consensus control means to design a networked interaction protocol such that all the agents reach an agreement on certain variables of common interest states asymptotically or in a finite time [14]. Due to potential applications of multiagent systems in broad areas such as distributed sensor networks, congestion

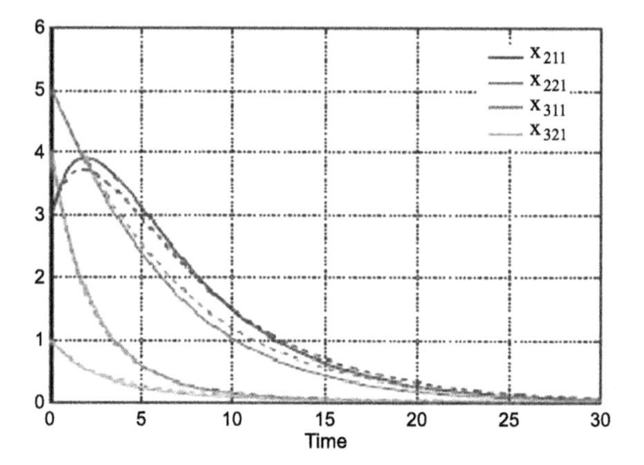

FIGURE 6.13 The trajectories of first state: Actual (solid), slow (dotted).

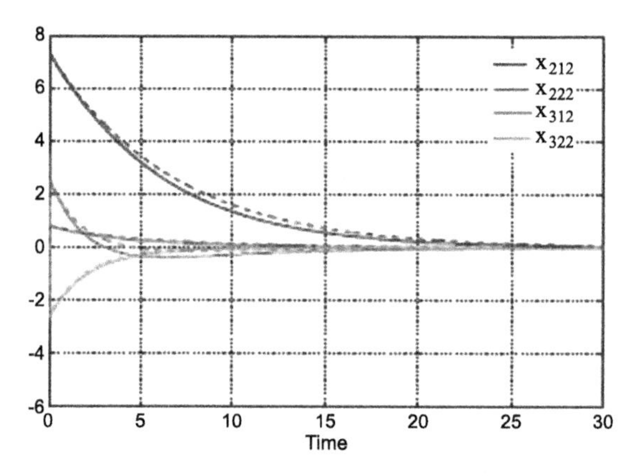

FIGURE 6.14 The trajectories of second state: Actual (solid), fast (dotted).

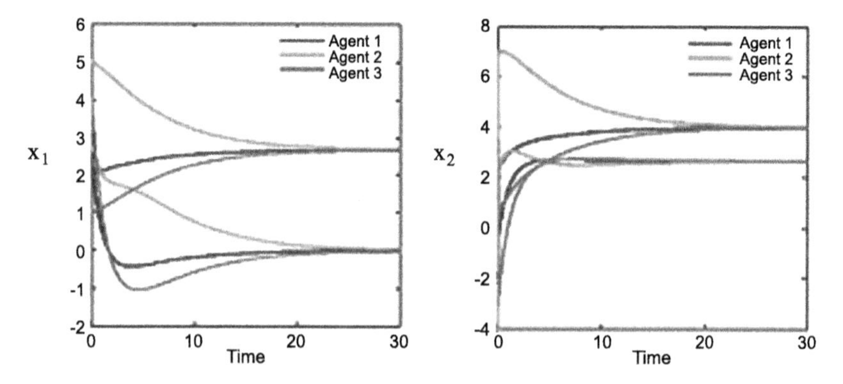

FIGURE 6.15 The overall system trajectories.

control, and formation flight, algorithms for consensus of multiagent systems is a field that has been studied by numerous researchers from various perspectives (see [15], [16], [17], [18], [19], [20] and references therein). Several results have appeared in recent literature that consider systems with different motion models and network interactions [21].

The resilient consensus problem for multiagent systems involves designing a control protocol such that the networks are robust to the effects of adversaries. [32] considered two cases of adversaries, namely, Byzantine and noncolluding faults. By considering Byzantine agents, they proved that if the connectivity of the communication graph is $2k + 1$, then a linear consensus network can be generically resilient to k concurrent faults, and $k + 1$ when consider noncolluding agents. [26] presented the adversarially robust consensus protocol (ARC-P) combining ideas from distributed computing and control consensus protocols. It shows that ARC-P solves the consensus problem in complete networks whenever there are more fault-free agents than adversarial agents. In [27], the authors extended the results to the discrete-time model and weakened the condition of full knowledge of the network topology. In addition to the above works, some researchers also considered linear iterative schemes in the presence of adversarial agents [33], [38].

Another important concern with respect to networked systems is the issue of communication delays. In practical networked multiagent systems delays are unavoidable in information acquisition and transmission, which should be taken into consideration when designing a consensus protocol. An initial study on this problem can be found in [31] using a frequency domain method, and a necessary and sufficient condition on the upper bound of time delays is provided under the assumption that all the delays are equal and time-invariant. [29] provided a stability analysis of a linear continuous-time system and showed the robustness of delays. The authors in [39] studied nonlinear time-delayed dynamics with a linear control law and a contraction theorem was applied to show consensus. [22] studied the continuous-time systems with transmission delays and provided a systematic way for analyzing the average consensus problem. In the presence of both delay and measurement noises, [14] proposed a consensus protocol based on the stochastic approximation theorem. Necessary and sufficient conditions were provided for mean square consensus and strong consensus. When we consider the quantization effect, [30] proved that the consensus can be achieved if we carefully design the dynamic quantizer.

In our work, we consider the consensus problem for discrete-time multiagent systems in the presence of both adversaries and transmission delays. We assume a directed delayed network in which all agents belong to two disjoint classes: loyal agents and adversarial ones; loyal agents will follow the protocol all the time, while, on the contrary, adversarial agents are rational and will attempt to alter the outcome so as to increase its utility. Every agent in the network, including the adversarial ones, sends the same information to all its neighbors, which appears to be realistic for control scenarios [33]. We consider the network-wide malicious adversary as the threat model. The malicious adversary has full knowledge of the networked system and may participate with the goal of leading the consensus process to an invalid value. To deal with this problem, we design an efficient consensus protocol while using only locally delayed information. The system allows for time-dependent communication patterns which are important when we take into account link failure and link creation, reconfigurable networks, and nearest neighbor coupling. Based on a proposed control protocol, sufficient conditions are obtained which guarantee that all loyal agents asymptotically achieve a consensus under both adversaries and delays.

6.4.1 Adversary model

We now set out to define the adversary model and the scope of the adversary we consider in the chapter. The adversary models can be classified as *crash failure* [35], *noncolluding* [32], [33], and *malicious* (or *Byzantine*) [37], [25], [42] agents, depending on their abilities. Crash failure agents behave correctly before crashing, but once they have crashed they will stop executing prematurely and do nothing. Noncolluding agents are unaware of the structure and state of the network and ignore the presence of other adversarial agents. Instead, both malicious and Byzantine agents have a complete knowledge of the networked system and possess unlimited sensing, communication, and computation capabilities. Compared with the Byzantine agents, the malicious agents simply lack the deceitful capacity to convey different information to different neighbors in the network. However, this type of deceit is usually not possible for the agents broadcasting their information to neighbors. Therefore, we consider malicious as an adversary model under the local broadcast assumption. Here, we assume that the total number of malicious agents is upper-bounded by a number $f \in \mathbb{Z}^+$ in each loyal agent's neighborhood. This is called the f-local malicious model [34], [27]. We provide the following formal definition.

Definition 6.5 (f-Local malicious model). A set $\mathcal{S} \subset \mathcal{V}$ of malicious agents is f-local malicious model if it contains at most f agents in the neighborhood of the other agents for all time instants k, that is, $|\mathcal{N}_i \cap \mathcal{S}| \leq f, \forall i \in \mathcal{V} \backslash \mathcal{S}, f \in \mathbb{Z}^+$.

6.4.2 Delay-robust resilient consensus

We shall formulate the consensus problem in a multiagent network system. Suppose that the network system under consideration consists of n dynamic agents, labeled from 1 to n. Each agent is regarded as a node in a directed graph, denoted by $\mathbb{G} = (\mathbb{V}, \mathbb{E})$. Without loss of generality, the agent set is partitioned into the set of p loyal agents $\mathcal{V}_l = \{1, 2, \ldots, p\}$ and the set of $n - p$ adversarial agents $\mathcal{V}_a = \{p+1, p+2, \ldots, n\}$.

For all agents $i \in \mathbb{V}_l$ and $j \in \mathcal{N}_i$, the information flow from agent j to agent i through edge (j, i) is assumed to suffer from a time delay $d_{i,j}$, which is an unknown positive integer and may be time-varying. The agent is assumed to be able to access its own instantaneous state value, i.e., $d_{i,i}=0$.

We assume that each loyal agent has the following dynamics:

$$\begin{cases} x_i(k+1)=x_i(k)+u_i(k), & i\in\mathcal{V}_l, \\ x_i(s)=\phi_i(s), & s=-\bar{d},-\bar{d}+1,\ldots,0, \end{cases} \tag{6.117}$$

where $x_i(k)\in\Re$ is the state value of agent i, $\phi_i(s)$ is a initial function, $\bar{d}$ is the upper delay bound to be determined, and $u_i(k)\in\Re$ is the control protocol. We will design the control protocol by using the local information such as neighbors' state values received through wireless channels such that two important conditions, safety and consensus, are satisfied. In most of the cases, transmission delay is not negligible in the control protocol design. We shall show that the proposed control protocol is robust to time delay.

Denote

$$M(k)=\max_{i\in\mathbb{V}_l,\theta=0,\ldots,\bar{d}} x_i(k-\theta),\quad m(k)=\min_{i\in\mathcal{V}_l,\theta=0,\ldots,\bar{d}} x_i(k-\theta), \tag{6.118}$$

where $\bar{d}=\sup_{k\geq 0}\max\{d_{j,i}(k), i\in\mathcal{N}_j, j\in\mathbb{V}\}$. Since time delays are uniformly bounded, we know that $\bar{d}<+\infty$. It is clear that $M(0)$ and $m(0)$ are the maximum and minimum values of the initial states among the loyal agents, respectively. Then, we define the problem to be studied specifically as follows.

Definition 6.6 (Delay-robust f-local resilient consensus). The discrete-time multiagent system (6.117) with bounded time-varying delays $\{d_{i,j}, i\in\mathcal{V}_l, j\in\mathcal{N}_i\}$ is said to achieve resilient consensus in the presence of f-local malicious agents if and only if

$$m(0)\leq \inf_{k\geq 0}\min_{i\in\mathcal{V}_l} x_i(k)\leq \sup_{k\geq 0}\max_{i\in\mathcal{V}_l} x_i(k)\leq M(0), \tag{6.119}$$

$$\lim_{k\to\infty}\left(x_i(k)-x_j(k)\right)=0,\quad \forall i,j\in\mathcal{V}_l. \tag{6.120}$$

In Definition 6.6, (6.119) guarantees that all the loyal agents are within the safe region all the time. This condition is equivalent to that of $\forall k\geq 0$, $m(k)\geq m(0)$, and $M(k)\leq M(0)$. On the other hand, consensus is guaranteed by (6.120).

In the sequel, we shall introduce a delay-robust f-local resilient consensus protocol first, and next, the convergence analysis is provided.

6.4.3 Delay-robust f-local resilient consensus protocol

In the proposed protocol, each agent i sorts its received (delayed) state values in a descending order. If there are fewer than f values strictly larger than its own value $x_i(k)$, then agent i removes all of these values by effectively setting their weights to zero. Otherwise, it removes precisely the largest f values in the sorted list. The same manipulation is applied to the smallest values in agent i's neighborhood. The similar protocols without considering transmission delay can be found in [42], [28], [24]. Let $\mathcal{R}_i(k)$ denote the set of agents whose values are removed by agent i at time instant k. The proposed consensus protocol is provided as follows:

$$u_i(k) = \sum_{j \in \mathcal{N}_i \mathcal{R}_i(k)} a_{i,j}(k)(x_j(k - d_{i,j}) - x_i(k)). \tag{6.121}$$

We assume that $\sum_{j \in \mathcal{N}_i} a_{i,j}(k) < 1$, which is very standard for discrete-time systems. According to the definition of $\mathcal{R}_i(k)$, we say that protocol (6.121) is with parameter f. Due to the existence of the adversary agents, $\mathcal{R}_i(k)$ is possibly time-varying even if the underlying network is fixed, which implies the control topology is dynamically switching. In the following, we shall provide the stability analysis of multiagent system (6.117) under protocol (6.121).

6.4.4 Stability analysis

In this section, we will prove that when the network topology satisfies some connectivity conditions, the safe condition (6.119) and consensus goal (6.120) can be achieved. In order to interpret the delayed value of the agents in the protocol (6.121), we introduce some virtual nodes $v_{i,j}$, which corresponds to node i with j steps time delay. It is clear that $v_{i,0} = i$. Denote $\mathcal{V}_{(i)} = \{v_{i,0}, \cdots, v_{i,\bar{d}}\}$. To begin with, we need to introduce the following topological properties [23], [27].

Definition 6.7 (Delay graph). Given a digraph $\mathcal{G}$ with node set $\mathcal{V}$ and edge set $\mathcal{E}$, the graph $\bar{\mathcal{G}}(k)$ is called delay graph corresponding to $\mathcal{G}$ if

1. the node set is $\bigcup_{i \in \mathcal{V}} \mathcal{V}(i)$;
2. the edge set is $\{(v_{i,j-1}, v_{i,j}), j = 1, \ldots, \bar{d}\} \bigcup \{(v_{i,d_{j,i}}, v_{j,0}) : \forall (i, j) \in \mathcal{E}\}$.

From Definition 6.7, we can see that there is one and only one path from the node to the virtual nodes representing its delayed values. Due to the time-varying delays, the delay graph is essentially dynamically switching.

Definition 6.8 (r-Reachable set). Given a digraph $\mathbb{G} = \{\mathbb{V}, \mathbb{E}\}$ and a nonempty subset $\mathbb{S} \subset \mathbb{V}$, we say $\mathbb{S}$ is an r-reachable set if $\exists_i \in \mathbb{S}$ such that $|\mathcal{N}_i \backslash \mathcal{S}| \geq r$, $r \in \mathbb{Z}^+$.

Observe that an r-reachable set $\mathcal{S}$ contains an agent that has at least r neighbors outside of $\mathcal{S}$ at time instant k, which can be intuitively understood as that there exists at least one agent in the set that can be influenced by a large number of agents from outside the set. The following definition of r-robust graph can generalize this notion of redundancy to the entire network.

Definition 6.9 (r-Robust graph). A digraph $\mathcal{G} = \{\mathcal{V}, \mathcal{E}\}$ is r-robust, with $r \in \mathbb{Z}^+$, if for every pair of nonempty, disjoint subsets of $\mathcal{V}$, at least one of the subsets is r-reachable.

Given an r-robust graph $\mathcal{G}$, let $\mathcal{G}'$ be the graph produced by removing up to t incoming edges of each agent in $\mathcal{G}(t < r)$. According to Definition 6.9, we can observe that $\mathcal{G}'$ is $(r - t)$-robust. Then, we have the following lemmas.

Lemma 6.5. *A graph contains a spanning tree if and only if $\mathcal{G}$ is 1-robust.*

Proof. **Necessity:** By contradiction, we assume that $\mathcal{G}$ contains a spanning tree and it is not 1-robust. According to Definition 6.9, there exist two disjoint subsets, $\mathcal{S}_1$ and $\mathcal{S}_2$, which do not have neighbors outside their own set. Then, there is no information flow between $\mathcal{S}_1$ and $\mathcal{S}_2$, which contradicts the definition of spanning tree.

Sufficiency: By contradiction, we assume that $\mathcal{G}$ does not contain a spanning tree. Denote A as the adjacency matrix of $\mathcal{G}$. According to [36], [40], A is decomposable, which means we can partition the node set into two subsets and no information exchange happens between them. This contradicts the definition of 1-robust. □

The same result without necessity part can be found in [42], [28]. It is noted that if a graph is r-robust it implies that it is 1-robust, and therefore contains a spanning tree.

Lemma 6.6. *[23] For system (6.117) with protocol (6.121), if the communication graph jointly contains a spanning tree, there is a constant x^*, depending only on initial values, for which*

$$\lim_{k \to \infty} x_i(k) = x^*, \tag{6.122}$$

where the limit is approached exponentially fast.

Now we are in the position to provide the main result.

Theorem 6.7. *Consider a discrete-time directed delayed network modeled by (6.123). Suppose each loyal agent updates its state according to consensus protocol (6.121) with parameter f. The delay-robust f-local resilient consensus can be achieved if the network $\mathcal{G}$ is $(2f + 1)$-robust.*

Proof. The proof is divided into two steps. The first step is to prove the safety condition (6.119) and the second step is to prove the consensus condition (6.120).

Step 1 (Safety): The closed-loop system can be written in the following form:

$$\begin{cases} x_i(k+1) = x_i(k) + \sum_{j \in \mathcal{N}_i \setminus \mathcal{R}_i(k)} a_{i,j}(k)(x_j(k - d_{i,j}) - x_i(k)), \\ x_i(s) = \phi_i(s), \quad s = -\bar{d}, -\bar{d} + 1, \ldots, 0. \end{cases} \tag{6.123}$$

According to the definition of $M(k)$ and $m(k)$ in (6.118), it follows from (6.123) that $\forall i \in \mathcal{V}_l$

$$\begin{aligned} x_i(k+1) &\leq x_i(k) + \sum_{j \in \mathcal{N}_i \backslash \mathcal{R}_i(k)} a_{i,j}(k)(M(k) - x_i(k)) \\ &= \alpha M(k) + (1-\alpha)x_i(k) \\ &\leq M(k), \end{aligned} \tag{6.124}$$

which implies that $M(k+1) \leq M(k)$. Here $\alpha = \sum_{j \in \mathcal{N}_i \backslash \mathcal{R}_i(k)} a_{i,j}(k) < 1$. Similarly, we can get $m(k+1) \geq m(k)$, which guarantees the safety condition (6.119).

Step 2 (Consensus): From consensus protocol (6.121), we can observe that each agent can remove up to $2f$ values from the data it received at each time step. When the values of some adversary agents are not in the top and bottom f values for some nodes, the removed state value from these nodes may be less than $2f$. In this case, the state values of the unremoved adversary agents can be expressed in terms of the linear combination of all the (delayed) state values of all the loyal nodes, i.e.,

$$x_i(k) = \sum_{j \in \mathcal{V}_l} \beta_{q,j}(k) x_j(k), \quad i \in \mathcal{V}_a \cap (\mathcal{N}_q \backslash \mathcal{R}_q(k)),$$

where $\beta_{q,j} \geq 0$ and $\sum_{j \in \mathcal{V}_l} \beta_{q,j} = 1$. It is noted that the state value of adversary agent i may be included in the protocol of a different loyal agent q and there may exist different (in fact, infinitely many) expressions of $x_i(k)$. We just arbitrarily choose one. To slightly abuse the notation, for each loyal agent q, which does not include adversary state information in the protocol at time instant k, we set $\beta_{q,j}(k) = 0$, $j \in \mathcal{V}_l$.

We consider the following augmented system:

$$\bar{X}(k+1) = \Theta(k)\bar{X}(k),$$

where

$$\bar{X}(k) = \begin{bmatrix} X(k) \\ X(k-1) \\ \vdots \\ X(k-\bar{d}) \end{bmatrix}, \quad X(k) = \begin{bmatrix} x_1(k) \\ x_2(k) \\ \vdots \\ x_p(k) \end{bmatrix}, \quad \Theta(k) = \Theta_1(k) + \Theta_2(k),$$

$$\Theta_1(k) = \begin{bmatrix} A_0 & A_1 & \cdots & A_{\bar{d}} \\ I & & & \\ & \ddots & & \\ & & I & \end{bmatrix}, \quad \Theta_2(k) = \begin{bmatrix} B & 0 & \cdots & 0 \\ 0 & 0 & \cdots & 0 \\ \vdots & \vdots & \ddots & \vdots \\ 0 & 0 & \cdots & 0 \end{bmatrix},$$

$$B^{i,j}(k) = \beta_{i,j}(k),$$

$$A_0^{j,q}(k) = \begin{cases} 1 - \sum_{q \in \mathcal{N}_j \backslash \mathcal{R}_j(k)} a_{j,q}(k), & j = q, \\ a_{j,q}(k)\delta_{0,d_{j,q}}, & j \neq q, \end{cases}$$

$$A_i^{j,q}(k) = a_{j,q}(k)\delta_{i,d_{j,q}}, \quad i = 1, \ldots, p.$$

It can be checked that Θ, Θ_1, and Θ_2 are valid adjacency matrices corresponding to some delay graphs as defined in Definition 6.7. Note that $\sum_{i=1,\ldots,\bar{d}} A_i(k)$ is the adjacency matrix corresponding to $\mathcal{G}_r$, which is $\mathcal{G}$ by removing some links according to protocol (6.121). By Lemma 6.5 we know that $\mathcal{G}_r$ is 1-robust and consequently contains a spanning tree. On the other hand, according to [41], we know that the delay graph corresponding to Θ_1, denoted by $\mathcal{G}_1$, contains a spanning tree if graph $\mathcal{G}_r$ contains a spanning tree. Meanwhile, the delay graph corresponding to Θ, denoted by $\mathcal{G}_0$, contains more edges than $\mathcal{G}_1$, which implies that $\mathcal{G}_0$ contains a spanning tree. Then, according to Lemma 6.6, the consensus can be achieved under protocol (6.121), which completes the proof. □

6.4.5 Simulation example 6.4

In this section, we illustrate the result derived in the above sections by numerical simulations using the MATLAB® tool. We consider a digraph with $n = 5$ agents as shown in Fig. 6.16. The initial values of the 5 agents are $x_1(0) = 1, x_2(0) = 2, x_3(0) = 3, x_4(0) = 4$, and $x_5(0) = 5$. Here, we assume that agent 3 is a malicious agent and designed as

$$x_3(k+1) = 0.9x_3(k) + 0.1u,$$

where the reference inputs u for the malicious agent is $u = 10$. The goal of agent 3 is to drive the loyal agent values outside of the range of their initial

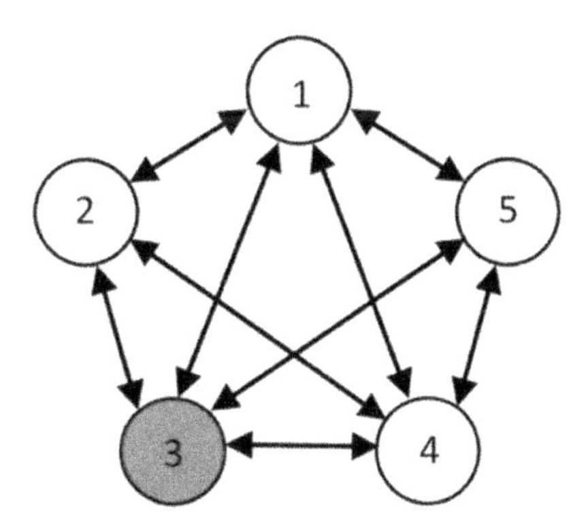

FIGURE 6.16 Directed graph with 5 agents.

values. To illustrate Theorem 6.7, let the communication graph be 3-robust in Fig. 6.16. By the above analysis, it can be seen that system (6.123) can sustain one malicious agent in the network under the 1-local malicious model. Here, we assume an upper delay bound $\bar{d} = 5$, the edge weight $a_{i,j} = \frac{0.8}{|\mathcal{N}_i \backslash \mathcal{R}_i|}$, and initial condition $\phi_i(s) = i$, $s = -\bar{d}, -\bar{d}+1, \ldots, 0$.

The results for this example are shown in Fig. 6.17. It can be seen that the malicious agent 3 attempts to drive the values of the loyal agents to a value of 10, but fails whenever the obtained control protocol (6.121) depicted in Fig. 6.17A. Furthermore, the result for the delay-free communication network of Fig. 6.16 is shown in Fig. 6.17B.

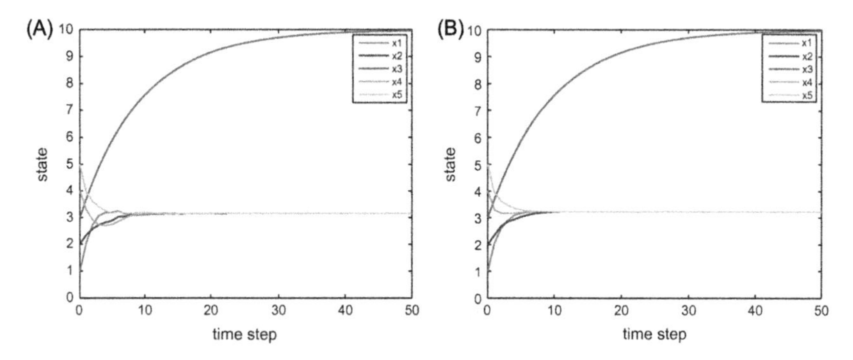

FIGURE 6.17 State trajectories of all agents with and without a transmission delay.

6.5 Notes

In this chapter, we have investigated event-triggered strategies for control of a class of linear discrete-time systems with fast–slow separation property. The plant is assumed to be ISS with respect to measurement errors and the control law is updated once a triggering condition involving the norm of a measurement error is violated. A sufficient condition for nontrivial interexecution times was derived. The results were also extended to a self-triggered formulation, where the next control updates are decided at the previous ones, thus relaxing the need for continuous monitoring of the measurement error.

Next, we focused on a class of singularly perturbed systems. A complete analysis and decentralized controller design for the synchronization of coupled singularly perturbed systems is accomplished by adapting the basic elements of discrete quasisteady state theory. The feedback controller uses only local information. The controller design is computationally oriented since it is obtained by solving some LMIs. Under mild assumptions the LMIs are always feasible.

Finally, we have considered the resilient consensus problem of discrete-time multiagent systems with transmission delays. An improved form of local filtering strategy for the loyal agents to resist adversaries in a directed delayed network has been provided. Under the condition that the time delays are uniformly bounded, a sufficient condition for the resilient consensus under the proposed protocol has been obtained, which is $(2f+1)$-robust.

References

[1] R. Horn, C. Johnson, Matrix Analysis, Cambridge University Press, Cambridge, 2005.
[2] T. Apostol, I. Makai, Mathematical Analysis, Addison-Wesley, Reading, Massachusetts, 1974.
[3] Y. Cao, W. Ren, Y. Li, Distributed discrete-time coordinated tracking with a time-varying reference state and limited communication, Automatica 45 (2009) 1299–1305.
[4] C. Ma, J. Zhang, Necessary and sufficient conditions for consensusability of linear multi-agent systems, IEEE Trans. Autom. Control 55 (2010) 1263–1268.
[5] H. Khalil, Nonlinear Systems, Prentice-Hall, NJ, 2002.
[6] H. Othman, N. Khraishi, M.S. Mahmoud, Discrete regulators with time-scale separation, IEEE Trans. Autom. Control 30 (3) (1985) 293–297.
[7] M.S. Mahmoud, Order reduction and control of discrete systems, IEE Proc. Part D, Control Theory Appl. 129 (4) (1982) 129–135.
[8] M.S. Mahmoud, Structural properties of discrete systems with slow and fast modes, Large Scale Syst. 3 (1982) 227–236.
[9] B. Litkouhi, H.K. Khalil, Multirate and composite control of two-time-scale discrete-time systems, IEEE Trans. Autom. Control AC-30 (7) (1985) 645–654.
[10] A.M. Memon, M.S. Mahmoud, An evaluation of self-triggering method for optimization of communication and control, IET Control Theory Appl. 10 (1) (January 2016) 76–83.
[11] M. Bhandari, D.M. Fulwani, R. Gupta, Event-triggered composite control of a two time scale system, IEEE Trans. Circuits Syst. II, Express Briefs 65 (4) (April 2018) 471–475.
[12] P. Sahu, M. Bhandari, D. Fulwani, Event-trigger control of discrete two-time scale system by leveraging its intrinsic properties, in: Proc. the 18th European Control Conference (ECC), Naples, Italy, 2019, pp. 3970–3975.
[13] M.S. Mahmoud, Event-based control of discrete two-time-scale systems, in: IECON 2017 – 43rd Annual Conference of the IEEE Industrial Electronics Society, Beijing, 2017, pp. 7217–7220.
[14] S. Liu, L. Xie, H. Zhang, Distributed consensus for multi-agent systems with delays and noises in transmission channels, Automatica 47 (5) (2011) 920–934.
[15] A. Jadbabaie, J. Lin, A.S. Morse, Coordination of groups of mobile autonomous agents using nearest neighbor rules, IEEE Trans. Autom. Control 48 (6) (2003) 988–1001.
[16] J. Cortés, F. Bullo, Coordination and geometric optimization via distributed dynamical systems, SIAM J. Control Optim. 44 (5) (2005) 1543–1574.
[17] F. Paganini, J. Doyle, S. Low, Scalable laws for stable network congestion control, in: Proc. the 40th IEEE Conference on Decision and Control, Orlando, FL, December 4–7, 2001, pp. 185–190.
[18] W. Ren, R.W. Beard, Consensus seeking in multiagent systems under dynamically changing interaction topologies, IEEE Trans. Autom. Control 50 (5) (2005) 655–661.
[19] J.A. Fax, R.M. Murray, Information flow and cooperative control of vehicle formations, IEEE Trans. Autom. Control 49 (9) (2004) 1465–1476.
[20] R. Vidal, O. Shakernia, S. Sastry, Formation control of nonholonomic mobile robots with omnidirectional visual servoing and motion segmentation, in: Proc. IEEE Int. Conference on Robotics and Automation (ICRA'03), Taipei, September 14–19, 2003, pp. 584–589.
[21] S. Liu, T. Li, L. Xie, M. Fu, J. Zhang, Continuous-time and sampled-data based average consensus with logarithmic quantizers, Automatica 49 (11) (2013) 3329–3336.
[22] B. Bliman, G. Ferrari-Trecate, Average consensus problems in networks of agents with delayed communications, Automatica 44 (8) (2008) 1985–1995.

[23] M. Cao, A.S. Morse, B.D.O. Anderson, Reaching a consensus in a dynamically changing environment: convergence rates, measurement delays, and asynchronous events, SIAM J. Control Optim. 47 (2) (2008) 601–623.

[24] A.A. Cárdenas, S. Amin, S. Sastry, Research challenges for the security of control systems, in: Proceedings of the 3rd Conference on Hot Topics in Security, San Jose, CA, July 28–August 1, 2008, pp. 1–6.

[25] A. Ichimura, M. Shigeno, A new parameter for a broadcast algorithm with locally bounded Byzantine faults, Inf. Process. Lett. 110 (12) (2010) 514–517.

[26] H.J. LeBlanc, X.D. Koutsoukos, Consensus in networked multi-agent systems with adversaries, in: Proceedings of the 14th International Conference on Hybrid Systems: Computation and Control, Chicago, IL, April 12–14, 2011, pp. 281–290.

[27] H.J. LeBlanc, H. Zhang, X. Koutsoukos, S. Sundaram, Resilient asymptotic consensus in robust networks, IEEE J. Sel. Areas Commun. 31 (4) (2013) 766–781.

[28] H.J. LeBlanc, H. Zhang, S. Sundaram, X. Koutsoukos, Consensus of multi-agent networks in the presence of adversaries using only local information, in: Proceedings of the 1st International Conference on High Confidence Networked Systems, Beijing, April 17–18, 2012, pp. 1–10.

[29] D. Lee, M.W. Spong, Agreement with non-uniform information delays, in: Proceedings of the 2006 American Control Conference, Minneapolis, MN, June 14–16, 2006, pp. 756–761.

[30] S. Liu, T. Li, L. Xie, Distributed consensus for multiagent systems with communication delays and limited data rate, SIAM J. Control Optim. 49 (6) (2011) 2239–2262.

[31] R. Olfati-Saber, R.M. Murray, Consensus problems in networks of agents with switching topology and time-delays, IEEE Trans. Autom. Control 49 (9) (2004) 1520–1533.

[32] F. Pasqualetti, A. Bicchi, F. Bullo, On the security of linear consensus networks, in: Proceedings of the 48th IEEE Conference on Decision and Control, 2009 Held Jointly with the 2009 28th Chinese Control Conference, Shanghai, December 16–18, 2009, pp. 4894–4901.

[33] F. Pasqualetti, A. Bicchi, F. Bullo, Consensus computation in unreliable networks: a system theoretic approach, IEEE Trans. Autom. Control 57 (1) (2012) 90–104.

[34] A. Pelc, D. Peleg, Broadcasting with locally bounded Byzantine faults, Inf. Process. Lett. 93 (3) (2005) 109–115.

[35] M. Raynal, Fault-tolerant Agreement in Synchronous Message-passing Systems, Synthesis Lectures on Distributed Computing Theory, vol. 1, 2010, pp. 1–189.

[36] E. Seneta, Non-negative Matrices and Markov Chains, Springer, New York, 2006.

[37] S. Sundaram, C.N. Hadjicostis, Distributed function calculation via linear iterative strategies in the presence of malicious agents, IEEE Trans. Autom. Control 56 (7) (2011) 1495–1508.

[38] N.H. Vaidya, L. Tseng, G. Liang, Iterative approximate Byzantine consensus in arbitrary directed graphs, in: Proceedings of the 2012 ACM Symposium on Principles of Distributed Computing, 2012, pp. 365–374.

[39] W. Wang, J.J. Slotine, Contraction analysis of time-delayed communications and group cooperation, IEEE Trans. Autom. Control 51 (4) (2006) 712–717.

[40] J. Wolfowitz, Products of indecomposable, aperiodic, stochastic matrices, Proc. Am. Math. Soc. 14 (5) (1963) 733–737.

[41] F. Xiao, L. Wang, State consensus for multi-agent systems with switching topologies and time-varying delays, Int. J. Control 79 (10) (2006) 1277–1284.

[42] H. Zhang, S. Sundaram, Robustness of information diffusion algorithms to locally bounded adversaries, in: Proceedings of the 2012 American Control Conference, Montréal, QC, June 27–29, 2012, pp. 5855–5861.

Chapter 7

Advanced approaches to multiagent coordination

7.1 Synchronization of stochastic dynamic networks

Complex dynamic networks (CDNs) consist of a great number of interconnected nodes, in which each node is a specific dynamical system and connects to its neighbors through a certain topological link [1]. Many kinds of real-world dynamical systems, including the World Wide Web, ecosystem networks, and social relationship networks, can be described by CDNs. Since the pioneer research work on scale-free networks [2], CDNs have been becoming a hot research topic in a variety of science and engineering fields, including mathematics, physics, engineering, and sociology [3], [4], [5]. In real-world CDNs, the synchronization of all dynamical nodes is an interesting and significant phenomenon, such as the synchronous transfer of signals in communication networks. The problem of complete synchronization has recently drawn attention due to its practical applications in secure communication [6], [7]. In this case, the boundedness appears to be an ideal candidate for evaluating the synchronization errors, in which the corresponding synchronization is known as the quasisynchronization or the bounded synchronization [8], [9].

Stochastic models have been of significant importance in reflecting more realistic dynamical behavior of complex networks to discuss the complete synchronization problems [10], [11]. The quasisynchronization issues of SCDNs under exogenous disturbances considered the $\mathcal{H}_\infty$ synchronization problem of coupled discrete time-varying stochastic complex networks over a finite horizon in [12].

Different from the common synchronization criterion in the mean square, the notion of quasisynchronization in probability takes into account both the error bound and the synchronization probability. A natural strategy is to design appropriate controllers for some nodes in SCDNs. The pinning control, which means that only part of the network nodes are directly controlled, has proven to be both economical and effective for the synchronization of SCDNs [13].

With the fast development of networked control technologies, event-triggered mechanisms are employed in the consensus of multiagent systems [14], [15] and synchronization of dynamical networks [16], [17]. Compared with the conventional time-triggered control, the event-triggered control (ETC) strategy takes into account the system behavior between the sampling periods,

Discrete Networked Dynamic Systems. https://doi.org/10.1016/B978-0-12-823698-7.00015-1

and is therefore more effective in reducing the burden of communication/actuation in control processes, while preserving satisfactory control performance.

In the sequel, we let a function $\gamma : \Re^+ \to \Re^+$ be a $\mathcal{K}$-function if it is continuous and strictly increasing and $\gamma(0) = 0$; it is a $\mathcal{K}_\infty$-function if it is a $\mathcal{K}$-function and also $\gamma(s) \to \infty$ as $s \to \infty$. A function $\beta : \Re^+ \times \Re^+ \to \Re^+$ is a $\mathcal{KL}$-function if, for each fixed $t \in \Re^+$, the function $\beta(., k)$ is a $\mathcal{K}$-function, and for each fixed $s \in \Re^+$, the function $\beta(s, .)$ is decreasing and $\beta(s, k) \to 0$ as $k \to \infty$. Let I_d be the identity function and $\psi \circ \phi$ represent the composition of two functions ϕ and ψ. Also, $(\Omega, \mathcal{F}, \mathcal{P})$ stands for a probability space relative to an increasing family $\{\mathcal{F}_k\}_{k \geq 0}$ of σ-algebras $\mathcal{F}_k \subset \mathcal{F}$ generated by $\{w(k)\}_k \geq 0$, where $k \in \mathbb{Z}^+$; $E\{.\}$ represents the mathematical expectation of random variables, and $\mathcal{L}^n_\infty$ denotes the set of all locally essentially bounded sequences $v : \mathbb{Z}^+ \to \Re^n$ with norm $|v|_\infty = \textbf{ess}\sup_{k \geq 0} |v(k)| < \infty$. In particular, we denote $\mathcal{N} = 1, 2, ..., N$.

7.1.1 Problem formulation

Consider the following SCDN, consisting of N identical nodes with nonlinear inner-coupling and exogenous disturbances:

$$x_i(k+1) = f(x_i(k)) + \sum_{j=1}^{N} \ell_{ij} \Gamma g(x_j(k)) + h(x_i(k))w(k) + u_i(k) + v_i(k), \quad i \in \mathcal{N}, \tag{7.1}$$

where $x_i(k) = [x_{i1}(k), x_{i2}(k), \cdots, x_{in}(k)]^t \in \Re^n$ is the state vector of node i, $u_i(k) = [u_{i1}(k), u_{i2}(k), \cdots, u_{in}(k)]^t \in \Re^n$ is the control input, $v_i(k) = [v_{i1}(k), v_{i2}(k), \cdots, v_{in}(k)]^t \in \Re^n$ denotes the exogenous disturbance belonging to $\mathcal{L}^n_\infty$, $w(k) \in \Re$ is a zero mean random sequence on the probability space $(\Omega, \mathcal{F}, \mathcal{P})$ satisfying $\mathbb{E}\{w^2(k)\} = 1$, and $f,\ g,\ h : \Re^n \to \Re^n$ are continuously nonlinear functions which denote, respectively, the node dynamics, the inner-coupling dynamics, and the intensity of stochastic noises. In addition, $\Gamma = [\gamma_{ij}]_{n \times n}$ is the inner-coupling matrix linking the subsystems and $L = [\ell_{ij}]_{N \times N}$ is the coupled configuration matrix denoting the topological structure of the network in which ℓ_{ij} are defined as follows: If there is a connection from node j to i, then $\ell_{ij} > 0$; otherwise, $\ell_{ij} = 0$ for $i \neq j$; and the diagonal elements of matrix L are defined as $\ell_{ii} = -\sum_{j=1, j \neq i}^{N} \ell_{ij}$, which means $\sum_{j=1}^{N} \ell_{ij} = 0$.

The dynamics of the unforced isolate node are described as

$$s(k+1) = f(s(k)) + h(s(k))w(k), \tag{7.2}$$

in which the solution $s(k) = [s_1(k), s_2(k), \cdots, s_n(k)]^t$ may be an equilibrium point, a periodic solution, or a chaotic orbit, and is viewed as the target state that will be tracked by the SCDN (7.1).

Remark 7.1. In the literature, the network nodes have been assumed to be linearly coupled (see [2], [4], [6], [8], [9]), and such an assumption does not seem to fully reflect the actual situation. In many networks, the dynamics of inner-couplings are much more complex than expected due to the difficulties in directly observing the node states. Therefore, we consider the information of nodes in the form of a nonlinear function with respect to the state. Furthermore, the coupled configuration matrix L is not necessarily symmetric, which implies that the corresponding topological graph is allowed to be either directed or undirected. Moreover, the nonidentical exogenous disturbances are taken into account for the network nodes. In addition, all the network nodes and the isolate node are subject to the stochastic noises with the same intensity function, which may not disappear even though the SCDN comes to complete synchronization.

In what follows, the pinning control strategy is applied to a small fraction of the nodes in the network. Without loss of generality, we let $\mathcal{N}_p = \{1, 2, ..., N\} \subset \mathcal{N}$ be the set of nodes which are selected to be pinned. Defining the synchronization error as

$$e_i(k) = x_i(k) - s(k), \quad i \in \mathcal{N}. \tag{7.3}$$

Then we consider the pinning state feedback controller as

$$u_i(k) = \begin{cases} K_i e_i(k), & i \in \mathcal{N}_p, \\ 0, & i \in \mathcal{N} \setminus \mathcal{N}_p, \end{cases} \tag{7.4}$$

in which $K_i \in \Re^{n \times n}$ represents the feedback gain matrix of the ith node. Specifically, if $\mathcal{N}_p = \mathcal{N}$, then all nodes are controlled.

In order to reduce the burden of the communications in the control loop, we introduce the event-triggered mechanism into the computation of control inputs and the updating of the actuator signals. For $i \in \mathcal{N}_p$, let $\{k_s^i : s \in \mathbb{Z}^+\}$ be the event-triggered instant sequence of the ith network node, which is determined iteratively through the following certain event generator rule:

$$k_{s+1}^i = \inf\{k \in \mathbb{Z}^+ : k > k_s^i, \text{and} |\delta_i(k)|^2 - \zeta |e_i(k_s^i)|^2 - \theta_i \geq 0\}, \tag{7.5}$$

where $\delta_i(k) = e_i(k_s^i) - e_i(k)$ for $k \in [k_s^i, k_{s+1}^i)$ denotes the measurement error. The event-triggered weight parameter ζ and the threshold parameters θ_i are nonnegative constants which are not zero at the same time and will be determined later.

The control input of the ith actuator is generated by a zero-order holder in the holding time interval, that is,

$$u_i(k) = K_i e_i(k_s^i), \quad k \in [k_s^i, k_{s+1}^i), \ i \in \mathcal{N}_p. \tag{7.6}$$

Taking into consideration that $e_i(k_s^i) = \delta_i(k) + e_i(k)$, we derive the event-triggered pinning controller as follows:

$$\begin{matrix} u_i(k) = K_i(\delta_i(k) + e_i(k)), & i \in \mathcal{N}_p, \\ 0, & 7i \in \mathcal{N} \setminus \mathcal{N}_p. \end{matrix} \tag{7.7}$$

By subtracting (7.2) from (7.1) we obtain the synchronization error system in the following form:

$$
\begin{aligned}
e_i(k+1) &= \tilde{f}(e_i(k)) + \sum_{j=1}^{N} l_{ij}\Gamma\tilde{g}(e_j(k)) + \tilde{h}(e_i(k))w(k) \\
&\quad + K_i(e_i(k)+\delta_i(k)) + v_i(k), \quad i \in \mathcal{N}_p, \\
e_i(k+1) &= \tilde{f}(e_i(k)) + \sum_{j=1}^{N} l_{ij}\Gamma\tilde{g}(e_j(k)) + \tilde{h}(e_i(k))w(k) \\
&\quad + v_i(k), \quad i \in \mathcal{N} \setminus \mathcal{N}_p,
\end{aligned} \tag{7.8}
$$

in which $\tilde{f}(e_i(k)) = f(x_i(k)) - f(s(k))$, $\tilde{g}(e_i(k)) = g(x_i(k)) - g(s(k))$ and $\tilde{h}(e_i(k)) = h(x_i(k)) - h(s(k))$. By using the properties of the Kronecker product, the system (7.8) is transformed into a compact vector form as follows:

$$
\begin{aligned}
e(k+1) &= F(e(k)) + (L \otimes \Gamma)G(e(k)) + H(e(k))w(k) \\
&\quad + Ke(k) + K\delta(k) + v(k),
\end{aligned} \tag{7.9}
$$

where

$$
\begin{aligned}
K &= \mathrm{diag}\{K_1, K_2, \cdots, K_l, 0, \cdots, 0\}, \\
e(k) &= \left[e_1^t(k), e_2^t(k), \cdots, e_N^t(k)\right]^t, \\
v(k) &= \left[v_1^t(k), v_2^t(k), \cdots, v_N^t(k)\right]^t, \\
\delta(k) &= [\delta_1^t(k), \delta_2^t(k), \cdots, \delta_l^t(k), 0, \cdots, 0]^t, \\
F(e(k)) &= \left[\tilde{f}^t(e_1(k)), \tilde{f}^t(e_2(k)), \cdots, \tilde{f}^t(e_N(k))\right]^T, \\
G(e(k)) &= \left[\tilde{g}^t(e_1(k)), \tilde{g}^t(e_2(k)), \cdots, \tilde{g}^t(e_N(k))\right]^t, \\
H(e(k)) &= \left[\tilde{h}^t(e_1(k)), \tilde{h}^t(e_2(k)), \cdots, \tilde{h}^t(e_N(k))\right]^t.
\end{aligned}
$$

Seeking to establish a workable criterion for the stochastic synchronization of SCDN (7.1), we introduce the following assumptions.

Assumption 7.1. For any $x,\ y \in \Re^n$, the nonlinear vector-value functions f, g satisfy

$$
\begin{aligned}
(f(x) - f(y) - U_{1f}(x-y))^t(f(x) - f(y) - U_{2f}(x-y)) \le 0, \\
(g(x) - g(y) - U_{1g}(x-y))^t(g(x) - g(y) - U_{2g}(x-y)) \le 0,
\end{aligned}
$$

where $U_{1f}, U_{2f}, U_{1g}, U_{2g} \in \Re^{n\times n}$ are known real matrices.

Assumption 7.2. The noise intensity function h satisfies the global Lipschitz condition in the sense that there exists a real matrix $U_h \in \Re^{n\times n}$ such that the inequality

$$(h(x) - h(y))^t (h(x) - h(y)) \le (x - y)^t U_h^t U_h (x - y)$$

holds for any $x,\ y \in \Re^n$.

The following definition will be used in the sequel.

Definition 7.1. For a given positive constant $\epsilon \in (0,\ 1)$, the synchronization error system (7.9) is said to be ultimately bounded in probability $1 - \epsilon$ if there exists a compact set $\mathcal{M} = \{x \in \mathbb{R}^{nN} : |x| \le r\}$ such that, for any initial value $e(0) \in \mathbb{R}^{nN}$, the solution $e(k)$ of system (7.9) satisfies

$$\mathcal{P}\{\lim_{k \to +\infty} \text{dist}(e(k), \mathcal{M}) = 0\} > 1 - \epsilon,$$

where $\text{dist}(x, \mathcal{M}) \triangleq \inf_{y \in \mathcal{M}} |x - y|$ denotes the distance from a point $x \in \mathbb{R}^{nN}$ to the compact set $\mathcal{M}$. The positive constant r is said to be the error bound in the probability.

Remark 7.2. Several definitions on the quasisynchronization and the bounded synchronization have been proposed for deterministic CDNs in [8], [9]. To this end, Definition 7.1 is introduced to evaluate the stochastic dynamical behaviors of the synchronization error system (7.9) when the time tends to infinity, which is different from the notion of transient consensus in probability in [14]. Furthermore, compared with those notions of stochastic synchronization in the mean square proposed in [12], Definition 7.1 provides a new perspective to investigate the synchronization performance for the SCDNs by resorting to the upper bound of the synchronization error and the probability of sample trajectories entering the bounded domain.

Definition 7.2. The SCDN (7.1) is said to be quasisynchronized to the isolate node (7.2) in probability $1 = \epsilon$ if the synchronization error dynamical system (7.9) is ultimately bounded in probability $1 - \epsilon$. In particular, the SCDN (7.1) is said to be completely synchronized to the dynamical system (7.2) in probability $1 - \epsilon$ if the error bound r satisfies $r = 0$.

7.1.2 Design results

A theoretical framework is first established for the synchronization error dynamics (7.9), by which the ultimately dynamical behaviors are explored in the probability sense.

Theorem 7.1. *For system (7.9), let the positive constant $\epsilon \in (0, 1)$ and the controller (7.7) be given. Assume that there are functions $V : \mathbb{R}^{nN} \to \mathbb{R}^+$, $\alpha_1, \alpha_2, \alpha \in \mathcal{K}_\infty$, $\eta \in \mathcal{K}$ and a scalar $\rho \ge 0$ such that: (1) $\alpha_1(|e|) \le V(e) \le \alpha_2(|e|)$ and (2) $\mathbb{E}\{V(e(k+1)) - V(e(k))\} \le -\alpha(\mathbb{E}V(e(k))) + \eta(|v(k)|) + \rho$.*

Then, the synchronization error dynamics (7.9) are ultimately bounded in probability $1-\epsilon$*, and there exist a function* $\hat{\gamma} \in \mathcal{K}$ *and a scalar* $d \geq 0$ *such that the error bound is equal to* $\hat{\gamma}(|v|_\infty) + d$.

Proof. For any initial condition $e_0 = e(0) \in \Re^{nN}$, we denote by $e(k)$ the solution of system (7.9). According to [20], without loss of generality, we assume that α satisfies $Id - \alpha \in \mathcal{K}$. Hence, it follows from condition (7.2) that

$$\begin{aligned}\mathbb{E}V(e(k+1)) &\leq \mathbb{E}V(e(k)) - \alpha(\mathbb{E}V(e(k)) + \eta(|v(k)|)) + \rho \\ &= (Id-\alpha)(\mathbb{E}V(e(k))) + \eta(|v(k)|) + \rho \\ &\leq (Id-\alpha)(\mathbb{E}V(e(k))) + \eta(|v|_\infty) + \rho. \end{aligned} \tag{7.10}$$

For any $\mu > 1$, let

$$\mathcal{B} \triangleq \{e \in \Re^{nN} : V(e) \leq \alpha^{-1}(\mu\eta(|v|_\infty) + \mu\rho)\}. \tag{7.11}$$

In what follows, by dividing the initial conditions into two cases of (7.1) $e_0 \in \mathcal{B}$ and (7.2) $e_0 \in \mathcal{R}^{nN} \setminus \mathcal{B}$, we will prove that, for any positive scalar $\epsilon \in (0,1)$, there exist $\gamma_\mu \in \mathcal{K}$ and $d_\mu \geq 0$ satisfying

$$\mathcal{P}\{\limsup_{k\to+\infty} |e(k)| \leq \gamma_\mu(|v|_\infty) + d_\mu\} > 1-\epsilon. \tag{7.12}$$

Case 1: We have $e_0 \in \mathcal{B}$.
It follows from $e_0 = e_0 \in \mathcal{B}$ that

$$\mathbb{E}V(e(0)) \leq \alpha^{-1}(\mu\eta(|v|_\infty) + \mu\rho). \tag{7.13}$$

Considering $Id - \alpha \in \mathcal{K}$, we have from (7.10)

$$\begin{aligned}\mathbb{E}V(e(1)) &\leq (Id-\alpha)(\mathbb{E}V(e(0))) + \eta(|v|_\infty) + \rho \\ &\leq (Id-\alpha)(\alpha^{-1}(\mu\eta(|v|_\infty) + \mu\rho)) + \eta(|v|_\infty) + \rho \\ &= \alpha^{-1}(\mu\eta(|v|_\infty) + \mu\rho) - \alpha \circ \alpha^{-1}(\mu\eta(|v|_\infty) + \mu\rho) + \eta(|v|_\infty) + \rho \\ &\leq \alpha^{-1}(\mu\eta(|v|_\infty) + \mu\rho) + (1-\mu)(\eta(|v|_\infty) + \rho) \\ &< \alpha^{-1}(\mu\eta(|v|_\infty) + \mu\rho). \end{aligned} \tag{7.14}$$

By induction, one has

$$\mathbb{E}V(e(k)) \leq \alpha^{-1}(\mu\eta(|v|_\infty) + \mu\rho), \quad k \in \mathbb{Z}^+. \tag{7.15}$$

Applying the Chebyshev inequality yields

$$\begin{aligned}&\mathcal{P}\left\{V(e(k)) \geq \frac{1}{\epsilon}\alpha^{-1}(\mu\eta(|v|_\infty) + \mu\rho)\right\} \\ &\qquad \leq \frac{\mathbb{E}V(e(k))\epsilon}{\alpha^{-1}(\mu\eta(|v|_\infty) + \mu\rho)} \leq \epsilon. \end{aligned} \tag{7.16}$$

Noting (7.11), we obtain

$$\mathcal{P}\left\{\alpha_1(|e(k)|) \geq \frac{\alpha^{-1}}{\epsilon}(\mu\eta(|v|_\infty) + \mu\rho)\right\}$$
$$\leq \mathcal{P}\left\{V(e(k)) \geq \frac{1}{\epsilon}\alpha^{-1}(\mu\eta(|v|_\infty) + \mu\rho)\right\}, \tag{7.17}$$

which implies that

$$\mathcal{P}\left\{|e(k)| \geq \alpha_1^{-1} \circ \frac{\alpha^{-1}}{\epsilon}(\mu\eta(|v|_\infty) + \mu\rho)\right\} \leq \epsilon. \tag{7.18}$$

This, together with the properties of the $\mathcal{K}$-function (see [22, Lemma 4.2]), means that

$$\mathcal{P}\left\{|e(k)| \geq \alpha_1^{-1} \circ \frac{\alpha^{-1}}{\epsilon}(2\mu\eta(|v|_\infty)) + \alpha_1^{-1} \circ \frac{\alpha^{-1}}{\epsilon}(2\mu\rho)\right\} \leq \epsilon. \tag{7.19}$$

By letting $\gamma_\mu \triangleq \alpha_1^{-1} \circ (\alpha^{-1}/\epsilon) \circ (2\mu\eta)$ and $d_\mu \triangleq \alpha_1^{-1} \circ (\alpha^{-1}/\epsilon)(2\mu\rho)$, we conclude that

$$\mathcal{P}\{|e(k)| < \gamma_\mu(|v|_\infty) + d_\mu\} > 1 - \epsilon \tag{7.20}$$

for $k \in \mathbb{Z}^+$, which leads to (7.12).

Case 2: We have $e_0 \in \mathbb{R}^{nN} \setminus \mathcal{B}$.

In this case, it is readily evident that $\mathbb{E}V(e(0)) > \alpha^{-1}(\mu\eta(|v|_\infty) + \mu\rho)$. If this holds true for all $k \in \mathbb{Z}^+$, then one deduces that

$$\eta(|v|_\infty) + \rho < \frac{1}{\mu}\alpha(\mathbb{E}V(e(k))). \tag{7.21}$$

Substituting (7.21) into (7.10) gives

$$\mathbb{E}V(e(k+1)) - \mathbb{E}V(e(k)) \leq \left(\frac{1}{\mu} - 1\right)\alpha(\mathbb{E}V(e(k))). \tag{7.22}$$

Keeping in mind that $\mu > 1$, it follows from the standard comparison lemma [21] that there exists a function $\tilde{\beta} \in \mathcal{KL}$ such that

$$\mathbb{E}V(e(k)) \leq \tilde{\beta}(V(e(0)), k), \qquad k \in \mathbb{Z}^+. \tag{7.23}$$

By using the Chebyshev inequality again, we have

$$\mathcal{P}\left\{V(e(k)) \geq \frac{1}{\epsilon}\tilde{\beta}(V(e(0)), k)\right\} \leq \frac{\mathbb{E}V(e(k))\epsilon}{\tilde{\beta}(V(e(0)), k)} \leq \epsilon, \tag{7.24}$$

which, together with the condition (7.1), results in

$$\mathcal{P}\left\{\alpha_1(|e(k)|) \geq \frac{\tilde{\beta}}{\epsilon}(V(e(0)), k)\right\}$$
$$\leq \mathcal{P}\left\{V(e(k)) \geq \tilde{\beta}(V(e(0)), k)\right\}. \quad (7.25)$$

By recalling the definition of the $\mathcal{KL}$-function, we deduce that

$$\mathcal{P}\left\{\alpha_1(|e(k)|) \geq \frac{\tilde{\beta}}{\epsilon}(\alpha_2(e(0)), k)\right\}$$
$$\leq \mathcal{P}\left\{\alpha_1(|e(k)|) \geq \frac{\tilde{\beta}}{\epsilon}(V(e(0)), k)\right\}, \quad (7.26)$$

which means that

$$\mathcal{P}\left\{|e(k)| \geq \alpha_1^{-1}\left(\frac{\tilde{\beta}}{\epsilon}(\alpha_2(e(0)), k)\right)\right\} \leq \epsilon. \quad (7.27)$$

Let $\beta(s,k) \triangleq \alpha_1^{-1}((\tilde{\beta}/\epsilon)(\alpha_2(s), k))$. We deduce that $\beta \in \mathcal{KL}$ in [22, Lemma 4.2]. Then, it follows from (7.27) that

$$\mathcal{P}\{|e(k)| < \beta(e(0), k)\} > 1 - \epsilon, \quad k \in \mathbb{Z}^+. \quad (7.28)$$

On the other hand, if $\mathbb{E}V(e(0)) > \alpha^{-1}(\mu\eta(|v|_\infty) + \mu\rho)$ but there is some $k > 0$ such that $\mathbb{E}V(e(k)) \leq \alpha^{-1}(\mu\eta(|v|_\infty) + \mu\rho)$, then we denote

$$k^* \triangleq \min\{k : \mathbb{E}V(e(k)) \leq \alpha^{-1}(\mu\eta(|v|_\infty) + \mu\rho)\} \quad (7.29)$$

and, when $k \in [0, k_\star - 1]$, one has

$$\mathbb{E}V(e(k)) > \alpha^{-1}(\mu\eta(|v|_\infty) + \mu\rho). \quad (7.30)$$

Invoking (7.10), we conclude that

$$\mathbb{E}V(e(k+1)) - \mathbb{E}V(e(k)) \leq \left(\frac{1}{\mu} - 1\right)\alpha(\mathbb{E}V(e(k))) \quad (7.31)$$

is true for $k \in [0, k_\star - 1]$. By similarity to the proof of (7.28), we obtain

$$\mathcal{P}\{|e(k)| < \beta(e(0), k)\} > 1 - \epsilon, \quad k \in [0, k^* - 1]. \quad (7.32)$$

When $k = k_\star$, it follows from the definition of $k_\star$ that

$$\mathbb{E}V(e(k^*)) \leq \alpha^{-1}(\mu\eta(|v|_\infty) + \mu\rho). \quad (7.33)$$

Similar to (7.15), we have

$$\mathbb{E}V(e(k)) \leq \alpha^{-1}(\mu\eta(|v|_\infty) + \mu\rho), \quad k \geq k^*. \tag{7.34}$$

Repeating the process from (7.16) to (7.20) leads to

$$\mathcal{P}\{\sup_{k \geq k^*} |e(k)| < \gamma_\mu(|v|_\infty) + d_\mu\} > 1 - \epsilon. \tag{7.35}$$

Then, (7.32) and (7.35) indicate that, for $k \in \mathbb{Z}^+$,

$$\mathcal{P}\{|e(k)| < \beta(e(0), k) + \gamma_\mu(|v|_\infty) + d_\mu\} > 1 - \epsilon. \tag{7.36}$$

From (7.28), (7.36), and $\beta \in \mathcal{KL}$, we derive that

$$\mathcal{P}\{\limsup_{k \to +\infty} |e(k)| \leq \gamma_\mu(|v|_\infty) + d_\mu\} > 1 - \epsilon \tag{7.37}$$

for $e_0 \in \Re^{nN} \setminus \mathcal{B}$, which means (7.12) still holds for **Case 2**.

We denote $\hat{\gamma} \triangleq \alpha_1^{-1} \circ (\alpha^{-1}/\epsilon) \circ (2\eta)$, $\hat{d} \triangleq \alpha_1^{-1} \circ (\alpha^{-1}/\epsilon)(2\rho)$, and $\mathcal{M} \triangleq \{x \in \mathbb{R}^{nN} : |x| \leq \hat{\gamma}(|v|_\infty) + \hat{d}\}$. By letting $\mu \to 1^+$,

$$\mathcal{P}\{\limsup_{k \to +\infty} |e(k)| \leq \hat{\gamma}(|v|_\infty) + \hat{d}\} > 1 - \epsilon, \tag{7.38}$$

which is followed by

$$\mathcal{P}\{\lim_{k \to +\infty} \text{dist}(e(k), \mathcal{M}) = 0\} > 1 - \epsilon. \tag{7.39}$$

The proof is now complete. □

Remark 7.3. Theorem 7.1 establishes a general framework for analyzing the dynamical behaviors of a class of nonlinear discrete-time stochastic systems with control input and exogenous disturbances. Sufficient conditions are given under which the sample trajectories of the synchronization error dynamics (7.9) are ultimately bounded by a positive constant in the probability sense. There are two contributions that should be highlighted. First, the compact set $\mathcal{M}$ depicts the final bounded area in which the sample trajectories of error states may dwell when the time evolves to infinity. This is quite different from the traditional complete synchronization previously studied in [13], which requires that the synchronization error approaches zero asymptotically. In addition, we utilize the positive scalar $1 - \epsilon$ to quantify the possibility that the error states enter into the bounded set $\mathcal{M}$ (instead of investigating the evolution properties for the error states in the mean square sense).

Remark 7.4. Theorem 7.1 can be regarded as a discrete-time counterpart of Theorem 7.2 in [23] and a generalization of [20, Lemma 3.5], in the case of stochastic systems. Hence, it is easy to apply Theorem 7.1 to study the input-to-state stability and the input-to-state practical stability in probability for the discrete-time stochastic systems with deterministic inputs.

Under Assumptions 7.1 and 7.2, we will establish several criteria to verify the stochastic synchronization between the SCDN (7.1) and the isolate node (7.2).

Theorem 7.2. *Under Assumptions 7.1 and 7.2, let constants* $\epsilon \in (0, 1)$, $\zeta \in [0, 1/2)$, $\theta_i \geq 0$ $(i \in \mathcal{N}_p)$ *and feedback gain matrices* $i \in \mathcal{N}_p$ *be given. The synchronization error dynamics (7.9) are ultimately bounded in probability* $1 - \epsilon$ *if there exist positive definite matrices* $P_i \in \mathbb{R}^{n\times n}$ $(i \in \mathcal{N})$ *and six positive constants* $\lambda_\star$, λ_i $(i = 1, 2, 3)$, σ_i $(i = 1, 2)$ *such that*

$$\frac{2\zeta}{1-2\zeta}\lambda_2 < \lambda_1 < \lambda^*, \tag{7.40}$$

$$\mathbf{P} < \lambda^* I_{Nn}, \tag{7.41}$$

$$\begin{pmatrix} \Pi & \mathbf{S}^t \\ \mathbf{S} & -\mathbf{P}^{-1} \end{pmatrix} < 0, \tag{7.42}$$

where

$$\begin{aligned}
\mathbf{P} &= diag\{\mathbf{P}_1, \mathbf{P}_2, \cdots, \mathbf{P}_N\}, \\
\mathbf{S} &= (K, I_{nN}, L \otimes \Gamma, K, I_{nN}), \\
\Pi_{11} &= I_N \otimes (\lambda^* U_h^t U_h - \sigma_1 \bar{U}_f - \sigma_2 \bar{U}_g + \lambda_1 I_n) - P, \\
\Pi_{12} &= \sigma_1 I_N \otimes \tilde{U}_f^t, \\
\tilde{\Pi}_{13} &= \sigma_2 I_N \otimes \tilde{U}_g^t, \\
\bar{U}_f &= U_{1f}^t U_{2f} + U_{2f}^t U_{1f}, \\
\bar{U}_g &= U_{1g}^t U_{2g} + U_{2g}^t U_{1g}, \\
\tilde{U}_f &= U_{1f} + U_{2f}, \\
\tilde{U}_g &= U_{1g} + U_{2g},
\end{aligned}$$

and

$$\Pi = \begin{bmatrix} \Pi_{11} & \Pi_{12} & \Pi_{13} & 0 & 0 \\ * & -2\sigma_1 I_{nN} & 0 & 0 & 0 \\ * & * & -2\sigma_2 I_{nN} & 0 & 0 \\ * & * & * & -\lambda_2 I_{nN} & 0 \\ * & * & * & * & -\lambda_3 I_{nN} \end{bmatrix}.$$

Proof. It follows from Assumption 7.1 that, for $i \in \mathcal{N}$,

$$
\begin{aligned}
&2\sigma_1(\tilde{f}^T(e_i(k))\tilde{f}(e_i(k)) - \tilde{f}^T(e_i(k))U_{2f}e_i(k) \\
&- e_i^T(k)U_{1f}^T\tilde{f}(e_i(k)) + e_i^T(k)U_{1f}^TU_{2f}e_i(k)) \le 0, \\
&2\sigma_2(\tilde{g}^T(e_i(k))\tilde{g}(e_i(k)) - \tilde{g}^T(e_i(k))U_{2g}e_i(k) \\
&- e_i^T(k)U_{1g}^T\tilde{g}(e_i(k)) + e_i^T(k)U_{1g}^TU_{2g}e_i(k)) \le 0,
\end{aligned}
$$

which further shows that

$$
\begin{aligned}
&2\sigma_1\tilde{f}^t(e_i(k))\tilde{f}(e_i(k)) + \sigma_1 e_i^t(k)\bar{U}_f e_i(k) \\
&- \sigma_1(\tilde{f}^t(e_i(k))\tilde{U}_f e_i(k) + e_i^t(k)\tilde{U}_f^t\tilde{f}(e_i(k))) \le 0, \qquad (7.43) \\
&2\sigma_2\tilde{g}^t(e_i(k))\tilde{g}(e_i(k)) + \sigma_2 e_i^t(k)\bar{U}_g e_i(k) \\
&- \sigma_2(\tilde{g}^t(e_i(k))\tilde{U}_g e_i(k) + e_i^t(k)\tilde{U}_g^t\tilde{g}(e_i(k))) \le 0. \qquad (7.44)
\end{aligned}
$$

By summing up (7.43) and (7.44) from $1 \to N$ with respect to i, respectively, we have

$$
\begin{aligned}
\mathcal{F}(k) \triangleq\ & 2\sigma_1\ F^t(e(k))F(e(k)) + \sigma_1\ e^t(k)(I_N \otimes \bar{U}_f)e(k) \\
&- \sigma_1(F^t(e(k))(I_N \otimes \tilde{U}_f)e(k)) \\
&- \sigma_1(e^t(k)(I_N \otimes \tilde{U}_f^t)F(e(k))) \le 0, \qquad (7.45) \\
\mathcal{G}(k) \triangleq\ & 2\sigma_2 G^T(e(k))G(e(k)) + \sigma_2 e^t(k)(I_N \otimes \bar{U}_g)e(k) \\
&- \sigma_2(G^t(e(k))(I_N \otimes \tilde{U}_g)e(k)) \\
&- \sigma_2(e^T(k)(I_N \otimes \tilde{U}_g^T)G(e(k))) \le 0. \qquad (7.46)
\end{aligned}
$$

Consider the Lyapunov function candidate,

$$
V(e(k)) = e^t(k)Pe(k) = \sum_{i=1}^{N} e_i^t(k)P_i e_i(k). \qquad (7.47)
$$

Computing the difference of $V(e(k))$ along the solution of synchronization error dynamics (7.9) and then taking the mathematical expectation, we obtain

$$
\begin{aligned}
\mathbb{E}\Delta V(e(k)) &= \mathbb{E}\{V(e(k+1)) - V(e(k))\} \\
&= \mathbb{E}\{[Ke(k) + F(e(k)) + (L \otimes \Gamma)G(e(k)) \\
&+ K\delta(k) + v(k)]^T P[Ke(k) + F(e(k)) \\
&+ (L \otimes \Gamma)G(e(k)) + K\delta(k) + v(k)]\} \\
&+ \mathbb{E}\{H^T(e(k))PH(e(k))\} - \mathbb{E}\{e^T(k)Pe(k)\}. \qquad (7.48)
\end{aligned}
$$

Considering Assumption 7.2, one concludes that

$$\begin{aligned}H^t(e(k))PH(e(k)) &= \sum_{i=1}^{N} \tilde{h}^t(e_i(k))P_i\tilde{h}(e_i(k)) \\ &\leq \lambda^* \sum_{i=1}^{N} \tilde{h}^t(e_i(k))\tilde{h}(e_i(k)) \\ &\leq \lambda^* \sum_{i=1}^{N} e_i^T(k)U_h^tU_he_i(k) \\ &= \lambda^* e^t(k)(I_N \otimes (U_h^tU_h))e(k). \end{aligned} \tag{7.49}$$

Substituting (7.48) into (7.49) leads to

$$\begin{aligned}\mathbb{E}\Delta V(e(k)) \leq\ & \mathbb{E}\{[Ke(k) + F(e(k)) + (L \otimes \Gamma)G(e(k)) \\ & + K\delta(k) + v(k)]^t P[Ke(k) + F(e(k)) \\ & + (L \otimes \Gamma)G(e(k)) + K\delta(k) + v(k)]\} \\ & + \mathbb{E}\{e^T(k)(I_N \otimes \lambda^* U_h^tU_h - P)e(k)\}. \end{aligned} \tag{7.50}$$

Defining

$$\xi^t(k) \triangleq [e^t(k),\ F^t(e(k)),\ G^t(e(k)),\ \delta^t(k),\ v^t(k)]^t, \tag{7.51}$$

it follows that

$$\begin{aligned}&\mathbb{E}[\Delta V(e(k)) + \lambda_1 e^t(k)e(k) - \lambda_2\delta^t(k)\delta(k) \\ &- \lambda_3 v^t(k)v(k) - \mathcal{F}(k) - \mathcal{G}(k)] \\ &\leq \mathbb{E}[\xi^t(k)\mathbf{S}^t bf\, P\mathbf{S}\xi(k) + \xi^t(k)\Pi\xi(k)]. \end{aligned} \tag{7.52}$$

According to condition (7.42), it follows from the Schur complement lemma that

$$\mathbb{E}[\xi^t(k)\mathbf{S}^T(I_N \otimes \mathbf{P})\mathbf{S}\xi(k) + \xi^t(k)\Pi\xi(k)] \leq 0. \tag{7.53}$$

Therefore, (7.52) together with (7.45) and (7.46) gives

$$\mathbb{E}\Delta V(e(k)) \leq -\lambda_1\mathbb{E}e^t(k)e(k) + \lambda_2\mathbb{E}\delta^t(k)\delta(k) + \lambda_3 v^t(k)v(k). \tag{7.54}$$

Note that the event generator rule (7.5) forces the measurement error to satisfy $|\delta_i(k)|^2 \leq \zeta|e_i(k_s^i)|^2 + \theta_i$. Bearing in mind that $e_i(k_s^i) = e_i(k) + \delta_i(k)$, we have

$$\begin{aligned}|\delta_i(k)|^2 &\leq \zeta|e_i(k) + \delta_i(k)|^2 + \theta_i \\ &\leq 2\zeta|e_i(k)|^2 + 2\zeta|\delta_i(k)|^2 + \theta_i, \end{aligned} \tag{7.55}$$

which implies that

$$\mathbb{E}\delta_i^t(k)\delta_i(k) \leq \frac{2\zeta}{1-2\zeta}\mathbb{E}e_i^t(k)e_i(k) + \theta_i \tag{7.56}$$

for $i \in \mathcal{N}_p$ and $k \in \mathbb{Z}^+$. Then, it follows from the definition of $\delta(k)$ that

$$\begin{aligned}
\mathbb{E}\delta^t(k)\delta(k) &= \sum_{i=1}^{N}\mathbb{E}\delta_i^t(k)\delta_i(k) = \sum_{i=1}^{l} t\mathbb{E}\delta_i^t(k)\delta_i(k) \\
&\leq \frac{2\zeta}{1-2\zeta}\sum_{i=1}^{l}\mathbb{E}e_i^t(k)e_i(k) + \sum_{i=1}^{l}\theta_i \\
&\leq \frac{2\zeta}{1-2\zeta}\mathbb{E}e^t(k)e(k) + \sum_{i=1}^{l}\theta_i.
\end{aligned} \tag{7.57}$$

By substituting (7.57) into (7.54), one has

$$\begin{aligned}
\mathbb{E}\Delta V(e(k)) \leq &-\left(\lambda_1 - \frac{2\zeta\lambda_2}{1-2\zeta}\right)\mathbb{E}e^t(k)e(k) \\
&+ \lambda_2\sum_{i=1}^{l}\theta_i + \lambda_3 v^t(k)v(k),
\end{aligned} \tag{7.58}$$

which, together with (7.40), (7.41), and (7.47), implies that

$$\mathbb{E}\Delta V(e(k)) \leq -\sigma\mathbb{E}V(e(k)) + \lambda_3|v(k)|^2 + \theta, \tag{7.59}$$

where

$$\sigma \triangleq (\lambda_1/\lambda^*) - (2\zeta\lambda_2/(1-2\zeta)\lambda^*) \in (0,1),$$

$$\theta \triangleq \lambda_2\sum_{i=1}^{l}\theta_i \geq 0.$$

Now, let $\alpha_1(s) \triangleq \lambda_{\min}(P)s^2$, $\alpha_2(s) \triangleq \lambda^* s^2$, $\alpha(s) \triangleq \sigma s$, $\eta(s) \triangleq \lambda_3 s^2$, and $\rho \triangleq \theta$. It follows easily from Theorem 7.1 that the synchronization error dynamics (7.9) are ultimately bounded in probability $1-\epsilon$ for a given $\epsilon \in (0,1)$. Moreover, we select

$$\hat{\gamma}(s) \triangleq \sqrt{\frac{2\lambda_3}{\epsilon\sigma\lambda_{\min}(P)}}s, \quad \hat{d} \triangleq \sqrt{\frac{2\theta}{\epsilon\sigma\lambda_{\min}(P)}} \tag{7.60}$$

to obtain the synchronization error bound $\hat{\gamma}(|v|_\infty) + \hat{d}$. The proof is complete. □

Now, we proceed to deal with the codesign problems of the control gain matrices and the event-triggered parameters for the SCDN (7.1) considered in this section.

Theorem 7.3. *Under Assumptions 7.1 and 7.2, let the positive constant* $\epsilon \in (0,1)$ *be given. The SCDN (7.1) is quasisynchronized to the isolate node (7.2) in probability* $1-\epsilon$ *if there exist positive definite matrices* $P_i \in \Re^{n\times n}$, $i \in \mathcal{N}$, *constant matrices* $Y_i \in \Re^{n\times n}$, $i \in \mathcal{N}_p$, *and six positive constants* $\lambda_\star$, λ_i $(i = 1,2,3)$, σ_i $(i = 1,2)$ *such that*

$$\lambda_1 < \lambda^*, \tag{7.61}$$

$$\mathbf{P} < \lambda^* I_{nN}, \tag{7.62}$$

$$\begin{pmatrix} \Pi & \tilde{S}^t \\ \tilde{\mathbf{S}} & -\mathbf{P} \end{pmatrix} < 0, \tag{7.63}$$

where

$$\begin{aligned} \mathbf{P} &= diag\{\mathbf{P}_1, \mathbf{P}_2, \cdots, \mathbf{P}_N\}, \\ \tilde{\mathbf{S}} &= (Y,\ \mathbf{P},\ (L \otimes \Gamma)\mathbf{P},\ Y,\ \mathbf{P}), \\ Y &= diag\{Y_1, Y_2, \cdots, Y_p, 0, \cdots, 0\}, \end{aligned}$$

and Π *is defined in Theorem 7.2. Moreover, the control gain matrices are designed as*

$$K_i = P_i^{-1} Y_i, \quad i \in \mathcal{N}_p, \tag{7.64}$$

with the event-triggered parameters satisfying

$$0 \le \zeta < \frac{\lambda_1}{2(\lambda_1 + \lambda_2)} \quad and \quad \theta_i \ge 0, \quad i \in \mathcal{N}_p. \tag{7.65}$$

The synchronization error bound is equal to

$$\sqrt{\frac{(2-4\zeta)\lambda_3\lambda^*}{\epsilon(\lambda_1 - 2\zeta(\lambda_1+\lambda_2))\lambda_{\min}(P)}}|v|_\infty + \sqrt{\frac{(2-4\zeta)\lambda_2\lambda^* \sum_{i=1}^{l}\theta_i}{\epsilon(\lambda_1 - 2\zeta(\lambda_1+\lambda_2))\lambda_{\min}(P)}}. \tag{7.66}$$

Proof. It suffices to prove that (7.41)–(7.42) hold under the conditions in Theorem 7.3. Premultiplying and postmultiplying (7.63) by

$$\{I_{nN}, I_{nN}, I_{nN}, I_{nN}, I_{nN}, P^{-1}\}$$

and noting that $Y_i = P_i K_i$ for $i \in \mathcal{N}_p$ and $2\zeta\lambda_2 < (1-2\zeta)\lambda_1$, we obtain (7.40)–(7.42) immediately. The synchronization error bound is readily accessible from (7.60). The proof is complete. □

Remark 7.5. Theorem 7.3 presents an effective design method such that the pinning control gain matrices K_i and the event-triggered parameters ζ, θ_i are codesigned to guarantee the desired synchronization of the SCDN (7.1) in the probability sense by a set of feasible solutions of inequalities (7.61)–(7.63). Moreover, it provides a quantitative calculation for the upper bound of the synchronization error in (7.66). Note that when the synchronization probability and the feasible solutions of inequalities (7.61)–(7.63) are given, the upper bound of the synchronization error increases as long as the event-triggered parameters increase, which means a lower updating frequency leads to a worse control performance.

Remark 7.6. Compared with the results in [3], Theorem 7.3 proposes a novel control approach that includes both the pinning strategy and the event-triggered mechanism. This method could be extended to the synchronization of duplex systems with star or other topologies [3].

By further designing the event-triggered parameters, we have the following corollary.

Corollary 7.1. *Under Assumptions 7.1 and 7.2, let the constants $\epsilon \in (0, 1)$ and $\kappa > 0$ be given. If there exist positive definite matrices $P_i \in \Re^{n\times n}$ $(i \in \mathcal{N}_p)$, constant matrices $Y_i \in \Re^{n\times n}$ $(i \in \mathcal{N}_p)$, and six positive constants $\lambda^\star$, λ_i $(i = 1, 2, 3)$, σ_i $(i = 1, 2)$ satisfying (7.61)–(7.63), then the SCDN (7.1) is quasisynchronized to the isolate node (7.2) in probability $1 - \epsilon$ and the error bound is equal to $\kappa|v|_\infty$ provided that the pinning control gain matrices are designed as*

$$K_i = P_i^{-1} Y_i, \quad i \in \mathcal{N}_p, \tag{7.67}$$

with the event-triggered parameters satisfying

$$0 \le \zeta < \frac{\lambda_1}{2(\lambda_1 + \lambda_2)}, \quad 0 \le \sum_{i=1}^{l} \theta_i \le \tilde{\kappa}|v|_\infty^2, \tag{7.68}$$

where

$$\tilde{\kappa} = \frac{(\kappa\sqrt{\epsilon(\lambda_1 - 2\zeta(\lambda_1 + \lambda_2))\lambda_{\min}(P)} - \sqrt{(2 - 4\zeta)\lambda_3\lambda^*})^2}{(2 - 4\zeta)\lambda_2\lambda^*}.$$

Remark 7.7. Corollary 7.1 provides a rather elegant estimation for the upper bound of the synchronization error by effectively designing the event-triggered parameters. To this end, for any $\epsilon \in (0,\ 1)$, the SCDN (7.1) and the isolate node (7.2) will be able to achieve the quasisynchronization in probability $-\epsilon$ while the synchronization error possesses a fixed upper bound $\kappa|v|_\infty$, if we choose the scalar $\kappa > 0$ appropriately.

When the nodes in the network are coupled linearly, that is, $g(x_j(k)) = x_j(k)$ for $j \in \mathcal{N}$, SCDN (7.1) reduces to

$$\begin{aligned} x_i(k+1) = {} & f(x_i(k)) + \sum_{j=1}^{N} l_{ij}\Gamma x_j(k) + h(x_i(k))w(k) \\ & + u_i(k) + v_i(k), \quad i \in \mathcal{N}. \end{aligned} \tag{7.69}$$

Without loss of generality, we also assume that the nonlinear dynamical function f and the noise intensity function h satisfy Assumptions 7.1 and 7.2, respectively.

The following results are readily derived from Theorem 7.3 and Corollary 7.1.

Theorem 7.4. *Let the constant $\epsilon \in (0,1)$ be given. The SCDN (7.69) is quasisynchronized to the isolate node (7.2) in probability $1-\epsilon$ if there exist positive definite matrices $P_i \in \Re^{n\times n}$ $(i \in \mathcal{N}_p)$, constant matrices $Y_i \in \Re^{n\times n}$ $(i \in \mathcal{N}_p)$, and five positive constants ϱ, $\lambda^\star$, and $\lambda_i i$ $(i = 1,2,3)$ such that*

$$\lambda_1 < \lambda^*, \tag{7.70}$$

$$P < \lambda^* I_n, \tag{7.71}$$

$$\begin{pmatrix} \bar{\Pi} & \bar{S}^T \\ \bar{S} & -P \end{pmatrix} < 0, \tag{7.72}$$

$$\begin{aligned} \bar{S} &= (Y + (L \otimes \Gamma)P, P, Y, P), \\ \bar{\Pi}_{11} &= I_N \otimes (\lambda^* U_h^t U_h - \varrho \bar{U}_f + \lambda_1 I_n) - P, \\ \bar{\Pi}_{12} &= \varrho I_N \otimes \tilde{U}_f^t, \\ \bar{\Pi} &= \begin{bmatrix} \bar{\Pi}_{11} & \bar{\Pi}_{12} & 0 & 0 \\ * & -2\varrho I_{nN} & 0 & 0 \\ * & * & -\lambda_2 I_{nN} & 0 \\ * & * & * & -\lambda_3 I_{nN}, \end{bmatrix} \end{aligned}$$

and P, Y, $\bar{U}_f$, and $\tilde{U}_f$ are defined in Theorem 7.3. Moreover, the pinning control gain matrices K_i $(i \in \mathcal{N}_p)$ are designed as (7.64) with the event-triggered parameters ζ, θ_i $(i \in \mathcal{N}_p)$ satisfying (7.65). The error bound is determined by (7.66).

Corollary 7.2. *Assume that there exist positive definite matrices $P_i \in \Re^{n\times n}$, constant matrices $Y_i \in \Re^{n\times n}$, and five positive constants ϱ, $\lambda^\star$, and λ_i $(i = 1,2,3)$ satisfying (7.70)–(7.72). Then, for any $\epsilon \in (0,1)$, the SCDN (7.69) is* *quasisynchronized to the isolate node (7.2) in probability* $1-\epsilon$ *and the error bound is equal to* $\kappa|v|_\infty$ *for* $\kappa>0$ *chosen appropriately, provided that the pinning control gain matrices* K_i $(i \in \mathcal{N}_p)$ *are designed as (7.67) with the event-triggered parameters satisfying (7.68).*

7.1.3 Simulation example 7.1

Consider a linear coupling SCDN (7.69) consisting of 10 identical nodes, in which the state vector of each node is $2-D$, that is, $N=10$ and $n=2$. The nonlinear dynamical function of the network nodes is

$$f(x_i(k)) = \begin{bmatrix} -0.5x_{i1}(k) + 0.25x_{i2}(k) + \tanh(0.05x_{i1}(k)) \\ 0.85x_{i2}(k) - \tanh(0.05x_{i1}(k) + 0.05x_{i2}(k)) \end{bmatrix}.$$

It is easy to verify that Assumption 7.1 holds with

$$U_{1f} = \begin{bmatrix} -0.5 & 0.25 \\ -0.05 & 0.8 \end{bmatrix}, \quad U_{2f} = \begin{bmatrix} -0.45 & 0.25 \\ 0 & 0.85 \end{bmatrix}.$$

Let the noise intensity and the exogenous disturbances be

$$h(x_i(k)) = \begin{bmatrix} 0.3 & 0 \\ 0 & 0.3 \end{bmatrix}\begin{bmatrix} x_{i1}(k) \\ x_{i2}(k) \end{bmatrix}, \quad v_i(k) = \begin{bmatrix} d\sin(k) \\ d\cos(k) \end{bmatrix},$$

which implies that Assumption 7.2 is true with

$$U_h = \begin{bmatrix} 0.3 & 0 \\ 0 & 0.3 \end{bmatrix}$$

and $|v|_\infty = |d|$. We choose the inner-coupling matrix $\Gamma = diag\{1,\ 1\}$ and we choose the coupled configuration matrix L to be as follows:

$$L = \begin{bmatrix} -0.7 & 0.2 & 0.3 & 0 & 0 & 0 & 0 & 0 & 0.2 & 0 \\ 0 & -0.6 & 0 & 0.25 & 0.25 & 0.05 & 0.05 & 0 & 0 & 0 \\ 0 & 0.1 & -0.35 & 0 & 0 & 0.1 & 0 & 0.1 & 0 & 0.05 \\ 0 & 0.1 & 0 & -0.2 & 0 & 0 & 0 & 0 & 0.1 & 0 \\ 0.05 & 0 & 0 & 0 & -0.25 & 0.1 & 0 & 0 & 0.1 & 0 \\ 0 & 0 & 0 & 0 & 0.05 & -0.10 & 0.05 & 0 & 0 & \\ 0 & 0.1 & 0 & 0.05 & 0 & 0 & -0.20 & 0.05 & 0 & \\ 0 & 0 & 0 & 0.15 & 0.05 & 0 & 0 & -0.35 & 0 & 0.15 \\ 0 & 0 & 0 & 0.05 & 0.05 & 0 & 0 & 0 & -0.2 & 0 \\ 0.05 & 0.1 & 0 & 0 & 0.1 & 0 & 0 & 0 & 0 & -0.25 \end{bmatrix}.$$

In the absence of the control strategy, the dynamical behaviors of nodes in the SCDN are no longer stable, which implies that the SCDN fails to synchronize with the desired motion of the unforced isolate node mainly due to the coupling effects. The state errors between the SCDN and the isolate node are depicted in Fig. 7.1, from which we see that the synchronization errors between the open-loop SCDN and the target node tend to infinity with time.

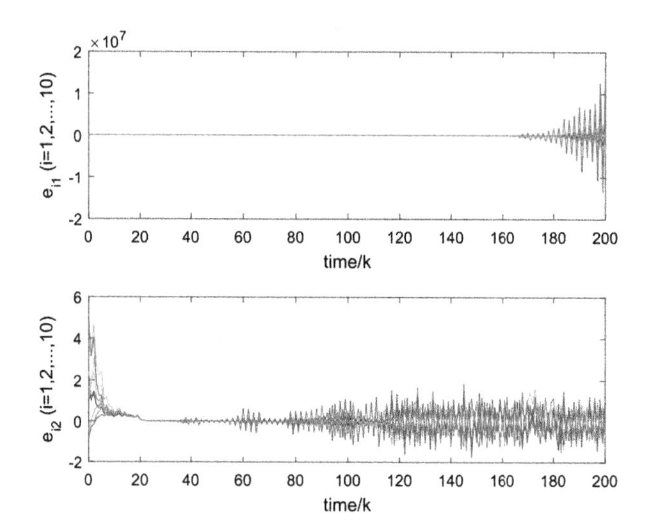

FIGURE 7.1 State errors between the isolate node and the SCDN without controller.

Next, we are interested in designing an event-triggered pinning controller with which the SCDN under consideration eventually achieves the quasisynchronization in probability with the isolate node. Assume that the synchronization probability is $\epsilon = 0.15$ and the first two nodes in the SCDN are chosen to be pinned. By solving the desired linear matrix inequalities (LMIs), a set of feasible solutions is obtained as follows:

$$P_1 = \begin{bmatrix} 0.1806 & -0.0174 \\ -0.0174 & 0.4297 \end{bmatrix}, \quad P_2 = \begin{bmatrix} 0.2758 & -0.0260 \\ -0.0260 & 0.4348 \end{bmatrix},$$

$$P_3 = \begin{bmatrix} 0.1944 & -0.0354 \\ -0.0354 & 0.6176 \end{bmatrix}, \quad P_4 = \begin{bmatrix} 0.4872 & -0.0795 \\ -0.0795 & 0.6300 \end{bmatrix},$$

$$P_5 = \begin{bmatrix} 0.3981 & -0.0663 \\ -0.0663 & 0.5960 \end{bmatrix}, \quad P_6 = \begin{bmatrix} 0.6618 & -0.1089 \\ -0.1089 & 0.5576 \end{bmatrix},$$

$$P_7 = \begin{bmatrix} 0.6401 & -0.0976 \\ -0.0976 & 0.6332 \end{bmatrix}, \quad P_8 = \begin{bmatrix} 0.1967 & -0.0359 \\ -0.0359 & 0.5448 \end{bmatrix},$$

$$P_9 = \begin{bmatrix} 0.4846 & -0.0796 \\ -0.0796 & 0.6779 \end{bmatrix}, \quad P_{10} = \begin{bmatrix} 0.4244 & -0.0682 \\ -0.0682 & 0.6565 \end{bmatrix},$$

$$Y_1 = \begin{bmatrix} 0.2075 & -0.0389 \\ -0.0109 & -0.0441 \end{bmatrix}, \; Y_2 = \begin{bmatrix} 0.3328 & -0.0683 \\ -0.0281 & -0.1863 \end{bmatrix},$$
$$\lambda^* = 0.0260, \quad \varrho = 1.8327,$$
$$\lambda_1 = 0.0028, \quad \lambda_2 = 2.9025, \quad \lambda_3 = 4.7984.$$

Moreover, the pinning feedback control gain matrix is obtained as

$$K_1 = \begin{bmatrix} 1.1506 & -0.2263 \\ 0.0211 & -0.1118 \end{bmatrix}, \quad K_2 = \begin{bmatrix} 1.2076 & -0.2898 \\ 0.0077 & -0.4459 \end{bmatrix}$$

and the upper bound of the event-triggered weight parameter is calculated as

$$\lambda_1 1/2(\lambda_1 + \lambda_2) = 4.8953 \times 10^{-4}.$$

It follows from Corollary 7.2 that the SCDN is quasisynchronized in probability to the isolate node and the synchronization probability equals 0.85.

For the purpose of simulation, we assume that $d = 0.001$ and $\zeta = \theta_1 = \theta_2 = 0.0001$, which implies that the synchronization error is upper-bounded by 0.5618. We let the total time interval be $[0,\ 200]$ and the initial values are chosen to be

$$x_1(0) = (0.5, -0.5)^t, \; x_2(0) = (-1.2, 0.3)^t, \; x_3(0) = (1.3, -0.8)^t,$$
$$x_4(0) = (-3.5, -0.9)^t, \; x_5(0) = (2.5, -2.5)^t, \; x_6(0) = (-1.9, 1.7)^t,$$
$$x_7(0) = (1.3, -2.8)^t, \; x_8(0) = (-0.5, 2.9)^t, \; x_9(0) = (2.3, -2.8)^t,$$
$$x_{10}(0) = (0.9, 0.4)T, \quad s(0) = (2, -2)^t.$$

The simulation results are presented in Figs. 7.2 and 7.3. Specifically, with the event-triggered pinning feedback controller, the dynamical behavior of the closed-loop synchronization error system is illustrated in Fig. 7.2, which implies that the sample trajectories of error states will eventually achieve the quasisynchronization in the probability sense. Under the event-triggered mechanism, the updating instants of the controllers are shown in Fig. 7.3, which indicates that the frequency of control updating has been reduced effectively.

If ζ and θ_i are selected to be zero, the ETC reduces to the traditional periodic-triggered control (PTC). The corresponding results on comparing the two different cases are quantitatively illustrated in Table 7.1. With the presence of ETC, the updating number of controllers reduces by more than 80% while the upper bound of synchronization errors increases by 0.5047, which means a lower updating frequency leads to a worse synchronization performance. Thus, in order to guarantee a certain control performance and reduce the updating frequency simultaneously, the event-triggered parameters should be properly fine-tuned to achieve a tradeoff between the control performance according to practical consideration and the design of the event-triggered scheme for energy saving.

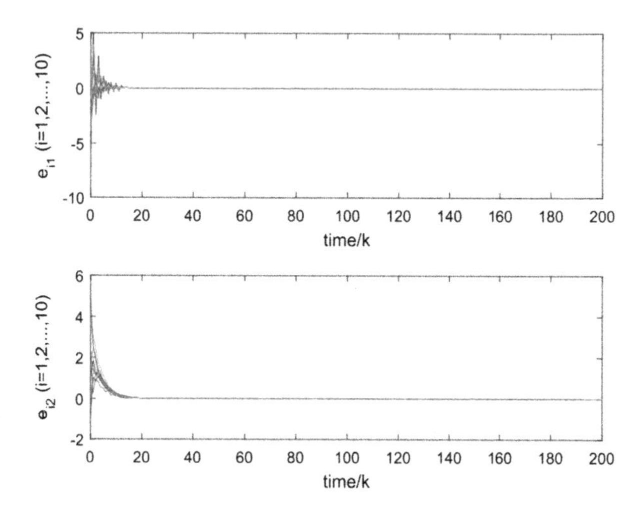

FIGURE 7.2 Synchronization errors of the closed-loop SCDN to the isolate node.

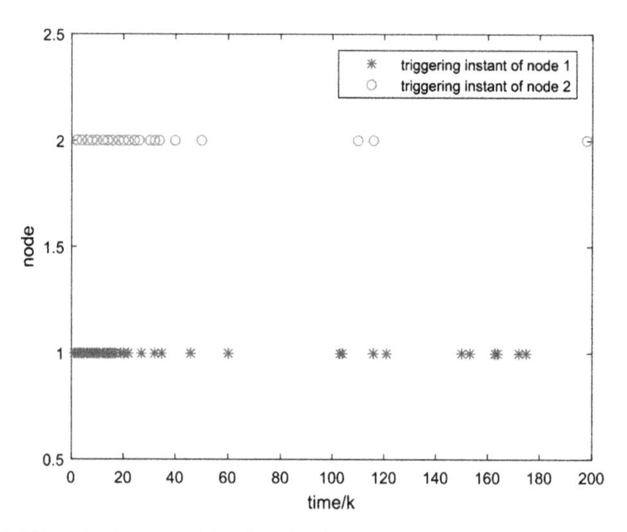

FIGURE 7.3 Triggering instants of the pinned nodes.

TABLE 7.1 Comparison of control performance between PTC and ETC.

Mechanism	Freq. node 1	Freq. node 2	Error bound
PTC	201	201	0.0571
ETC	35	29	0.5618
Perform. Percent	−82.6	−85.6	0.5047

7.2 Observer-based consensus protocols

This section considers the consensus of linear multiagent systems with directed communication topologies. Available results are found in [18], [19] for static consensus protocols based on relative states of neighboring agents. Here, we look at an observer-type consensus protocol based only on relative output measurements of neighboring agents. A decomposition approach is utilized here to convert the consensus of a multiagent system, whose communication topology has a spanning tree, into the stability of a set of matrices with the same dimension as a single agent. The final consensus value reached by the agents is derived and the notion of discrete-time consensus region is introduced and analyzed.

In the sequel, we recall from Chapter 1 that Γ_m denotes the set of all directed graphs with m nodes such that each graph contains a directed spanning tree, and we let $\Gamma_{\leq\delta}$ $(0 < \delta < 1)$ denote the set of all directed graphs containing a directed spanning tree, whose eigenvalues different from 1 lie in the disk of radius δ centered at the origin.

7.2.1 Problem formulation

Consider a network of $\mathbb{N}$ identical agent with linear discrete-time setting, where the dynamics of the ith agent are described by

$$
\begin{aligned}
x_i(k+1) &= Ax_i(k) + Bu_i(k), \\
y_i(k) &= Cx_i(k), \quad i = 1,\ 2,\ \ldots,\ \mathbb{N}.
\end{aligned}
\tag{7.73}
$$

In system (7.73), $x_i(k) \in \Re^n$ is the state vector, $y_i(k) \in \Re^p$ is the measured output, and $u_i(k) \in \Re^q$ is the control input. The matrices $A \in \Re^{n\times n}$, $B \in \Re^{n\times m}$, $C \in \Re^{p\times n}$ are real and known constant matrices. In the sequel, we assume that the pairs $(A,\ B)$–$(A,\ C)$ are stabilizable-detectable.

The communication topology among agents is represented by a directed graph $\mathbb{G} = (\mathbb{V},\ \mathbb{E})$, where $\mathbb{V} = \{1,\ \ldots,\ \mathbb{N}\}$ is the set of nodes (agents) and $\mathbb{E} = \mathbb{V} \times \mathbb{V}$ is the set of edges. An edge $(i,\ j)$ in graph $\mathbb{G}$ means that agent j can obtain information from agent i, but not conversely.

At each time instant, the information available to agent i is the relative measurements of other agents with respect to itself, given by

$$
\xi_i(k) = \sum_{j=1}^{\mathbb{N}} d_{ij}[y_i(k) - y_j(k)],
\tag{7.74}
$$

where $\mathbb{D} := diag\{d_{11},\ \ldots,\ d_{\mathbb{NN}}\}$ is the row-stochastic matrix associated with graph $\mathbb{G}$. A distributed observer-type consensus protocol can be described by

$$v_i(k+1) = [A + BK]v_i(k) + L\Bigg(\sum_{j=1}^{\mathbb{N}} d_{ij}C[v_i(k) - v_j(k)] - \xi(k)\Bigg),$$

$$u_i(k) = Kv_i(k), \tag{7.75}$$

where $v_i(k) \in \Re^n$ is the protocol state, $i = 1, 2, ..., \mathbb{N}$, and $L \in \Re^{n \times n}$ and $K \in \Re^{p \times n}$ are feedback gain matrices to be determined. In (7.75), the term

$$\sum_{j=1}^{\mathbb{N}} d_{ij}C[v_i(k) - v_j(k)]$$

denotes the information exchanges between the protocol of agent i and those of its neighboring agents. It is observed that the protocol (7.75) maintains the same communication topology as the agents in (7.73).

Introducing the augmentation

$$\eta_i^t = [x_i^t, \ v_i^t], \ \eta^t = [\eta_1^t, ..., \eta_{\mathbb{N}}^t],$$

combining (7.73) and (7.75), we obtain the following closed-loop system:

$$\eta(k+1) = [(I_{\mathbb{N}} \otimes \mathcal{A}) + ((I_{\mathbb{N}} - \mathbb{D}) \otimes \mathcal{H})]\eta(k),$$

$$\mathcal{A} = \begin{bmatrix} A & BK \\ 0 & A + BK \end{bmatrix}, \ \mathcal{H} = \begin{bmatrix} 0 & 0 \\ -LC & LC \end{bmatrix}. \tag{7.76}$$

Definition 7.3. Given agents (7.73), the protocol (7.75) is said to solve the consensus problem if

$$||x_i(k) = x_m(k)|| \to 0 \text{ as } k \to \infty, \ \ \forall\, i, \ m = 1, 2, \ldots, \mathbb{N}. \tag{7.77}$$

A decomposition approach to the consensus problem of network (7.77) is provided by the following theorem.

Theorem 7.5. *For any $\mathbb{G} \in \Gamma_{\mathbb{N}}$, the agents in (7.73) reach consensus under protocol (7.75) if all the matrices $A + BK$, $A + (1 - \lambda_i)LC$, $i = 2, \ldots \mathbb{N}$, are Schur stable, where λ_i, $i = 2, \ldots \mathbb{N}$, denote the eigenvalues of $\mathbb{D}$ located in the open unit disc.*

Proof. For any $\mathbb{G} \in \Gamma_{\mathbb{N}}$, it is known that 0 is a simple eigenvalue of $I_{\mathbb{N}} - \mathbb{D}$ and the other eigenvalues lie in the open unit disk centered at $1 + j\,0$ in the complex plane, where $j = \sqrt{-1}$. We let $s^t \in \Re^{1 \times \mathbb{N}}$ be the left eigenvector of $I_{\mathbb{N}} - \mathbb{D}$ associated with the eigenvalue 0, satisfying $x^t\,\mathbf{1} = 1$ with $\mathbf{1} \in \Re^p$ denoting the vector with all entries equal to one. Now, letting $\zeta \in \Re^{2\mathbb{N}n \times 2\mathbb{N}n}$, we have

$$\begin{aligned}\zeta(k) &= \eta(k) - [(\mathbf{1}s^t) \otimes I_{2n}]\eta(k) \\ &= [(I_{\mathbb{N}} - \mathbf{1}s^t) \otimes I_{2n}]\eta(k),\end{aligned} \tag{7.78}$$

which satisfies $(s^t \otimes I_{2n})\ \zeta = 0$. It is readily seen that 0 is a simple eigenvalue of $I_{\mathbb{N}} - \mathbf{1}s^t$ with $\mathbf{1}$ as its right eigenvector, and 1 is another eigenvalue with multiplicity $\mathbb{N} - 1$. It follows from (7.78) that $\zeta = 0$ if and only if $\eta_1 = \eta_2 = \ldots = \eta_{\mathbb{N}}$, that is, the consensus problem can be cast into the Schur stability of vector ζ, which evolves according to the following dynamics:

$$\zeta(k+1) = [(I_{\mathbb{N}} \otimes \mathcal{A}) + ((I - \mathbb{D}) \otimes \mathcal{H})]\eta(k). \tag{7.79}$$

Proceeding, we let $\mathcal{Y} \in \Re^{\mathbb{N}\times(\mathbb{N}-1)}$, $\mathcal{W} \in \Re^{(\mathbb{N}-1)\times\mathbb{N}}$, $\mathcal{V} \in \Re^{\mathbb{N}\times\mathbb{N}}$, and we let the upper triangular $\Delta \in \Re^{(\mathbb{N}-1)\times(\mathbb{N}-1)}$ be such that

$$\mathcal{V} = [\mathbf{1} - \mathcal{Y}],\ \mathcal{V}^{-1} = \begin{bmatrix} s^t \\ \mathcal{W} \end{bmatrix},$$

$$\mathcal{V}^{-1}(I_{\mathbb{N}} - \mathbb{D})\mathcal{V} = J = \begin{bmatrix} 0 & 0 \\ 0 & \Delta \end{bmatrix}, \tag{7.80}$$

where the diagonal entries of Δ are the nonzero eigenvalues of $(I_{\mathbb{N}} - \mathbb{D})$. We define the state transformation

$$\phi(k) = [\mathcal{V}^{-1} \otimes I_{2n}]\zeta(k),\ \phi^t = [\phi_1^t,\ \ldots,\ \phi_{\mathbb{N}}^t].$$

Then, (7.79) can be represented in terms of ϕ as follows:

$$\phi(k+1) = [(I_{\mathbb{N}} \otimes \mathcal{A}) + (J \otimes \mathcal{H})]\phi(k). \tag{7.81}$$

As to ϕ_1, it can be seen from (7.79) that

$$\phi_1(k+1) = [s^t \otimes I_{2n}]\zeta(k) \equiv 0. \tag{7.82}$$

It is worth noting that the elements of the state matrix of (7.81) are either block-diagonal or block upper-triangular. Hence, ϕ_i, $i = 2, \ldots \mathbb{N}$, converge asymptotically to zero if and only if the $\mathbb{N} - 1$ subsystems along the diagonal, that is,

$$\phi_i(k+1) = [\mathcal{A} + (1 - \lambda_i)\mathcal{H})]\phi_i(k), \tag{7.83}$$

are Schur stable. It is straightforward to verify that matrices $[(\mathcal{A} + \lambda_i\mathcal{H})]$ are similar to

$$\begin{bmatrix} A + (1-\lambda_i)LC & 0 \\ -(1-\lambda_i)LC & A + BK \end{bmatrix}.$$

Therefore, the Schur stability of the matrices $A + BK$, $A + (1 - \lambda_i)LC$, $i = 2, \ldots, \mathbb{N}$, is equivalent to that the state ζ of (7.79) converges asymptotically to zero, implying that consensus is achieved. □

Remark 7.8. The importance of this theorem lies in that it converts the consensus problem of a large-scale and therefore very high-dimensional multiagent network under the observer-type protocol (7.75) to the stability of a set of matrices with the same dimension as a single agent, thereby significantly reducing the computational complexity. The directed communication topology $\mathbb{G}$ is only assumed to have a directed spanning tree. The effects of the communication topology on the consensus problem are characterized by the eigenvalues of the corresponding row-stochastic matrix D, which may be complex, rendering the matrices to be complex-valued in Theorem 7.5.

Remark 7.9. The observer-type consensus protocol (7.75) can be seen as an extension of the conventional observer-based controller for a single system to one for multiagent systems. The classical *separation principle* of the conventional observer-based controllers still holds in this multiagent setting. Moreover, the protocol (7.75) is based only on relative output measurements between neighboring agents.

Theorem 7.6. *Consider the multiagent network (7.76) with a communication topology $\mathbb{G} \in \Gamma_{\mathbb{N}}$. If protocol (7.75) satisfies Theorem 7.6, then*

$$\begin{aligned} x_i(k) &\to \vartheta_i \stackrel{\Delta}{=} (r^t \otimes A^k)[x_1^t(0), \cdots, x_{\mathbb{N}}^t(0)]^t, \\ v_i(k) &\to 0, \quad i = 1, 2, \cdots, N, \quad as\ k \to \infty. \end{aligned} \tag{7.84}$$

Proof. It can be easily established in the manner of [46]. □

7.3 Event-based tracking control

Recent years have witnessed an increasing interest in the study of the coordination problem for multiagent systems. In this area, the consensus of multiagent systems is a basic and important problem, which has many applications in the fields of biology, computer science, and control engineering. During the past decades, the study of the consensus problem has attracted much attention for systems of single-integrator kinematics [44], double-integrator dynamics [46], fractional-order dynamics [36], and high-order dynamics [43] based on the continuous-time models [36] or discrete-time models [46].

At the same time, several scenarios have also been addressed, such as networks with switching topologies and time-varying delays [38], asynchronous consensus, finite-time consensus [40], and so on [42]. More specially, consensus with a dynamic leader is called a consensus tracking problem [41]. The tracking control has been widely used in many practical applications, such as unmanned aerial vehicle formation, target tracking in sensor networks, and so on [39].

7.3.1 Introduction

Decentralized consensus control is currently facilitated by recent technological advances on computing and communication resources. Each agent can be equipped with a small embedded microprocessor, which will be responsible for collecting the information from neighboring nodes and actuating the controller updates according to some ruling [37]. One of the most important aspects in the implementation of decentralized consensus algorithms is the communication and control actuation schemes. In order to reduce the communication burden and meanwhile save the communication resource for the network, the data-sampled control for continuous-time multiagent systems has been investigated in which the sampled signals of the agents should be instantly transmitted through the network.

In order to overcome the conservatism of data-sampled control, the event-triggered control strategy [37] provides a useful way of determining when the sampled signal is sent out. In [45], the control actuation is updated whenever a certain error becomes large enough to overtake the norm of the state.

In the following, we focus our attention on the design of a reasonable discrete event-triggered communication scheme and its application to the tracking control design of multiagent systems in order to save the limited resource and reduce the communication burden while preserving the desired performance. During the event-triggered tracking control process, it is assumed that every agent (or the leader) releases its state information to its neighbors only when the state error is above a specified threshold and each agent's local controller is updated by using the latest information received from neighboring agents.

Note that to realize the tracking control, the reference state is available to only a subset of the agents followed, and only the communication between the agent and its local neighbors is needed; therefore, the designed control is essentially distributed.

7.3.2 Problem statement

Suppose that agent i takes the following dynamics:

$$x_i(k+1) = x_i(k) + Tu_i(k), \ i = 1, 2, \cdots, n, \tag{7.85}$$

where the integer k is the discrete-time index, T is the sampling period which is assumed to be given a priori, and $x_i(k) \in \Re$ and $u_i(k) \in \Re$, respectively, represent the agent's state and the control input of the ith agent at time $t = kT$. In this section, the consensus tracking problem will be considered, where the time-varying reference state is denoted by $x^r(k)$. If the ith agent can access the leader's state information, then $a_{(i,n+1)} > 0$; otherwise, $a_{(i,n+1)} = 0$.

In what follows, for simplicity, we use (k) to represent (kT), for example, $x_i(k) = x_i(kT)$, $x^r(k) = x^r(kT)$, etc.

Definition 7.4. We say the solution of a dynamic system is uniformly ultimately bounded (UUB) if for a compact set U of R^n and for all $x(t_0) = x_0 \in U$, there exists an $\varepsilon > 0$ and a number $T(\varepsilon, x_0)$ such that $||x(t)|| < \varepsilon$ for all $t > t_0 + T$.

Assumption 7.3. The communication graph between the agents and the leader is fixed and directed.

Assumption 7.4. No data loss and transmission delay occur in the network communication.

The following consensus algorithm was shown to be able to guarantee the tracking of states of agents (7.85) with a time-varying reference state $x^r(k)$:

$$u_i(k) = \frac{1}{T\sum_{j=1}^{n+1} a_{(i,j)}} \sum_{j=1}^{n} a_{(i,j)}(x_j(k) - x_j[k-1] - T\gamma\{x_i(k) - x_j(k)\})$$
$$+ \frac{a_{(i,n+1)}}{T\sum_{j=1}^{n+1} a_{(i,j)}}(x^r(k) - x^r[k-1] - T\gamma\{x_i(k) - x_r(k)\}), \tag{7.86}$$

where γ is a positive gain.

Remark 7.10. The tracking controller (7.86) was assumed to be realized based on a time-triggered scheme with a given sampling period T, which means that the state sampled at the current time kT for each agent i and the leader should be transmitted to its neighbors in order to realize the algorithm (7.86). In order to save the communication energy of the multiagent network, in the following section, the event-triggered scheme is first introduced to realize the tracking controller (7.86), where the events are triggered for each agent i when the norm of the measurement errors exceeds certain dynamic thresholds presented by some trigger functions.

The realizing process for the algorithm goes as follows: All agents compute their trigger functions based on the past and current sampled signals; if the trigger condition is fulfilled for one agent, the agent will broadcast its actual measurement value to its neighbors. The time when the agent sends the measurement value of the state out to its neighbors is called the release time. Each agent's controller $u_i(k)$ is updated by evaluation using the latest information from its neighbors.

In order to model the event triggers for agents, here we denote the release times for the ith agent and its neighbors by s_m^i and s_m^j $(m = 0, 1, 2, \cdots, \ j \in N_i)$ and for the leader by s_m^r $(m = 0, 1, 2, \cdots)$.

The broadcast states for agents and the leader can be described by the following piece-wise constant functions:

$$\tilde{x}_i(k) = x_i(s_m^i), k \in (s_m^i, s_{m+1}^i), \tag{7.87}$$
$$\tilde{x}r(k) = xr(s_m^r), k \in (s_m^r, s_{m+1}^r), \tag{7.88}$$

where $s_m^i - s_{m+1}^i$ is an integer and $s_m^i T - s_{m+1}^i T$ is a multiple of T.

We define the trigger functions for agent i as $f_i(\cdot)$, $i = 1, 2, \cdots, n$, and the trigger function for the leader as $g(\cdot)$. For agent i, the states $x_j(k)$, $j \in N_i$, are unknown; only $x_i(k)$ and $\tilde{x}_i(k)$ and the broadcast values $\tilde{x}_j(k)$ of the neighbors $j \in N_i$ are available. An event for agent i about the state is triggered only when the trigger condition

$$f_i(k, x_i(k), \tilde{x}_i(k)) < 0, \; i = 1, 2, \cdots, n, \tag{7.89}$$

is violated, while the state of the leader is triggered as soon as

$$g(k, x^r(k), \tilde{x}^r(k)) < 0 \tag{7.90}$$

is violated.

The consensus algorithm with the tracking controller (7.86) under an event-triggered scheme (called event-triggered consensus algorithm) or (event-triggered tracking control) can be described as

$$\begin{aligned} u_i(k) &= \frac{1}{T\sum_{j=1}^{n+1} a_{(i,j)}} \sum_{j=1}^{n} a_{(i,j)}(\tilde{x}_j(k) - \tilde{x}_j[k-1] - T\gamma\{\tilde{x}_i(k) - \tilde{x}_j(k)\}) \\ &+ \frac{a_{(i,n+1)}}{T\sum_{j=1}^{n+1} a_{(i,j)}}(\tilde{x}^r(k) - \tilde{x}^r[k-1] - T\gamma\{\tilde{x}_i(k) - \tilde{x}^r(k)\}) \end{aligned} \tag{7.91}$$

$$\begin{aligned} &= \frac{1}{T\sum_{j=1}^{n+1} a_{(i,j)}} \sum_{j=1}^{n} a_{(i,j)}(x_j[s^j_{m_j}(k)] - xj[s^j_{m_j}(k-1)] \\ &- T\gamma xi[s^i_m] - xj[s^j_{m_j}(k)]) \\ &+ \frac{a_{(i,n+1)}}{T\sum_{j=1}^{n+1} a_{(i,j)}}(x^r[s^r_{n(k)}] - x^r[s^r_{n(k-1)}] - T\gamma\{x_i[s^i_m] - x^r[s^r_n(k)]\}), \\ &k \in [s^i_m, s^i_{m+1}), \end{aligned} \tag{7.92}$$

where

$$\begin{aligned} m_j(k) &\triangleq arg \min_{p \in N: s^j_p \le k} \{k - s^j_p\}, \\ n(k) &\triangleq \mathbf{arg} \min_{q \in N: s^r_q \le k} \{k - s^r_q\}, \; k \in [s^i_m, s^i_{m+1}), \end{aligned}$$

and $N = \{1, 2, 3, \cdots\}$, $\{s^j_p, \; p \in \mathbb{Z}^+\}$ represents the set of release times for the jth agent before time k and $\{s^r_q, q \in Z^+\}$ represents the set of release times for the leader before time k.

Remark 7.11. Note that $s^i_{m+1} - s^i_m$ in (7.92) may be larger than T. Moreover, in the time interval $[s^i_m, s^i_{m+1})$, the events triggered by the neighbors of the ith agent and the leader may occur, which means that $s^j_{m_j(k)}$ and $s^r_{n(k)}$ may change for $k \in [s^i_m, s^i_{m+1})$. Therefore, $u_i(k)$ in (7.92) may be time-varying for $k \in [s^i_m, s^i_{m+1})$ depending on the variations of $s^j_{m_j(k)}$ and $s^r_{n(k)}$.

7.3.3 Design results

The tracking problem for system (7.85) under the event-triggered consensus algorithm (7.92) will be investigated. Some arrangement is needed. We define the state error between the current sampled time and the last release time of the ith agent and the leader, respectively, as

$$\begin{aligned} e_i(k) &= x_i[s^i_m] - x_i(k),\ k \in [s^i_m, s^i_{m+1}), \\ e_j(k) &= x_j[s^j_m] - x_j(k),\ k \in [s^j_m, s^j_{m+1}), \end{aligned} \tag{7.93}$$

and

$$e^r(k) = x^r[s^r_m] - x^r(k),\ k \in [s^r_m, s^r_{m+1}), \tag{7.94}$$

where $m = 0, 1, \cdots$. Then $u_i(k)$ in (7.92) can be rewritten as

$$\begin{aligned} u_i(k) = & \frac{1}{T\sum_{j=1}^{n+1} a_{(i,j)}} \sum_{j=1}^{n} a_{(i,j)}(x_j(k) + e_j(k) - x_j[k-1] - e_j[k-1] \\ & - T\gamma x_i(k) + e_i(k) - x_j(k) - e_j(k)) \\ & + \frac{a_{(i,n+1)}}{T\sum_{j=1}^{n+1} a_{(i,j)}}(x^r(k) + e^r(k) - x^r[k-1] - e^r[k-1] \\ & - T\gamma\{x_i(k) + e_i(k) - x_r(k) - e^r(k)\}). \end{aligned} \tag{7.95}$$

Furthermore, we define the tracking error between the ith agent and the leader as $\delta_i(k) \triangleq x_i(k) - x^r(k)$. Then

$$x_i[k+1] - x_i(k) = \delta_i[k+1] + x^r[k+1] - \delta_i(k) - x^r(k) \tag{7.96}$$

and

$$\begin{aligned} u_i(k) = & \frac{1}{T\sum_{j=1}^{n+1} a_{(i,j)}} \sum_{j=1}^{n} a_{(i,j)}(\delta_j(k) + x^r(k) \\ & + e_j(k) - \delta_j[k-1] - x^r[k-1] - e_j[k-1] \\ & - \gamma\{\delta_i(k) + x^r(k) + e_i(k) - \delta_j(k) - x^r(k) - e_j(k)\}) \\ & + \frac{a_{(i,n+1)}}{T\sum_{j=1}^{n+1} a_{(i,j)}}(x^r(k) + e^r(k) - x^r[k-1] \\ & - e^r[k-1] - T\gamma\{\delta_i(k) + e_i(k) - e^r(k)\}). \end{aligned} \tag{7.97}$$

Substituting (7.96) and (7.97) into (7.85) and making some arrangements, we obtain, $k \in [s^i_m, s^i_{m+1})$,

$$\begin{aligned} \delta_i[k+1] = & 2x^r(k) - x^r[k+1] - x^r[k-1] \\ & + \frac{a_{(i,n+1)}}{\sum_{j=1}^{n+1} a_{(i,j)}}\{e^r(k) + T\gamma e^r(k) - e^r[k-1]\} \end{aligned}$$

$$
\begin{aligned}
&+(1-T\gamma)\delta_i(k)-T\gamma e_i(k)\\
&+\frac{1}{\sum_{j=1}^{n+1}a_{(i,j)}}\sum_{j=1}^{n}a_{(i,j)}[(1+T\gamma)(\delta_j(k)+e_i(k))\\
&-\delta_j[k-1]-e_j[k-1]]. \qquad (7.98)
\end{aligned}
$$

In order to write (7.98) in a matrix form, we introduce the following matrix and augmented variable definitions:

$$
\begin{aligned}
B &= diag\{a_{(1,n+1)},\cdots,a_{(n,n+1)}\},\\
D &= diag\{\sum_{j=1}^{n+1}a_{(1,j)},\cdots,\sum_{j=1}^{n+1}a_{(n,j)}\},\\
\Delta[k+1] &= (\delta_1[k+1],\cdots,\delta_n[k+1])^\top,\\
e(k) &= (e_1(k),\cdots,e_n(k))^\top,\\
\theta^r(k) &= 2x^r(k)-x^r[k+1]-x^r[k-1],\\
\beta^r(k) &= e^r(k)+T\gamma e^r(k)-e^r[k-1].
\end{aligned}
$$

Therefore, (7.98) can be rewritten as

$$
\begin{aligned}
\Delta(k+1) =\ &\theta^r(k)1_n+D^{-1}B\beta^r(k)1n+(1-T\gamma)I_n\Delta(k)-T\gamma e(k)\\
&+D^{-1}A(1+T\gamma)\Delta(k)+D^{-1}A(1+T\gamma)e(k)\\
&-D^{-1}A\Delta[k-1]-D^{-1}Ae[k-1] \qquad (7.99)
\end{aligned}
$$

and then

$$
\begin{bmatrix}\Delta[k+1]\\ \Delta(k)\end{bmatrix}=\tilde{A}\begin{bmatrix}\Delta(k)\\ \Delta[k-1]\end{bmatrix}+\tilde{B}\begin{bmatrix}e(k)\\ e[k-1]\end{bmatrix}+\tilde{C}\begin{bmatrix}\theta^r(k)1_n\\ \beta^r(k)1_n\end{bmatrix}, \qquad (7.100)
$$

$$
\begin{aligned}
\tilde{A} &= \begin{bmatrix}(1-T\gamma)I_n+(1+T\gamma)D^{-1}A-D^{-1}A & -D^{-1}A\\ I_n & 0\end{bmatrix},\\
\tilde{B} &= \begin{bmatrix}(1+T\gamma)D^{-1}A-T\gamma In-D^{-1}A & -D^{-1}A\\ 0 & 0\end{bmatrix},\\
\tilde{C} &= \begin{bmatrix}I_n & D^{-1}B\\ 0 & 0\end{bmatrix}. \qquad (7.101)
\end{aligned}
$$

We define the augmented variables

$$
Y(k)=\begin{bmatrix}\Delta(k)\\ \Delta[k-1]\end{bmatrix},\ \omega(k)=\begin{bmatrix}e(k)\\ e[k-1]\end{bmatrix},\ z(k)=\begin{bmatrix}\beta^r(k)1_n\\ \theta^r(k)1_n]\end{bmatrix},
$$

and then (7.100) becomes

$$Y[k+1] = \tilde{A}Y(k) + \tilde{B}\omega(k) + \tilde{C}z(k). \tag{7.102}$$

Successive iteration on (7.102) yields

$$Y(k) = \tilde{A}^k Y(0) + \sum_{s=1}^{k} \tilde{A}^{k-s}\{\tilde{B}\omega(s-1) + \tilde{C}z(s-1)\}. \tag{7.103}$$

Convergence analysis of system (7.102) will be carried out based on the solution (7.103). For this purpose, the following lemma is needed.

Lemma 7.1. *([46]) Assume that the leader has a directed path to all agents from* $1, \cdots, n$ *and let* λ_i *be the eigenvalue of* $D^{-1}A$*. Then the condition*

$$\tau_i \triangleq 2|1-\lambda_i|2\{2[1-\mathbf{Re}(\lambda_i)] - |1-\lambda_i|2\}/(|1-\lambda_i|^4 + 4[\mathbf{Im}(\lambda_i)]^2) > 0$$

holds and $\mathbf{Re}(\cdot)$ *and* $\mathbf{Im}(\cdot)$ *denote, respectively, the real and imaginary parts of a number. If positive scalars* T *and* γ *satisfy*

$$T\gamma < \min\{1, \min_{i=1,\dots,n} \tau_i\}, \tag{7.104}$$

then $\tilde{A}$*, defined in (7.101), has all eigenvalues within the unit circle.*

The bound for the tracking error is estimated under the control (7.92) with the event-triggered scheme.

Theorem 7.7. *Assume that the leader has a directed path to all agents from* 1 *to* n *and its states* $x^r(k)$ *satisfy* $|(x^r(k) - x^r[k-1])/T| \le \bar{\xi}$ *(that is, the changing rate of* $x^r(k)$ *is bounded), and*

$$\|e_i[s]\| \le \alpha_1 + c_1 e^{-\beta_1 sT},\ i = 1, 2, \cdots, n, \tag{7.105}$$

$$\|e^r[s]\| \le \alpha_2 + c_2 e^{-\beta_2 sT}, \tag{7.106}$$

where $\alpha_j \ge 0,\ \beta_j \ge 0,\ c_j \ge 0,\ j = 1, 2,$ *are some constants,* $s \in \mathbb{Z}^+$*. If positive scalars* γ *and* T *satisfy (7.104), under the control algorithm (7.92), the infinite norm of the solution of (7.102) is UUB by*

$$\begin{aligned} ||Y(k)||_\infty \le\ & [2\alpha_1(1+T\gamma) + 2c_1(1+T\gamma)e^{-\beta_1 T} + b_1] \\ & \times ||(I_{2n} - \tilde{A})^{-1}||_\infty, \quad (k \to \infty), \\ b_1 =\ & \max\{2T\bar{\xi}, (2+T\gamma)(\alpha_2 + c2e^{-\beta_2 T})\}. \end{aligned} \tag{7.107}$$

Proof. From the definition of $||\cdot||_\infty$, it is readily seen from condition (7.106) that

$$||\omega[s]||_\infty \le \alpha_1 + c_1 e^{-\beta_1}(s-1)T. \tag{7.108}$$

Furthermore, it is easy to see from the definitions of $\tilde{B}$, $\tilde{C}$, and $z[s]$ that

$$||\tilde{B}||_\infty \leq 2(1+T\gamma),$$

and

$$\begin{aligned}||\tilde{C}z[s]||_\infty &\leq \max\{2T\bar{\xi}, (2+T\gamma)(\alpha_2+c_2e^{-\beta_2 T})\}\\ &\triangleq b_1\end{aligned} \tag{7.109}$$

Under the assumptions of the theorem, it follows from Lemma 7.1 that $\tilde{A}$ has all eigenvalues within the unit circle. According to [47], $||\tilde{A}|| < 1$. Therefore, the following relation can be deduced:

$$\lim_{k\to\infty} ||\sum_{s=0}^{k-1} \tilde{A}^s||_\infty \leq ||(I_{2n}-\tilde{A})^{-1}||_\infty. \tag{7.110}$$

Combining (7.103), (7.106), and (7.108)–(7.110), according to [46], it can be concluded that

$$\begin{aligned}||Y(k)|_\infty &\leq ||\tilde{A}^k Y[0]||_\infty + ||\sum_{s=1}^{k} \tilde{A}^{k-s}\tilde{B}\omega[s-1]||_\infty\\ &+||\sum_{s=1}^{k} \tilde{A}^{k-s}\tilde{C}z[s-1]||_\infty\\ &\leq ||\tilde{A}^k Y[0]||_\infty + (\alpha_1+c1e^{-\beta_1 T})||\tilde{B}||_\infty||\sum_{s=1}^{k} \tilde{A}^{k-s}||_\infty\\ &+b_1||\sum_{s=1}^{k} \tilde{A}^{k-s}||_\infty\\ &\leq ||\tilde{A}^k||_\infty||Y[0]||_\infty + 2(1+T\gamma)(\alpha_1+c1e^{-\beta_1 T})||\sum_{s=1}^{k} \tilde{A}^{k-s}||_\infty\\ &+b_1||\sum_{s=1}^{k} \tilde{A}^{k-s}||_\infty\\ &\leq [2\alpha_1(1+T\gamma)+2c_1(1+T\gamma)e^{-\beta_1 T}+b_1]||(I_{2n}-\tilde{A})^{-1}||_\infty\end{aligned}$$

as $k \to \infty$, which ends the proof. □

Remark 7.12. From the definition of $||\cdot||_\infty$ and inequality (7.107), we have

$$\begin{aligned}|\delta_i(k)| &\leq [2\alpha_1(1+T\gamma)+2c_1(1+T\gamma)e^{-\beta_1 T}+b_1]\\ &\times ||(I_{2n}-\tilde{A})^{-1}||_\infty\end{aligned}$$

as $k \to \infty, i = 1, 2, \cdots, n$, which means the tracking error between the ith agent and the leader is ultimately bounded.

Remark 7.13. For all agents and the leader, we define the event-triggered functions, respectively, as

$$f_i(s, |e_i[s]|) \triangleq |e_i[s]| - (\alpha_1 + c1e^{-\beta_1 sT}), \; i = 1, 2, \cdots, n, \tag{7.111}$$

$$g(s, |e^r[s]|) \triangleq |e^r[s]| - (\alpha_2 + c_2 e^{-\beta_2 sT}). \tag{7.112}$$

The events are triggered for the ith agent and the leader when $f_i(s, |e_i[s]|) > 0$ and $g(s, |e^r[s]|) > 0$, respectively. Under the above event-triggered schemes, the conditions in (7.105)–(7.106) can be guaranteed for all $s \in \mathbb{Z}^+$, which can be concluded from the following analysis: For agent i, $\forall s \in \mathbb{Z}^+$, there exists one interval $[s_l^i, s_{l+1}^i)$, such that $s \in [s_l^i, s_{l+1}^i)$, and in the time interval (s_l^i, s_{l+1}^i) no event has occurred.

Case 1: We have $s = s_p^i$, that is, an event is triggered for agent i at time s. According to (7.93), it can be obtained that $e_i[s] = 0$, so the condition (7.105) is guaranteed.

Case 2: We have $s \in (s_l^i, s_{l+1}^i)$. Because no event has occurred in this time interval, it can be concluded that the trigger function in (7.111) satisfies $f_i(s, e_i[s]) < 0$. So the condition in (7.105) can be guaranteed. Also, (7.106) can be guaranteed by using a similar analysis method as above. From the structure of event triggers (7.111) and (7.112), it can also be seen that the triggered mechanism used in each agent is decentralized and therefore realizable.

Remark 7.14. The discrete exponentially decreasing threshold $\alpha + ce^{-\beta kT}$ provides a very flexible event-triggered control strategy for multiagent systems. The parameter α can be used to adjust the state errors' convergence region. The parameter c can be tuned in such a way that the events are not too dense for small times kT. The parameter β can be used to determine the speed of convergence. For small times kT, the event times depend dominantly on c, so the density of events does not increase with decreasing α. For larger times kT, the density does not increase with decreasing c either.

In Theorem 7.7, the parameters β_1 and β_2 in (7.106) and (7.107), respectively, are assumed to be constant. In the following, revised versions of (7.106) and (7.107) are proposed by setting β_1 and β_2 as time-varying functions which can lead to a smaller upper bound for $||Y(k)||_\infty$ compared with that in Theorem 7.7.

To show this fact, the following lemma is needed.

Lemma 7.2. *If function* $f(k) = kC_k^m a^{k-m}$, *where* $0 < a < 1$, *and* m *is a finite positive integer, then* $\lim_{k \to +\infty} f(k) = 0$.

Proof. It can be easily concluded based on the L'Hopital rule in [48]. □

Theorem 7.8. *Assume that the leader has a direct path to all agents from* $1 \cdots n$ *and its state* $x^r(k)$ *satisfies* $|(x^r(k) - x^r[k-1])/T| \leq \bar{\xi}$ *(i.e., the changing rate of* $x^r(k)$ *is bounded), and*

$$|e_i[s]| \leq \alpha_1 + c_1 e^{-\beta_1(s)sT}, \ i = 1, 2, \cdots, n, \tag{7.113}$$

$$|e^r[s]| \leq \alpha_2 + c_2 e^{-\beta_2(s)sT}, \tag{7.114}$$

where

$$\beta_1(s) = -\frac{ln\rho(\tilde{A})s}{sT}, \quad \beta_2(s) = -\frac{ln\rho(\tilde{A})s}{sT}, \tag{7.115}$$

and $\alpha_j \geq 0$, $c_j \geq 0$, $j = 1, 2$, *are some constants,* $s \in Z^+$. *If positive scalars* γ *and* T *satisfy (7.104), under the control algorithm (7.92), the infinite norm of the solution of (12) is UUB by*

$$||Y(k)||_\infty \leq [2\alpha_1(1 + T\gamma) + b_2]||(I_{2n} - \tilde{A})^{-1}||_\infty \ (k \to \infty), \tag{7.116}$$

where $b_2 = \max\{2T\bar{\xi}, \alpha_2(2 + T\gamma)\}$.

Proof. It follows from (7.103) that

$$||Y(k)||_\infty \leq ||\tilde{A}^k Y[0]||_\infty + ||\sum_{s=1}^{k} \tilde{A}^{k-s}\tilde{B}\omega[s-1]||_\infty + ||\sum_{s=1}^{k} \tilde{A}^{k-s}\tilde{C}z[s-1]||_\infty.$$

Using the conditions in (7.113)–(7.115), we can easily show that

$$||\omega[s-1]||_\infty \leq \alpha_1 + c_1\rho(\tilde{A})^{s-1}$$

and

$$||\tilde{C}z[s-1]||_\infty \leq \max\{2T\bar{\xi}, (2 + T\gamma)(\alpha_2 + c_2\rho(\tilde{A})^{s-1})\} \triangleq b(s). \tag{7.117}$$

Then, according to [46], it can be concluded that

$$||Y(k)||_\infty \leq ||\tilde{A}^k||_\infty ||Y[0]||_\infty + \alpha_1 ||\tilde{B}||_\infty ||\sum_{s=1}^{k} \tilde{A}^{k-s}||_\infty + c_1||\tilde{B}||_\infty \sum_{s=1}^{k} ||\tilde{A}^{k-s}||_\infty \rho(\tilde{A})^{s-1} + ||\sum_{s=1}^{k} \tilde{A}^{k-s} b(s)||_\infty. \tag{7.118}$$

As shown in [49], there exists an invertible matrix P such that $\tilde{A}$ is similar to a Jordan canonical matrix J, i.e., $P^{-1}\tilde{A}P = J = diag\{J_1, J_2, \cdots, J_l\}$, where $J_s, s = 1, 2, \cdots, l$, are upper-triangular Jordan blocks, whose principal diagonal elements are the eigenvalues of $\tilde{A}$. Then, for the third term on the right hand side of inequality (7.118), we have

$$
\begin{aligned}
& c_1||\tilde{B}||_\infty \sum_{s=1}^{k} ||\tilde{A}^{k-s}||_\infty \rho(\tilde{A})^{s-1} \\
& \leq 2(1+T\gamma)\sum_{s=1}^{k} ||\tilde{A}^{k-s}||_\infty \rho(\tilde{A})^{s-1} \\
& = 2(1+T\gamma)\sum_{s=1}^{k} ||(PJP^{-1})^{k-s}||_\infty \rho(\tilde{A})^{s-1} \\
& \leq 2(1+T\gamma)\cdot c\sum_{s=1}^{k} ||J||_\infty^{k-s} \rho(\tilde{A})^{s-1} \\
& \leq 2c(1+T\gamma)\sum_{s=1}^{k}[\rho(\tilde{A})^{k-s} + C_{k-s}^{1}\rho(\tilde{A})^{k-s-1} + \cdots \\
& = C_{k-s}^{m}\rho(\tilde{A})^{k-s-m+1}]\rho(\tilde{A})^{s-1} \\
& \leq 2c(1+T\gamma)[k\rho(\tilde{A})^{k-1} + kC_k^1\rho(\tilde{A})^{k-2} + \cdots + kC_k^m\rho(\tilde{A})^{k-m}],
\end{aligned}
$$

where $c \triangleq ||P||_\infty \cdot ||P-1||_\infty$ and m is the maximum order of $J_s, s = 1, 2, \cdots, l$. According to Lemma 7.2 and the above analysis, it can be easily obtained that

$$
c_1||\tilde{B}||_\infty \sum_{s=1}^{k} ||\tilde{A}_\infty^{k-s}\rho(\tilde{A})^{s-1} \to 0, (k \to \infty). \tag{7.119}
$$

By a similar proof process for (7.119) and using (7.110), the following relation can be deduced:

$$
||\sum_{s=1}^{k} \tilde{A}^{k-s}b(s)||_\infty \leq b_2||(I_{2n} - \tilde{A})^{-1}||_\infty, (k \to \infty), \tag{7.120}
$$

where $b_2 = \max\{2T\bar{\xi}, \alpha_2(2+T\gamma)\}$.

Since all the eigenvalues of $\tilde{A}$ are within the unit circle, as stated in Lemma 7.1, we can obtain $\lim_{k\to\infty}\tilde{A}^k = 0_{2n\times 2n}$. Combining (7.118)–(7.120), we obtain

$$
||Y(k)||_\infty \leq [2(1+T\gamma)\alpha_1 + b_2]||(I_{2n} - \tilde{A})^{-1}||_\infty, (k \to \infty). \tag{7.121}
$$

This ends the proof. □

Remark 7.15. We define the trigger functions for each agent i and the leader, respectively, as follows:

$$f_i(s, |e_i[s]|) \triangleq |e_i[s]| - (\alpha_1 + c_1 e^{-\beta_1(s)sT}),\ i = 1, 2, \cdots, n, \tag{7.122}$$

$$g(s, |e^r[s]|) \triangleq |e^r[s]| - (\alpha_2 + c_2 e^{-\beta_2(s)sT}). \tag{7.123}$$

In view of the foregoing analysis, it is known that, under the event-triggered schemes with trigger functions above, the conditions in (7.113) and (7.114) can be guaranteed for all $s \in \mathbb{Z}^+$.

Remark 7.16. Similarly to Remark 7.12, inequality (7.116) yields

$$|\delta_i(k)| \le 2[(1 + T\gamma)\alpha_1 + b_2]||(I_{2n} - \tilde{A})_{\infty}^{-1}$$

as $k \to \infty$, $i = 1, 2, \cdots, n$. Obviously, under the new event-triggered scheme, the resulting upper bound for the tracking error $\delta_i(k)$ is smaller than that under the event-triggered scheme in Remark 7.14. However, due to the introduction of the time-varying parameters β_1 and β_2, the relatively heavier computation is needed when using the event-triggered scheme in Remark 7.16.

Remark 7.17. If we choose $\alpha_i = 0$, $c_i = 0$ $(i = 1, 2)$ in Theorems 7.7 and 7.8, then the event-triggered communication scheme proposed in this section reduces to the time-triggered scheme which has been studied in [46], and the upper bound of tracking errors between agents and the leader is $2T\bar{\xi}||(I_{2n} - \tilde{A})^{-1}||_{\infty}$.

7.3.4 Simulation example 7.2

A numerical example with four agents and a time-varying reference satisfying the same communication graph is considered to validate the main results of this section (see Fig. 7.4). If agent j is a neighbor of agent i, we let $a_{(i,j)} = 1$; otherwise, $a_{(i,j)} = 0$. Then, for this example, the corresponding adjacency matrix is

$$A = \begin{bmatrix} 0 & 1 & 0 & 0 \\ 1 & 0 & 1 & 0 \\ 1 & 0 & 0 & 0 \\ 0 & 1 & 1 & 0 \end{bmatrix}.$$

Since agent 3 can have access to the leader, $a_{(3,5)} = 1$, $a_{(1,5)} = 0$, $a_{(2,5)} = 0$, and $a_{(4,5)} = 0$. The reference state is chosen as $x^r(k) = \sin(kT) + kT$. In the following, we will study this example by using the asynchronous communication scheme, that is, by using the developed event-triggered scheme.

The initial states of the four agents are chosen as

$$[x_1[0], x_2[0], x_3[0], x_4[0]] = [2, 1, -1, -3].$$

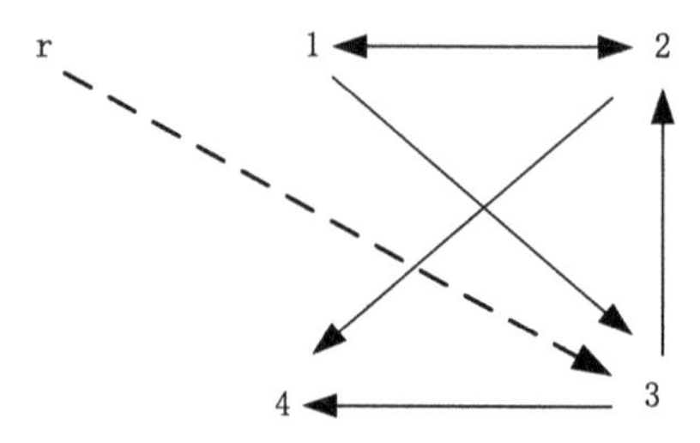

FIGURE 7.4 Directed graph for four agents with a leader.

Without loss of generality, suppose that

$$[x_1[-1], x_2[-1], x_3[-1], x_4[-1]] = [0, 0, 0, 0].$$

Let $T = 0.1$ and $\gamma = 3$. By simple computation, it can be seen that the condition (7.104) holds.

In the following simulation, two cases for the selection of α_j, c_j, and β_j $(j = 1, 2)$ will be considered, which correspond to different types of trigger thresholds.

Case 1: $\alpha_1 = 0.01, \alpha_2 = 0.04, c_1 = 0.1, c_2 = 0.5, \beta_1 = 0.5, \beta_2 = 0.5$. Under the tracking control (7.92) with the event-triggered schemes in Remark 7.5, the dynamic responses for the states $x_i(k)$ and the tracking errors $\delta_i(k) = x_i(k) - x^r(k)$, $i = 1, \cdots, 4$, are shown in Figs. 7.5 and 7.6, respectively. By simple computation according to Table 7.2, we can obtain that only 80.68% of sampled states of all agents and the leader need to be sent out to their neighbors.

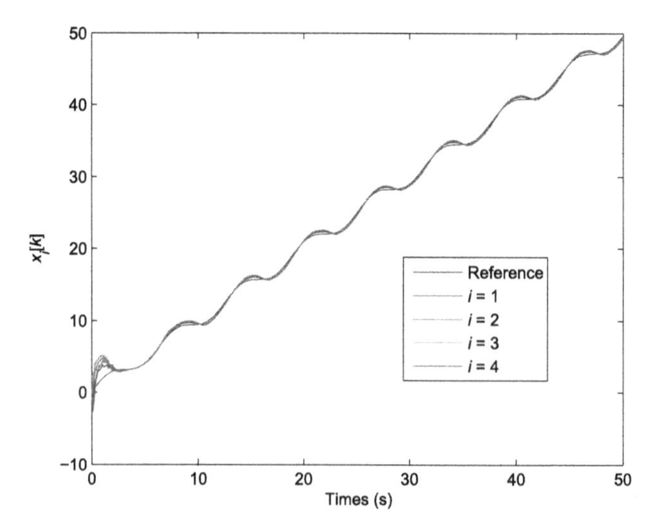

FIGURE 7.5 State responses of system (7.85) in Case 1.

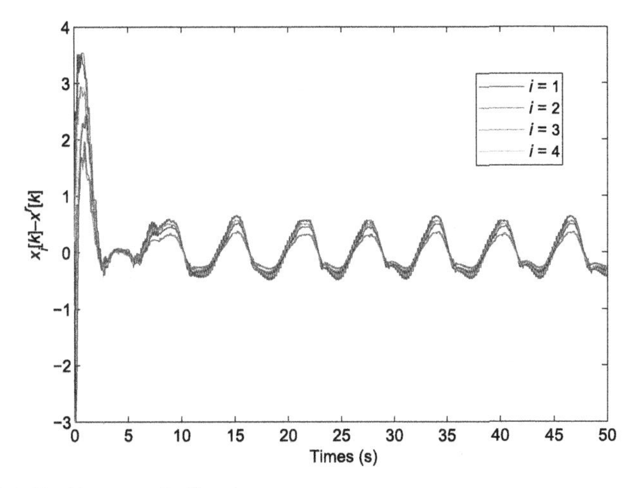

FIGURE 7.6 Tracking errors in Case 1.

TABLE 7.2 Sample and release times for Case 1.

	Leader	Agent 1	Agent 2	Agent 3	Agent 4
Sample times	500	500	500	500	500
Release times	354	411	416	427	409

TABLE 7.3 Sample and release times for Case 2.

	Leader	Agent 1	Agent 2	Agent 3	Agent 4
Sample times	500	500	500	500	500
Release times	356	412	409	420	399

Case 2: α_j and c_j $(j = 1, 2)$ are the same as in Case 1, but the parameters β_1 and β_2 are chosen as $\beta_1(s) = -ln\varrho(\tilde{A})s/0.1s$, $\beta_2(s) = -ln\rho(\tilde{A})s/0.1s$, $s \in Z^+$ and $\rho(\tilde{A}) = 0.9405$. Under the tracking control (7.92) with the event-triggered schemes in Remark 7.7, the dynamic responses for the states $x_i(k)$ and the tracking errors $\delta_i(k) = x_i(k) - x^r(k)$, $i = 1, \cdots, 4$, are shown in Figs. 7.7 and 7.8, respectively. By simple computation according to Table 7.3, we can obtain that only 79.84% of sampled states of all the agents and the leader need to be sent out to their neighbors.

Case 3: $\alpha_j = 0$ and $c_j = 0$ $(j = 1, 2)$. Under the tracking control (7.86) with the time-triggered scheme, the dynamic responses for the states $x_i(k)$ and the tracking errors $\delta_i(k) = x_i(k) - x^r(k)$, $i = 1, \cdots, 4$, are shown in Figs. 7.9 and 7.10, respectively.

As indicated above, only around 80% of sampled states of all agents and the leader need to be transmitted through the network for **Cases 1** and **2** in order to

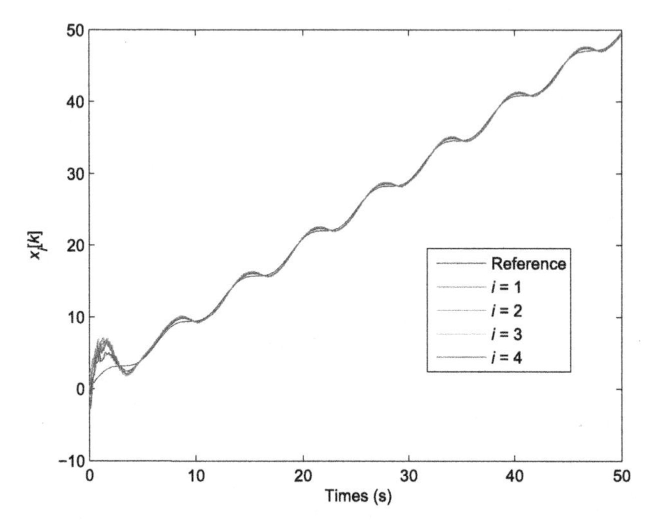

FIGURE 7.7 State responses of system (7.1) in Case 2.

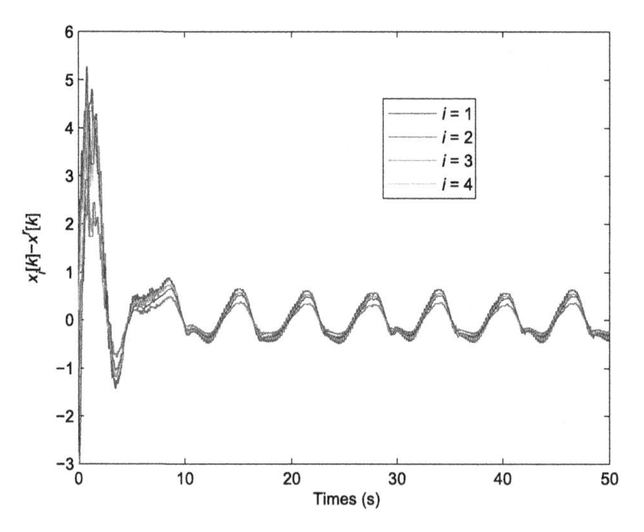

FIGURE 7.8 Tracking errors in Case 2.

guarantee the tracking performance. Therefore, around 20% of communication resource is saved. According to the simulation, when the time is longer than 30 s, it can be computed that the bound of the tracking error for **Case 1** is 0.6497 and for **Case 2** it is 0.6396. It is clear that $0.6396 < 0.6497$, which is consistent with the foregoing analysis and validates the foregoing theoretical results.

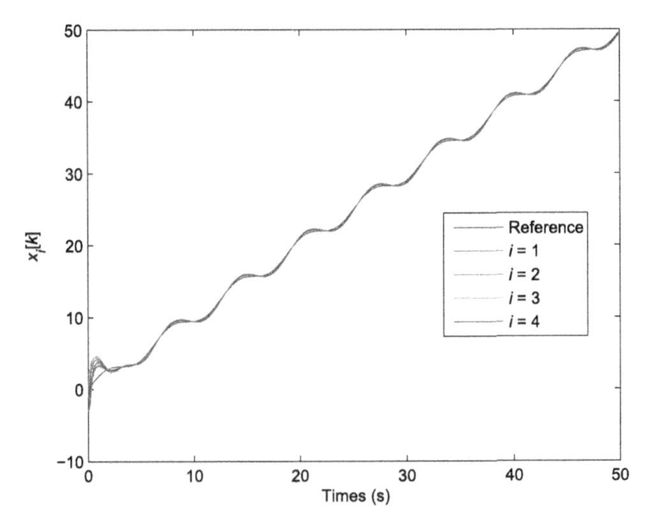

FIGURE 7.9 State responses of system (7.1) in Case 3.

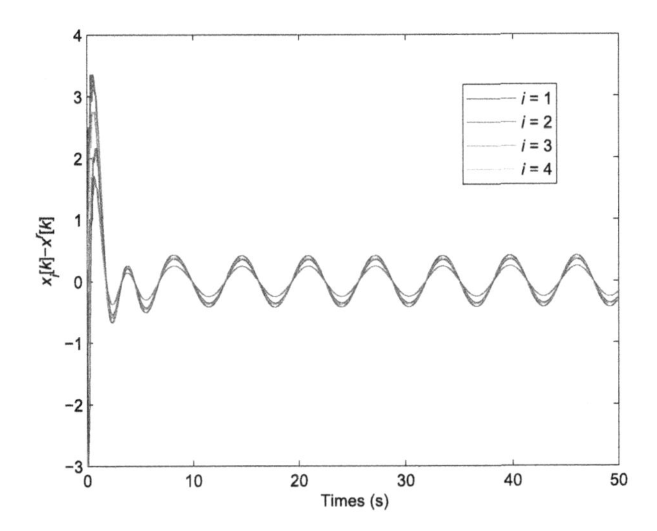

FIGURE 7.10 Tracking errors in Case 3.

7.4 Robust output regulation

We learned that an important issue of cooperative control is that the agents should be able to reach a common trajectory based on a consensus protocol under the shared information in the presence of information exchange and dynamically changing interaction topologies. One possible approach is to design a distributed algorithm by which all the agents reach an agreement on their states or outputs, using local information exchange under the communication topologies.

In the sequel, an output regulation concept is introduced to characterize the output consensus of leader following multiagent systems. To this end, output regulation theory is to control one or more than one plant so that their outputs can track a reference signal (and/or reject a disturbance) produced by an exosystem. In this section, we study discrete-time multiagent systems where every subsystem is identical and uncertain. Each agent also has different disturbance. We develop an internal model with n adjustable parameters and construct a distributed consensus protocol, which not only makes the subsystem stable regardless of external disturbance but also regulates the errors to converge to zero.

A directed path is a sequence of edges in a directed graph of the form $(v_1, v_2), (v_2, v_3), \cdots$, where $v_i \in \mathbb{V}$. A directed graph contains a directed spanning tree if there exists at least one agent which is called root node that has a directed path to every other agent.

From matrix theory, it follows that given $A \in \Re^{n \times n}$ and for $\lambda_1, \lambda_2, \cdots, \lambda_n$ being the eigenvalues of A, for $c \in \mathbf{C}$, $\lambda_1 + c$, $\lambda_2 + c$, $\cdots$, $\lambda_n + c$ are the eigenvalues of $A + cI_n$. The following initial result is recalled.

Lemma 7.3. *For any real symmetric matrix $Q > 0$, there exists a real symmetric matrix $\check{Q} > 0$ such that $Q = \check{Q}\check{Q}$.*

Proof. From the expression of Q, there is a real orthogonal matrix U such that

$$Q = U \Lambda_Q U^t = U \sqrt{\Lambda_Q} U^t U \sqrt{\Lambda_Q} U^t,$$

with $\lambda_{qi} > 0$, $i = 1, 2, \cdots, N$, being the eigenvalues of Q and

$$\Lambda_Q = \mathbf{diag}\{\lambda_{q1}, \cdots, \lambda_{qN}\}, \ \sqrt{\Lambda_Q} = \mathbf{diag}\{\sqrt{\lambda_{q1}}, \cdots, \sqrt{\lambda_{qN}}\}.$$

Let $\check{Q} = U \sqrt{\Lambda_Q} U^t$; then the proof is completed. □

A restatement of the *Gershgorin lemma* is presented below.

Lemma 7.4. *Let $A = [a_{ij}] \in \Re^{n \times n}$,*

$$R_i^{\circ}(A) \equiv \sum_{j=1,\ j \neq i}^{n} \left| a_{ij} \right|, \ 1 \leq i \leq n,$$

and denote the deleted absolute row sums of A. Then all the eigenvalues of A are located in the union of n discs

$$\bigcup_{i=1}^{N} \{z \in \mathbf{C} : |z - a_{ii}| \leq R_i'(A)\} \equiv \mathbf{G}(A).$$

7.4.1 Network model

Let the digraph $\bar{\mathbb{G}} = (\bar{\mathbb{V}}, \bar{\mathbb{E}}, \bar{\mathbb{A}})$ of order $N+1$ consist of a vertex set $\bar{\mathbb{V}} = (v_0, v_1, \cdots, v_N)$ and let v_0 be defined as the exosystem; $a_{i0} > 0$ if agent i can receive information from the exosystem; otherwise, $a_{i0} = 0$; $\bar{\mathcal{L}}$ is defined as the Laplacian matrix of subgraph $\bar{\mathbb{G}}$ except node v_0. The adjacency matrix of digraph $\bar{\mathbb{G}}$ is obtained as

$$\bar{\mathbb{A}} = \begin{pmatrix} 0 & \mathbf{0} \\ \mathbb{A}_0 \mathbf{1}_N & \bar{\mathbb{A}}_s \end{pmatrix}, \tag{7.124}$$

in which $\mathbb{A}_0 = \mathbf{diag}\{a_{10}, a_{20}, \cdots, a_{N0}\}$ and $\bar{\mathbb{A}}_s$ is the adjacency matrix of subgraph $\bar{\mathbb{G}}_s$ with vertex set $\bar{\mathbb{V}}_s = (v_1, v_2, \cdots, v_N)$. Then the Laplacian matrix $\bar{\mathbb{L}}$ of digraph $\bar{\mathbb{G}}$ is

$$\bar{\mathbb{L}} = \begin{pmatrix} 0 & \mathbf{0} \\ -\mathbb{A}_0 \mathbf{1}_N & \mathbb{A}_0 + \bar{\mathbb{L}}_s \end{pmatrix}, \tag{7.125}$$

in which $\bar{\mathbb{L}}_s = \bar{\mathbb{D}}_s - \bar{\mathbb{A}}_s$ is the Laplacian matrix of subgraph $\bar{\mathbb{G}}_s$, and $\bar{\mathbb{D}}_s = \mathbf{diag}\{\bar{d}_1, \bar{d}_2, \cdots, \bar{d}_N\}$ and $\bar{\mathbb{A}}_s$ are the degree matrix and adjacency matrix of subgraph $\bar{\mathbb{G}}_s$, respectively.

Lemma 7.5. *All the eigenvalues of $\mathbb{H} = \mathbb{A}_0 + \bar{\mathbb{L}}_s$ defined in (7.125) have positive real parts if and only if the digraph $\bar{\mathbb{G}}$ contains a spanning tree and node v_0 as its root.*

A group of N discrete-time linear agents with uncertain dynamics are described by

$$\begin{aligned} x_i(k+1) &= \bar{A} x_i(k) + \bar{B} u_i(k) + \bar{E}_i \omega(k), \quad x_i(0) = x_{i0}, \\ y_i(k) &= \bar{C} x_i(k), \quad i = 1, 2, \cdots, N, \end{aligned} \tag{7.126}$$

where $x_i \in \Re^n$ is the state, $y_i \in \Re^p$ is the output, $u_i \in \Re^q$ denotes the consensus protocol to be designed which depends on agent i and its neighbors, and $\bar{E}_i \omega$ denotes the external disturbance of the ith agent. The matrices $\bar{A}$, $\bar{B}$, $\bar{C}$, and $\bar{E}_i$ are uncertain and they could be written as follows:

$$\begin{aligned} \bar{A} &= A + \Delta A, \quad \bar{B} = B + \Delta B, \\ \bar{C} &= C + \Delta C, \quad \bar{E}_i = E_i + \Delta E_i, \end{aligned} \tag{7.127}$$

where ΔA, ΔB, ΔC, and ΔE_i are the perturbations of nominal matrices A, B, C, and E_i, respectively. It is convenient to identify the system uncertainties with a **vec** form,

$$\Delta = \begin{bmatrix} \mathbf{vec}(\Delta A, \Delta B, \Delta E_1, \cdots, \Delta E_N) \\ \mathbf{vec}(\Delta C) \end{bmatrix} \in \Re^{n(m+n+p+Nq)}.$$

System (7.126) with $\Delta = 0$ is called the nominal system. In addition, the reference inputs and the disturbances can be lumped together as the following exosystem:

$$\begin{aligned} \omega(k+1) &= S\,\omega(k), \;\; \omega(0) = \omega_0, \\ y_r(k) &= F\,\omega(k), \end{aligned} \tag{7.128}$$

with $y_r \in \Re^p$ being the reference signal. Thus, the consensus output regulation objectives may be prescribed in terms of the tracking errors $e_i \in \Re^p$ as the following form:

$$e_i(k) = y_i(k) - y_r(k) = \bar{C}x_i(k) - F\omega(k), \quad i = 1, 2, \cdots, N. \tag{7.129}$$

7.4.2 Consensus protocol

We consider that each agent can communicate with its neighbor agents and exosystem by sharing their output states. The output-coupling variable relationship between agent i and its neighbors $j \in \mathcal{N}_i$ is defined as follows:

$$\delta_i(k) = \sum_{j \in \mathcal{N}_i} a_{ij}(y_i(k) - y_j(k)) + a_{i0}(y_i(k) - y_r(k)), \quad i = 1, 2, \cdots, N. \tag{7.130}$$

The internal model technique which will be employed to synthesize protocols has the important advantage that the parameters of the internal model are not sensitive to the controlled systems. It is also necessary to solve the output regulation for uncertain multiagent systems. Thus, the internal model servocompensator $z_i \in \Re^{p(s_m)}$ under the network topology is built as follows:

$$z_i(k+1) = G_1 z_i(k) + G_2 \phi_i \delta_i(k), \quad i = 1,\ 2,\ \cdots, N, \tag{7.131}$$

with $\phi_i \in \Re^+$, $i = 1, 2, \cdots, N$. The pair of matrices (G_1, G_2) is said to incorporate a p-copy internal model of matrix S with

$$\begin{aligned} G_1 &= \textbf{block diag}\{\beta_1, \beta_2, \cdots, \beta_p\}, \\ G_2 &= \textbf{block diag}\{\sigma_1, \sigma_2, \cdots, \sigma_p\}, \end{aligned}$$

for all $i = 1,\ \cdots,\ p$, and β_i is a constant square matrix; σ_i is a constant column vector such that $(\beta_i,\ \sigma_i)$ is controllable and the minimal polynomial of S divides the characteristic polynomial of β_i.

To solve the regulator problem, we give the following distributed consensus protocols:

$$u_i = K_1 \phi_i \left[\sum_{j \in \mathcal{N}_i} a_{ij}(x_i - x_j) + a_{i0} x_i \right] + K_2 z_i, \tag{7.132}$$

with $i = 1, 2, \cdots, N$, $K_1 \in \Re^{q \times n}$ and $K_2 \in \Re^{q \times p(s_m)}$ being the gain matrices to be designed later.

Let $x = [x_1^t, x_2^t, \cdots, x_N^t]^t$, $z = [z_1^t, z_2^t, \cdots, z_N^t]^t$, and $\tilde{\omega} = \mathbf{1}_N \otimes \omega$. Then using protocol (7.132), the uncertain discrete-time linear dynamics (7.126) and the internal model servocompensator (7.131) can be written as

$$\begin{aligned} x(k+1) &= (I_N \otimes \bar{A} + \Phi \mathbb{H} \otimes \bar{B} K_1) x(k) + (I_N \otimes \bar{B} K_2) z(k) + \bar{E} \tilde{\omega}(k), \\ z(k+1) &= (\Phi \mathbb{H} \otimes G_2 \bar{C}) x(k) + (I_N \otimes G_1) z(k) - (\Phi \mathbb{H} \otimes G_2 F) \tilde{\omega}(k), \end{aligned} \tag{7.133}$$

where

$$\begin{aligned} \Phi &= \mathbf{diag}\{\phi_1, \phi_2, \cdots, \phi_N\}, \\ \bar{E} &= \mathbf{block\ diag}\{\bar{E}_1, \bar{E}_2, \cdots, \bar{E}_N\}, \\ (\Phi \mathcal{A}_0 \otimes G_2 F)\tilde{\omega} &= (\Phi \mathbb{H} \otimes G_2 F)\tilde{\omega}. \end{aligned}$$

Also, let $\zeta = [x^t, z^t]^t$. Then the closed loop system can be written in a compact form as

$$\zeta(k+1) = \bar{A}_c \zeta(k) + \bar{E}_c \tilde{\omega}(k), \tag{7.134}$$

where

$$\bar{A}_c = \begin{bmatrix} I_N \otimes \bar{A} + \Phi \mathcal{H} \otimes \bar{B} K_1 & I_N \otimes \bar{B} K_2 \\ \Phi \mathcal{H} \otimes G_2 \bar{C} & I_N \otimes G_1 \end{bmatrix},$$

$$\bar{E}_c = \begin{bmatrix} \bar{E} \\ -\Phi \mathcal{H} \otimes G_2 F \end{bmatrix}.$$

Now it is fair to state *the robust cooperative output regulation problem of discrete-time multiagent systems is solved if there is the consensus protocol (7.132) such that*

1) *The nominal closed-loop system matrix*

$$A_{c0} = \begin{bmatrix} I_N \otimes A + \Phi \mathbb{H} \otimes B K_1 & I_N \otimes B K_2 \\ \Phi \mathbb{H} \otimes G_2 C & I_N \otimes G_1 \end{bmatrix}$$

is Schur.

2) *There exists an open neighborhood W of $\Delta = 0$. For the closed-loop system (7.134), the local tracking error $e_i(k) \to 0$ as $k \to \infty$, for any initial conditions $x_{i0} \in \Re^n$, $\omega_0 \in \Re^s$.*

The output regulation problem is solvable subject to the following assumptions:

A1) *the pair* (A, B) *is stabilizable;*
A2) *S has no eigenvalues in the interior of the unit circle in the z-plane;*
A3) *for all* $\lambda \in \sigma(G_1)$,

$$\textbf{rank}\begin{bmatrix} A-\lambda I_n & B \\ C & 0 \end{bmatrix} = n+p;$$

A4) *digraph* $\bar{\mathcal{G}}$ *contains a spanning tree and node* v_0 *as its root.*

Remark 7.18. If all the eigenvalues of S lay in the interior of the unit circle in the z-plane, the trajectories of ω will decay exponentially to zero and not affect the output regulation. Without loss of generality, we assume that all the eigenvalues lay on or outside the unit circle.

Lemma 7.6. *If Assumptions A1–A3 hold, the pair* (G_1, G_2) *incorporates a p-copy internal model of matrix S. Let*

$$\mathsf{A} = \begin{bmatrix} A & 0 \\ G_2 C & G_1 \end{bmatrix}, \quad \mathsf{B} = \begin{bmatrix} B \\ 0 \end{bmatrix}.$$

Then pair (A, B) *is stabilizable.*

Lemma 7.7. *Under Assumption A2, assume* (G_1, G_2) *incorporates a p-copy internal model of S. If the matrix equation*

$$\Pi_2 S = G_1 \Pi_2 + G_2 \Omega$$

has a solution Π_2, *then* $\Omega = 0$.

Lemma 7.8. *If* (A, B) *is stabilizable and the gain matrix K is defined as*

$$K = -(\mathsf{B}^t P \mathsf{B})^{-1} \mathsf{B}^t P \mathsf{A},$$

then

$$\rho = \frac{1}{\lambda_{\max}[\check{Q}^{-1} \mathsf{A}^t P \mathsf{B}(\mathsf{B}^t P \mathsf{B})^{-1} \mathsf{B}^t P \mathsf{A} \check{Q}^{-1}]}$$

with $P = P^t > 0$, $Q = Q^t > 0$, *and* $Q = \check{Q}\check{Q}$ *for symmetric matrix* $\check{Q} > 0$. *Matrix* $\mathsf{A} + s\mathsf{B}K$, $s \in \mathbf{C}$, *is stable if and only if s lies in the stability region*

$$\Psi = \{s \in \mathbf{C} : |s-1|^2 < \rho\}. \tag{7.135}$$

Proof. Since (A, B) is stabilizable, for any $Q = Q^t > 0$, the discrete-time algebraic Riccati equation

$$\mathrm{A}^t P \mathrm{A} - P - \mathrm{A}^t P \mathrm{B}(\mathrm{B}^t P \mathrm{B})^{-1} \mathrm{B}^t P \mathrm{A} + Q = 0 \tag{7.136}$$

has a unique solution $P = P^t > 0$.

Invoking the Lyapunov stability theorem, for $P > 0$, $\mathrm{A} + s\mathrm{B}K$ is stable for some K if and only if

$$(\mathrm{A} + s\mathrm{B}K)^* P(\mathrm{A} + s\mathrm{B}K) - P < 0.$$

Since the gain matrix K is defined as $K = -(\mathrm{B}^t P\mathrm{B})^{-1}\mathrm{B}^T P\mathrm{A}$, it follows that

$$\begin{aligned}
&(\mathrm{A} + s\mathrm{B}K)^* P(\mathrm{A} + s\mathrm{B}K) - P \\
&= \mathrm{A}^t P\mathrm{A} + s^* K^t \mathrm{B}^t P\mathrm{A} + s\mathrm{A}^t P\mathrm{B}K + ss^* K^t \mathrm{B}^t P\mathrm{B}K - P \\
&= \mathrm{A}^t P\mathrm{A} + (-s - s^* + ss^*)\mathrm{A}^t P\mathrm{B}(\mathrm{B}^t P\mathrm{B})^{-1}\mathrm{B}^t P\mathrm{A} - P \\
&= \mathrm{A}^t P\mathrm{A} - \mathrm{A}^t P\mathrm{B}(\mathrm{B}^t P\mathrm{B})^{-1}\mathrm{B}^t P\mathrm{A} - P \\
&+ (1 - s - s^* + ss^*)\mathrm{A}^t P\mathrm{B}(\mathrm{B}^t P\mathrm{B})^{-1}\mathrm{B}^t P\mathrm{A} \\
&= -Q + |s-1|^2 \mathrm{A}^T P\mathrm{B}(\mathrm{B}^T P\mathrm{B})^{-1}\mathrm{B}^T P\mathrm{A} < 0.
\end{aligned} \tag{7.137}$$

From Lemma 7.8, there exists a symmetric matrix $\check{Q}$ such that $Q = \check{Q}\check{Q}$ and (7.137) is rewritten as

$$-I_n + |s-1|^2\, \check{Q}^{-1}\mathrm{A}^t P\mathrm{B}(\mathrm{B}^t P\mathrm{B})^{-1}\mathrm{B}^t P\mathrm{A}\check{Q}^{-1} < 0.$$

This provides the stability region as

$$\Psi = \{s \in \mathbf{C} : |s-1|^2 < \rho\}, \tag{7.138}$$

with $\rho = \frac{1}{\lambda_{\max}[\check{Q}^{-1}\mathrm{A}^t P\mathrm{B}(\mathrm{B}^t P\mathrm{B})^{-1}\mathrm{B}^t P\mathrm{A}\check{Q}^{-1}]}$ as desired.

Now consider that s lies in the region Ψ, and if λ_{ri}, $i = 1, 2, \cdots, n$, are the eigenvalues of $\check{Q}^{-1}\mathrm{A}^t P\mathrm{B}(\mathrm{B}^t P\mathrm{B})^{-1}\mathrm{B}^t \times P\mathrm{A}\check{Q}^{-1}$, there is $\frac{1}{|s-1|^2} > \frac{1}{\rho} \geq \lambda_{ri}$, that is,

$$\lambda_{ri}\, |s-1|^2 - 1 < 0. \tag{7.139}$$

By Lemma 7.3, $\lambda_{ri}\, |s-1|^2 - 1$ are the eigenvalues of

$$-I_n + |s-1|^2\, \check{Q}^{-1}\mathrm{A}^t P\mathrm{B}(\mathrm{B}^t P\mathrm{B})^{-1}\mathrm{B}^t P\mathrm{A}\check{Q}^{-1}$$

and (7.139) implies that

$$-I_n + |s-1|^2\, \check{Q}^{-1}\mathrm{A}^t P\mathrm{B}(\mathrm{B}^T P\mathrm{B})^{-1}\mathrm{B}^t P\mathrm{A}\check{Q}^{-1} < 0. \tag{7.140}$$

Pre- and postmultiplying (7.140) by $\check{Q}$ yields

$$- Q + |s-1|^2 \mathrm{A}^t P\mathrm{B}(\mathrm{B}^t P\mathrm{B})^{-1}\mathrm{B}^t P\mathrm{A} \\ + (\mathrm{A} + s\mathrm{B}K)^* P(\mathrm{A} + s\mathrm{B}K) - P < 0.$$

Then $\mathrm{A} + s\mathrm{B}K$ is stable and the proof is completed. □

Theorem 7.9. *Under Assumptions A1–A4, if* $\phi_i = \frac{1}{a_{i0}+\bar{d}_i}$, $K = \begin{bmatrix} K_1 & K_2 \end{bmatrix} = -(\mathrm{B}^t P\mathrm{B})^{-1}\mathrm{B}^t P\mathrm{A}$, *and* $\sigma(\Phi\mathbb{H}) \subset \Psi$, *then the consensus robust output regulation problem can be solved by the distributed consensus protocol (7.132).*

Proof. Let

$$T_1 = \begin{bmatrix} I_{Nn} & 0 \\ 0 & (\Phi\mathbb{H}) \otimes I_{p(s_m)} \end{bmatrix}. \tag{7.141}$$

The nominal closed-loop system matrix A_{c0} can be transformed into

$$\begin{aligned} \Lambda_1 &= T_1^{-1} A_{c0} T_1 \\ &= \begin{bmatrix} I_N \otimes A + \Phi\mathcal{H} \otimes BK_1 & \Phi\mathcal{H} \otimes BK_2 \\ I_N \otimes G_2 C & I_N \otimes G_1 \end{bmatrix} \\ &= I_N \otimes \mathrm{A} + \Phi\mathcal{H} \otimes \mathrm{B}K, \end{aligned} \tag{7.142}$$

with

$$K = \begin{bmatrix} K_1 & K_2 \end{bmatrix}.$$

Denote the eigenvalues of $\Phi\mathbb{H}$ by λ_i, $i = 1, \cdots, N$. There exists a unitary matrix U such that $U^{-1}\Phi\mathcal{H}U = \mathcal{T} = [t_{ij}]$ is upper-triangular, with diagonal entries $t_{ii} = \lambda_i$, $i = 1, \cdots, N$. Let

$$T_2 = \begin{pmatrix} U \otimes I_n & 0 \\ 0 & U \otimes I_{p(s_m)} \end{pmatrix}. \tag{7.143}$$

Then

$$\begin{aligned} \Lambda_2 &= T_2^{-1} \Lambda_1 T_2 \\ &= \begin{bmatrix} I_N \otimes A + \mathcal{T} \otimes BK_1 & \mathcal{T} \otimes BK_2 \\ I_N \otimes G_2 C & I_N \otimes G_1 \end{bmatrix} \\ &= I_N \otimes \mathrm{A} + \mathcal{T} \otimes \mathrm{B}K. \end{aligned} \tag{7.144}$$

Due to the nature of matrix Λ_2, being either block-diagonal or block upper-triangular, Λ_2 is Schur if and only if $\mathrm{A} + \lambda_i \mathrm{B} K$, $i = 1, \cdots, N$, is Schur. By the *Gershgorin disc theorem*, all the eigenvalues λ_i, $i = 1, \cdots, N$, of $\Phi\mathbb{H}$ are located in the union of N discs

$$\bigcup_{i=1}^{N} \{z \in \mathbf{C} : \left|z - \phi_i(\bar{d}_i + a_{i0})\right| \leq \phi_i \bar{d}_i\}. \tag{7.145}$$

If $\phi_i = \frac{1}{a_{i0}+\bar{d}_i}$, then (7.145) can be expressed as

$$\bigcup_{i=1}^{N} \left\{z \in \mathbf{C} : |z-1| \leq \frac{\bar{d}_i}{a_{i0}+\bar{d}_i}\right\}, \tag{7.146}$$

which represents that the eigenvalues of $\Phi\mathbb{H}$ span the neighborhood of 1. By Lemmas 7.6 and 7.8, the stability region of $\mathrm{A} + s\mathrm{B}K$ is given by

$$\Psi = \left\{s \in \mathbf{C} : |s-1|^2 < \rho\right\}. \tag{7.147}$$

If all the eigenvalues λ_i, $i = 1, \cdots, N$, are located in Ψ, that is, $\sigma(\Phi H) \subset \Psi$, then $\mathrm{A} + \lambda_i \mathrm{B} K$, $i = 1, \cdots, N$, is Schur. Therefore, the nominal closed-loop system matrix A_{c0} is Schur.

For each $\Delta \in \mathbf{W}$, where $\mathbf{W}$ is an open neighborhood of $\Delta = 0$ such that $\bar{A}_c$ is Schur and under Assumption A2, $\lambda_i(S) + \lambda_j(\bar{A}_c) \neq 0$, $i = 1, \cdots, s$, $j = 1, \cdots, N(n+p(s_m))$. There exists a unique solution $\Pi \in \Re^{N(n+p(s_m))\times N(n+p(s_m))}$ of

$$\Pi S = \bar{A}_c \Pi + \bar{E}_c(1_N \otimes I_s). \tag{7.148}$$

We let

$$\Pi = \begin{bmatrix} \Pi_1 \\ \Pi_2 \end{bmatrix}, \quad \Pi_1 \in \Re^{Nn\times Nn}, \quad \Pi_2 \in \Re^{Np(s_m)\times Np(s_m)}. \tag{7.149}$$

It follows that (7.148) can be expressed as

$$\begin{aligned} \Pi_1 S &= (I_N \otimes \bar{A} + \Phi\mathbb{H} \otimes \bar{B} K_1)\Pi_1 \\ &\quad + (I_N \otimes B K_2)\Pi_2 + \bar{E}(1_N \otimes I_s), \\ \Pi_2 S &= (\Phi\mathbb{H} \otimes G_2 \bar{C})\Pi_1 + (I_N \otimes G_1)\Pi_2 \\ &\quad - (\Phi\mathbb{H} \otimes G_2 F)(1_N \otimes I_s) \\ &= (I_N \otimes G_2)(\Phi\mathbb{H} \otimes I_q)((I_N \otimes \bar{C})\Pi_1 - 1_N \otimes F) \\ &\quad + (I_N \otimes G_1)\Pi_2. \end{aligned} \tag{7.150}$$

Since $(I_N \otimes G_1, \; I_N \otimes G_2)$ incorporates a pN-copy internal model of S, under Lemma 7.7 and by the invertibility of $(\Phi\mathbb{H} \otimes I_q)$, one gets

$$(I_N \otimes \bar{C})\Pi_1 - 1_N \otimes F = 0. \tag{7.151}$$

Introducing $\hat{\zeta} = \zeta - \Pi\omega$ and evaluating the difference of $\hat{\zeta}$ yields

$$\hat{\zeta}(k+1) = \bar{A}_c\zeta(k) + \bar{E}_c(1_N \otimes I_s)\omega(k) - \Pi S\omega = \bar{A}_c\hat{\zeta}(k). \tag{7.152}$$

We define $e = [e_1^t, e_2^t, \cdots, e_N^t]^t$. The tracking errors (7.130) have the following form:

$$\begin{aligned}
e &= \left[(I_N \otimes \bar{C})x - (1_N \otimes F)\right]\omega \\
&= \begin{bmatrix} I_N \otimes \bar{C} & 0 \end{bmatrix}\zeta - (1_N \otimes F)\omega \\
&= \begin{bmatrix} I_N \otimes \bar{C} & 0 \end{bmatrix}(\hat{\zeta} + \Pi\omega) - (1_N \otimes F)\omega \\
&= \begin{bmatrix} I_N \otimes \bar{C} & 0 \end{bmatrix}\hat{\zeta} + ((I_N \otimes \bar{C})\Pi_1 - 1_N \otimes F)\omega.
\end{aligned}$$

From (7.151), one gets

$$e = \begin{bmatrix} I_N \otimes \bar{C} & 0 \end{bmatrix}\hat{\zeta}.$$

Since $\bar{A}_c$ is Schur, $\hat{\zeta} \to 0$, and $k \to \infty$, we get

$$\lim_{k\to\infty} e = \lim_{k\to\infty}\begin{bmatrix} I_N \otimes \bar{C} & 0 \end{bmatrix}\hat{\zeta} = 0.$$

That is, the solution of the consensus robust output regulation problem is completed. □

Remark 7.19. For the system dynamics (7.126), if $\Delta = 0$, the robust output regulation problem for uncertain discrete-time multiagent systems becomes an output regulation problem for certain systems. It is crucial to emphasize that for discrete-time multiagent systems, it is quite hard to find a gain K to make the closed-loop system matrix in the form of $\mathrm{A} + \mathbb{H} \otimes \mathrm{B}K$ stable without considering the relationship between system matrix A and Laplacian matrix $\mathbb{H}$ of the information graph. More importantly, parameters ϕ_i are used to regulate all the eigenvalues of $\mathbb{H}$ to lie in the neighborhood of 1, which can reduce the conservatism.

7.4.3 Simulation example 7.3

An example is now given to validate the effectiveness of the theoretical result. Consider a network of discrete-time multiagent systems described by (7.126) with the exosystem in the form of (7.128) with $\omega = (\omega_1, \omega_2, \omega_3)^t$ and

$$S = \begin{bmatrix} 0 & 0 & 0 \\ 1 & 0 & -1 \\ 0 & 1 & 0 \end{bmatrix}, \quad F = \begin{bmatrix} 1 & 0 & 0 \end{bmatrix}.$$

By the exosystem matrix S, we find a pair of matrices

$$G_1 = \begin{bmatrix} 0 & 1 & 0 \\ -1 & 0 & -1 \\ 0 & 0 & 0 \end{bmatrix}, \quad G_2 = \begin{bmatrix} 0 \\ 0 \\ 1 \end{bmatrix}$$

which incorporate a 1-copy internal model of S. It is easy to check that Assumptions A1–A3 hold.

The system topology is described by the digraph shown in Fig. 7.1 with $\bar{\mathbb{V}} = (v_0, v_1, \cdots, v_7)$ and the adjacent weighted values are also shown in Fig. 7.11. Obviously, digraph $\bar{\mathbb{G}}$ contains a spanning tree and node 0 as its root. The parametric matrix $\Phi =$ **block diag**$\{1, 1, 1, 1/4, 1/2, 1, 1/4\}$. By some simple calculations, the stability region is shown as

$$\Psi = \{s \in \mathbf{C} : |s - 1|^2 < 0.2981\}.$$

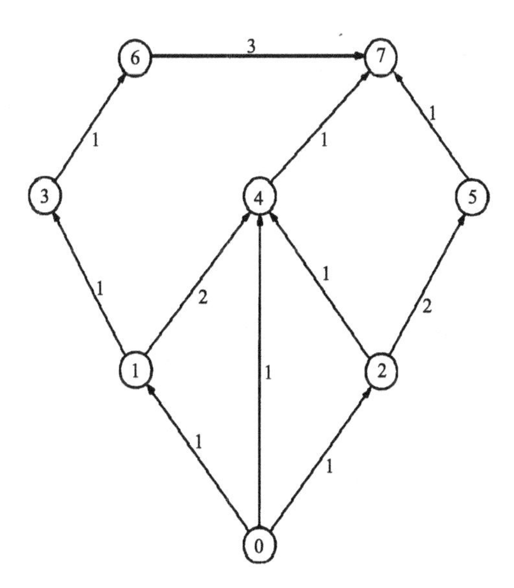

FIGURE 7.11 Communication topology.

The control gain $K_1 = \begin{bmatrix} -0.4514 & 0 \end{bmatrix}$, and $K_2 = \begin{bmatrix} 0 & -0.5486 & 0 \end{bmatrix}$ which can make $\bar{A}_c$ stable in an open neighborhood of $\Delta = 0$. Thus, we obtain the simulation results as shown in Fig. 7.12. The regulated errors converge to the origin asymptotically.

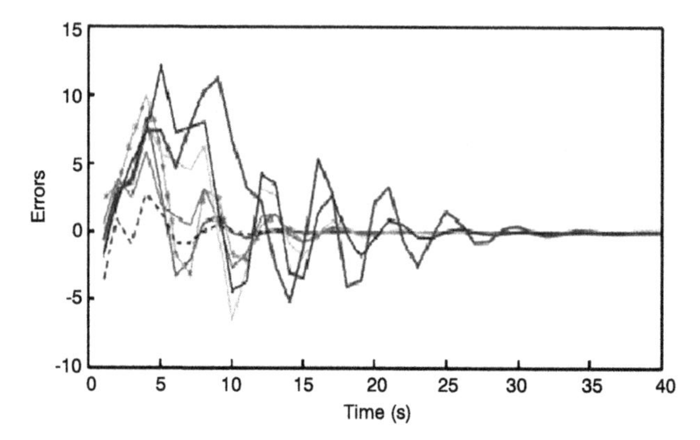

FIGURE 7.12 Regulated state errors.

7.5 Distributed networked control

We learned before that distributed control systems (DCSs) include shared communication networks to connect all components including controllers, sensors, and actuators. In other words, DCSs connect physicals parts with the cyber parts to allow distributed remote/estimation control tasks [24], [25], and [26].

Networked control systems implement a periodic control scheme at the early stage based on the sampled-data system theory which implements periodic signal regardless of the behavior of the system. Small sampling periods were used in this scheme to guarantee the desired performance in the worst circumstances. This way of design leads to huge redundant sampled data even in the steady state period where there is no need to carry out all these communications, causing a waste of system resources.

ETC is proposed to moderate the waste of system resources by minimizing communication of signals. In this scheme, the signal is transmitted based on the occurrences of an event causing the required change in the control behavior. As a result, the system resources are utilized while maintaining the desired performance of the system.

7.5.1 Introduction

Distributed networked control systems (DNCSs) have attracted acute attention in academic and industrial research since they are considered in several fields, including control systems, computer science, and communication networks. The DNCS consists of a number of simple interacting subsystems, and each subsystem uses a shared communication network to exchange its signals with other systems units. To reduce these communications, the ETC scheme is widely applied in DNCSs; some examples are given in [34] and [35].

Most of the event triggering literature considers the state feedback control. However, it is known that the full state is not available in many industrial applications. So, it is required to implement an output-feedback controller in such scenarios. Event-triggered dynamic output-feedback control was implemented in [27] to build an event-triggered scheme. The L_∞ gain and stability were presented using LMI conditions. An output feedback configuration was designed by [28] based on event triggering using the Lyapunov function between the sampling events. This work was extended by [29] for acyclically interconnected systems. Some studies, such as [30] and [31], applied event-based Kalman filtering for estimating the states from the event-sampled measured output. Other examples of output feedback event triggering control were presented in [32] and [33].

7.5.2 Problem statement

In the sequel, we study a distributed networked system (DNS) with N agents/subsystems. Each subsystem consists of physical and cyber components and focuses on the output feedback scenario. This implies partial information about the states is available, and local observers are designed to estimate the other states. The observer estimated states can be shared to the appropriate neighboring subsystem according to the overall control objectives of the distributed system. It is assumed that the agents have interacting dynamics, and thus require dynamic information from the neighboring agents.

The approach adopted hereinafter considers the output feedback of DNSs where sensors, actuators, and controllers are connected over shared communication networks. The distributed event triggering depends on local information at the agents and thus relaxes the synchronization requirement among the agents.

7.5.3 Distributed feedback scheme

This section shows how to implement the distributed output feedback scheme for linear DNSs. The notion of ℓ_2-stability is used for stability analysis. Consider the DNS with individual agent i dynamics presented as

$$\begin{aligned} x_i(k+1) &= A_{ii}x_i(k) + B_i u_i(k) + \sum_{j\in D_i} A_{ij}x_j(k) + \Gamma_i w_i(k), \\ y_i(k) &= C_{ii}x_i(k) + \sum_{j\in Zi} C_{ij}x_j(k). \end{aligned} \tag{7.153}$$

The structure of the distributed observer is written as

$$\tilde{x}_i(k) = A_{ii}\tilde{x}_i(k) + B_i u_i(k) + L_i[y_i - \tilde{y}_i],$$

$$\tilde{y}_i(k) = C_{ii}\tilde{x}_i(k) + \sum_{j \in Z_i} C_{ij}\tilde{x}_j(k), \tag{7.154}$$

where x_i, y_i, and u_i are physical variables representing physical coupling, $\tilde{x}_i$ and $\tilde{y}_i$ are observer variables, and $\hat{y}_i$ is a networked variable. Note that $\tilde{x}$ is used to represent the locally observed variable, while $\hat{x}$ is used to represent the variable received over the communication network.

The interacting dynamics of the DNCS require that the control action at the local subsystem should be calculated based on the locally observed dynamics as well as on the information dynamics of the interacting subsystem. Thus, the controller at the subsystem i is given by

$$u_i(k) = K_{ii}\hat{x}_i(k) + \sum_{j \in Z_i} K_{ij}\hat{x}_j(k), \tag{7.155}$$

where K_{ii}, and K_{ij} are the control gain matrices corresponding to the states $\tilde{x}_i(k)$ and $\hat{x}_i(k)$, respectively, in the interacting set $\mathbb{Z}_i$. We assume homogeneous agents, that is, the individual dynamics of each agent are of the same structure. Therefore, we can combine the overall DNCS system into the following notation for unified analysis and design purposes. The overall system is represented by

$$x(k+1) = Ax(k) + Bu(k) + \Gamma w(k), \tag{7.156}$$

$$y(k) = Cx(k), \tag{7.157}$$

where $x = [x_1^t, \dots, x_N^t]^t$, $u = [u_1^t, \dots, u_N^t]^t$, $w = [w_1^t, \dots, w_N^t]^t$, and $\hat{x} = [\hat{x}_1^t, \dots, \hat{x}_N^t]^t$. The overall distributed observer system is given by

$$\tilde{x}(k+1) = A_d\tilde{x}(k) + Bu(k) + L[y - \tilde{y}], \tag{7.158}$$

$$\tilde{y}(k) = C\tilde{x}(k), \tag{7.159}$$

where $x = [x_1^t, \dots, x_N^t]^t$, $u = [u_1^t, \dots, u_N^t]^t$, $w = [w_1^t, \dots, w_N^t]^t$, and $\hat{x} = [\hat{x}_1^t, \dots, \hat{x}_N^t]^t$, and the distributed protocols take the following form:

$$u(k) = K\hat{x}(k), \tag{7.160}$$

where $u = [u_1^t, \dots, u_N^t]^t$, $\tilde{x} = [\tilde{x}_1^t, \dots, \tilde{x}_N^t]^t$, $\hat{x} = [\hat{x}_1^t, \dots, \hat{x}_N^t]^t$, $K = (K_1, \dots, K_N)$.

In the sequel, we let an error $e_i = [k_0, \infty) \to \Re^n$ be defined as

$$e_i(k) \triangleq \tilde{x}_i(k) - \hat{x}_i(k), \quad \forall k \geq k_0. \tag{7.161}$$

This represents the error between the system's current state estimate and its last transmitted value. Thus the event condition-based inequality is expressed as

$$\|e\| \leq \alpha\|\hat{x}\|, \ \alpha > 0. \tag{7.162}$$

Remark 7.20. In the sequel, our strategy is to communicate the outputs of the system which reduces the communication burden and network traffic. We assume knowledge about the states and use local observers at the local subsystems for estimating the other states. The event triggering uses the observed states instead of the exact measurements but the controller here is designed based only on the broadcast states.

7.5.4 Reliable observer-based protocol

An observer-based protocol, as presented in (7.158), (7.159), and (7.160), is to be designed to render the distributed system (7.156) stable. We will thoroughly investigate the stability analysis and observer-based design for the interconnected agents in the form of (7.153). The results are summarized below.

Theorem 7.10. *Consider the DNS represented by (7.158), (7.159), and (7.160). Given scalars γ, ϵ, and $\alpha \in R^+$, assume that there exist positive definite block-diagonal matrices $\Omega_1, \Omega_2, \Omega_3, P, P_0 > 0$ and X_2, X_2, Y_1, Y_2 that satisfy the following inequalities:*

$$\begin{bmatrix} -\Omega_1 & 0 & 0 & 0 & A^t P \\ 0 & -\Omega_2 & 0 & 0 & X_1^t \\ 0 & 0 & -\Omega_3 & 0 & Y_1^t \\ * & * & * & -\epsilon I & \Gamma^t P \\ * & * & * & * & -P \end{bmatrix} \leq 0, \tag{7.163}$$

$$\begin{bmatrix} -P + \Omega_1 & Y_2^t \\ * & -P_0 \end{bmatrix} \leq 0, \tag{7.164}$$

$$\begin{bmatrix} -P_0 - \alpha I + \Omega_2 & A_d^t P_0 & X_2^t & Y_2^t \\ * & -3P_0 & 2P_0 & -2P_0 \\ * & * & -\epsilon I & 2P_0 \\ * & * & * & -\epsilon I \end{bmatrix} \leq 0, \tag{7.165}$$

$$\begin{bmatrix} -(1-\alpha)I + \Omega_3 & X_2^t \\ * & -P_0 \end{bmatrix} \leq 0. \tag{7.166}$$

The controller and observer gains are given by $K = (B^\dagger)PX_1$ and $L = P_0 Y_2 (C^\dagger)$.

Proof. Consider the Lyapunov function $V(k) = x^t(k)Px(k) + \tilde{x}^t(k)P_0\tilde{x}(k)$. Along the solutions of (7.158) and (7.160), we get

$$\begin{aligned}
\Delta V(k) &= V(k+1) - V(k) \\
&= x^t(k+1)Px(k+1) - x^t(k)Px(k) \\
&+ \tilde{x}^t(k+1)P_0\tilde{x}(k+1) - \tilde{x}^t(k)P_0\tilde{x}(k) \\
&= \left[x^t(k)A^t + u^t(k)B^t + w^t(k)\Gamma^t\right]P \\
&\times \left[Ax(k) + Bu(k) + \Gamma w(k)\right] - x^t(k)Px(k) \\
&+ \left[\tilde{x}^t(k)A_d^t + u^t(k)B^t + x^t(k)C^tL^t - \tilde{x}^t(k)C^tL^t\right]P_0 \\
&\times \left[A\tilde{x}(k) + Bu(k) + LCx(k) - LC\tilde{x}(k)\right] \\
&- \tilde{x}^t(k)P_0\tilde{x}(k).
\end{aligned} \tag{7.167}$$

Using (7.161), the protocol $u(k) = K\hat{x}(k)$, and $\hat{x}(k) = \tilde{x}(k) - e(k)$ and implementing the event condition-based inequality (7.162), the Lyapunov difference can be expressed as

$$\begin{aligned}
\Delta V(k) &= x^t(k)\left[A^tPA - P + C^tL^tP_0LC\right]x(k) \\
&+ \tilde{x}(k)\left[K^tB^tPBK + K^tB^tP_0BK\right. \\
&+ A_d^tP_0A_d - 2A_d^tP_0LC - P_0 + C^tL^tP_0LC \\
&\left.+ 2A_d^tP_0BK - C^tL^tP_0BK\right]\tilde{x}(k) \\
&+ e^t(k)\left[K^tB^tPBK + K^tB^tP_0BK\right]e(k) \\
&+ w^t(k)\Gamma^tP\Gamma w(k) \\
&- 2\tilde{x}^t(k)\left[K^tB^tPBK + K^tB^tP_0BK\right]e(k) \\
&+ 2x^t(k)\left[A^tPBK + C^tL^tP_0BK\right]\tilde{x}(k) \\
&- 2x^t(k)\left[A^tPBK + C^tL^tP_0BK\right]e(k) \\
&+ 2x^t(k)A^tP\Gamma w(k) + 2\tilde{x}^t(k)K^tB^tP\Gamma w(k) \\
&- 2e^t(k)K^tB^tP\Gamma w(k) \\
&- 2\tilde{x}^t(k)\left[A_d^tP_0BK - C^tL^tP_0BK\right]e(k) \\
&+ 2\tilde{x}^t(k)\left[A_d^tP_0LC - C^tL^tP_0LC\right]x(k) \\
&+ e^t(k)e(k) - \alpha e^t(k)e(k) + \alpha e^t(k)\tilde{x}(k) \\
&- \alpha\tilde{x}^t(k)\tilde{x}(k) + \alpha\tilde{x}^t(k)e(k).
\end{aligned} \tag{7.168}$$

In turn, this can be written into the compact form

$$\Delta V(k) \le \zeta^t\Xi\zeta, \quad \zeta = [x^t, \tilde{x}^t, e^t, w^t]^t,$$

$$\Xi = \begin{bmatrix} \Xi_1 & \Xi_2 & \Xi_3 & A^t P\Gamma \\ \bullet & \Xi_4 & \Xi_5 & K^t B^t P\Gamma \\ \bullet & \bullet & \Xi_6 & -K^t B^t P\Gamma \\ \bullet & \bullet & \bullet & \Gamma^t P\Gamma \end{bmatrix},$$

$$\begin{aligned}
\Xi_1 &= A^t PA - P + C^t L^t P_0 LC, \\
\Xi_2 &= A^t PBK + C^t L^t P_0 BK + C^t L^t P_0 A_d - C^t L^t P_0 LC, \\
\Xi_3 &= -A^t PBK - C^t L^t P_0 BK, \\
\Xi_4 &= K^t B^t PBK + K^t B^t P_0 BK \\
&\quad + A_d^t P_0 A_d - 2A_d^t P_0 LC - P_0 + C^t L^t P_0 LC \\
&\quad + 2A_d^t P_0 BK - C^t L^t P_0 BK - \alpha I, \\
\Xi_5 &= -K^t B^t PBK - K^t B^t P_0 BK - A_d^t P_0 BK \\
&\quad + C^t L^t P_0 BK + \alpha I, \\
\Xi_6 &= K^t B^t PBK + K^t B^t P_0 BK + (1-\alpha) I, \qquad (7.169)
\end{aligned}$$

where $\bullet$ stands for symmetric terms. A sufficient condition for stability is that $\Delta V(k) \leq 0$, which in turn is guaranteed by $\Xi \leq 0$; Ξ could be rewritten as

$$\Omega + \Theta P \Theta^t,$$

$$\Theta = \begin{bmatrix} A & BK & -BK & \Gamma \end{bmatrix}^t, \qquad (7.170)$$

$$\Omega = \text{diag}\{-\Omega_1 \; -\Omega_2 \; -\Omega_3 \; -\epsilon I\}, \qquad (7.171)$$

where

$$-\Omega_1 = -P + C^t L^t P_0 LC, \qquad (7.172)$$

$$\begin{aligned}
-\Omega_2 &= A_d^t P_0 A_d + K^t B^t P_0 BK + C^t L^t P_0 LC \\
&\quad + 2A_d^t P_0 BK - 2A_d^t P_0 LC - C^t L^t P_0 BK \\
&\quad - P_0 - \alpha I, \qquad (7.173)
\end{aligned}$$

$$-\Omega_3 = K^t B^t P_0 BK + (1-\alpha) I. \qquad (7.174)$$

Using the Schur complement, (7.171) could be rewritten in the form of (7.163) with $X_1 = K^t B^t P$ and $Y_1 = K^t B^t P_0$. Also, with $X_2 = Y_2 = L^t C^t P$ and $Y_2 = L^t C^t P_0$ it is straightforward to obtain LMIs (7.164) and (7.166) from (7.172) and (7.174), respectively. Finally, LMI (7.165) could be obtained by formulating (7.173) as follows:

$$\Omega_2 = \begin{bmatrix} P_0 A_d \\ P_0 BK \\ P_0 LC \end{bmatrix}^t \begin{bmatrix} P_0^{-1} & P_0^{-1} & -P_0^{-1} \\ P_0^{-1} & P_0^{-1} & -0.5P_0^{-1} \\ -P_0^{-1} & -0.5P_0^{-1} & P_0^{-1} \end{bmatrix} \begin{bmatrix} P_0 A_d \\ P_0 BK \\ P_0 LC \end{bmatrix} - (P_0 + \alpha I).$$

□

7.5.5 Design procedure

The problem of designing the event-triggered output feedback controller is solved in Theorem 7.10 as an LMI feasibility problem. The procedure can be stated in the following steps:

Decentralized event triggering scheme

1. Identify the A, A_d, B, C, and D matrices of the overall distributed system.
2. Choose $\alpha > 0$.
3. Find a feasible point for the set of LMIs (7.163), (7.164), (7.165), (7.166).
4. If stopping criteria are satisfied, exit. Otherwise choose another $\alpha > 0$, and go to step 2.
5. Calculate the controller and observer gains by $K = (B^{\dagger})PX_1$ and $L = P_0 Y_2 (C^{\dagger})$.

7.5.6 Simulation example 7.4

The effectiveness of the developed method is illustrated by considering one of the common applications of cyber-physical systems, i.e., the quadruple-tank process. As shown in Fig. 7.13, the system consists of four tanks (two upper and two lower) and our objective is to control the level in the lower two tanks with

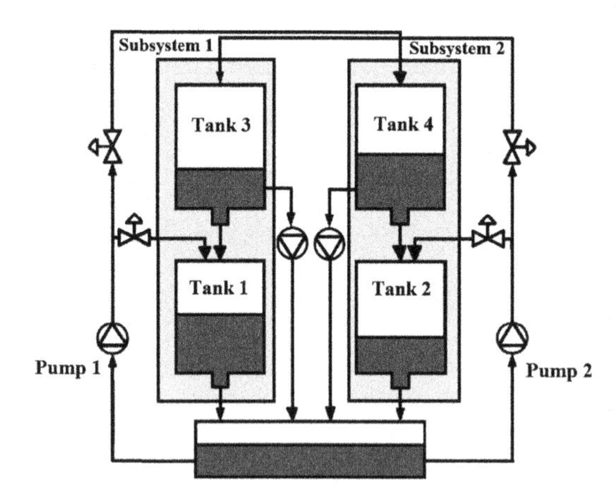

FIGURE 7.13 Schematic diagram of a quadruple-tank system.

two pumps. The process has two inputs (input voltages to the pumps) and two outputs (voltages from level measurement devices). Following are the system matrices for the linearized discrete-time state-space model of the system:

$$A_{11} = \begin{bmatrix} 0.9998 & 0 \\ 0 & 0.9998 \end{bmatrix}, \quad B_1 = \begin{bmatrix} 0.6359(10)^{-3} \\ 0.4559(10)^{-3} \end{bmatrix},$$

$$A_{12} = \begin{bmatrix} 0 & 0.0003 \\ 0 & 0 \end{bmatrix}, \qquad C_{11} = \begin{bmatrix} 1 & 0 \end{bmatrix},$$

$$A_{22} = \begin{bmatrix} 0.9999 & 0 \\ 0 & 0.9997 \end{bmatrix}, \quad B_2 = \begin{bmatrix} 0.488(10)^{-3} \\ 0.6279(10)^{-3} \end{bmatrix},$$

$$A_{21} = \begin{bmatrix} 0 & 0.0002 \\ 0 & 0 \end{bmatrix}, \qquad C_{22} = \begin{bmatrix} 1 & 0 \end{bmatrix}.$$

The gains of the controller and observer were obtained using YALMIP as follows:

$$K = \begin{bmatrix} -20.3236 & -14.5800 & 0.0248 & 0.0571 \\ 0.0546 & 0.0257 & -15.0746 & -19.4836 \end{bmatrix},$$

$$L = \begin{bmatrix} 0.0028 & 0 \\ 0 & 0 \\ 0 & 0.0008 \\ 0 & 0 \end{bmatrix}. \tag{7.175}$$

Fig. 7.14 shows the asymptotic stability of all states. The control signals for the subsystems, which vary before the system approaches its equilibrium, are shown in Fig. 7.15. Fig. 7.16 shows the absolute error between the subsystems' outputs and the observers' outputs.

The smooth behavior of the illustrated example shows the effectiveness of the developed approach.

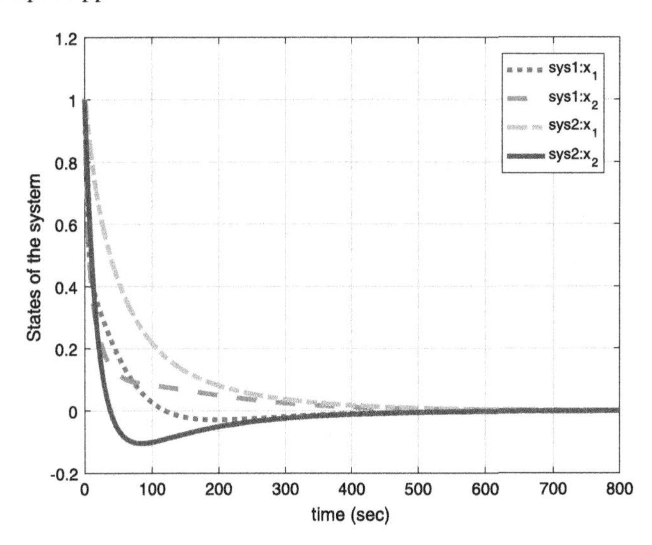

FIGURE 7.14 States of the subsystems.

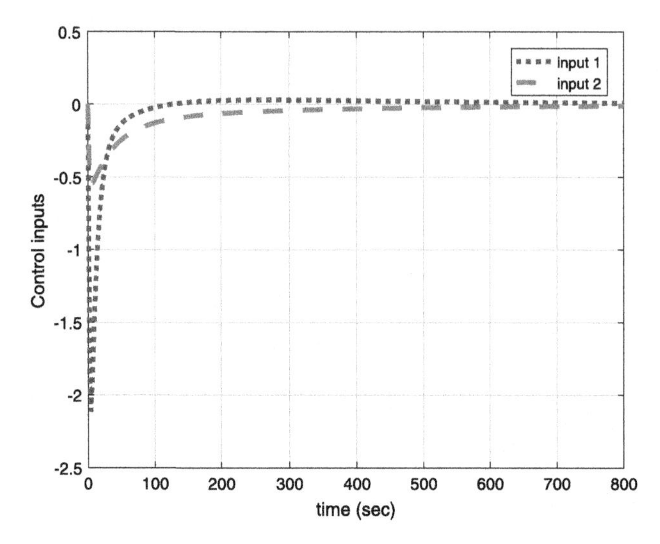

FIGURE 7.15 Signals of the inputs.

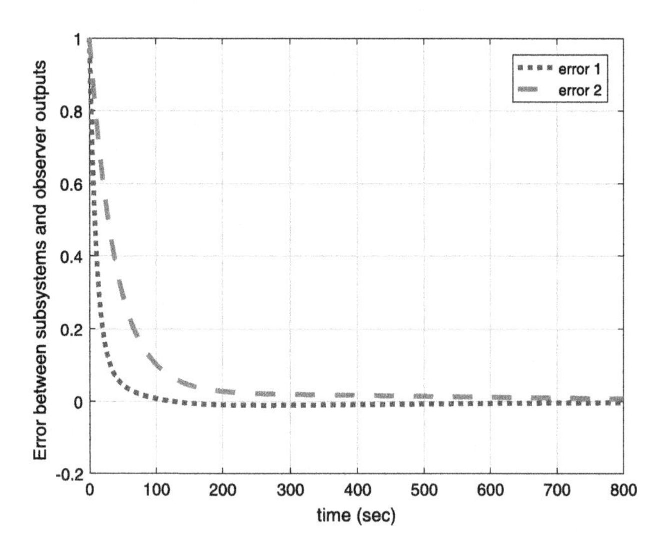

FIGURE 7.16 Error between subsystems' and observers' output.

7.6 Notes

In this chapter, the stochastic synchronization issues have been considered for a class of discrete-time complex dynamical networks with nonlinear coupling and exogenous disturbances. The notion of quasisynchronization in probability, which takes both the error bound and the synchronization probability into account, has been introduced to evaluate the synchronization quality of the coupled networks. The pinning feedback control strategy has been used to control part of the nodes and drive the whole SCDN asymptotically to synchronize the target motion. The event-triggered scheme has also been utilized to reduce the burden of updating of controller. Finally, a numerical example has been provided to show the power of theoretical approaches proposed in this chapter. The criteria proposed in this chapter provide some new insights for the possible applications in the real world, since the theoretical advance of SCDNs is a prerequisite and a significant step to brain science, and the quasisynchronization in probability of chaotic semiconductor lasers could be applied to secure communication in noisy channels. Moreover, with the rapid development of networked control technologies, the event-triggered pinning control strategy can be applied to the formation of unmanned aerial vehicles, attitude synchronization of spacecraft, and distributed sensor networks. We have also investigated the stability of networked systems.

Then, an observer-type protocol based on the relative outputs of neighboring agents has been proposed, which can be seen as an extension of the traditional observer-based controller for a single system to one for multiagent systems. The consensus of high-dimensional multiagent systems with directed communication topologies can be converted into the stability of a set of matrices with the same low dimension as that of a single agent. The notion of discrete-time consensus region has been introduced and analyzed. For neurally stable agents, an algorithm has been presented to construct a protocol having a bounded consensus region in the form of the open unit disk. Moreover, for unstable agents, another algorithm has also been proposed to construct a protocol having an origin-centered disk as its consensus region. The consensus algorithms have been further applied to solve formation control problems of multiagent systems.

To reduce the communication burden of multiagent networks, event-triggered schemes have been introduced to study the tracking problem for a class of discrete-time multiagent systems with a time-varying reference state where the agents communicate with their local neighbors at discrete-time instants. The control actuation updates considered in this chapter were event-driven, depending on certain measurement errors with respect to the states of agents and the leader. A numerical example was presented to demonstrate the effectiveness of the theoretical results. Throughout, the communication graph was assumed to be fixed and no communication delay was taken into account for agents. Future study of the tracking problem will involve extending the proposed approach to more general dynamic models such as systems with switching topologies and with communication time delays.

References

[1] S.H. Strogatz, Exploring complex networks, Nature 410 (Mar. 2001) 268–276.

[2] A.-L. Barabási, R. Albert, Emergence of scaling in random networks, Science 286 (5439) (1999) 509–512.

[3] Y. Li, X. Wu, J.-A. Lu, J. Lü, Synchronizability of duplex networks, IEEE Trans. Circuits Syst. II, Express Briefs 63 (2) (Feb. 2016) 206–210.

[4] X.F. Wang, G. Chen, Complex networks: small-world, scale-free and beyond, IEEE Circuits Syst. Mag. 3 (1) (Sep. 2003) 6–20.
[5] J. Zhou, J. Chen, J. Lu, J. Lü, On applicability of auxiliary system approach to detect generalized synchronization in complex network, IEEE Trans. Autom. Control 62 (7) (Jul. 2017) 3468–3473.
[6] A. Arenas, D.G.J. Kurths, Y. Moreno, C. Zhou, Synchronization in complex networks, Phys. Rep. 469 (3) (2008) 93–153.
[7] X. Mu, Y. Chen, Synchronization of delayed discrete-time neural networks subject to saturated time-delay feedback, Neurocomputing 175 (Jan. 2016) 293–299.
[8] W.L. He, F. Qian, J. Lam, G.R. Chen, Q.-L. Han, J. Kurths, Quasi-synchronization of heterogeneous dynamic networks via distributed impulsive control: error estimation, optimization and design, Automatica 62 (Dec. 2015) 249–262.
[9] J. Zhao, D.J. Hill, T. Liu, Global bounded synchronization of general dynamical networks with nonidentical nodes, IEEE Trans. Autom. Control 57 (10) (Oct. 2012) 2656–2662.
[10] B. Shen, Z. Wang, X. Liu, Sampled-data synchronization control of dynamical networks with stochastic sampling, IEEE Trans. Autom. Control 57 (10) (Oct. 2012) 2644–2650.
[11] W. Zhang, Z. Wang, Y. Liu, D. Ding, F.E. Alsaadi, Event-based state estimation for a class of complex networks with time-varying delays: a comparison principle approach, Phys. Lett. A 381 (1) (2017) 10–18.
[12] B. Shen, Z. Wang, X. Liu, Bounded $\mathcal{H}_\infty$ synchronization and state estimation for discrete time-varying stochastic complex networks over a finite horizon, IEEE Trans. Neural Netw. 22 (1) (Jan. 2011) 145–157.
[13] J. Lu, J. Kurths, J. Cao, N. Mahdavi, C. Huang, Synchronization control for nonlinear stochastic dynamical networks: pinning impulsive strategy, IEEE Trans. Neural Netw. Learn. Syst. 23 (2) (Feb. 2012) 285–292.
[14] D. Ding, Z. Wang, B. Shen, G. Wei, Event-triggered consensus control for discrete-time stochastic multi-agent systems: the input-to-state stability in probability, Automatica 62 (Dec. 2015) 284–291.
[15] Q. Liu, Z. Wang, X. He, D.H. Zhou, Event-based $\mathcal{H}_\infty$ consensus control of multi-agent systems with relative output feedback: the finite-horizon case, IEEE Trans. Autom. Control 60 (9) (Sep. 2015) 2553–2558.
[16] L. Sheng, Z. Wang, L. Zou, F. Alsaadi, Event-based $\mathcal{H}_\infty$ state estimation for time-varying stochastic dynamical networks with state- and disturbance-dependent noises, IEEE Trans. Neural Netw. Learn. Syst. 28 (10) (Oct. 2017) 2382–2394.
[17] A. Wang, T. Dong, X. Liao, Event-triggered synchronization strategy for complex dynamical networks with the Markovian switching topologies, Neural Netw. 74 (Jun. 2016) 52–57.
[18] C.Q. Ma, J.F. Zhang, Necessary and sufficient conditions for consensusability of linear multi-agent systems, IEEE Trans. Autom. Control 55 (2010) 1263–1268.
[19] J.H. Wang, D.Z. Cheng, X.M. Hu, Consensus of multi-agent linear dynamic systems, Asian J. Control 10 (2008) 144–155.
[20] Z.-P. Jiang, Y. Wang, Input-to-state stability for discrete-time nonlinear systems, Automatica 37 (6) (Jun. 2001) 857–869.
[21] Z.-P. Jiang, Y. Wang, A converse Lyapunov theorem for discrete-time systems with disturbances, Syst. Control Lett. 45 (1) (2002) 49–58.
[22] H.K. Khalil, Nonlinear Systems, Prentice-Hall, Upper Saddle River, NJ, USA, 2002.
[23] S.J. Liu, J.F. Zhang, Output-feedback control of a class of stochastic nonlinear systems with linearly bounded unmeasurable states, Int. J. Robust Nonlinear Control 18 (6) (2008) 665–687.
[24] J. Qiu, H. Gao, S.X. Ding, Recent advances on fuzzy-model-based nonlinear networked control systems: a survey, IEEE Trans. Ind. Electron. 63 (2) (2015) 1207–1217.
[25] C. Peng, F. Li, A survey on recent advances in event-triggered communication and control, Inf. Sci. 457 (2018) 113–125.
[26] M.S. Mahmoud, M.M. Hamdan, Fundamental issues in networked control systems, IEEE/CAA J. Autom. Sin. 5 (5) (2018) 902–922.

[27] M.S. Mahmoud, M. Sabih, M. Elshafei, Event-triggered output feedback control for distributed networked systems, ISA Trans. 60 (2016) 294–302.
[28] J. Almeida, C. Silvestre, A.M. Pascoal, Observer based self-triggered control of an acyclic interconnection of linear plants, in: Proc. the 51st IEEE Conference on Decision and Control (CDC), 2012, pp. 7553–7558.
[29] J. Almeida, C. Silvestre, A.M. Pascoal, Self-triggered output feedback control of linear plants in the presence of unknown disturbances, IEEE Trans. Autom. Control 59 (11) (2014) 3040–3045.
[30] L. Li, M. Lemmon, Event-triggered output feedback control of finite horizon discrete-time multi-dimensional linear processes, in: Proc. the 49th IEEE Conference on Decision and Control (CDC), 2010, pp. 3221–3226.
[31] M. Lemmon, et al., Weakly coupled event triggered output feedback system in wireless networked control systems, Discrete Event Dyn. Syst. 2 (24) (2014) 247–260.
[32] D. Lehmann, J. Lunze, Event-based output-feedback control, in: Proc. the 19th Mediterranean Conference on Control & Automation (MED), 2011, pp. 982–987.
[33] H. Yu, P.J. Antsaklis, Event-triggered output feedback control for networked control systems using passivity: achieving ℓ_2 stability in the presence of communication delays and signal quantization, Automatica 49 (1) (2013) 30–38.
[34] M. Andreasson, D.V. Dimarogonas, H. Sandberg, K.H. Johansson, Distributed control of networked dynamical systems: static feedback, integral action and consensus, IEEE Trans. Autom. Control 59 (7) (2014) 1750–1764.
[35] D.V. Dimarogonas, E. Frazzoli, K.H. Johansson, Distributed event-triggered control for multi-agent systems, IEEE Trans. Autom. Control 57 (5) (2011) 1291–1297.
[36] Y. Cao, Y. Li, W. Ren, Y. Chen, Distributed coordination of networked fractional-order systems, IEEE Trans. Syst. Man Cybern., Part B, Cybern. 40 (2010) 362–370.
[37] D. Dimarogonas, E. Frazzoli, Distributed event-triggered control strategies for multi-agent systems, in: 47th Annual Allerton Conference on Communication, Control and Computing, Monticello, IL, USA, 2009, pp. 906–910.
[38] L. Fang, P. Antsaklis, Information consensus of asynchronous discrete-time multi-agent systems, in: IEEE Proceedings of the 2005 American Control Conference, Portland, USA, 2005, pp. 1883–1888.
[39] J. Hu, G. Chen, H.X. Li, Distributed event-triggered tracking control of leader-follower multi-agent systems with communication delays, Inst. Inform. Theory Autom. AS CR 47 (2011) 630–643.
[40] S. Khoo, L. Xie, Z. Man, Robust finite-time consensus tracking algorithm for multirobot systems, IEEE/ASME Trans. Mechatron. 14 (2009) 219–228.
[41] Z. Meng, W. Ren, Y. Cao, Z. You, Leaderless and leader-following consensus with communication and input delays under a directed network topology, IEEE Trans. Syst. Man Cybern., Part B, Cybern. 41 (2011) 75–88.
[42] W. Ren, R. Beard, Distributed Consensus in Multi-Vehicle Cooperative Control: Theory and Applications, Springer, Berlin, 2008.
[43] F. Xiao, L. Wang, Consensus problems for high-dimensional multi-agent systems, IET Control Theory Appl. 1 (2007) 830–837.
[44] Y. Sun, L. Wang, Consensus of multi-agent systems in directed networks with nonuniform time-varying delays, IEEE Trans. Autom. Control 54 (2009) 1607–1613.
[45] P. Tabuada, Event-triggered real-time scheduling of stabilizing control tasks, IEEE Trans. Autom. Control 52 (2007) 1680–1685.
[46] Y. Cao, W. Ren, Y. Li, Distributed discrete-time coordinated tracking with a time-varying reference state and limited communication, Automatica 4 (6) (2009) 1299–1305.
[47] R. Horn, C. Johnson, Matrix Analysis, Cambridge University Press, Cambridge, 2005.
[48] T. Apostol, I. Makai, Mathematical Analysis, Addison-Wesley, Reading, Massachusetts, 1974.
[49] M.S. Mahmoud, Resilient L_2/L_∞ filtering of polytopic systems with state-delays, IET Control Theory Appl. 1 (1) (2007) 141–154.

Chapter 8

State estimation techniques

8.1 Asynchronous multirate multismart sensors

In asynchronous multisensor systems, data missing and delays are likely to occur when performing state estimation at different rates. In this chapter, an algorithm to process both the delayed and missing measurements is developed, under the assumption that the data arrive between two consecutive sampling times. Local estimation is then implemented using a fusion algorithm. This study extends the currently available state estimation for asynchronous multirate multisensor systems that only considers random measurements missing with the ability to remove delay effects. Simulation experiments are performed to illustrate the effectiveness of the developed theory.

8.1.1 Introduction

It becomes increasingly apparent that in multisensor systems there is a possibility to incorporate all of the measurements coming from these sensors to get a better estimation of the dynamical system. In this regard, state fusion is the process to combine information coming from different sources of state estimates. Formerly, sensor could only send the raw data, such as the value of its measurement, but with the advancement of electronics, the new generation of sensors, so called *smart sensors*, can do more complex tasks, such as signal conditioning, calibration, diagnostic tests, and communication. The smart sensor transforms the raw sensor signal to a standardized digital representation, checks and calibrates the signal, and finally transmits this digital signal to the outside world via a standardized interface using a standardized communication protocol.

Applications for multisensor data fusion are widespread. Military applications include automated target recognition (for example, for smart weapons), guidance for autonomous vehicles, remote sensing, battlefield surveillance, and automated threat recognition systems, such as identification-friend-foe-neutral systems. Nonmilitary applications include monitoring of manufacturing processes, condition-based maintenance of complex machinery, robotics, and medical applications.

Although there is a vast body of research related to multisensor fusion, only little work has been done in asynchronous multirate multisensor systems. The general asynchronous sensor-to-sensor linear distributed estimation problem is solved in [1,2,4]. On the theoretical level, the results of [1] solve three problems

Discrete Networked Dynamic Systems. https://doi.org/10.1016/B978-0-12-823698-7.00016-3

in one step: fusing asynchronous tracks without synchronization, dealing with out-of-sequence tracks, and accounting for latent tracks. In [5], the authors presented an exact state update equation for such out-of-sequence problem using retrodictated state estimation. Recently in [6], the authors considered the problem of removal of out-of-sequence (actually, "old") measurements from tracks. The optimal solution and a one-step solution have been presented using the information filter (IF)-based equivalent measurement.

The development of a multisensor optimal information fusion criterion weighted by matrices in the linear minimum variance sense is presented in [7]. Based on this optimal fusion criterion, a general multisensor optimal information fusion decentralized Kalman filter with a two-layer fusion structure is given for discrete-time linear stochastic control systems with multiple sensors and correlated noises. Some related works are reported in [8,9]. Recent results on multisensor data fusion can be found in [10,11].

The problem of state estimation using multiple sensors with incomplete measurement are found in several works [12–16]. In [12] Kalman filtering was studied with intermittent observations using modeling of the arrival of the observation as a random process. Later, state estimation for asynchronous multisensor systems was developed in [10]. In their approach, measurement losses are detected based on statistical hypotheses.

It appears from the published results that data missing and delays are likely to occur in asynchronous multisensor systems when performing state estimation at different rates. In this chapter, an algorithm to process both the delayed and missing measurements is developed, under the assumption that the data arrive between two consecutive sampling times. Local estimation is then implemented using a fusion algorithm. Looked at it in this light, our study extends the currently available state estimation for asynchronous multirate multisensor systems that only consider random measurements missing with the ability to remove delay effects.

On related research fronts, a pair-wise decomposition and coordinated control for complex interconnected systems with any information structure constraints is proposed in [17] based on a permuted inclusion principle. The authors in [18] introduced swarm intelligence as a relatively novel field in order to address the study of the collective behaviors of systems made by many components that are coordinated using decentralized controls and self-organization.

In what follows, we use a priori estimation to handle the case of measurement losses. The optimality of the fusion algorithm is also presented, neglecting the effect of cross-covariance among the local state estimation errors. In this regard, our approach extends the work of [10] by allowing delays to occur between two consecutive sampling times and by assuming that the measurement loss detection is done by communication protocols using data validity checking mechanisms.

8.1.2 Problem formulation

In the sequel, the following assumptions are recalled:

1. Fusion centers are assumed to have the capability to determine measurement loss resulting from communication mechanisms.
2. Sensor measurements are time-stamped.
3. Sensor sampling rates are assumed to be integer-number multiplications from the fastest rate.
4. Delay can happen on measurements, but it is assumed here that the sensor delay is distributed uniformly and less than the unit sampling time.
5. It is assumed that the sensor does not compute the estimation locally; the estimation process is performed at the fusion center.

Suppose that there are N sensors at different sampling rates. Because we allow sensors to measure asynchronously, the measurement is only assumed to come at the same number during a period, where the time distance between two measurements is not uniform.

For every measurement coming from these sensors that is received in the fusion center, there is a corresponding estimation, based solely on one sensor that is taken as the so-called *virtual sensor* (VS). Suppose that at the VS, the last estimate update is at t_{k_i}, and the next estimate time is $t_{k_i+1} = t_{k_i} + T_{si}$, where T_{si} is a constant sampling time at the VS chosen based on the sampling rate of the corresponding sensor. Suppose that during the time interval $[t_{k_i}, t_{k_i+1}]$, a number of asynchronous measurements arrive at the fusion center. Because of communication delays, some of the measurements are generated during the time interval $[t_{k_i}, t_{k_i+1}]$ and some may have been generated during the interval $[t_{k_i-1}, t_{k_i}]$ and arrive during the interval $[t_{k_i}, t_{k_i+1}]$.

Before the VS estimation process takes place, some of the measurements may be removed. If more than one measurement generated by the same sensor is received, only the most recent measurement is used for fusion. If only one measurement is received from a given sensor platform that was taken during the time interval $[t_{k_i-1}, t_{k_i}]$, a decision has to be made whether to keep such track or not. This mainly depends on the data rate of the sensor, the sensor accuracy, and how old that measurement is. The measurements that will be kept for track fusion are called valid measurements. Because of the difference in communication delays, some of the validated tracks may arrive out-of-sequence, which is considered as measurement loss.

8.1.3 Lossy channel

A lossy channel modeled by a stochastic process $\gamma_k \in \{0, 1\}, \forall k \in \mathbb{N}$ with the understanding that $\gamma_k = 1$ signifies that the measurement y_k sent at time k reaches its destination and that $\gamma_k = 0$ when it does not. Whether the measurement has reached the sensor or failed is assumed to be known to the sensor from the network data validity checking mechanism. It is generally assumed that the dropout process $\gamma(k)$ is statistically independent of x_0, w_k, and v_k. This estimate can be computed recursively using the following time-varying Kalman filter:

$$\begin{aligned}\hat{x}(0|-1) &= 0,\\ \hat{x}(k|k-1) &= A_i\hat{x}(k|k),\\ \hat{x}(k|k) &= \hat{x}(k|k-1) + \gamma(k)K(k)(y(k) - C\hat{x}(k|k-1)).\end{aligned} \tag{8.1}$$

8.1.4 Dealing with delay

As the VS in the central data fusion reads the measurement from each sensor in a uniform time step t_{k_i}, each measurement received could be the measurement from t_ζ, where $t_{k_i} < t_\zeta < t_{k_i+1}$. Since state estimation from each sensor will be counted at t_{k_i+1}, by incorporating the measurements that arrive before that, it should be an intermediate propagation from the former time sampling instance t_{k_i}. Considering that the measurement time stamp is at t_ζ, from the former explanation we obtain

$$\epsilon = \frac{t_\zeta - t_{k_i}}{T_{s_i}}, \quad \tau = 1 - \epsilon, \tag{8.2}$$

where it is assumed, for simplicity in exposition, that the dynamics of the system are linear time-invariant and the system modeling takes place in the continuous-time domain. Then it is discretized in view of the fact that the measurements are obtained in discrete time. Then we have

$$\begin{aligned}\dot{x}(t) &= A_c x(t) + w_c(t),\\ x_i(k+1) &= A_i x_i(k) + w_i(k), \text{ where } A_i = e^{A_c\, T_{s_i}}.\end{aligned} \tag{8.3}$$

The propagation is taken in two steps: initially from t_{k_i} to t_ζ and then from t_ζ to t_{k_i+1}. Then we have

$$\begin{aligned}&\text{Step 1 at time } k_i + \epsilon:\\ &\quad \hat{x}_i(k_i+\epsilon|k_i) = A_i^\epsilon \hat{x}_i(k_i|k_i),\\ &\quad P_i(k_i+\epsilon|k_i) = A_i^\epsilon P_i(k_i|k_i) A_i^{\epsilon,T} + Q_{i,\epsilon},\\ &\quad K_i(k_i+\epsilon) = P_i(k_i+\epsilon|k_i)C_i^T\left[C_i P_i(k_i+\epsilon|k_i)C_i^T + R_{i,\epsilon}\right],\\ &\quad \hat{x}_i(k_i+\epsilon|k_i+\epsilon) = \hat{x}_i(k_i+\epsilon|k_i) + \gamma_i(k_i)K_i(k_i+\epsilon)\tilde{y}(k_i+\epsilon),\\ &\quad \tilde{y}(k_i+\epsilon) = (y(k_i+\epsilon) - C_i\hat{x}_i(k_i+\epsilon|k));\\ &\text{Step 2 at time } k_i + 1:\\ &\quad \hat{x}_i(k_i+1|k_i+\epsilon) = A_i^\tau \hat{x}_i(k_i+\epsilon|k_i+\epsilon).\end{aligned} \tag{8.4}$$

Here both $Q_{i,\epsilon}$ and $R_{i,\epsilon}$ are the process and measurement noises of the corresponding sensors related to step ϵ. While $R_{i,\epsilon}$ will be readily set to R_i since there is no lumped effect on the measurement, $Q_{i,\epsilon}$ on the other hand will be different from Q_i.

Easy inspection of (8.4) shows that it essentially reduces to the ordinary Kalman filter, when $\epsilon \to 0$. When measurement losses occur, it follows that (8.4) reduces to one-step a priori state estimation leading to $\hat{x}_i(k_i+1) = A_i\hat{x}_i(k_i)$. At this stage, one might have doubt of whether $A^\epsilon \in \mathbb{R}$. This can be easily answered by carefully looking into (8.3) and $A^\epsilon = e^{A_c \epsilon T_{s_i}}$.

Let us assume that there is an integer $p > 0$ such that one time sampling of the corresponding sensor can be divided into p parts, and there is also an integer $m > 0$ such that $\epsilon = m/p$. Based on this fact, one can write an expression for the state propagation in step ϵ as follows:

$$\begin{aligned} x_i(k_i) &= \mathbf{x}_i(pk_i), \\ x_i(k_i+\epsilon) &= \mathbf{x}_i(pk_i+m) \\ &= \mathbf{A}_i\mathbf{x}_i(pk_i+m-1) + \mathbf{w}_i(pk_i+m-1) \\ &= \mathbf{A}_i^m\mathbf{x}_i(pk_i) + \sum_{j=0}^{m-1}\mathbf{A}_i^j\mathbf{w}_i(pk_i+m-1-j) \\ &= \mathbf{A}_i^m\mathbf{x}_i(pk_i) + \mathbf{w}_{i,m}(pk_i) \\ &= A^\epsilon x_i(k_i) + w_{i,\epsilon}(k_i). \end{aligned} \tag{8.5}$$

Compared with the original Kalman filter, only one additional computation has to be computed here, no matter how big the value p is. Higher p values do not affect the computational effort at every step. Rather, p only affects the way of computing A^ϵ and the process noise variance $Q_{i,\epsilon}$ at step ϵ. One can then express $Q_{i,\epsilon}$ as follows:

$$Q_{i,\epsilon} = \sum_{j=0}^{m-1}\mathbf{A}_i^j\mathbf{Q}\mathbf{A}_i^{j,T} = Q_i - \sum_{j=m}^{p}\mathbf{A}_i^j\mathbf{Q}\mathbf{A}_i^{j,T}. \tag{8.6}$$

Interestingly, we solve (8.6) by approximation as follows:

$$\begin{aligned} Q_i &= \sum_{j=0}^{p}\mathbf{A}_i^j\mathbf{Q}\mathbf{A}_i^{j,T}, \\ \lambda_M(Q_i) &= \lambda_M(\sum_{j=0}^{p-1}\mathbf{A}_i^j\mathbf{Q}\mathbf{A}_i^{j,T}) \end{aligned}$$

$$
\begin{aligned}
&\simeq \lambda_M(\mathbf{Q})\lambda_M(\sum_{j=0}^{p-1}\mathbf{A}_i^j\mathbf{A}_i^{j,T}) \\
&\simeq \lambda_M(\mathbf{Q})\sum_{j=0}^{p-1}\lambda_M(\mathbf{A}_i^j\mathbf{A}_i^{j,T}).
\end{aligned} \tag{8.7}
$$

This leads to

$$
\lambda_M(\mathbf{Q}) \simeq \frac{\lambda_M(Q_i)}{\sum_{j=0}^{p-1}\lambda_M(\mathbf{A}_i^j\mathbf{A}_i^{j,T})},
$$

which in turn yields

$$
\lambda_M(Q_{i,\epsilon}) \simeq \frac{\lambda_M(Q_i)\sum_{j=0}^{m-1}\lambda_M(\mathbf{A}_i^j\mathbf{A}_i^{j,T})}{\sum_{j=0}^{p-1}\lambda_M(\mathbf{A}_i^j\mathbf{A}_i^{j,T})}.
$$

Thus, we have

$$
\frac{\lambda_M(Q_{i,\epsilon})}{\lambda_M(Q_i)} \simeq \left(\frac{\sum_{j=0}^{m-1}\lambda_M(\mathbf{A}_i^j\mathbf{A}_i^{j,T})}{\sum_{j=0}^{p-1}\lambda_M(\mathbf{A}_i^j\mathbf{A}_i^{j,T})}\right).
$$

Letting

$$
\delta(m,p) \equiv \frac{\lambda_M(Q_{i,\epsilon})}{\lambda_M(Q_i)},
$$

we finally arrive at

$$
Q_{i,\epsilon} \simeq \delta(m,p)\, U_{m,p} Q_i U_{m,p}^T, \quad U_{m,p} = \frac{\mathbf{A}_i^m}{|\mathbf{A}_i^m|}. \tag{8.8}
$$

The problem therefore turns into the selection of an appropriate p for an acceptable delay tolerance. As soon as p is determined, $\delta(m,p)$ can be counted. For example, if p is 64, then one has a collection of 64 values of δ. These values can be counted once at initialization, and then they can be used at every time step. One can even have more precise values of $Q_{i,\epsilon}$ using a simple well-known Newton–Raphson method, using $Q_{i,\epsilon}$ obtained in (8.8) as initial guess. All procedures for state estimation using delayed measurements described hereafter are named *intermediate step Kalman filter* (ISKF). Fig. 8.1 shows a comparison between approximated results of $\delta_{m,p}$ processed with (8.8) along with the associated true value. Rough estimates can be taken as $Q_{i,\epsilon} \simeq \epsilon Q_i$, as given in [5].

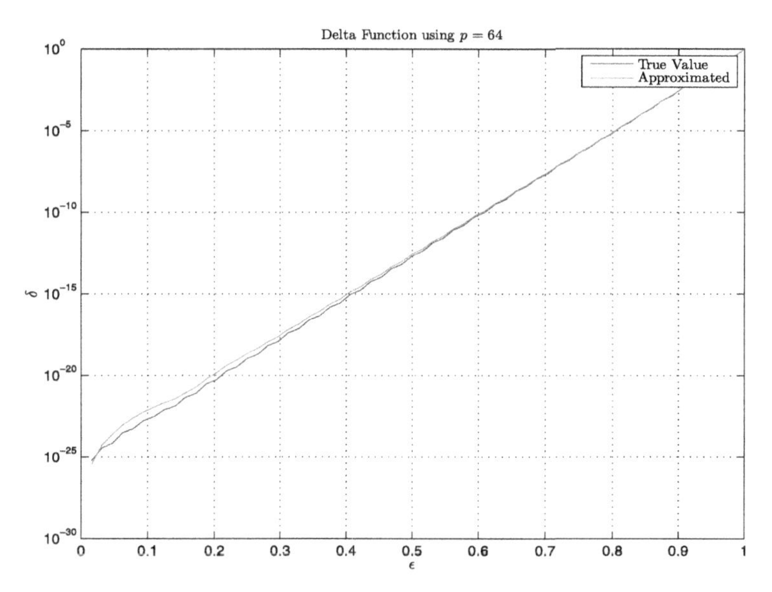

FIGURE 8.1 Comparison of approximated and true δ.

8.1.5 Fusion algorithm

When an estimate of the state is available, the problem now turns to the combination of these different estimations to get the optimal result. In the sequel, an approach proposed in [10] is incorporated effectively. The dynamics of the system are first defined on the lowest sampling time (VS) by

$$\begin{aligned} x(k+1) &= Ax(k) + w(k), \\ y_i(k_i) &= \gamma_i(k_i)C_i x_i(k_i) + v_i(k_i), \quad i = 1, 2, \cdots N, \end{aligned} \tag{8.9}$$

where nonsubindexed variables belong to the lowest sampling time sensor that the system dynamics are defined with, A is the system matrix, and w is supposed to be zero mean white Gaussian noise with variance Q. Also y_i is the measurement observed by sensor i, x_i is the state whose dynamics are described by A_i, which will be defined subsequently, C_i is the measurement matrix, and v_i is supposed to be zero mean white Gaussian noise with variance R_i, $i = 1, 2, \cdots, N$.

In the sequel, it is assumed that each sensor follows

$$n_i = \frac{S_N}{S_i}, n_i \in \mathbb{N}, \tag{8.10}$$

where S_i is the sampling rate on sensor i and S_N is the sampling rate at the highest sensor sampling rate. Equivalently to the VS,

$$n_i = \frac{T_{si}}{T_s}, \; n_i \in \mathbb{N}, \tag{8.11}$$

where T_s is the sampling time of the lowest sampling time VS and T_{si} is the VS i sampling time.

Fused estimation based on this series of sensors can then be computed every T_s. Obviously, the fused estimation $\hat{x}(k|k)$ is no more than a weighted estimation coming from each sensor $\hat{x}_i(k|k)$. One can intuitively think that the estimation from sensor i that has less error variance should be weighted more, compared with those that have higher error variance. It can be easily shown that the foregoing weighting selection of state estimations coming from the corresponding VS gives the optimal result in the sense of error variance under the assumption that there is no cross-covariance among VS estimations [10]. So, we have

$$\begin{aligned}
\hat{x}(k|k) &= \sum_{i=1}^{N} \alpha_i(k)\hat{x}_{N|i}(k|k), \\
P(k|k) &= \left[\sum_{i=1}^{N} P_{N|i}^{-1}(k|k)\right]^{-1}, \\
\alpha_i(k) &= P(k|k)P_{N|i}^{-1}(k|k),
\end{aligned} \tag{8.12}$$

where $\hat{x}_{N|i}(k|k)$ is the state estimation at the highest sampling rate based on estimation from VS i and $P_{N|i}(k|k)$ is its error covariance, with both being given as follows:

$$\hat{x}_{N|i}(k|k) = \begin{cases} [\sum_{m=0}^{n_i-1} A^-m]^{-1}\hat{x}_i(l|l), & k = n_i l, \\ n_i A^p \left[\sum_{m=0}^{n_i-1} A^-m\right]^{-1} \hat{x}_i(l|l), & k = n_i l + p, \end{cases} \tag{8.13}$$

$$P_{N|i}(k|k) = \begin{cases} n_i^2[\sum_{m=0}^{n_i-1} A^-m]^{-1} P_i(l|l)[\sum_{m=0}^{n_i-1} A^-m]^{-1,T}, & k = n_i l, \\ n_i^2 A^p[\sum_{m=0}^{n_i-1} A^-m]^{-1} P_i(l|l)[\sum_{m=0}^{n_i-1} A^-m]^{-1,T} A^{p,T} & \\ \quad + \sum_{m=0}^{p-1} A^-m^{-1} A^l Q A^{l,T}(l|l), & k = n_i l + p. \end{cases} \tag{8.14}$$

The terms $P_i(l|l)$ and $\hat{x}_i(l|l)$ in (8.14) and (8.13) are computed using the ISKF as described in Section 8.1.4, with system matrix A_i and process noise covariance Q_i in the corresponding sensor i defined as

$$A_i = A^{n_i},$$
$$Q_i = \frac{1}{n_i^2}\left[\sum_{m=0}^{n_i-1} A^{-m}\right]\left[\sum_{m=0}^{n_i-1} A^{m} Q A^{m,T}\right]\left[\sum_{m=0}^{n_i-1} A^{-m}\right]^t. \tag{8.15}$$

From (8.12) it can be verified that

$$P(k|k) \leq P_{N|i}(k|k), \tag{8.16}$$

which means that the fused estimation error generated from different estimation sensors is always less than or equal to the estimation error of each sensor.

In what follows, the proposed algorithm will be tested with a system using a kinetic model using the discrete Wiener acceleration process (DWAP) as in [2]. Given the true system as

$$x(k+1) = \begin{bmatrix} 1 & T_s & T_s^2/2 \\ 0 & 1 & T_s \\ 0 & 0 & 1 \end{bmatrix} x(k) + \Gamma w(k),$$
$$Q = \sigma_w^2 \, \Gamma \, \Gamma^t,$$
$$\Gamma = \begin{bmatrix} T_s^2/2 \\ T_s \\ 1 \end{bmatrix}. \tag{8.17}$$

Here $T_s = 0.01$ and $\sigma_w = 0.8$. The simulation experiments are divided into two parts: In the first part, the ISKF will be compared with the ordinary Kalman filter. In the second part, the fusion algorithm based on the ISKF is implemented.

8.1.6 Simulation example 8.1

The purpose here is to demonstrate single-sensor simulations. Given a position sensor, with its VS $T_s = 5.12$ s and with measurement noise of 0.9. At the beginning, the measurement loss probability is assumed zero to allow comparison between the ISKF and the Kalman filter. Next, the measurement loss probability is set as 0.3. Several selections of delay partition are presented here to demonstrate the effect of the ISKF using different delay partitions and make comparisons with the Kalman filter.

In the simulation results, the following notations are applied:

1. Black dotted line represents the measurement, and blue dotted line (dark gray in print version) represents the measurements that arrive at the fusion center (see the legend on Figs. 8.2 and 8.3).
2. Kalman estimation tends to follow the delayed measurement.

3. The ISKF, despite the small partition number, like in Fig. 8.2, is able to move state estimation closer to the true value.

Simulation results of single-sensor estimations without measurement loss and different partitions are displayed in Figs. 8.2–8.5.

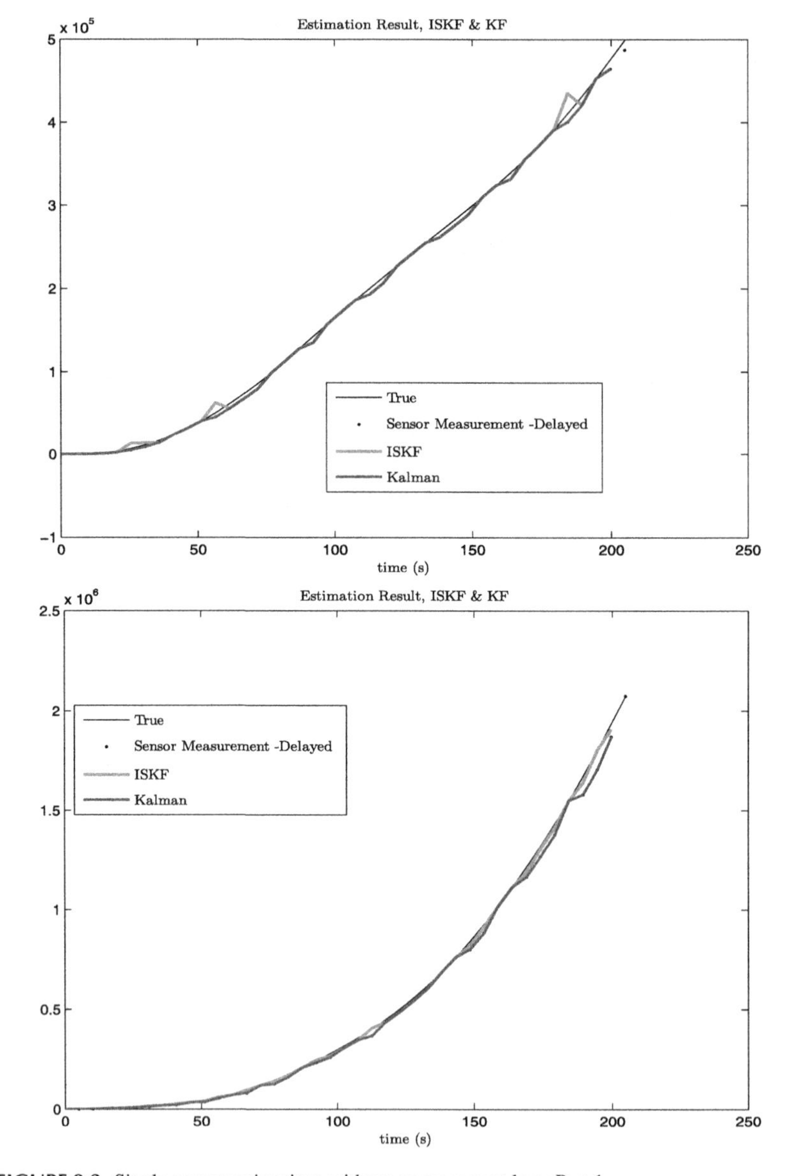

FIGURE 8.2 Single-sensor estimations without measurement loss: Part 1.

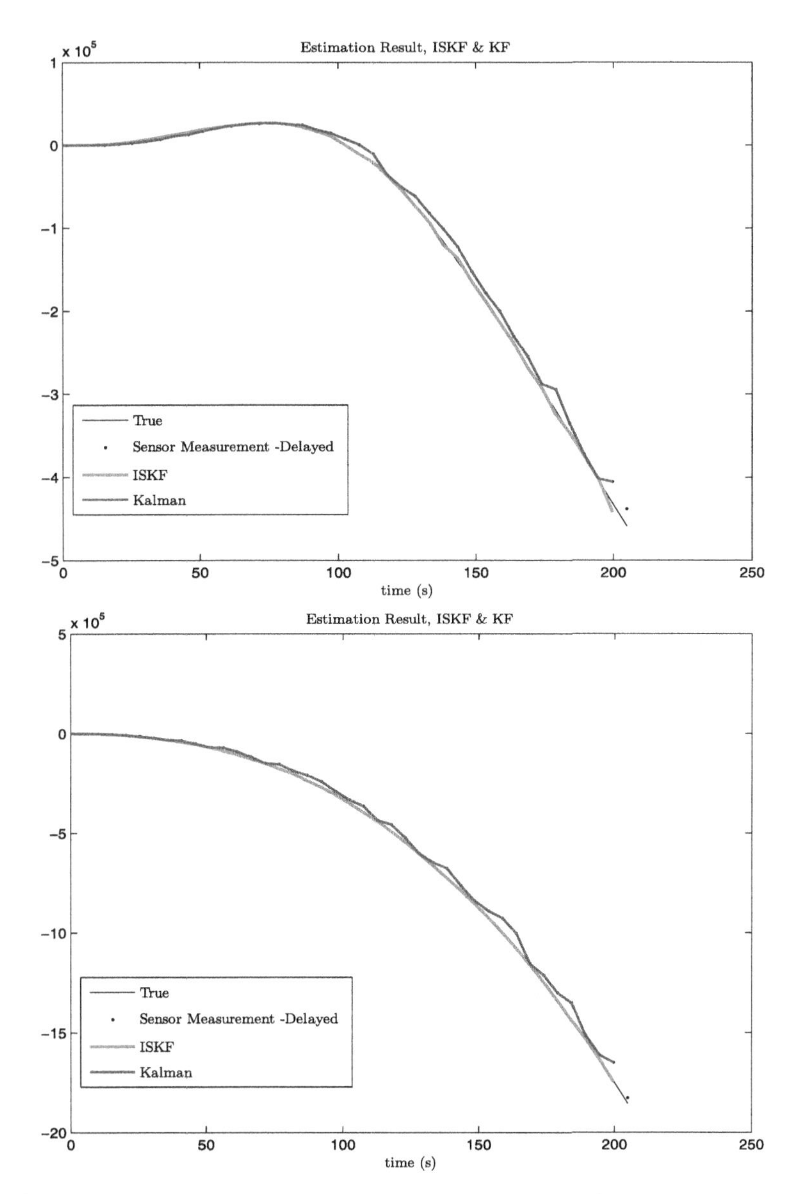

FIGURE 8.3 Single-sensor estimations without measurement loss: Part 2.

Remark 8.1. The following points pertaining to the simulation results are in order:

- When there is no loss in measurements, the ISKF is shown to be able to give nearly the "nondelayed" estimate from delayed measurements. We observe that the more delay partitions are used in ISKF, the lower is the estimation

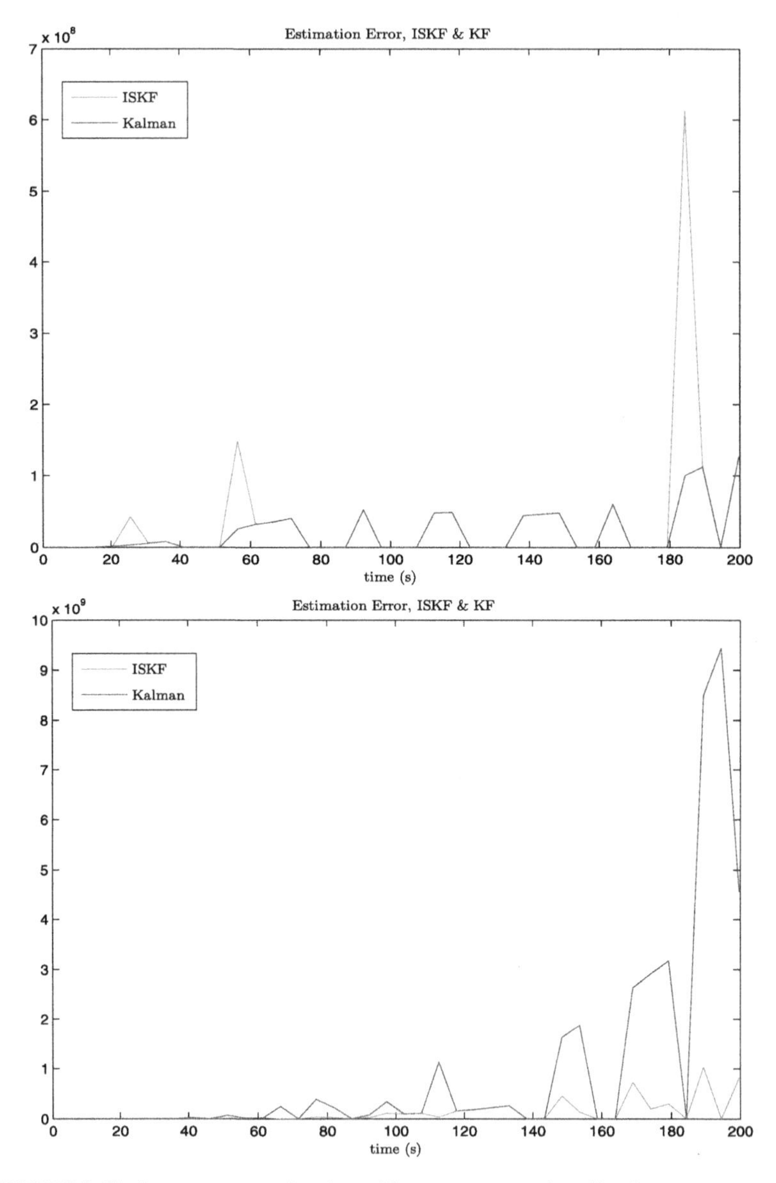

FIGURE 8.4 Single-sensor error estimations without measurement loss: Part 1.

error that is obtained. It can be verified that, if the delay partition number is equal to one, the ISKF reduces to the Kalman filter, however with its estimations being delayed. It can be seen clearly that Kalman filter estimation follows delayed estimations, where the ISKF nearly follows true measurements, especially if the partition number becomes higher (see Figs. 8.2–8.5).

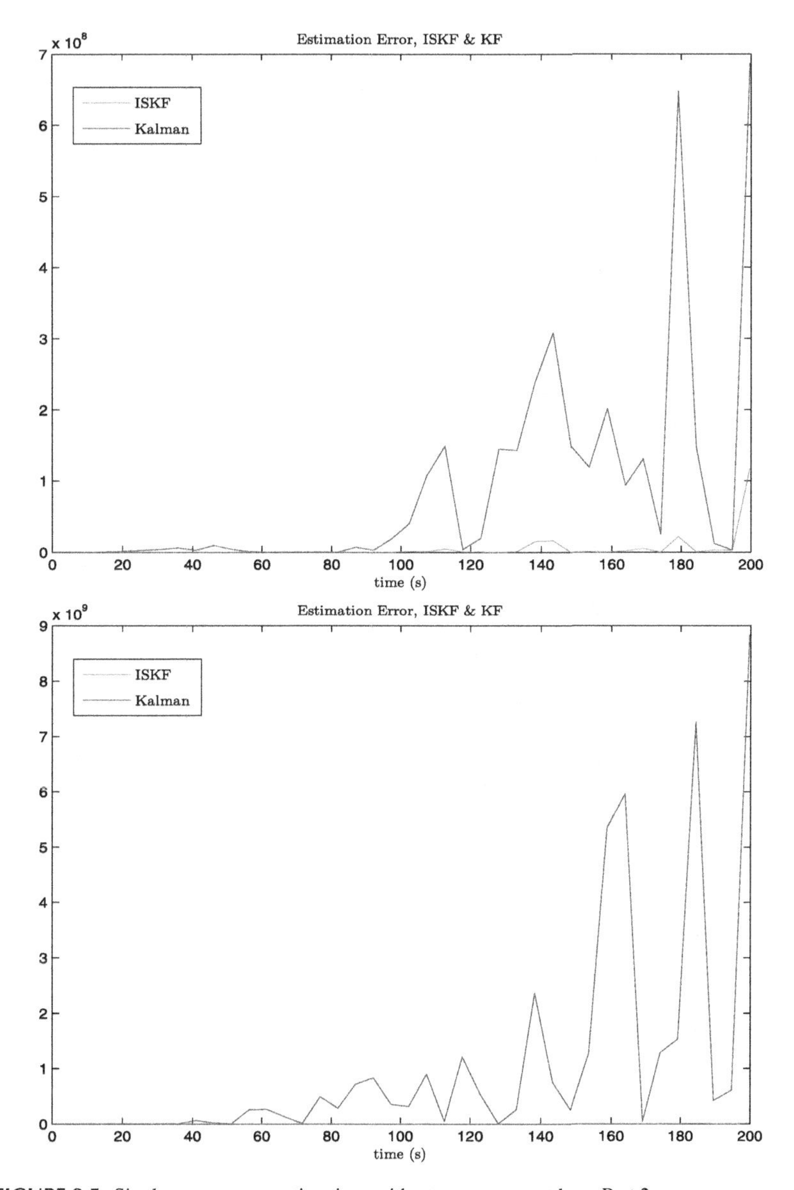

FIGURE 8.5 Single-sensor error estimations without measurement loss: Part 2.

- When measurement loss occurs, the ISKF tends to give a better a priori estimation compared with the Kalman filter. This can be readily seen on several sharp peaks on Kalman filter estimation results. Measurement loss is the main cause of the occurrence of such sharp peaks on Kalman estimation. When measurements are lossy, Kalman filter propagates its state estimation using its last estimate that already had an error due to delayed measurement. Upon propagation, the state estimate tends to go to false directions, until the arrival of the next measurement. The peak becomes higher if measurement loss happens in consecutive measurements. Similar things also happen in the ISKF, except that the estimate is closer to the true value before the loss occurs. In such case, although it also propagates in the wrong direction, the state estimation peak is not as high as in Kalman filter estimations. The peak even reduces significantly if the number of partitions becomes higher (see Figs. 8.6 and 8.9).

- As noted earlier, since higher p values do not influence the computational effort, one can easily select the highest number that is reasonable. The difference will only be how many matrices will be stored at the initialization phase of the algorithm in addition to the convergence of the Newton–Raphson algorithm used to compute $Q_{i,\epsilon}$ and the precision of $Q_{i,\epsilon}$ matrices that can be reached.

The corresponding simulation results of single-sensor estimations with measurement loss and different partitions are displayed in Figs. 8.6–8.9.

8.1.7 Simulation example 8.2

In this section, we illustrate multiple-sensor simulations. Using the same model as in Section 8.1.6, a fusion algorithm is simulated using position, velocity, and acceleration sensors. The highest sampling rate is that of the acceleration sensor, with $T_s = 5.12$ at its VS, drop probability $= 0.6$, and noise variance $= 0.1$. Velocity sensors have half the sensor sampling rate, and 0.3 drop probability, with 1.9 noise variance. The position sensor has the slowest sensor, at one-third of the sampling rate of the acceleration sensor, and 0.1 drop probability, with 2.8 noise variance.

It is expected that the fusion algorithm can overcome the measurement loss problem, by considering that during the event of measurement loss in one sensor, other sensors still get the measurement, so the peak arising in the estimation pattern due to the measurement loss can be essentially removed. It is also expected that variance of the state estimation error would be less than the state estimation error arising from single-sensor measurements.

Simulation results of multisensor estimations with measurement loss and different partitions are displayed in Figs. 8.10–8.12.

Remark 8.2. Estimations arising from single measurements (VS estimations) using the ISKF (red line, light gray in print version) generally perform better than using a Kalman filter (blue line, dark gray in print version). This can be seen in Figs. 8.10 and 8.11. Estimations of position and velocity based on the position and velocity sensors based on the ISKF are closer to the true values compared with Kalman filter estimation. ISKF estimation also has fewer peaks on each VS estimation compared with the Kalman filter.

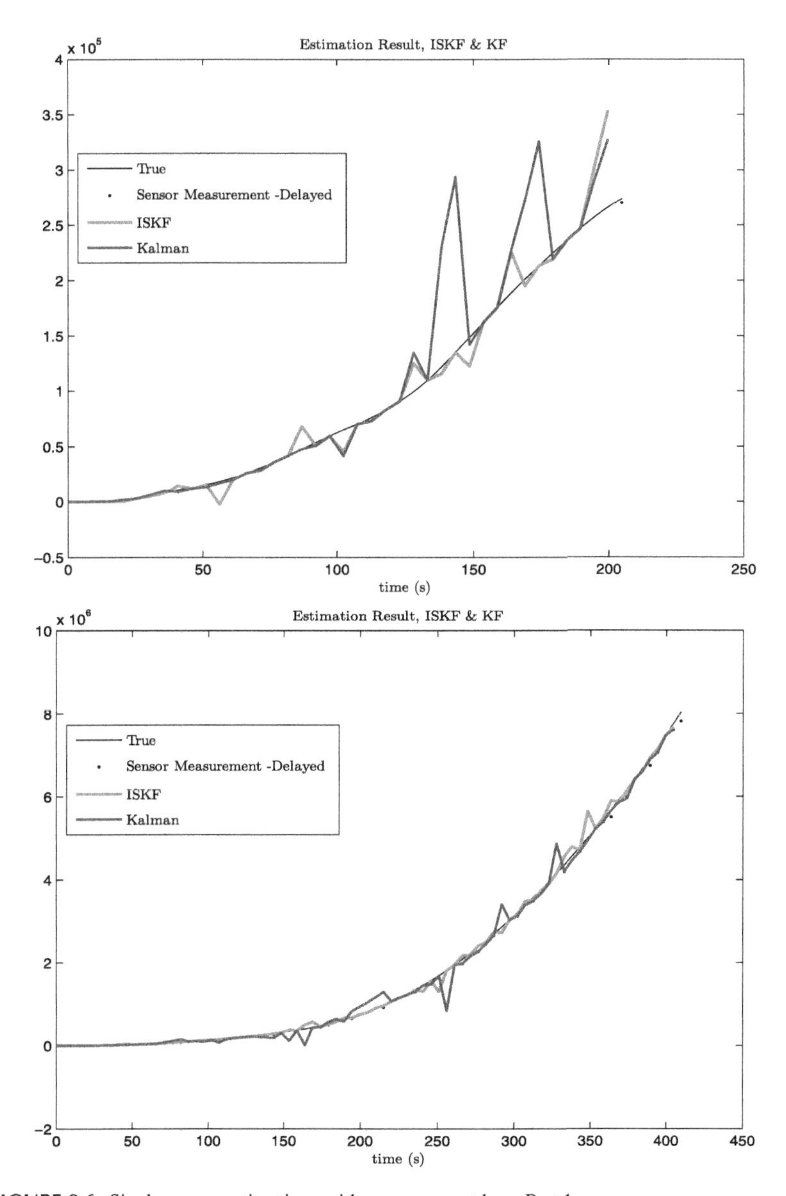

FIGURE 8.6 Single-sensor estimations with measurement loss: Part 1.

8.2 Nonlinear state estimation

This section concerns the nonlinear $\mathbb{H}_\infty$ estimation/filtering problem for sensor networks with repeated scalar nonlinearities and multiple probabilistic packet losses. The class of nonlinear systems is represented by a discrete-time, state-

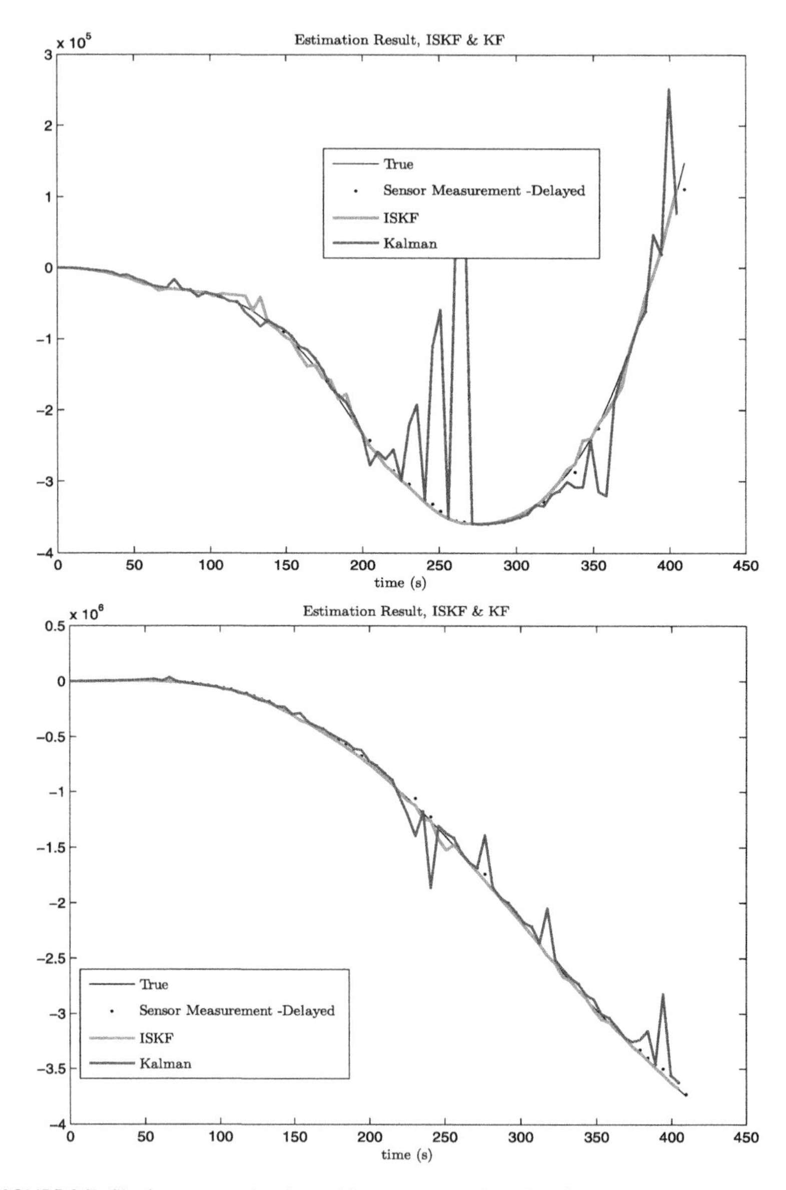

FIGURE 8.7 Single-sensor estimations with measurement loss: Part 2.

space model involving repeated scalar nonlinearities that cover several types of frequently investigated nonlinearities as special cases. A number of stochastic variables, all of which are mutually independent but satisfy a certain probabilistic distribution in the interval [0, 1], are introduced to account for the packet dropout phenomena occurring in the channels from the original system to the

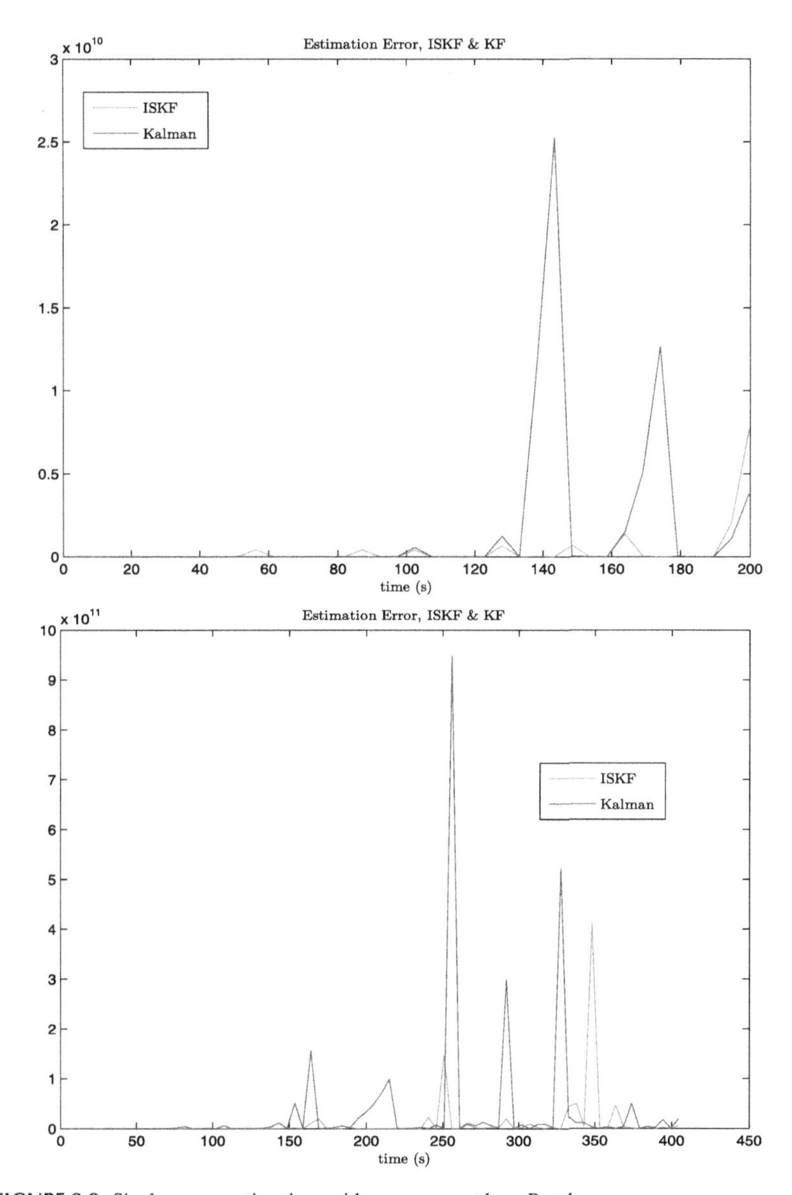

FIGURE 8.8 Single-sensor estimations with measurement loss: Part 1.

networked sensors. The concept of average $\mathbb{H}_\infty$ index is first introduced to measure the overall performance of the sensor networks. Then, by utilizing available measurement information from not only each individual sensor but also its neighboring sensors according to a given topology, stability analysis is

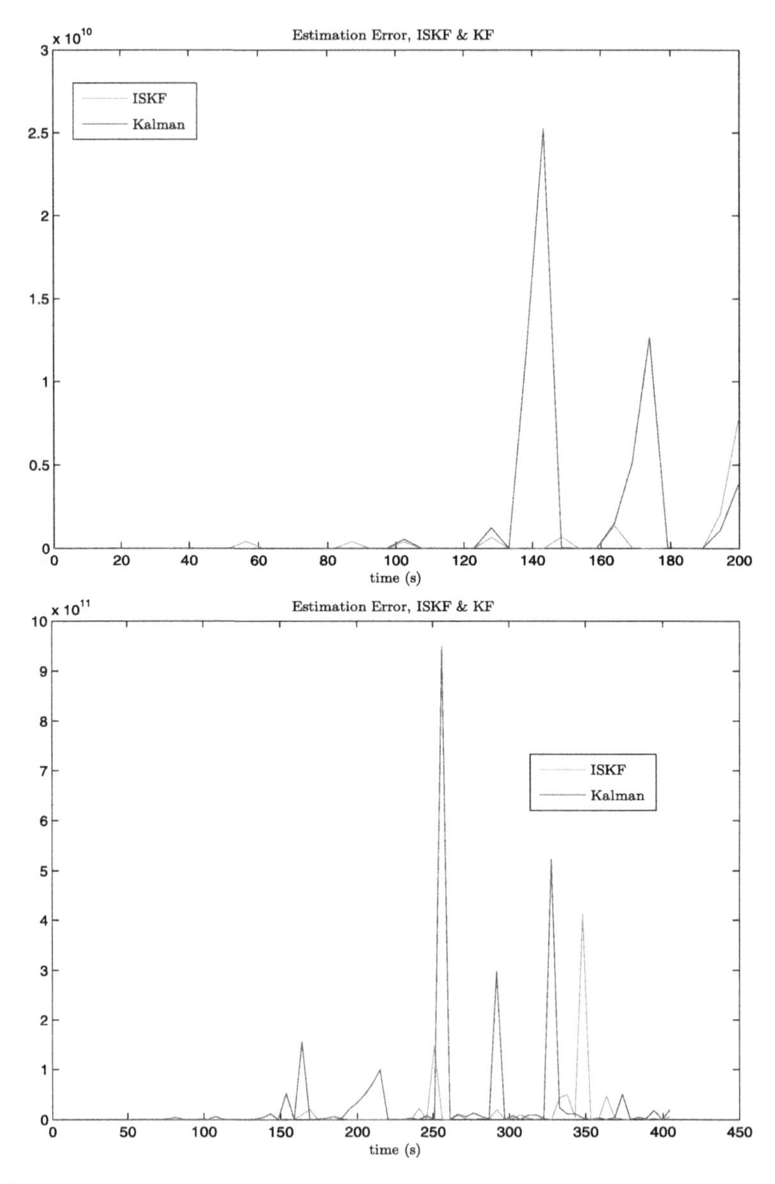

FIGURE 8.9 Single-sensor estimations with measurement loss: Part 2.

carried out to obtain sufficient conditions for ensuring stochastic stability as well as the prescribed average $\mathbb{H}_\infty$ performance constraint. The determination of the parameters of the distributed filters is attained through the feasibility of a convex optimization problem.

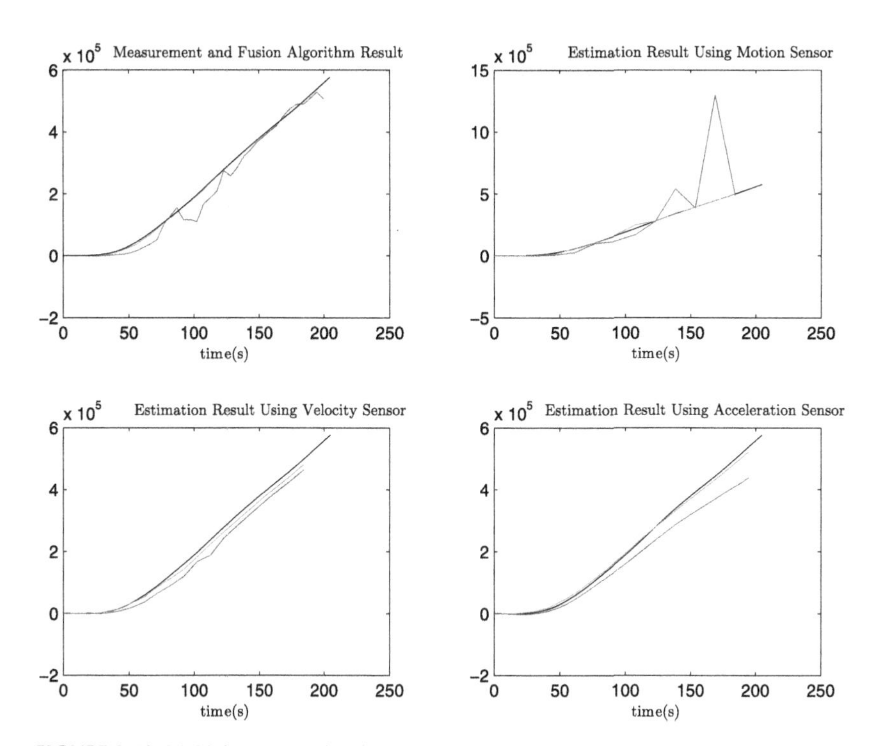

FIGURE 8.10 Multiple-sensor estimation: position state.

8.2.1 Introduction

In the 2010s, sensor networks have been attracting increasing attention from many researchers in different disciplines owing to the extensive applications of sensor networks in many areas, including surveillance, environment monitoring, information collection, industrial automation, and wireless networks [27].

A sensor network typically consists of a large number of sensor nodes and also a few control nodes, all of which are distributed over a spatial region. The distributed filtering or estimation, as an important issue for sensor networks, has been an area of active research for many years. Different from the traditional filtering for a single sensor [28], the information available for the filter algorithm on an individual node of the sensor network is not only from its own measurement but also from its neighboring sensors' measurements according to the given topology. As such, the objective of filtering based on a sensor network can be achieved in a distributed yet collaborative way. It is noted that one of the main challenges for distributed filtering lies in the handling of the complicated coupling issues between one sensor and its neighboring sensors.

In recent years, the distributed filtering problem for sensor networks has received considerable research interest, and a lot of research results are available

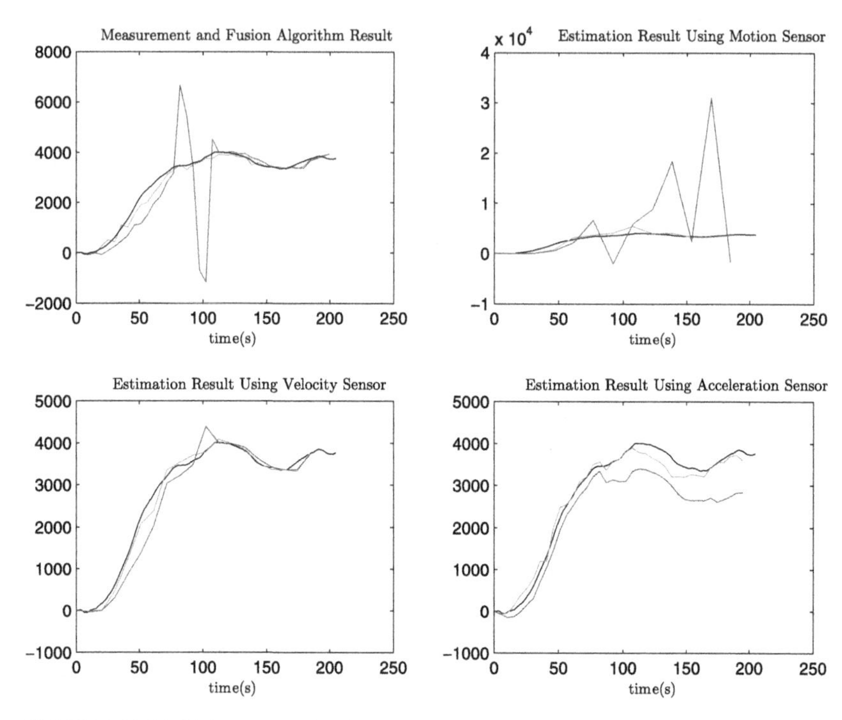

FIGURE 8.11 Multiple-sensor estimation: velocity state.

in the literature (see [3,29]). Most of the reported distributed filter algorithms for sensor networks have thus far been mainly based on the traditional Kalman filtering theory, which requires exact information about the plant model. In the presence of unavoidable parameter drifts and external disturbances, a desired distributed filtering scheme should be made as robust as possible. However, the robust performance of the available distributed filters has not yet been thoroughly studied, and this inevitably restricts the application potential in practical engineering. Therefore, it is of great significance to introduce the $\mathbb{H}_\infty$ performance requirement with the hope to enhance the disturbance rejection attenuation level of designed distributed filters. Note that some initial efforts have been made to address the robustness issue. Very recently, a new distributed $\mathbb{H}_\infty$ consensus performance has been proposed in [30] to quantify the consensus degree over a finite horizon, the distributed filtering problem has been addressed for a class of linear time-varying systems in the sensor network, and the filter parameters have been designed recursively by resorting to the difference linear matrix inequalities. In [32], an $\mathcal{H}_\infty$-type performance measure of disagreement between adjacent nodes of the network has been included and a robust filtering approach has been proposed to design the distributed filters for uncertain plants. On the other hand, since nonlinearities are ubiquitous in practice, it is neces-

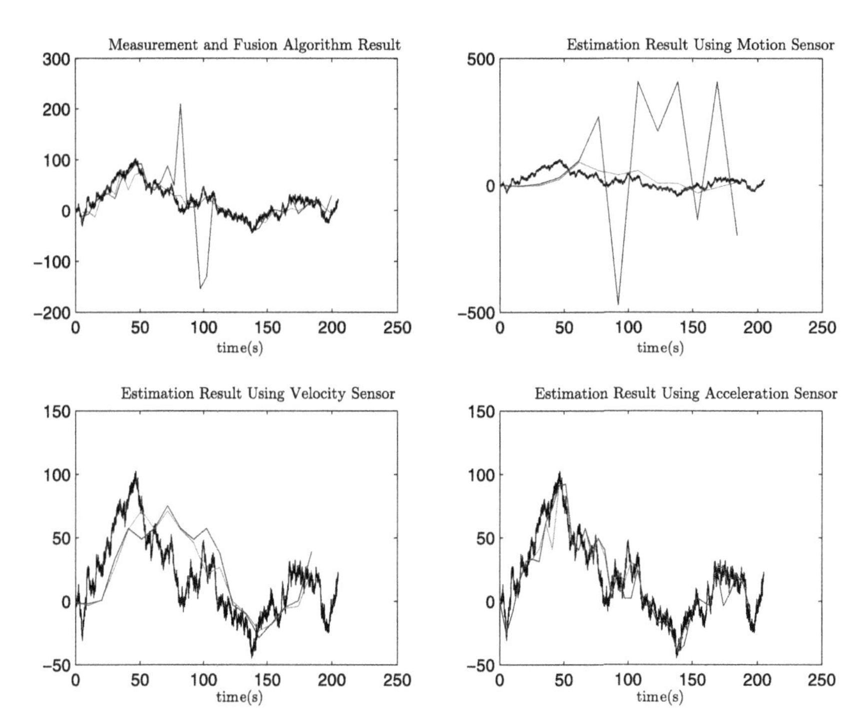

FIGURE 8.12 Multiple-sensor estimation: acceleration state.

sary to consider the distributed filtering problem for target plants described by nonlinear systems.

A sensor network possesses spatially distributed autonomous sensors to monitor physical conditions in cooperative environments where wireless communication is typically adopted among the sensors. As the network scale increases, distributed filtering in a sensor network unavoidably suffers from the limited communication induced by the channel bandwidth constraints. Among others, the network-induced data packet loss phenomenon takes place quite often in wireless communication channels between nodes because of fading and interference. In the most relevant literature, the binary random distribution model has been widely employed due to its simplicity in describing the packet loss phenomenon. In such a model, probability 0 is usually used to represent the missing of an entire signal and probability 1 denotes the intactness, that is, there is no signal missing at all.

Recently, due to various reasons such as sensor aging and temporal sensor failure, the traditional model of binary random distribution has been generalized in [31] to account for the partial data missing problem where the missing probability is defined over the interval [0, 1].

8.2.2 Problem formulation

The filtering problem in sensor networks is shown in Fig. 8.13, where each sensor can derive the information from both the plant and its neighboring sensors according to the sensor network topology. The information received by sensor i from the plant is transmitted via communication cables that are of limited capacity, and may therefore face data packet losses.

In the sequel, we assume that the N sensor nodes are distributed in space according to a fixed network topology represented by a directed graph $\mathbb{G} = (\mathbb{V}, \xi, \mathbb{A})$ of order N with the set of nodes $\mathbb{V} = 1, 2, \ldots, N$, the set of edges $\xi \in \mathbb{V} \times \mathbb{V}$, and the weighted adjacency matrix $\mathbb{A} = [a_{ij}]$ with nonnegative adjacency element a_{ij}. An edge of G is denoted by the ordered pair (i, j). The adjacency elements associated with the edges of the graph are positive, that is, $a_{ij} > 0 \Leftrightarrow (i, j) \in \mathbb{E}$, which means that sensor i can obtain information from sensor j.

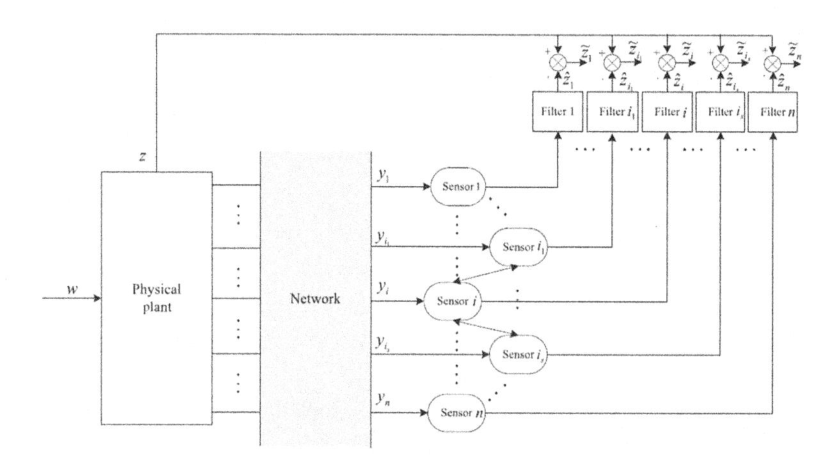

FIGURE 8.13 A sensor network filtering problem.

Also, we assume that $a_{ii} = 1$ for all $i \in \mathbb{V}$, and therefore (i, i) can be regarded as an additional edge. The set of neighbors of node $i \in V$ plus the node itself are denoted by $N_i = \{j \in \mathbb{V} : (i, j) \in \mathbb{E}\}$. With reference to Fig. 8.13, the plant is described by the following discrete-time nonlinear system:

$$
\begin{aligned}
x(k+1) &= A\, f(x(k)) + B\omega(k), \\
z(k) &= M\, f(x(k)),
\end{aligned}
\tag{8.18}
$$

where $x(k) \in \Re^{n_x}$ represents the state vector which cannot be observed directly, $z(k) \in \Re^{n_z}$ is the output to be estimated, $\omega(k) \in \Re^{n_w}$ is the disturbance input belonging to $\ell_2[0, \infty)$, and f is a nonlinear function. In the sequel, for the vector

$x = [x_1\ x_2\ \dots\ x_{n_x}]^t$, we denote

$$f(x) := [g(x_1)\ g(x_2)\ \dots\ g(x_{n_x})]^t \tag{8.19}$$

and $g : \Re \to \Re$ is a piece-wise continuous identical nonlinearity acting on every component of the state x as specified in the following assumption [33].

Assumption 8.1. The nonlinear function $g : R \to R$ in system (8.18) satisfies

$$\forall a, b \in R \quad |g(a) + g(b)| \leq |a + b|. \tag{8.20}$$

The model (8.18) is called a system with repeated scalar nonlinearity since it is analogous to an upper NFT with respect to a repeated scalar block (see [33–35]). It is implied from Assumption 8.1 that f encapsulates some typical classes of nonlinearities, such as:

- the standard saturation

$$\begin{cases} s, & |s| \leq 1, \\ sgn(s), & |s| > 1; \end{cases}$$

- the hyperbolic tangent function that has been extensively used for activation function in neural networks;
- the sine function.

The target plant (8.18) can be applied in many areas, including marketing and production control systems and n-stand cold rolling mills and manufacturing systems for decision making [19].

The N sensors are modeled by

$$y_i(k) = \beta_i(k)\, C_i\, f(x(k)) + D_i\, \omega(k), \quad i = 1, 2, \dots, n, \tag{8.21}$$

where $y_i(k) \in \Re^{n_y}$ is the measurement output measured by sensor i from the plant and $\beta_i(k)$ $(i = 1, 2, \dots, N)$ are N unrelated and independent stochastic variables. It is assumed that $\beta_i(k)$ has the probabilistic density function $q_i(s)$ $(i = 1,\ 2, \dots,\ N)$ on the interval $[0, 1]$ with mathematical expectation μ_i and variance σ_i^2. Moreover, the matrices mentioned above, that is, A, B, C_i, D_i, and M, are all known matrices with appropriate dimensions.

Remark 8.3. In model (8.21), the stochastic variable $\beta_i(k)$ corresponding to the ith sensor is introduced to cater for the probabilistic packet loss problem and can take values on the interval $[0,\ 1]$, and is therefore more general than the traditional Bernoulli variables.

The key point in the distributed state estimator design for sensor networks is the fusion of the information available for the filter on the sensor node i both from the sensor i itself and from its neighbors. The following estimator structure is adopted on sensor node i:

$$\begin{aligned}\hat{x}_i(k+1) &= \sum_{j\in N_i} a_{ij} K_{ij}\, f(\hat{x}_j(k)) + \sum_{j\in N_i} a_{ij} H_{ij} y_j(k),\\ \hat{z}_i(k) &= M\, f(\hat{x}_i(k)),\end{aligned} \tag{8.22}$$

where $\hat{x}_i(k) \in \Re^{n_x}$ is the state estimate on sensor node i and $\hat{z}_i(k) \in \Re^{n_z}$ is the estimate of $z(k)$ from the filter on sensor node i. Here, matrices K_{ij} and H_{ij} are the parameters of estimator on node i to be determined.

Now, we denote

$$\begin{aligned}&\hat{x}(k) = [\hat{x}_1^t(k)\ \hat{x}_2^t(k)\ \dots\ \hat{x}_n^t(k)]^t,\\ &\bar{x}(k) = \mathbf{1}_N \otimes x(k), \quad \bar{C}_i = (e_i e_i^t)\otimes C_i,\\ &\tilde{z}(k) = [(z(k)-\hat{z}_1(k))^t\ (z(k)-\hat{z}_2(k))^t\ \dots\ (z(k)-\hat{z}_n(k))^t]^t,\\ &C_\mu = diag\{\mu_1 C_1, \mu_2 C_2, \dots, \mu_n C_n\},\\ &\bar{A} = I_n \otimes A, \quad \bar{M} = I_n \otimes M,\\ &\bar{B} = I_N \otimes B, \quad \bar{D} = [D_1^t\ D_2^t\ \dots\ D_n^t]^t,\\ &\tilde{M} = \begin{bmatrix} \bar{M} & -\bar{M}\end{bmatrix}, \quad \tilde{A} = \begin{bmatrix} \bar{A} & 0\\ \bar{H}C_\mu & \bar{K}\end{bmatrix},\\ &\tilde{C}_i = \begin{bmatrix} 0 & 0\\ \bar{H}\bar{C}_i & 0\end{bmatrix}, \quad \tilde{D} = \begin{bmatrix} \bar{B}^t & (\bar{H}\bar{D})^t\end{bmatrix}^t,\end{aligned} \tag{8.23}$$

where

$$\begin{aligned}&\bar{K} = [\bar{K}_{ij}]_{n\times n}, \quad \text{with}\\ &\bar{K}_{ij} = \begin{cases} a_{ij}K_{ij}, & i = 1,2,\dots,n,\ j \in N_i,\\ 0, & i = 1,2,\dots,n,\ j \neq N_i,\end{cases}\\ &\bar{H} = [\bar{H}_{ij}]_{n\times n}\\ &\bar{H}_{ij} = \begin{cases} a_{ij}H_{ij}, & i = 1,2,\dots,n,\ j \in N_i,\\ 0, & i = 1,2,\dots,n,\ j \neq N_i.\end{cases}\end{aligned} \tag{8.24}$$

Obviously, since $a_{ij} = 0$ when $j \neq N_i$, $\bar{K}$ and $\bar{H}$ are two sparse matrices that can be expressed as

$$\bar{K} \in T_{n_x \times n_x}, \qquad \bar{H} \in T_{n_x \times n_y}, \tag{8.25}$$

where $T_{p\times q} = \{\bar{U} = [U_{ij}] \in \Re^{Np\times Nq} | U_{ij} \in \Re^{p\times q};\ U_{ij} = 0 \text{ if } j \neq N_i\}$.

Introducing $\tilde{x}(k) = [\tilde{x}^t(k) \quad \hat{x}^t(k)]^t$, the following augmented system which governs the filtering dynamics for the sensor network is obtained:

$$\tilde{x}(k+1) = \left[\tilde{A} + sum_{i=1}^{n} (\beta_i(k) - \mu_i)\tilde{C}_i \right] f(\tilde{x}(k)) + \tilde{D}w(k),$$
$$\tilde{z}(k) = \tilde{M} f(\tilde{x}(k)). \tag{8.26}$$

Next, we introduce the following definitions.

Definition 8.1. The solution $\tilde{x}(k) = 0$ of the augmented system in (8.26) with $\omega(k) \equiv 0$ is said to be stochastically stable if, for any $\varepsilon > 0$, there exists a $\delta > 0$ such that

$$\mathbb{E}\{\|\tilde{x}(k)\|\} < \varepsilon, \tag{8.27}$$

whenever $k \in \mathbb{Z}^+$ and $\|\tilde{x}(0)\| < \delta$.

Definition 8.2. A square matrix $P := [p_{ij}] \in \Re^{N \times N}$ is called diagonally dominant if for all $i = 1, \ldots, N$

$$\forall i, |p_{ii}| \geq \sum_{j \neq i} |p_{ij}|. \tag{8.28}$$

The objective now is to design a filter of the form in (8.22) on each node i of the sensor network for system (8.18). This is achieved by finding the filter parameters K_{ij} and H_{ij} such that the following two requirements are satisfied simultaneously:

- (stochastic stability) the zero solution of the augmented system (8.26) with $w(k)0$ is stochastically stable;
- (average $\mathbb{H}_\infty$ performance) under the zero initial condition, for a given disturbance attenuation level $\gamma > 0$ and all nonzero $\omega(k)$, the filtering error $\tilde{z}(k)$ from (8.26) satisfies the following condition, which will be called in the sequel *average $\mathbb{H}_\infty$ performance constraint*:

$$J = \frac{1}{n}\|\tilde{z}\|_{E_2}^2 - \gamma^2\|w\|_2^2 < 0, \tag{8.29}$$

where

$$\|\tilde{z}\|_{E_2} = \mathbb{E}\left\{ \sqrt{\sum_{k=0}^{\infty} \tilde{z}^t(k)\tilde{z}(k)} \right\}, \quad \|\omega\|_2 = \sqrt{\sum_{k=0}^{\infty} \omega^t(k)\omega(k)}. \tag{8.30}$$

Remark 8.4. It is significant to note that the average $\mathcal{H}_\infty$ performance presented in (8.29) is a generalization of the classical $\mathbb{H}_\infty$ performance constraint. It means that the average energy gains (from the average energy of all disturbances on the system and sensor networks to the average energy of all estimation errors) for filters in the sensor network should be less than a given disturbance attenuation level γ.

8.2.3 Analytic results

In this section, let us investigate both the distributed $\mathbb{H}_\infty$ filtering performance and filter design problems for system (8.18) with n sensors whose topology is determined by the given graph $\mathbb{G} = (\mathbb{V}, \mathbb{E}, \mathbb{A})$. The following lemmas will be needed in establishing our main results.

Lemma 8.1. *[35] Let f be defined as in (8.19). If $P > 0$ is diagonally dominant, then the following inequality holds for any vector ζ:*

$$f^T(\zeta)Pf(\zeta) \leq \zeta^T P\zeta. \tag{8.31}$$

Lemma 8.2. *If there exist a Lyapunov function $V(\tilde{x}(k))$ and a function $\phi(r) \in \Xi$ satisfying the following conditions:*

$$V(0) = 0, \tag{8.32}$$

$$\phi(\|\tilde{x}(k)\|) \leq V(\tilde{x}(k)), \tag{8.33}$$

$$\mathbb{E}\{V(\tilde{x}(k+1))\} - \mathbb{E}\{V(\tilde{x}(k))\} < 0, \quad k \in I^+, \tag{8.34}$$

then the solution $\tilde{x}(k) = 0$ of the filtering augmented system in (8.26) with $w(k) \equiv 0$ is stochastically stable.

The following theorem gives a sufficient condition under which the stochastic repeated scalar nonlinear system governed by (8.26) is stochastically stable in the sense of Definition 8.1 with $\mathbb{H}_\infty$ performance constraint given in (8.29).

Theorem 8.1. *Consider the repeated scalar nonlinear system (8.18) with given filter parameters K_{ij}, H_{ij} and a prescribed $\mathcal{H}_\infty$ index $\gamma > 0$. If there exists a positive diagonally dominant matrix $P > 0$ satisfying*

$$\phi = \begin{bmatrix} \begin{bmatrix} \tilde{A}^t P\tilde{A} + \sum_{i=1}^n \sigma_i^2 \tilde{C}_i^t P\tilde{C}_i \\ -P + \frac{1}{n}\tilde{M}^t\tilde{M} \end{bmatrix} & \bullet \\ \tilde{D}^t P\tilde{A} & \tilde{D}^t P\tilde{D} - \gamma^2 I \end{bmatrix} < 0, \tag{8.35}$$

then the filtering dynamics in (8.26) are stochastically stable and also satisfy the $\mathbb{H}_\infty$ performance constraint (8.29).

Proof. We choose the following Lyapunov function for system (8.26):

$$V(\tilde{x}(k)) = \tilde{x}^T(k)P\tilde{x}(k), \tag{8.36}$$

with $P > 0$ being a matrix to be determined. Then, along the trajectory of system (8.26) with $w(k) = 0$, we have

$$\mathbb{E}\left\{\tilde{x}^t(k+1)P\tilde{x}(k+1) - \tilde{x}^t(k)P\tilde{x}(k)\right\} =$$

$$\mathbb{E}\left\{((\tilde{A} + \sum_{i=1}^n (\beta_i(k) - \mu_i)\tilde{C}_i)f(\tilde{x}(k)))^t\right.$$

$$\times P\big(\tilde{A}+\sum_{i=1}^{n}(\beta_i(k)-\mu_i)\tilde{C}_i\big)f(\tilde{x}(k))-\tilde{x}^t(k)P\tilde{x}(k)\Big\}=$$

$$\mathbb{E}\Big\{f^t(\tilde{x}(k))\Big(\tilde{A}^T P\tilde{A}+\sum_{i=1}^{n}\sigma_i^2\tilde{C}_i^t P\tilde{C}_i\Big)f(\tilde{x}(k))-\tilde{x}^t(k)P\tilde{x}(k)\Big\}. \quad (8.37)$$

According to Lemma 8.1, we have

$$\begin{aligned}&\mathbb{E}\{\Delta V(\tilde{x}(k))\}\\&\le\mathbb{E}\Big\{f^T(\tilde{x}(k))\Big(\tilde{A}^t P\tilde{A}+\sum_{i=1}^{n}\sigma_i^2\tilde{C}_i^t P\tilde{C}_i-P\Big)f(\tilde{x}(k))\Big\}.\end{aligned} \quad (8.38)$$

Note that (8.35) in Theorem 8.1 implies that

$$\tilde{A}^t P\tilde{A}+\sum_{i=1}^{n}\sigma_i^2\tilde{C}_i^t P\tilde{C}_i-P<0,$$

which results in

$$\mathbb{E}\{V(\tilde{x}(k+1))\}-E\{V(\tilde{x}(k))\}<0.$$

Taking $\phi(r)=\lambda_{\min}(P)\Re^2$ such that $\phi(r)\in\Xi$, wc obtain

$$\begin{aligned}\phi(\|\tilde{x}(k)\|)&=\lambda_{\min}(P)\|\tilde{x}(k)\|^2\\&=\lambda_{\min}(P)\tilde{x}^t(k)\tilde{x}(k)\le\tilde{x}^t(k)P\tilde{x}(k)=V(\tilde{x}(k)),\end{aligned}$$

which satisfies (8.33). Considering $V(0)=0$, it follows readily from Lemma 8.2 that the filtering dynamics in (8.26) with $w(k)=0$ are stochastically stable.

In order to establish the $\mathbb{H}_\infty$ performance, we introduce the following index:

$$\begin{aligned}\bar{J}=&\ \mathbb{E}\{V(\tilde{x}(k+1))\}+E\Big\{\frac{1}{n}\tilde{z}^t(k)\tilde{z}(k)\Big\}-\gamma^2\omega^t(k)\omega(k)\\&-\mathbb{E}\big\{f^t(\tilde{x}(k))Pf(\tilde{x}(k))\big\}.\end{aligned} \quad (8.39)$$

From (8.26), one has

$$\begin{aligned}\bar{J}=&\ \mathbb{E}\Big\{\Big[\Big(\tilde{A}+\sum_{i=1}^{n}(\beta_i(k)-\mu_i)\tilde{C}_i\Big)f(\tilde{x}(k))+\tilde{D}\omega(k)\Big]^t\\&\times P\Big[\Big(\tilde{A}+\sum_{i=1}^{n}(\beta_i(k)-\mu_i)\tilde{C}_i\Big)f(\tilde{x}(k))+\tilde{D}\omega(k)\Big]\\&+f^T(\tilde{x}(k))\Big(\frac{1}{n}\tilde{M}^T(k)\tilde{M}-P\Big)f(\tilde{x}(k))-\gamma^2\omega^t(k)\omega(k)\Big\}\\&=\mathbb{E}\{\eta^t(k)\Phi\eta(k)\},\end{aligned} \quad (8.40)$$

where $\eta(k) = [f^t(\tilde{x}(k))\ \omega^t(k)w(k)]^t$. It follows from (8.35) in Theorem 8.1 that $\bar{J} < 0$.

By summing up (8.39) on both sides from 0 to ∞ with respect to k and taking mathematical expectation, we arrive at

$$\sum_{k=0}^{\infty} \bar{J} = \sum_{k=0}^{\infty} \mathbb{E}\{V(\tilde{x}(k+1))\} - \sum_{k=0}^{\infty} \mathbb{E}\{f^t(\tilde{x}(k))Pf(\tilde{x}(k))\} + \frac{1}{n}\|\tilde{z}\|_{E_2}^2 - \gamma^2\|\omega\|_2^2 < 0. \tag{8.41}$$

Hence, the $\mathbb{H}_\infty$ performance index defined in (8.29) is given by

$$J = \sum_{k=0}^{\infty} \bar{J} - \Big(\sum_{k=0}^{\infty} E\{V(\tilde{x}(k+1))\} - \sum_{k=0}^{\infty}\{f^T(\tilde{x}(k))Pf(\tilde{x}(k))\}\Big). \tag{8.42}$$

By Lemma 8.1 and noting that $\bar{J} < 0$ together with the zero initial condition, we have

$$\begin{aligned} J &\le \sum_{k=0}^{\infty} \bar{J} - \Big(\sum_{k=0}^{\infty} E\{V(\tilde{x}(k+1))\} - \sum_{k=0}^{\infty} E\{\tilde{x}^T(k)P\tilde{x}(k)\}\Big) \\ &\le \sum_{k=0}^{\infty} \bar{J} - E\{VV(\tilde{x}(\infty))\} < 0, \end{aligned}$$

which is equivalent to (8.29), and the proof is complete. □

We now direct our attention to the problem of designing distributed $\mathbb{H}_\infty$ estimators for the repeated scalar nonlinear systems. One way to bypass the issue of dilating a diagonally dominant positive definite matrix P, that is, letting P remain in its original form (instead of being partitioned) in solving the estimator design problems, we rewrite the parameters $\tilde{A}$, $\tilde{D}$, and $\tilde{C}_i$ in Theorem 8.1 in the following form:

$$\tilde{A} = \tilde{A}_0 + EK_hR_1, \quad \tilde{D} = \tilde{D}_0 + EK_hR_2, \quad \tilde{C}_i = EK_hR_{3i}, \tag{8.43}$$

where

$$\tilde{A}_0 = \begin{bmatrix} \bar{A} & 0 \\ 0 & 0 \end{bmatrix}, \quad E = \begin{bmatrix} 0 \\ I \end{bmatrix}, \quad R_1 = \begin{bmatrix} 0 & I \\ C_\mu & 0 \end{bmatrix},$$

$$R_{3i} = \begin{bmatrix} 0 & 0 \\ \bar{C}_i & 0 \end{bmatrix}, \quad \tilde{D}_0 = \begin{bmatrix} \bar{B}^t & 0 \end{bmatrix}^t,$$

$$R_2 = \begin{bmatrix} 0 & \bar{D}^t \end{bmatrix}^t, \quad K_h = \begin{bmatrix} \bar{K} & \bar{H} \end{bmatrix}. \tag{8.44}$$

Observe that the matrix K_h consists of all desired estimator parameters. By the Schur complement and noting (8.43), it follows that (8.35) is equivalent to

$$\begin{bmatrix} \frac{1}{n}\tilde{M}^t(k)\tilde{M} - P & \bullet & \bullet & \bullet \\ 0 & -\gamma^2 I & \bullet & \bullet \\ \tilde{A}_0 + EK_hR_1 & \tilde{D}_0 + EK_hR_2 & -P^{-1} & \bullet \\ \bar{X} & 0 & 0 & -\bar{P}^{-1} \end{bmatrix} < 0, \tag{8.45}$$

where

$$\bar{X} = [\sigma_1 \Re_{31}^t (EK_h)^t \ \ldots \ \sigma_n \Re_{3n}^t (EK_h)^t]^t, \ \bar{P}^{-1} = I_n \otimes P^{-1}. \tag{8.46}$$

The existence condition of the desired distributed $\mathbb{H}_\infty$ estimators is provided by the following theorem.

Theorem 8.2. *Given a positive scalar $\gamma > 0$, the filtering dynamics in (8.26) are stochastically stable and satisfy the distributed $\mathcal{H}_\infty$ performance constraint (8.29) if there exist matrices $0 < P := [p_{ij}] \in \Re^{2nn_x \times 2nn_x}$, $0 < Q := [q_{ij}] \in \Re^{2nn_x \times 2nn_x}$, $L > 0$, $\bar{K} \in T_{n_x \times n_x}$, and $\bar{H} \in T_{n_x \times n_y}$ satisfying*

$$\begin{bmatrix} \frac{1}{n}\tilde{M}^T(k)\tilde{M} - P & \bullet & \bullet & \bullet \\ 0 & -\gamma^2 I & \bullet & \bullet \\ \tilde{A}_0 + EK_hR_1 & \tilde{D}_0 + EK_hR_2 & -L & \bullet \\ \bar{X} & 0 & 0 & -\bar{L} \end{bmatrix} < 0, \tag{8.47}$$

$$p_{ii} - \sum_{j \neq i} (p_{ij} + 2q_{ij}) \geq 0, \tag{8.48}$$

$$q_{ij} \geq 0 \, \forall \, i \neq j, \tag{8.49}$$

$$p_{ij} + q_{ij} \geq 0 \, \forall \, i \neq j, \tag{8.50}$$

$$P \, L = I, \tag{8.51}$$

where K_h is defined in (8.44) and

$$\bar{L} = I_n \otimes L. \tag{8.52}$$

Moreover, if the above inequalities (with one equality constraint) are feasible, the desired filter parameters K_{ij} and H_{ij} ($i = 1, 2, \ldots, n$, $j \in N_i$) can be derived from (8.24).

Proof. By letting $P^{-1} = L$, inequality (8.45) follows from (8.47) immediately. In addition, from (8.48)–(8.50), we have

$$p_{ii} \geq \sum_{j\neq i}(p_{ij} + 2q_{ij}) = \sum_{j\neq i}(|p_{ij} + q_{ij}| + |-q_{ij}|) \geq \sum_{j\neq i}|p_{ij}|,$$

which guarantees the positive definite matrix P to be diagonally dominant. This completes the proof. □

Due to the presence of the matrix equality in Theorem 8.2, condition (8.47) will give rise to a nonconvex feasible set. In the following, a CCL method is proposed to solve the problem.

In [35], it is reported that the basic idea in the CCL algorithm is that if $\Omega(P, L) = \begin{bmatrix} P & I \\ I & L \end{bmatrix} \geq 0$ *is feasible in the* $2nn_x \times 2nn_x$ *matrix variables* $L > 0$ *and* $P > 0$, *then* $tr(PL) \geq 2nn_x$, *and* $tr(PL) = 2nn_x$ *if and only if* $PL = I$.

Based on such a statement, we put forward the following CCL problem instead of the original nonconvex feasibility problem formulated in Theorem 8.2.

The following design procedure is proposed to implement the distributed $\mathbb{H}_\infty$ filters for the target plant with repeated scalar nonlinearity and random packet losses.

8.2.4 Design algorithm

(S1) Given the $\mathbb{H}_\infty$ performance index γ and the positive weighting matrix R, find a feasible set $(P_{(0)}, L_{(0)}, Q_{(0)}, \bar{K}_{(0)}, \bar{H}_{(0)})$ satisfying (8.47), (8.50), and $\Omega(P, L) \geq 0$. Set $q = 0$. If there are none, exit.

(S2) Solve the problem min **trace**$(PL_{(q)} + P_{(q)}L)$ using (8.47), (8.50), and $\Omega(P, L) \geq 0$.

(S3) Substitute the obtained matrix variables $(P, L, Q, \bar{K}, \bar{H})$ into (8.35). If condition (8.35) is satisfied with $|tr(PL) - 2nn_x| < \delta$ for some sufficiently small scalar $\delta > 0$, then output the feasible solutions $(P, L, Q, \bar{K}, \bar{H})$ and exit.

(S4) If $q > N$, where N is the maximum number of iterations allowed, exit.

(S5) Set $q = q + 1$, $(P_{(q)},$

$$L_{(q)}, Q_{(q)}, \bar{K}_{(q)}, \bar{H}_{(q)}) = (P, L, Q, \bar{K}, \bar{H})$$

and go to Step S2.

8.2.5 Simulation example 8.3

To demonstrate the applicability of the proposed filtering techniques, following [19], we consider a factory that produces two kinds of products ($j = 1, 2$)

sharing common resources and raw materials, like, for instance, color TV and black/white TV or PC and laptop computer. During the kth period (quarter or season), we define:

$s_j(k)$, amount of sales of product j;
$a_j(k)$, advertisement cost spent for product j;
$i_j(k)$, amount of inventory of product j; and
$m_j(k)$, production of product j,

where $j = 1, 2$. Let

$$x(k) = \begin{bmatrix} s_1(k) \\ s_2(k) \\ i_1(k) \\ i_2(k) \end{bmatrix}, \quad u_k = \begin{bmatrix} m_1(k+1) \\ m_2(k+1) \\ a_1(k) \\ a_2(k) \end{bmatrix}.$$

Then, the effect of advertisements on sales in the marketing process and the interlink between inventory and production in the production process can be expressed dynamically by the following form:

$$x(k+1) = Af(x(k)) + Bw(k) + Eu(k),$$

where $f(x(k))$, which accounts for the state-dependent perturbations, is a saturation nonlinearity function with the following form:

$$f(x(k)) := \begin{bmatrix} g(x_1(k)) & g(x_2(k)) & g(x_3(k)) & g(x_4(k)) \end{bmatrix}^t, \tag{8.53}$$

$$g(x_i(k)) = \begin{cases} x_i(k), & -1 \le x_i(k) \le 1, \\ 1, & x_i(k) > 1, \\ -1, & x_i(k) < -1. \end{cases} \tag{8.54}$$

Let us assume that no production and advertisement are available in this example, and hence $E = 0$. Also, we assume that the information transmission for the measured amounts of sales and inventory of the products is conducted through a sensor network which is subject to possible packet losses. The sensor network shown in Fig. 8.14 is represented by a directed graph $\mathbb{G} = (\mathbb{V}, \mathbb{E}, \mathbb{A})$ with the set of nodes $\mathbb{V} = \{1, 2, 3, 4\}$, the set of edges

$$\mathbb{E} = \{(1,1), (1,4), (2,1), (2,2), (2,3), (3,1), (3,3), (4,3), (4,4)\},$$

and the following adjacency matrix:

$$\mathbb{A} = \begin{bmatrix} 1 & 0 & 0 & 1 \\ 1 & 1 & 1 & 0 \\ 1 & 0 & 1 & 0 \\ 0 & 0 & 1 & 1 \end{bmatrix}.$$

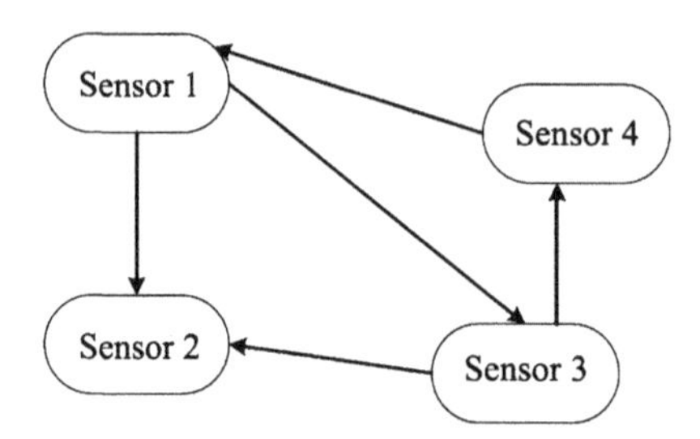

FIGURE 8.14 Topological structure of the sensor network.

To apply the proposed distributed filtering scheme, we consider a specific example of the problem discussed above. The corresponding system parameters are listed as follows:

$$A = \begin{bmatrix} 0.4 & 0 & 0 & 0 \\ 0 & 0.5 & 0 & 0 \\ -0.4 & 0 & 0.6 & 0 \\ 0 & -0.5 & 0 & 0.6 \end{bmatrix}, \quad B = \begin{bmatrix} 0 & 0.3 & 0 & 0 \\ 0.2 & 0 & 0 & 0 \\ 0 & -0.4 & 0 & 0 \\ -0.2 & 0 & 0 & 0 \end{bmatrix},$$
$$C_1 = \begin{bmatrix} 1 & 0 & 0 & 0 \end{bmatrix}, C_2 = \begin{bmatrix} 0 & 1 & 0 & 0 \end{bmatrix},$$
$$C_3 = \begin{bmatrix} 1 & 1 & 0 & 0 \end{bmatrix}, C_4 = \begin{bmatrix} 1 & 0 & 1 & 0 \end{bmatrix},$$
$$D_1 = \begin{bmatrix} 0.5 & 0 & 0 & 0 \end{bmatrix}, D_2 = \begin{bmatrix} 0.2 & 0.1 & 0 & 0 \end{bmatrix},$$
$$D_3 = \begin{bmatrix} 0 & 0 & 0.2 & 0 \end{bmatrix}, D_4 = \begin{bmatrix} 0.1 & 0 & 0 & 0.1 \end{bmatrix},$$
$$M = \begin{bmatrix} 0 & 1 & 0 & 0 \end{bmatrix}, E = 0.$$

Let the probabilistic density functions of $\beta_1(k)$, $\beta_2(k)$, $\beta_3(k)$, and $\beta_4(k)$ in [0, 1] be described by

$$q_1(s_1) = \begin{cases} 0, & s_1 = 0, \\ 0.1, & s_1 = 0.5, \\ 0.9, & s_1 = 1, \end{cases} \quad q_2(s_2) = \begin{cases} 0.1, & s_2 = 0, \\ 0.1, & s_2 = 0.5, \\ 0.8, & s_2 = 1, \end{cases}$$
$$q_3(s_3) = \begin{cases} 0, & s_3 = 0, \\ 0.2, & s_3 = 0.5, \\ 0.8, & s_3 = 1, \end{cases} \quad q_4(s_4) = \begin{cases} 0, & s_4 = 0, \\ 0.4, & s_4 = 0.5, \\ 0.6, & s_4 = 1. \end{cases} \tag{8.55}$$

It follows that the expectations and variances are calculated as $\mu_1 = 0.95$, $\mu_2 = 0.85$, $\mu_3 = 0.9$, $\mu_4 = 0.8$, $\sigma_1^2 = 0.15$, $\sigma_2^2 = 0.32$, $\sigma_3^2 = 0.2$, and $\sigma_4^2 = 0.245$. The disturbance attenuation level and the positive definite matrices are given as

$\gamma = 1.2$ and $R_1 = R_2 = R_3 = R_4 = 2I_8$, respectively. The purpose is to design a distributed $\mathbb{H}_\infty$ filter in order to estimate the amount of sales of a product. By applying Theorem 8.2 with help from the design procedure, we can obtain the following parameters of the desired distributed filters:

$$K_{11} = \begin{bmatrix} 0.1137 & -0.1625 & 0.0158 & -0.0531 \\ -0.1915 & 0.4749 & -0.0877 & -0.1100 \\ 0.2716 & -0.1200 & 0.3754 & -0.2617 \\ -0.0188 & 0.0202 & 0.1044 & 0.6081 \end{bmatrix},$$

$$K_{14} = \begin{bmatrix} 0.2439 & 0.0750 & -0.0961 & -0.0938 \\ 0.0458 & 0.1037 & 0.0736 & 0.0247 \\ -0.2167 & 0.3131 & -0.1808 & -0.0713 \\ 0.0102 & -0.1909 & 0.0602 & 0.0352 \end{bmatrix},$$

$$K_{21} = \begin{bmatrix} -0.0852 & 0.0193 & 0.0280 & -0.0121 \\ -0.1592 & 0.0814 & 0.1600 & 0.0569 \\ 0.0110 & 0.0287 & 0.2255 & 0.0179 \\ -0.0409 & -0.0854 & -0.0536 & 0.0567 \end{bmatrix},$$

$$K_{22} = \begin{bmatrix} 0.5380 & -0.0176 & -0.0566 & 0.1805 \\ -0.2530 & 0.4950 & -0.0272 & -0.0026 \\ -0.3970 & 0.0877 & 0.4407 & -0.2361 \\ 0.3020 & 0.0255 & -0.0732 & 0.4410 \end{bmatrix},$$

$$K_{23} = \begin{bmatrix} 0.0175 & 0.0216 & -0.0046 & 0.0136 \\ 0.1703 & -0.0971 & -0.1353 & 0.1793 \\ 0.2161 & -0.0226 & -0.1292 & 0.1740 \\ -0.0519 & 0.0367 & 0.0420 & -0.1516 \end{bmatrix},$$

$$K_{31} = \begin{bmatrix} 0.0690 & -0.0771 & -0.1142 & 0.0979 \\ 0.2841 & 0.0122 & 0.0060 & -0.0909 \\ 0.2946 & 0.0726 & 0.0485 & -0.0467 \\ -0.1500 & -0.0464 & 0.1253 & 0.1185 \end{bmatrix},$$

$$K_{33} = \begin{bmatrix} 0.3091 & 0.0055 & 0.1363 & 0.0433 \\ -0.0245 & 0.6476 & 0.0959 & 0.0404 \\ 0.3798 & 0.3682 & 0.4620 & 0.2560 \\ 0.2610 & -0.0723 & 0.0302 & 0.6325 \end{bmatrix},$$

$$K_{43} = \begin{bmatrix} -0.1472 & 0.0150 & 0.0801 & 0.0331 \\ -0.1411 & -0.0687 & 0.1451 & -0.2717 \\ -0.2472 & -0.2496 & 0.3624 & -0.1996 \\ -0.0143 & -0.0310 & -0.0192 & 0.0628 \end{bmatrix},$$

$$K_{44} = \begin{bmatrix} 0.5910 & 0.0532 & -0.1676 & -0.1281 \\ -0.0105 & 0.5823 & -0.0889 & -0.0401 \\ -0.3628 & 0.1010 & 0.4647 & 0.0711 \\ -0.1753 & 0.0505 & 0.0249 & 0.4736 \end{bmatrix},$$

$$H_{11} = \begin{bmatrix} 0.0373 \\ -0.1116 \\ -0.8248 \\ 0.1619 \end{bmatrix}, H_{14} = \begin{bmatrix} 0.1336 \\ -0.0340 \\ 0.5162 \\ 0.0074 \end{bmatrix},$$

$$H_{21} = \begin{bmatrix} 0.0278 & -0.0884 & -0.0630 & 0.0845 \end{bmatrix}^t,$$

$$H_{22} = \begin{bmatrix} 0.3096 & -0.3339 & -0.6263 & 0.4021 \end{bmatrix}^t,$$

$$H_{23} = \begin{bmatrix} 0.0180 & 0.0047 & 0.0066 & -0.0897 \end{bmatrix}^t,$$

$$H_{31} = \begin{bmatrix} 0.6074 & 0.0560 & -0.8735 & 0.0408 \end{bmatrix}^t,$$

$$H_{23} = \begin{bmatrix} 0.0180 & 0.0047 & 0.0066 & -0.0897 \end{bmatrix}^t,$$

$$H_{33} = \begin{bmatrix} -0.2468 & -0.0439 & 0.2344 & 0.1118 \end{bmatrix}^t,$$

$$H_{43} = \begin{bmatrix} -0.0235 & -0.0300 & -0.1328 & 0.0966 \end{bmatrix}^t,$$

$$H_{44} = \begin{bmatrix} 0.1702 & 0.0700 & 0.4929 & -0.0385 \end{bmatrix}^t.$$

For simulation purposes, we first assume that $w(k) = 0$ and the initial conditions $x(0) = [-0.4\ 0.5\ 0\ 0.2]^t$ and $\hat{x}_i(0) = [0\ 0.1\ 0\ 0]^t$ $(i = 1, 2, 3, 4)$. Fig. 8.15 shows the output $z(k)$ and its estimates from the filters 1, 2, 3, and 4, and Fig. 8.16 gives the simulation results

of estimation error $\tilde{z}(k)$, which confirms that the augmented system in (8.26) is stochastically stable. To further illustrate the performance of the designed distributed filters, we now assume the zero initial conditions and set the exogenous disturbance input to be $w(k) = \exp(-k/40)$. The simulation results are shown in Figs. 8.17 and 8.18. Fig. 8.17 plots the output $z(k)$ and its estimates from the filters 1, 2, 3, and 4. Fig. 8.18 shows the estimation errors $z(k) - \hat{z}_i(k)$ $(i = 1, 2, 3, 4)$. It can be calculated that $\frac{\frac{1}{n}\|\tilde{z}\|_{E_2}}{\|w\|_2} = 0.8933$, which stays below the prescribed upper bound $\gamma = 1.2$.

8.3 Distributed filtering with saturation

In this section, we examine the distributed $\mathbb{H}_\infty$ estimation problem for a class of nonlinear systems with randomly occurring sensor saturation (ROSS) and successive packet dropouts over networks. To this end, we consider the esti-

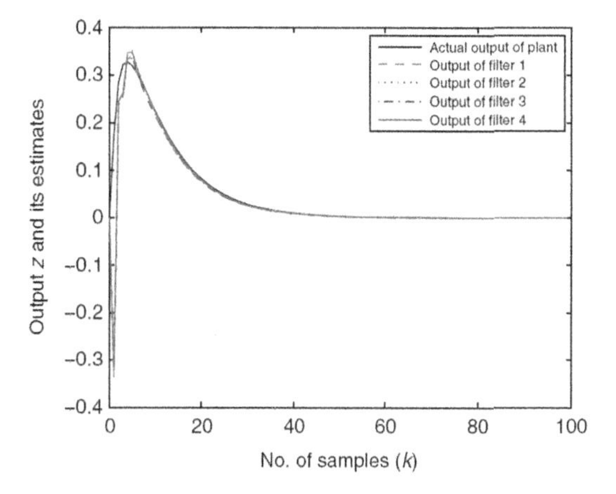

FIGURE 8.15 Output $z(k)$ and its estimates when $w(k) = 0$.

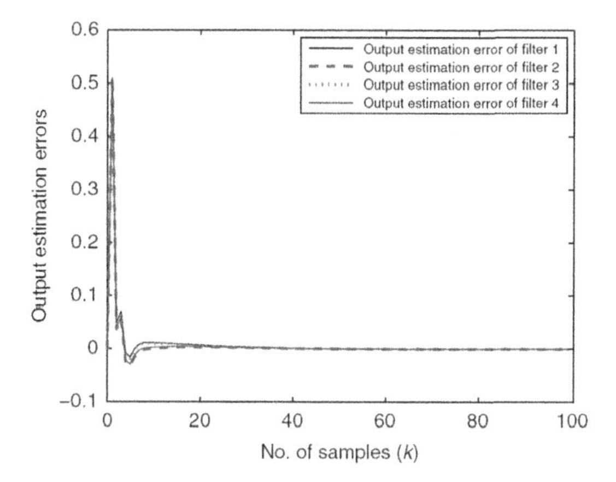

FIGURE 8.16 Filtering errors $z(k) - \hat{z}_i(k)$ when $w(k) = 0$ $(i = 1, 2, 3, 4)$.

mator configuration with n sensors, where each sensor can receive information from both the plant and its neighboring sensors. The information received by sensor i from the plant is transmitted via communication cables which are of limited capacity and, therefore, may suffer from the phenomena of ROSS and packet dropouts. On the other hand, sensor i can also obtain information from its neighboring sensors according to the topology of the sensor network.

In a similar way, we assume that the N sensor nodes are distributed in space according to a fixed network topology represented by a directed graph $\mathbb{G} = (\mathbb{V}, \mathbb{E}, \mathbb{A})$ of order n with the set of nodes $\mathcal{V} = 1, 2, \ldots, n$, the set of edges $\mathbb{E} \in \mathbb{V} \times \mathbb{V}$, and the weighted adjacency matrix $\mathbb{A} = [a_{ij}]$ with nonnegative adjacency element a_{ij}. An edge of $\mathbb{G}$ is denoted by the ordered pair (i, j). The adjacency elements associated with the edges of the graph are positive (i.e., $a_{ij} > 0 \Leftrightarrow$

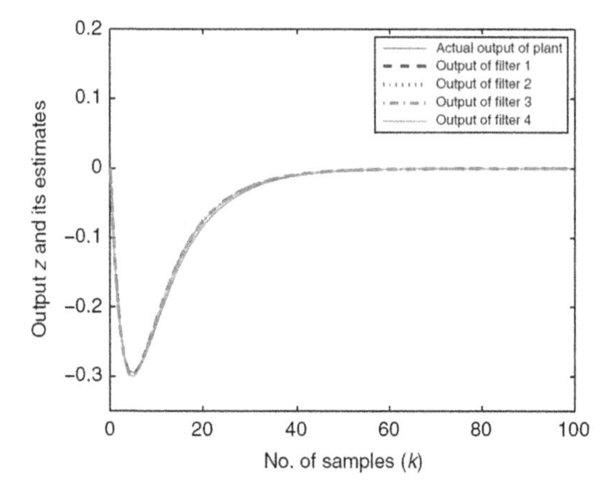

FIGURE 8.17 Output $z(k)$ and its estimates.

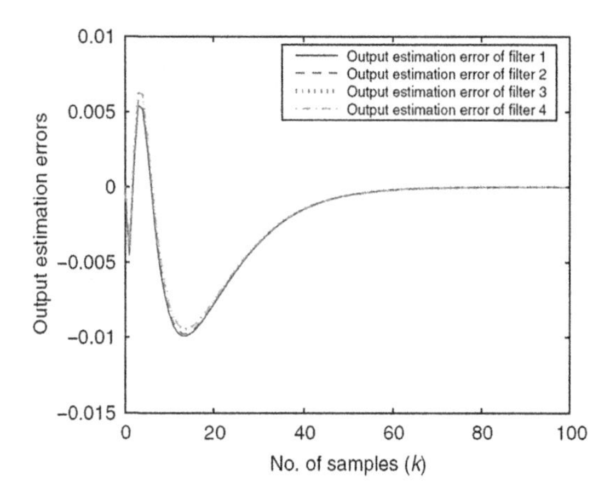

FIGURE 8.18 Filtering errors $z(k) - \hat{z}_i(k)$ $(i = 1, 2, 3, 4)$.

$(i, j) \in \mathbb{E}$), which means that sensor i can obtain information from sensor j. Also, we assume that $a_{ii} = 1$ for all $i \in \mathbb{V}$; therefore, (i, i) can be regarded as an additional edge. The set of neighbors of node $i \in \mathbb{V}$ plus the node itself are denoted by $\mathbb{N}_i = \{j \in \mathbb{V} : (i, j) \in \mathcal{E}\}$.

8.3.1 Problem formulation

Consider a discrete-time nonlinear system

$$\begin{aligned} x(k+1) &= A\,x(k) + f(x(k)) + Bw(k), \\ z(k) &= M\,x(k), \end{aligned} \tag{8.56}$$

where $x(k) \in \Re^{n_x}$ represents the state vector which cannot be observed directly, $z(k) \in \Re^{n_z}$ is the output to be estimated, and $w(k) \in \Re^{n_w}$ is the disturbance input belonging to $l_2[0, \infty)$. The nonlinear function $f(x(\cdot))$ satisfies the following condition:

$$\|f(x(\cdot))\|^2 \le \|Ex(\cdot)\|^2, \tag{8.57}$$

where E is a known constant matrix.

In this section, the n sensors with both saturation and packet dropouts are modeled by

$$\begin{aligned} y_i(k) &= \beta_i(k)\sigma(C_i x(k)) + (1-\beta_i(k))\gamma_i(k)C_i x(k) + (1-\beta_i(k))(1-\gamma_i(k)) \\ &\times y_i(k-1) + D_i v(k), \quad i = 1, 2, \ldots, n, \end{aligned} \tag{8.58}$$

where $y_i(k) \in \Re^{n_y}$ is the measurement output measured by sensor i from the plant and $\beta_i(k)$ and $\gamma_i(k) (i = 1, 2, \ldots, n)$ are Bernoulli-distributed white sequences taking values on 0 and 1 with

$$\begin{cases} \text{Prob}\{\beta_i(k) = 1\} = \bar{\beta}_i, \\ \text{Prob}\{\beta_i(k) = 0\} = 1 - \bar{\beta}_i \end{cases} \quad \text{and} \quad \begin{cases} \text{Prob}\{\gamma_i(k) = 1\} = \bar{\gamma}_i, \\ \text{Prob}\{\gamma_i(k) = 0\} = 1 - \bar{\gamma}_i, \end{cases}$$

respectively, where $\bar{\beta}_i, \bar{\gamma}_i \subset [0, 1]$ are known constants. Throughout the chapter, the stochastic variables $\beta_i(k)$ and $\gamma_i(k)$ are independent mutually in all i $(1 \le i \le n)$.

The saturation function $\sigma(\cdot) : \Re^{n_y} \mapsto \Re^{n_y}$ is defined as

$$\sigma(\vartheta) = \left[\sigma_1^T(\vartheta_1) \; \sigma_2^T(\vartheta_2) \; \cdots \; \sigma_{n_y}^T(\vartheta_{n_y})\right]^T, \tag{8.59}$$

with $\sigma_i(\vartheta_i) = \text{sign}(\vartheta_i)\min\{\vartheta_{i,\max}, |\vartheta_i|\}$, where $\vartheta_{i,max}$ is the ith element of the vector $\vartheta_{\max}$, the saturation level.

Definition 8.3. [20] A nonlinearity $\Psi : \Re^m \mapsto \Re^m$ is said to satisfy a sector condition if

$$(\Psi(\vartheta) - \mathcal{H}_1\vartheta)^T(\Psi(\vartheta) - \mathcal{H}_2\vartheta) \le 0, \quad \forall \vartheta \in \Re^{n_y} \tag{8.60}$$

for some real matrices $\mathcal{H}_1, \mathcal{H}_2 \in \Re^{n_y \times n_y}$, where $\mathcal{H} = \mathcal{H}_2 - \mathcal{H}_1$ is a symmetric positive definite matrix. In this case, we say that Ψ belongs to the sector $[\mathcal{H}_1, \mathcal{H}_2]$.

As in Refs. [21], [22], [24], assuming that there exist diagonal matrices $\underline{L}_i$ and $\bar{L}_i$ such that $0 \le \underline{L}_i < I \le \bar{L}_i$, the saturation function $\sigma(Cix(k))$ in (8.58) can be decomposed into a linear and a nonlinear part as

$$\sigma(C_i x(k)) = \underline{L}_i C_i x(k) + \Psi(C_i x(k)), \tag{8.61}$$

where $\Psi(C_i x(k))$ is a nonlinear vector-valued function satisfying a sector condition with $\mathcal{H}_1 = 0$ and $\mathcal{H}_2 = L_i$, which can be described as follows:

$$\Psi^T(C_i x(k))(\Psi(C_i x(k)) - L_i C_i x(k)) \leq 0, \tag{8.62}$$

where $L_i = \bar{L}_i - \underline{L}_i$.

Remark 8.5. The proposed measurement model in (8.58) provides a novel unified framework to account for the phenomena of both ROSS and successive packet dropouts. The stochastic variable $\beta_i(k)$ characterizes the random nature of sensor saturation and, on the other hand, the stochastic variable $\gamma_i(k)$ describes possible successive packet dropouts. It can be seen clearly from (8.58) that, if $\beta_i(k) = 1$, model (8.58) is reduced to the one with saturation; if $\beta_i(k) = 0$ and $\gamma_i(k) = 1$, model (8.58) specializes to the one with neither saturation nor packet dropouts (that is, the sensor i works normally); if $\beta_i(k) = 0$ and $\gamma_i(k) = 0$, model (8.58) degenerates to the one with successive packet dropouts (i.e., the latest measurement received for sensor i will be used if the current measurement is lost during transmissions). Therefore, model (8.58) is comprehensive in that it takes into account both the probabilistic sensor saturation and probabilistic successive packet dropouts in an environment of sensor networks.

As is well known, a key point in designing distributed filters for sensor networks is how to fuse the information available for the filter on the sensor node i both from the sensor i itself and from its neighbors. Keeping such a fact in mind, the following filter structure is adopted in this chapter on sensor node i:

$$\begin{cases} \hat{x}_i(k+1) = \sum_{j \in \mathcal{N}_i} a_{ij} K_{ij} \hat{x}_j(k) + \sum_{j \in \mathcal{N}_i} a_{ij} H_{ij} y_j(k), \\ \hat{z}_i(k) = M \hat{x}_i(k), \end{cases} \tag{8.63}$$

where $\hat{x}_i(k) \in \Re^{n_x}$ is the state estimate on sensor node i and $\hat{z}_i(k) \in \Re^{n_z}$ is the estimate of $z(k)$ from the filter on sensor node i. Here, matrices K_{ij} and H_{ij} are the filter parameters on node i to be determined. The initial values of filters $\hat{x}_i(0)$ $(i = 1, 2, \ldots, n)$ are assumed to be known vectors.

Remark 8.6. The filter structure in (8.63) establishes the communications between sensor node i and its neighboring nodes, in which the sensor nodes are distributed over a spatial region. It is worth mentioning that (8.63) represents a quite general filter model structure. To see this, assuming that there is no communication between sensor node i and its neighboring nodes, the filter (8.63) can be reduced to

$$\hat{x}_i(k+1) = K_{ii} \hat{x}_i(k) + H_{ii} y_i(k), \tag{8.64}$$

which has been widely adopted for filter design in the literature.

For convenience of later analysis, we denote

$$\begin{aligned}
&\hat{x}(k)=\begin{bmatrix}\hat{x}_1^T(k) & \hat{x}_2^T(k) & \cdots & \hat{x}_n^T(k)\end{bmatrix}^T, \ \bar{x}(k)=\mathbf{1}_n\otimes x(k),\ \bar{C}_i=(e_ie_i^T)\otimes C_i,\\
&\hat{z}(k)=\begin{bmatrix}\hat{z}_1^T(k) & \hat{z}_2^T(k) & \cdots & \hat{z}_n^T(k)\end{bmatrix}^T, \ \bar{z}(k)=\mathbf{1}_n\otimes z(k),\ \bar{B}=\mathbf{1}_n\otimes B,\\
&\tilde{y}(k)=\begin{bmatrix}y_1^t(k) & y_2^t(k) & \cdots & y_n^t(k)\end{bmatrix}^t, \ \bar{A}=I_n\otimes A,\ \bar{M}=I_n\otimes M,\\
&\tilde{D}=\begin{bmatrix}D_1^T & D_2^T & \cdots & D_n^T\end{bmatrix}^T, \ g(\bar{x}(k))=\mathbf{1}_n\otimes f(x(k)),\\
&\bar{\Lambda}_\gamma=\mathrm{diag}\{\bar{\gamma}_1,\bar{\gamma}_2,\ldots,\bar{\gamma}_n\},\ \ \bar{\Lambda}_\beta=\mathrm{diag}\{\bar{\beta}_1,\bar{\beta}_2,\ldots,\bar{\beta}_n\},\\
&\tilde{C}=\mathrm{diag}\{C_1,C_2,\ldots,C_n\},\ \ \tilde{L}=\mathrm{diag}\{\underline{L}_1,\underline{L}_2,\ldots,\underline{L}_n\},\\
&\hat{L}=\mathrm{diag}\{L_1,L_2,\ldots,L_n\},\ \ \bar{E}_n=In\otimes E^TE,
\end{aligned} \tag{8.65}$$

where

$$\begin{aligned}
&\bar{K}=[\bar{K}_{ij}]_{n\times n}\text{ with }\bar{K}_{ij}=\begin{cases}a_{ij}K_{ij}, & i=1,2,\ldots,n,\ \ j\in\mathcal{N}_i,\\ 0, & i=1,2,\ldots,n,\ \ j\notin\mathcal{N}_i,\end{cases}\\
&\bar{H}=[\bar{H}_{ij}]_{n\times n}\text{ with }\bar{H}_{ij}=\begin{cases}a_{ij}H_{ij}, & i=1,2,\ldots,n,\ \ j\in\mathcal{N}_i,\\ 0, & i=1,2,\ldots,n,\ \ j\notin\mathcal{N}_i.\end{cases}
\end{aligned} \tag{8.66}$$

Obviously, since $a_{ij}=0$ when $j\notin\mathcal{N}_i$, $\bar{K}$ and $\bar{H}$ are two matrices that can be expressed as

$$\bar{K}\in\mathcal{T}_{n_x\times n_x},\quad \bar{H}\in\mathcal{T}_{n_x\times n_y}, \tag{8.67}$$

where $\mathcal{T}_{p\times q}=\{\bar{U}=[U_{ij}]\in\Re^{np\times nq}|U_{ij}\in R^{p\times q},\ U_{ij}=0\text{ if }j\notin\mathcal{N}_i\}$.

Letting $\eta(k)=[\bar{x}^t(k)\ \ \hat{x}^t(k)\tilde{y}^t(k-1)]^t$ and $\overrightarrow{z}(k)=\bar{z}(k)-\hat{z}(k)$, the following augmented system is obtained that governs the networked filtering dynamics:

$$\begin{aligned}
\eta(k+1)=&\ \overrightarrow{A}\eta(k)+\overrightarrow{g}(\overrightarrow{H}\eta(k))+\hat{H}\sigma(\tilde{C}\overrightarrow{H}\eta(k))+\overrightarrow{B}\overrightarrow{w}(k)\\
&+\sum_{i=1}^{n}(\beta_i(k)-\bar{\beta}_i)\overrightarrow{E}_i\sigma(\tilde{C}\overrightarrow{H}\eta(k))\\
&+\sum_{i=1}^{n}((1-\beta_i(k))\gamma_i(k)-(1-\bar{\beta}_i)\bar{\gamma}_i)\overrightarrow{E}_i\tilde{C}\overrightarrow{H}\eta(k)\\
&+\sum_{i=1}^{n}((1-\beta_i(k))(1-\gamma_i(k))-(1-\bar{\beta}_i)(1-\bar{\gamma}_i))\overrightarrow{E}_i\overrightarrow{F}\eta(k),\\
\overrightarrow{z}(k)=&\ \overrightarrow{M}\eta(k),
\end{aligned} \tag{8.68}$$

where

$$\vec{A} = \begin{bmatrix} \bar{A} & 0 & 0 \\ \bar{H}(I-\bar{\Lambda}_\beta)\bar{\Lambda}_\gamma\tilde{C} & \bar{K} & \bar{H}(I-\bar{\Lambda}_\beta)(I-\bar{\Lambda}_\gamma) \\ (I-\bar{\Lambda}_\beta)\bar{\Lambda}_\gamma\tilde{C} & 0 & (I-\bar{\Lambda}_\beta)(I-\bar{\Lambda}_\gamma) \end{bmatrix},$$

$$\vec{g}(\vec{H}\eta(k)) = \begin{bmatrix} g(\bar{x}(k)) \\ 0 \\ 0 \end{bmatrix},$$

$$\vec{M} = \begin{bmatrix} \bar{M} & -\bar{M} & 0 \end{bmatrix}, \; E_i = e_i e_i^T, \; \vec{H} = \begin{bmatrix} I & 0 & 0 \end{bmatrix},$$

$$\vec{F} = \begin{bmatrix} 0 & 0 & I \end{bmatrix}, \; \vec{B} = \begin{bmatrix} \bar{B} & 0 \\ 0 & \bar{H}\tilde{D} \\ 0 & \tilde{D} \end{bmatrix}, \; \vec{w}(k) = \begin{bmatrix} w(k) \\ v(k) \end{bmatrix},$$

$$\vec{E}_i = \begin{bmatrix} 0 \\ \bar{H}E_i \\ E_i \end{bmatrix}, \; \hat{H} = \begin{bmatrix} 0 \\ \bar{H}\bar{\Lambda}_\beta \\ \bar{\Lambda}_\beta \end{bmatrix}. \tag{8.69}$$

Also, from (8.57), (8.61), and (8.62), we have

$$\vec{g}^T(\vec{H}\eta(k))\vec{g}(\vec{H}\eta(k)) \le \eta^T(k)\vec{H}^t\bar{E}_n\vec{H}\eta(k), \tag{8.70}$$

$$\sigma(\tilde{C}\vec{H}\eta(k)) = \tilde{L}\tilde{C}\vec{H}\eta(k) + \|(\tilde{C}\vec{H}\eta(k)), \tag{8.71}$$

$$\Psi^T(\tilde{C}\vec{H}\eta(k))\left(\Psi(\tilde{C}\vec{H}\eta(k)) - \hat{L}\tilde{C}\vec{H}\eta(k)\right) \le 0, \tag{8.72}$$

where

$$\sigma(\tilde{C}\vec{H}\eta(k)) := [\sigma^t(C_1x(k)) \; \sigma^t(C_2x(k)) \; \cdots \; \sigma^T(C_nx(k))]^t,$$
$$\Psi(\tilde{C}\vec{H}\eta(k)) := [\Psi^t(C_1x(k)) \; \Psi^T(C_2x(k)) \; \cdots \; \Psi^T(C_nx(k))]^t.$$

Here, the notations σ and Ψ have been slightly abused to denote the vector-valued saturation functions and vector nonlinear functions of different dimensions, respectively.

Before proceeding further, we introduce the following definition.

Definition 8.4. The augmented system in (8.68) is said to be exponentially mean square stable if, with $\vec{w}(k) = 0$, there exist constants $\delta > 0$ and $0 < \kappa < 1$ such that

$$\mathbb{E}\{\|\eta(k)\|^2\} \le \delta\kappa^k\mathbb{E}\{\|\eta(0)\|^2\}, \;\; \forall\eta(0) \in \Re^n, \;\; k \in I^+.$$

Our aim in the sequel is to design a filter of the form in (8.63) on each node i of the sensor network for system (8.56). In other words, we are going to find the filter parameters K_{ij} and H_{ij} such that the following two requirements are satisfied simultaneously:

- Exponentially mean square stability. The zero solution of the augmented system (8.68) with $\vec{w}(k) = 0$ is exponentially mean square stable.
- $\mathbb{H}_\infty$ performance. Under zero initial conditions, for a given disturbance attenuation level $\gamma > 0$ and all nonzero $\vec{w}(k)$, the filtering error $\vec{z}(k)$ from (8.68) satisfies the following condition:

$$\sum_{k=0}^{\infty} \mathbb{E}\{\|\vec{z}(k)\|^2\} \leq \gamma^2 \sum_{k=0}^{\infty} \|\vec{w}(k)\|^2. \tag{8.73}$$

8.3.2 Design results

In this section, we investigate both the filter analysis and design problems for the distributed $\mathbb{H}_\infty$ filtering of system (8.56) with n sensors whose topology is determined by the given graph $\mathbb{G} = (\mathbb{V}, \mathbb{E}, \mathbb{A})$. The following lemma will be needed in establishing our main results.

Lemma 8.3. *[26] Let $P = diag\{P_1, P_2, \ldots, P_n\}$ with $P_i \in R^{p\times p}(1 \leq i \leq)$ being invertible matrices. If $X = PW$ for $W \in \Re^{np\times nq}$, then we have $W \in \mathcal{T}_{p\times q} \Leftrightarrow X \in \mathcal{T}_{p\times q}$.*

The following theorem gives a sufficient condition under which the augmented system (8.68) is exponentially mean square stable in the sense of Definition 8.4 with $\mathbb{H}_\infty$ performance constraint given in (8.73).

Theorem 8.3. *For given filter parameters K_{ij}, H_{ij} and a prescribed $\mathbb{H}_\infty$ index $\gamma > 0$, the filtering dynamics in (8.68) are exponentially mean square stable and also satisfy the $\mathbb{H}_\infty$ performance constraint (8.73) if there exist a positive definite matrix $P > 0$ and positive scalars ϵ_1 and ϵ_2 satisfying*

$$\Sigma = \begin{bmatrix} \Sigma_{11} & \bullet & \bullet & \bullet \\ P(\vec{A} + \hat{H}\tilde{L}\tilde{C}\vec{H}) & P - \epsilon_2 I & \bullet & \bullet \\ \Sigma_{31} + \epsilon_1 \hat{L}\tilde{C}\vec{H} & \hat{H}^t P & \Sigma_{33} - \epsilon_1 I & \bullet \\ \vec{B}^t P(\vec{A} + \hat{H}\tilde{L}\tilde{C}\vec{H}) & \vec{B}^t P & \vec{B}^t P\hat{H} & \vec{B}^t P\vec{B} - \gamma^2 I \end{bmatrix} < 0, \tag{8.74}$$

where

$$\Sigma_{11} = \bar{\Upsilon}_{11} + \epsilon_2 \vec{H}^T \bar{E}_n \vec{H} + \vec{M}^T \vec{M}, \ \Sigma_{31} = \bar{\Upsilon}_{31} + \hat{H}^T P(\vec{A} + \hat{H}\tilde{L}\tilde{C}\vec{H}),$$

$$\Sigma_{33} = \bar{\Upsilon}_{33} + \hat{H}^T P\hat{H}, \ \bar{\Upsilon}_{31} = \sum_{i=1}^{n} \varphi_i^2 \vec{E}_i^T P\vec{E}_i \tilde{L}\tilde{C}\vec{H}, \ \phi_i^2 = \bar{\beta}_i(1 - \bar{\beta}_i)\bar{\gamma}_i,$$

$$
\begin{aligned}
\bar{\Upsilon}_{11} &= (\vec{A} + \hat{H}\tilde{L}\tilde{C}\vec{H})^T P(\vec{A} + \hat{H}\tilde{L}\tilde{C}\vec{H}) \\
&\quad + \sum_{i=1}^{n} (2\bar{\phi}_i^2 + \hat{\phi}_i^2 + \hat{\varphi}_i^2)\vec{F}^T \vec{E}_i^T P \vec{E}_i \vec{F} \\
&\quad + \sum_{i=1}^{n} (\bar{\phi}_i^2 + \phi_i^2 + \varphi_i^2)\vec{H}^T \tilde{C}^T \tilde{L}^T \vec{E}_i^T P \vec{E}_i \tilde{L}\tilde{C}\vec{H} - P \\
&\quad + \sum_{i=1}^{n} (2\phi_i^2 + \hat{\phi}_i^2 + \bar{\varphi}_i^2)\vec{H}^T \tilde{C}^T \vec{E}_i^T P \vec{E}_i \tilde{C}\vec{H}, \\
\Upsilon_{33} &= \sum_{i=1}^{n} (\phi_i^2 + \bar{\phi}_i^2 + \varphi_i^2)\vec{E}_i^T P \vec{E}_i, \quad \bar{\varphi}_i^2 = (1-\bar{\beta}_i)\bar{\gamma}_i - (1-\bar{\beta}_i)^2\bar{\gamma}_i^2, \\
\varphi_i^2 &= \bar{\beta}_i(1-\bar{\beta}_i), \quad \hat{\varphi}_i^2 = (1-\bar{\beta}_i)(1-\bar{\gamma}_i) - (1-\bar{\beta}_i)^2(1-\bar{\gamma}_i)^2, \\
\bar{\phi}_i^2 &= \bar{\beta}_i(1-\bar{\beta}_i)(1-\bar{\gamma}_i), \quad \hat{\phi}_i^2 = \bar{\gamma}_i(1-\bar{\beta}_i)^2(1-\bar{\gamma}_i).
\end{aligned}
$$

Proof. Choose the following Lyapunov function for system (8.68):

$$V(\eta(k)) = \eta^T(k)P\eta(k); \tag{8.75}$$

the difference of the Lyapunov function is described as follows:

$$\Delta V(\eta(k)) = \mathbb{E}\{V(\eta(k+1))|\eta(k)\} - V(\eta(k)).$$

Then, along the trajectory of system (8.68) with $\vec{w}(k) = 0$, we have

$$
\begin{aligned}
&\mathbb{E}\{\Delta V(\eta(k))\} \\
&= \mathbb{E}\eta^t(k+1)P\eta(k+1) - \eta^t(k)P\eta(k) \\
&= \mathbb{E}\left\{\left[\vec{A}\eta(k) + \vec{g}(\vec{H}\eta(k)) + \hat{H}\sigma(\tilde{C}\vec{H}\eta(k)) + \sum_{i=1}^{n}(\beta_i(k) - \bar{\beta}_i)\vec{E}_i\right.\right. \\
&\times \sigma(\tilde{C}\vec{H}\eta(k)) + \sum^{n} i = 1((1-\beta_i(k))\gamma_i(k) - (1-\bar{\beta}_i)\bar{\gamma}_i)\vec{E}_i\tilde{C}\vec{H}\eta(k) \\
&\left. + \sum_{i=1}^{n}((1-\beta_i(k))(1-\gamma_i(k)) - (1-\bar{\beta}_i)(1-\bar{\gamma}_i))\vec{E}_i\vec{F}\eta(k)\right]^t \\
&\times P\left[\vec{A}\eta(k) + \vec{g}(\vec{H}\eta(k)) + \hat{H}\sigma(\tilde{C}\vec{H}\eta(k)) + \sum_{i=1}^{n}(\beta_i(k) - \bar{\beta}_i)\right. \\
&\times \vec{E}_i\sigma(\tilde{C}\vec{H}\eta(k)) + \sum_{i=1}^{n}((1-\beta_i(k))\gamma_i(k) - (1-\bar{\beta}_i)\bar{\gamma}_i)\vec{E}_i\tilde{C}\vec{H}\eta(k)
\end{aligned}
$$

$$+\sum_{i=1}^{n}((1-\beta_i(k))(1-\gamma_i(k))-(1-\bar{\beta}_i)(1-\bar{\gamma}_i))\overrightarrow{E}_i\overrightarrow{F}\eta(k)\Bigg]$$
$$-\eta^T(k)P\eta(k)\Bigg\}. \tag{8.76}$$

By noting (8.76), it can be obtained that

$$\begin{aligned}\mathbb{E}\{\Delta V(\eta(k))\} = \mathbb{E}\Bigg\{&[(\overrightarrow{A}+\hat{H}\tilde{L}\tilde{C}\overrightarrow{H})\eta(k)+\hat{H}\Psi(\tilde{C}\overrightarrow{H}\eta(k))+\overrightarrow{g}(\overrightarrow{H}\eta(k))]^T P\\ &\times[(\overrightarrow{A}+\hat{H}\tilde{L}\tilde{C}\overrightarrow{H})\eta(k)+\hat{H}\Psi(\tilde{C}\overrightarrow{H}\eta(k))+\overrightarrow{g}(\overrightarrow{H}\eta(k))]\\ &+\sum_{i=1}^{n}\hat{\varphi}_i^2\eta^t(k)\overrightarrow{F}^T\overrightarrow{E}_i^tP\overrightarrow{E}_i\overrightarrow{F}\eta(k)+\sum_{i=1}^{n}\varphi_i^2[\tilde{L}\tilde{C}\overrightarrow{H}\eta(k)\\ &+\Psi(\tilde{C}\overrightarrow{H}\eta(k))]^t\overrightarrow{E}_i^tP\overrightarrow{E}_i[\tilde{L}\tilde{C}\overrightarrow{H}\eta(k)+\Psi(\tilde{C}\overrightarrow{H}\eta(k))]\\ &-2\sum_{i=1}^{n}\phi_i^2[\tilde{L}\tilde{C}\overrightarrow{H}\eta(k)+\Psi(\tilde{C}\overrightarrow{H}\eta(k))]^t\overrightarrow{E}_i^tP\overrightarrow{E}_i\tilde{C}\overrightarrow{H}\eta(k)\\ &-2\sum_{i=1}^{n}\bar{\phi}_i^2[\tilde{L}\tilde{C}\overrightarrow{H}\eta(k)+\Psi(\tilde{C}\overrightarrow{H}\eta(k))]^t\overrightarrow{E}_i^tP\overrightarrow{E}_i\overrightarrow{F}\eta(k)\\ &+\sum_{i=1}^{n}\bar{\varphi}_i^2\eta^t(k)\overrightarrow{H}^t\tilde{C}^t\overrightarrow{E}_i^tP\overrightarrow{E}_i\tilde{C}\overrightarrow{H}\eta(k)-\eta^t(k)P\eta(k)\\ &-2\sum_{i=1}^{n}\hat{\phi}_i^2\eta^T(k)\overrightarrow{H}^t\tilde{C}^t\overrightarrow{E}_i^TP\overrightarrow{E}_i\overrightarrow{F}\eta(k)\Bigg\}.\end{aligned} \tag{8.77}$$

From the elementary inequality $2a^Tb \le a^Ta + b^Tb$, we have

$$\begin{aligned}&-2\sum_{i=1}^{n}\phi_i^2[\tilde{L}\tilde{C}\overrightarrow{H}\eta(k)+\|(\tilde{C}\overrightarrow{H}\eta(k))]^t\overrightarrow{E}_i^tP\overrightarrow{E}_i\tilde{C}\overrightarrow{H}\eta(k)\\ &\le\sum_{i=1}^{n}\phi_i^2[\eta^t(k)\overrightarrow{H}^t\tilde{C}^t\tilde{L}^t\overrightarrow{E}_i^tP\overrightarrow{E}_i\tilde{L}\tilde{C}\overrightarrow{H}\eta(k)\\ &+2\eta^t(k)\overrightarrow{H}^t\tilde{C}^t\overrightarrow{E}_i^tP\overrightarrow{E}_i\tilde{C}\overrightarrow{H}\eta(k)\\ &+\Psi^t(\tilde{C}\overrightarrow{H}\eta(k))\overrightarrow{E}_i^tP\overrightarrow{E}_i\Psi(\tilde{C}\overrightarrow{H}\eta(k))],\end{aligned} \tag{8.78}$$

$$\begin{aligned}&-2\sum_{i=1}^{n}\bar{\phi}_i^2[\tilde{L}\tilde{C}\overrightarrow{H}\eta(k)+\|(\tilde{C}\overrightarrow{H}\eta(k))]^t\overrightarrow{E}_i^tP\overrightarrow{E}_i\overrightarrow{F}\eta(k)\\ &\le\sum_{i=1}^{n}\bar{\phi}_i^2[\eta^t(k)\overrightarrow{H}^t\tilde{C}^t\tilde{L}^T\overrightarrow{E}_i^tP\overrightarrow{E}_i\tilde{L}\tilde{C}\overrightarrow{H}\eta(k)\end{aligned}$$

$$+ 2\eta^t(k)\overrightarrow{F}^t\overrightarrow{E}_i^t P\overrightarrow{E}_i\overrightarrow{F}\eta(k)$$
$$+ \Psi^t(\tilde{C}\overrightarrow{H}\eta(k))\overrightarrow{E}_i^t P\overrightarrow{E}i\|(\tilde{C}\overrightarrow{H}\eta(k))], \tag{8.79}$$

$$\begin{aligned}
&-2\sum_{i=1}^{n}\hat{\phi}_i^2\eta^t(k)\overrightarrow{H}^t\tilde{C}^t\overrightarrow{E}_i^t P\overrightarrow{E}_i\overrightarrow{F}\eta(k)\\
&\le \sum_{i=1}^{n}\hat{\phi}_i^2[\eta^t(k)\overrightarrow{H}^t\tilde{C}^t\overrightarrow{E}_i^t P\overrightarrow{E}_i\tilde{C}\overrightarrow{H}\eta(k)\\
&+ \eta^t(k)\overrightarrow{F}^t\overrightarrow{E}_i^t P\overrightarrow{E}_i\overrightarrow{F}\eta(k)],
\end{aligned} \tag{8.80}$$

which result in

$$\mathbb{E}\{\Delta V(\eta(k))\} \le \mathbb{E}\{\xi^T(k)\bar{\Gamma}\xi(k)\},$$

where

$$\begin{aligned}
\xi(k) &= \begin{bmatrix}\eta^T(k) & \overrightarrow{g}^T(\overrightarrow{H}\eta(k)) & \Psi^T(\tilde{C}\overrightarrow{H}\eta(k))\end{bmatrix}^T,\\
\bar{\Gamma} &= \begin{bmatrix}\bar{\Upsilon}_{11} & * & * \\ P(\overrightarrow{A} + \hat{H}\tilde{L}\tilde{C}\overrightarrow{H})P & * & \\ \Sigma_{31} & \hat{H}^T P & \Sigma_{33}\end{bmatrix}.
\end{aligned} \tag{8.81}$$

Moreover, it follows from (8.70) and (8.72) that

$$\begin{aligned}
\mathbb{E}\{\Delta V(\eta(k))\} &\le \mathbb{E}\{\xi^T(k)\bar{\Gamma}\xi(k) - \epsilon_1\Psi^T(\tilde{C}\overrightarrow{H}\eta(k))(\Psi(\tilde{C}\overrightarrow{H}\eta(k)) - \hat{L}\tilde{C}\overrightarrow{H}\eta(k))\\
&\quad - \epsilon_2(\overrightarrow{g}^T(\overrightarrow{H}\eta(k))\overrightarrow{g}(\overrightarrow{H}\eta(k)) - \eta^T(k)\overrightarrow{H}^T\bar{E}_n\overrightarrow{H}\eta(k))\}\\
&= \mathbb{E}\{\xi^T(k)\Gamma\xi(k)\},
\end{aligned}$$

where

$$\Gamma = \begin{bmatrix}\bar{\Upsilon}_{11} + \epsilon_2\overrightarrow{H}^T\bar{E}_n\overrightarrow{H} & * & * \\ P(\overrightarrow{A} + \hat{H}\tilde{L}\tilde{C}\overrightarrow{H}) & P - \epsilon_2 I & * \\ \Sigma_{31} + \epsilon_1\hat{L}\tilde{C}\overrightarrow{H} & \hat{H}^T P & \bar{\Upsilon}_{33} + \hat{H}^T P\hat{H} - \epsilon_1 I\end{bmatrix}. \tag{8.82}$$

We can obtain from (8.74), by considering the third leading principal submatrix, that $\Gamma < 0$ and, subsequently,

$$\mathbb{E}\{\Delta V(\eta(k))\} \le -\lambda_{\min}(-\Gamma)\|\xi(k)\|^2.$$

Finally, we can confirm from [25] that the augmented filtering system (8.68) is exponentially mean square stable.

To establish the $\mathbb{H}_\infty$ performance, we assume zero initial conditions and introduce the following index:

$$\begin{aligned}
&\mathbb{E}\{\Delta V(\eta(k))\} + \mathbb{E}\{\|\vec{z}(k)\|^2\} - \gamma^2\|\vec{w}(k)\|^2 \\
&\quad = \mathbb{E}\{\xi^t(k)\bar{\Gamma}\xi(k)\} + \vec{w}^t(k)\vec{B}^t P\vec{B}\vec{w}(k) \\
&\qquad + 2\vec{w}^T(k)\vec{B}^T P[(\vec{A} + \hat{H}\tilde{L}\tilde{C}\vec{H})\eta(k) + \hat{H}(\tilde{C}\vec{H}\eta(k)) + \vec{g}(\vec{H}\eta(k))] \\
&\qquad + \eta^t(k)\vec{M}^t\vec{M}\eta(k) - \gamma^2\vec{w}^t(k)\vec{w}(k) \\
&\quad = \mathbb{E}\{\hat{\xi}^t(k)\bar{\Sigma}\hat{\xi}(k)\},
\end{aligned}$$

where

$$\hat{\xi}(k) = \begin{bmatrix} \eta^t(k) & \vec{g}^t(\vec{H}\eta(k)) & \Psi^t(\tilde{C}\vec{H}\eta(k)) & \vec{w}^T(k) \end{bmatrix}^T,$$

$$\bar{\Sigma} = \begin{bmatrix} \bar{\Upsilon}_{11} + \vec{M}^t\vec{M} & \bullet & \bullet & \bullet \\ P(\vec{A} + \hat{H}\tilde{L}\tilde{C}\vec{H}) & P & \bullet & \bullet \\ \Sigma_{31} & \hat{H}^t P & \bar{\Upsilon}_{33} + \hat{H}^t P\hat{H} & \bullet \\ \vec{B}^t P(\vec{A} + \hat{H}\tilde{L}\tilde{C}\vec{H}) & \vec{B}^t P & \vec{B}^t P\hat{H} & \vec{B}^t P\vec{B} - \gamma^2 I \end{bmatrix}.$$

Again, it follows from the constraints (8.70) and (8.72) that

$$\begin{aligned}
&\mathbb{E}\{\Delta V(\eta(k))\} + \mathbb{E}\left\{\|\vec{z}(k)\|^2\right\} - \gamma^2\|\vec{w}(k)\|^2 \\
&\leq \mathbb{E}\{\hat{\xi}^t(k)\bar{\Sigma}\hat{\xi}(k) - \epsilon_1\Psi^T(\tilde{C}\vec{H}\eta(k))(\Psi(\tilde{C}\vec{H}\eta(k)) - \hat{L}\tilde{C}\vec{H}\eta(k)) \\
&- \epsilon_2(\vec{g}^t(\vec{H}\eta(k))\vec{g}(\vec{H}\eta(k)) - \eta^t(k)\vec{H}^t\bar{E}_n\vec{H}\eta(k))\} \\
&= \mathbb{E}\{\hat{\xi}^t(k)\Sigma\hat{\xi}(k)\}.
\end{aligned}$$

Furthermore, we can see from (8.74) in Theorem 8.4 that

$$\mathbb{E}\{\Delta V(\eta(k))\} + \mathbb{E}\{\|\vec{z}(k)\|^2\} - \gamma^2\|\vec{w}(k)\|^2 \leq 0$$

for all nonzero $\vec{w}(k)$.

By considering zero initial conditions, it follows from the above inequality that

$$\sum_{k=0}^{\infty}\mathbb{E}\{\|\vec{z}(k)\|^2\} \leq \gamma^2\sum_{k=0}^{\infty}\|\vec{w}(k)\|^2,$$

which is equivalent to (8.73), and the proof is now complete. □

Having conducted the filtering performance analysis in Theorem 8.3, we are now in a position to deal with the problem of designing distributed $\mathbb{H}_\infty$

filters. The solution to the distributed $\mathbb{H}_\infty$ filtering problem with both ROSS and successive packet dropouts is obtained by the following theorem.

Theorem 8.4. *Let a positive scalar $\gamma > 0$ be given. For the nonlinear system (8.56) and sensors (8.58) with both ROSS and successive packet dropouts, the filtering dynamics in (8.68) is exponentially mean square stable and satisfies the H_∞ performance constraint (8.73) if there exist positive constant scalars ϵ_1 and ϵ_2, positive definite matrices $S > 0$, $Q_i > 0$ $(i = 1, 2, \ldots, n)$, and $R > 0$, and matrices $X \in \mathcal{T}_{n_x \times n_x}$ and $Y \in \mathcal{T}_{n_x \times n_y}$ satisfying*

$$\begin{bmatrix} -S + \epsilon_2 \bar{E}_n + \bar{M}^T \bar{M} & * & * & * \\ \Pi_{21} & \Pi_{22} & * & * \\ \Pi_{31} & \Pi_{32} & -\bar{S} & * \\ \Pi_{41} & \Pi_{42} & 0 & -\tilde{S} \end{bmatrix} < 0, \tag{8.83}$$

where

$$\Pi_{21} = \begin{bmatrix} (-\bar{M}^T \bar{M})^T & 0| & 0 & 0 & 0| & (\epsilon_1 \hat{L} \tilde{C})^T & 0 & 0 \end{bmatrix}^T, \quad \Theta_I = \mathbf{1}_n \otimes I,$$

$$\Pi_{22} = diag\{-Q + \bar{M}^T \bar{M}, -R, -\epsilon_2 I, -\epsilon_2 I, -\epsilon_2 I, -\epsilon_1 I, -\gamma_2 I, -\gamma_2 I\},$$

$$\Pi_{31} = \begin{bmatrix} S\bar{A} \\ Y[(I - \bar{\Lambda}_\beta)\bar{\Lambda}_\gamma + \bar{\Lambda}_\beta \tilde{L}]\tilde{C} \\ R[(I - \bar{\Lambda}_\beta)\bar{\Lambda}_\gamma + \bar{\Lambda}_\beta \tilde{L}]\tilde{C} \end{bmatrix}, \quad \Pi_{32} = \begin{bmatrix} \bar{\Pi}_{311} & \bar{S} & \bar{\Pi}_{313} \end{bmatrix},$$

$$\bar{\Pi}_{311} = \begin{bmatrix} 0 & 0 \\ X & Y(I - \bar{\Lambda}_\beta)(I - \bar{\Lambda}_\gamma) \\ 0 & R(I - \bar{\Lambda}_\beta)(I - \bar{\Lambda}_\gamma) \end{bmatrix}, \quad \bar{\Pi}_{313} = \begin{bmatrix} 0 & S\bar{B} & 0 \\ Y\bar{\Lambda}_\beta & 0 & Y\tilde{D} \\ R\bar{\Lambda}_\beta & 0 & R\tilde{D} \end{bmatrix},$$

$$\Pi_{41} = \begin{bmatrix} (\Lambda_\varphi \mathcal{W} \Theta_L)^T & (\Lambda_{2\phi} \mathcal{W} \Theta_C)^T & 0 & (\Lambda_\phi \mathcal{W} \Theta_L)^T & 0 \end{bmatrix}^T,$$

$$\Theta_C = \mathbf{1}_n \otimes \tilde{C},$$

$$\Pi_{42} = \begin{bmatrix} 0 & \bar{\Pi}_{412}| & 0 & 0 & 0| & \bar{\Pi}_{416} & 0 & 0 \end{bmatrix},$$

$$\bar{\Pi}_{412} = \begin{bmatrix} 0 & 0 & (\Lambda_{2\bar{\phi}} \mathcal{W} \Theta I)^T & 0 & 0 \end{bmatrix}^T,$$

$$\bar{\Pi}_{416} = \begin{bmatrix} (\Lambda_\varphi \mathcal{W} \Theta_I)^T & 0 & 0 & 0 & (\Lambda_\phi \mathcal{W} \Theta_I)^T \end{bmatrix}^T, \quad \bar{S} = diag\{S, Q, R\},$$

$$\Theta_{Hi} = \begin{bmatrix} 0 & (YE_i)^T & (RE_i)^T \end{bmatrix}^T, \quad Q = diag\{Q_1, Q_2, \ldots, Q_n\}, \quad \tilde{S} = I_{5n} \otimes \bar{S},$$

$$\Lambda_\varphi = diag\{\varphi_1 I, \varphi_2 I, \ldots, \varphi_n I\}, \quad \mathcal{W} = diag\{\Theta_{H1}, \Theta_{H2}, \ldots, \Theta_{Hn}\},$$

$$\Lambda_\phi = diag\left\{\sqrt{\phi_1^2 + \bar{\phi}_1^2} I, \sqrt{\phi_2^2 + \bar{\phi}_2^2} I, \ldots, \sqrt{\phi_n^2 + \bar{\phi}_n^2} I\right\}, \quad \Theta_L = \mathbf{1}_n \otimes \tilde{L}\tilde{C},$$

$$\Lambda_{2\phi} = diag\left\{\sqrt{2\phi_1^2 + \hat{\phi}_1^2 + \bar{\varphi}_1^2 I}, \sqrt{2\phi_2^2 + \hat{\phi}_2^2 + \bar{\varphi}_2^2 I}, \ldots, \sqrt{2\phi_n^2 + \hat{\phi}_n^2 + \bar{\varphi}_n^2 I}\right\},$$
$$\Lambda_{2\bar{\phi}} = diag\left\{\sqrt{2\bar{\phi}_1^2 + \hat{\phi}_1^2 + \hat{\varphi}_1^2 I}, \sqrt{2\bar{\phi}_2^2 + \hat{\phi}_2^2 + \hat{\varphi}_2^2 I}, \ldots, \sqrt{2\bar{\phi}_n^2 + \hat{\phi}_n^2 + \hat{\varphi}_n^2 I}\right\} \tag{8.84}$$

and the other parameters are defined in (8.65). Moreover, if the above inequality is feasible, two matrices $\bar{K}$ and $\bar{H}$ are given as follows:

$$\bar{K} = Q^{-1}X, \quad \bar{H} = Q^{-1}Y. \tag{8.85}$$

Therefore, the desired filter parameters K_{ij} and H_{ij} ($i = 1, 2, \ldots, n$, $j \in \mathcal{N}_i$) can be obtained from (8.66).

Proof. By setting $P = \text{diag}\{S, Q, R\}$, applying the Schur complement lemma [23], and noting (8.69), it can be seen that (8.74) is equivalent to

$$\begin{bmatrix} -S + \epsilon_2 \bar{E}_n + \bar{M}^T \bar{M} & * & * & * \\ \Pi_{21} & \Pi_{22} & * & * \\ \hat{\Pi}_{31} & \hat{\Pi}_{32} & -\bar{S} & * \\ \hat{\Pi}_{41} & \hat{\Pi}_{42} & 0 & -\tilde{S} \end{bmatrix} < 0, \tag{8.86}$$

where

$$\hat{\Pi}_{31} = \begin{bmatrix} S\bar{A} \\ Q\bar{H}[(I - \bar{\Lambda}_\beta)\bar{\Lambda}_\gamma + \bar{\Lambda}_\beta \tilde{L}]\tilde{C} \\ R[(I - \bar{\Lambda}_\beta)\bar{\Lambda}_\gamma + \bar{\Lambda}_\beta \tilde{L}]\tilde{C} \end{bmatrix},$$
$$\hat{\Pi}_{32} = \begin{bmatrix} \hat{\Pi}_{311} & \bar{S} & \hat{\Pi}_{313} \end{bmatrix},$$
$$\hat{\Pi}_{311} = \begin{bmatrix} 0 & 0 \\ Q\bar{K} & Q\bar{H}(I - \bar{\Lambda}_\beta)(I - \bar{\Lambda}_\gamma) \\ 0 & R(I - \bar{\Lambda}_\beta)(I - \bar{\Lambda}_\gamma) \end{bmatrix},$$
$$\hat{\Pi}_{313} = \begin{bmatrix} 0 & S\bar{B} & 0 \\ Q\bar{H}\bar{\Lambda}_\beta & 0 & Q\bar{H}\tilde{D} \\ R\bar{\Lambda}_\beta & 0 & R\tilde{D} \end{bmatrix},$$
$$\hat{\Pi}_{41} = \begin{bmatrix} (\Lambda_\varphi \hat{\mathcal{W}} \Theta_L)^T & (\Lambda_{2\phi} \hat{\mathcal{W}} \Theta_C)^t & 0 & (\Lambda_\phi \hat{\mathcal{W}} \Theta_L)^T & 0 \end{bmatrix}^t,$$
$$\hat{\Pi}_{42} = \begin{bmatrix} 0 & \hat{\Pi}_{412}| & 0 & 0 & 0| & \hat{\Pi}_{416} & 0 & 0 \end{bmatrix},$$
$$\hat{\Pi}_{412} = \begin{bmatrix} 0 & 0 & (\Lambda_2 \bar{\phi} \hat{\mathcal{W}} \Theta_I)^t & 0 & 0 \end{bmatrix}^t,$$

$$\hat{\Pi}_{416} = \left[(\Lambda_{\varphi}\hat{\mathcal{W}}\Theta_I)^t \quad 0 \quad 0 \quad 0 \quad (\Lambda_{\phi}\hat{\mathcal{W}}\Theta_I)^t \right]^t,$$
$$\hat{\mathcal{W}} = \mathbf{diag}\{\hat{\Theta}_{H1}, \hat{\Theta}_{H2}, \ldots, \hat{\Theta}_{Hn}\},$$
$$\hat{\Theta}_{Hi} = \left[0 \quad (Q\bar{H}E_i)^t \quad (RE_i)^t \right]^t. \tag{8.87}$$

Letting $Q = \text{diag}\{Q_1, Q_2, \ldots, Q_n\}$ and noting $Q\bar{K} = X$ and $Q\bar{H} = Y$, we can obtain (8.83) readily. In addition, from Lemma 8.3, it follows that $\bar{K} \in \mathcal{T}_{n_x \times n_x}$ and $\bar{H} \in \mathcal{T}_{n_x \times n_y}$, which completes the proof of this theorem. □

Remark 8.7. It is well known that the main difficulties in designing distributed filters in sensor networks lie in the tight coupling among sensors in terms of both time and space. In this chapter, the filter parameters K_{ij} and H_{ij} ($i = 1, 2, \ldots, n$, $j \in N_i$) are "assembled" to matrices $\bar{K}$ and $\bar{H}$, which should meet the constraints (8.67). Then, by Lemma 8.3, we can derive the conditions that $X \in \mathcal{T}_{n_x \times n_x}$ and $Y \in \mathcal{T}_{n_x \times n_y}$ are required to satisfy. Consequently, the distributed filters can be designed effectively.

8.3.3 Simulation example 8.4

In this section, we present a simulation example to illustrate the effectiveness of the proposed distributed filter design scheme for nonlinear systems with both ROSS and successive packet dropouts over sensor networks.

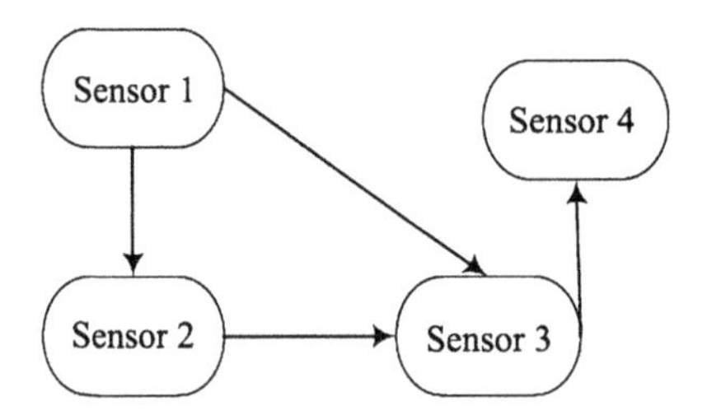

FIGURE 8.19 Topological structure of the sensor network.

The sensor network shown in Fig. 8.19 is represented by a directed graph $\mathbb{G} = (\mathbb{V}, \mathbb{E}, \mathbb{A})$ with the set of nodes $\mathbb{V} = \{1, 2, 3, 4\}$, the set of edges

$$\mathbb{E} = \{(1,1), (2,1), (2,2), (3,1), (3,2), (3,3), (4,3), (4,4)\},$$

and the adjacency matrix

$$\mathbb{A} = \begin{bmatrix} 1 & 0 & 0 & 0 \\ 1 & 1 & 0 & 0 \\ 1 & 1 & 1 & 0 \\ 0 & 0 & 1 & 1 \end{bmatrix}.$$

The nonlinear discrete system considered is modeled by (8.56) with the following parameters:

$$A = \begin{bmatrix} -0.6 & 0.2 \\ 0 & -0.8 \end{bmatrix}, \quad B = \begin{bmatrix} 0.5 & 1 \end{bmatrix}^T, \quad M = \begin{bmatrix} 0.1 & 0.1 \end{bmatrix},$$

and the nonlinear function $f(x(k))$ is selected as

$$f(x(k)) = \begin{bmatrix} \frac{0.2x_1(k)}{2x_2^2(k)+1} & 0.1\sin(x_1(k))x_2(k) \end{bmatrix}^T.$$

It is easy to see that the constraint (8.57) can be met with $E = \mathbf{diag}\{0.2, 0.15\}$. Consider the sensors with both ROSS and successive packet dropouts described by (8.58) with the following parameters:

$$C_1 = \begin{bmatrix} 0.1 & 0 \end{bmatrix}, \quad C_2 = \begin{bmatrix} 0.2 & 0.1 \end{bmatrix}, \quad C_3 = \begin{bmatrix} 0.5 & 0.7 \end{bmatrix}, \quad C_4 = \begin{bmatrix} 0.1 & 0.2 \end{bmatrix},$$
$$D_1 = 1, \quad D_2 = 0.5, \quad D_3 = 0.7, \quad D_4 = 0.5.$$

In this example, the probabilities are taken as $\bar{\beta}_1 = 0.9$, $\bar{\beta}_2 = 0.8$, $\bar{\beta}_3 = 0.85$, $\bar{\beta}_4 = 0.7$ and $\bar{\gamma}_1 = 0.9$, $\bar{\gamma}_2 = 0.8$, $\bar{\gamma}_3 = 0.7$, $\bar{\gamma}_4 = 0.6$. Take the saturation level as $\vartheta_{\max} = 0.3$, and other parameters are chosen as $\underline{L}_1 = 0.3$, $\underline{L}_2 = 0.4$, $\underline{L}_3 = 0.2$, $\underline{L}_4 = 0.1$, $L_1 = 0.7$, $L_2 = 0.6$, $L_3 = 0.8$, $L_4 = 0.9$. By solving (8.83) and (8.85) in Theorem 8.4, we can obtain the following parameters of the desired distributed filters:

$$K_{11} = \begin{bmatrix} 0.2997 & 0.2511 \\ 0.1260 & 0.1092 \end{bmatrix}, \quad K_{21} = \begin{bmatrix} 0.2238 & 0.1939 \\ 0.1512 & 0.1349 \end{bmatrix},$$

$$K_{22} = \begin{bmatrix} 0.2599 & 0.2554 \\ 0.1854 & 0.1812 \end{bmatrix}, \quad K_{31} = \begin{bmatrix} 0.2570 & 0.2254 \\ -0.0387 & 0.0405 \end{bmatrix},$$

$$K_{32} = \begin{bmatrix} 0.3055 & 0.2995 \\ 0.0622 & 0.0593 \end{bmatrix}, \quad K_{33} = \begin{bmatrix} -0.0533 & -0.0537 \\ 0.4484 & 0.4423 \end{bmatrix},$$

$$K_{43} = \begin{bmatrix} 0.3239 & 0.3177 \\ 0.0723 & 0.0694 \end{bmatrix}, \quad K_{44} = \begin{bmatrix} -0.0488 & -0.0491 \\ 0.5097 & 0.5051 \end{bmatrix},$$

$$H_{11} = \begin{bmatrix} 0.0844 & 0.1755 \end{bmatrix}^t, \quad H_{21} = \begin{bmatrix} 0.0070 & 0.0604 \end{bmatrix}^t,$$

$$H_{22} = \begin{bmatrix} 0.1151 & 0.1578 \end{bmatrix}^t, \quad H_{31} = \begin{bmatrix} -0.1054 & 0.1529 \end{bmatrix}^t,$$

$$H_{32} = \begin{bmatrix} 0.0761 & 0.2111 \end{bmatrix}^t, \quad H_{33} = \begin{bmatrix} -0.0407 & 0.0249 \end{bmatrix}^t,$$

$$H_{43} = \begin{bmatrix} 0.0113 & -0.0017 \end{bmatrix}^t, \quad H_{44} = \begin{bmatrix} 0.2516 & -0.0414 \end{bmatrix}^t,$$

and the optimal performance index given in (8.73) is $\gamma^* = 1.0214$. In the simulation, the exogenous disturbance inputs are selected as $w(k) = \exp(-0.2k)\sin(k)$ and $v(k) = [\sin(10k+1)]/(3k+1)$. The initial conditions are $x(0) = [0.4\ \ 0.2]^t$ and $\hat{x}_i(0) = [0\ \ 0]^t$ $(i = 1, 2, 3, 4)$.

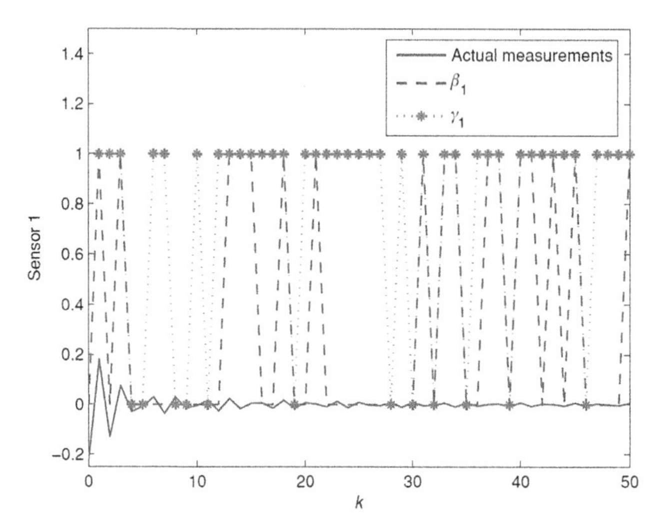

FIGURE 8.20 Measurements from Sensor 1.

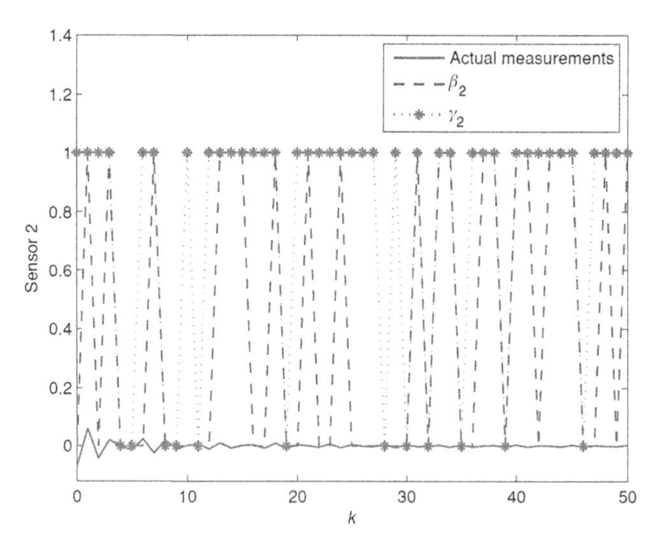

FIGURE 8.21 Measurements from Sensor 2.

Simulation results are shown in Figs. 8.20–8.24. In Figs. 8.20–8.23 the actual measurements and the binary signals β and γ for Sensors 1 to 4 are depicted.

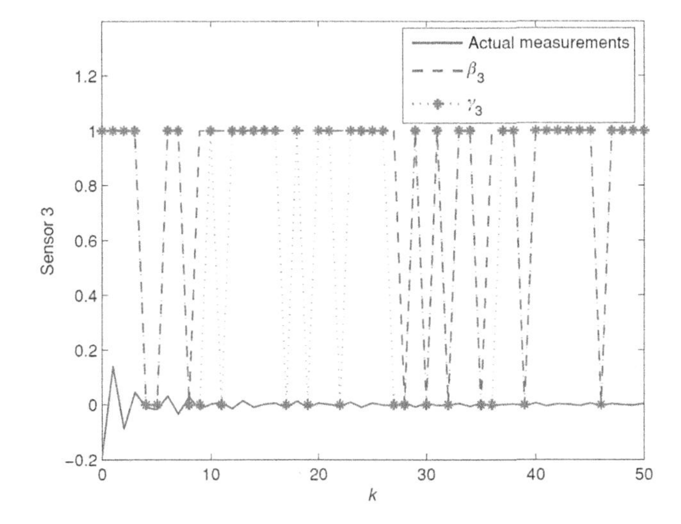

FIGURE 8.22 Measurements from Sensor 3.

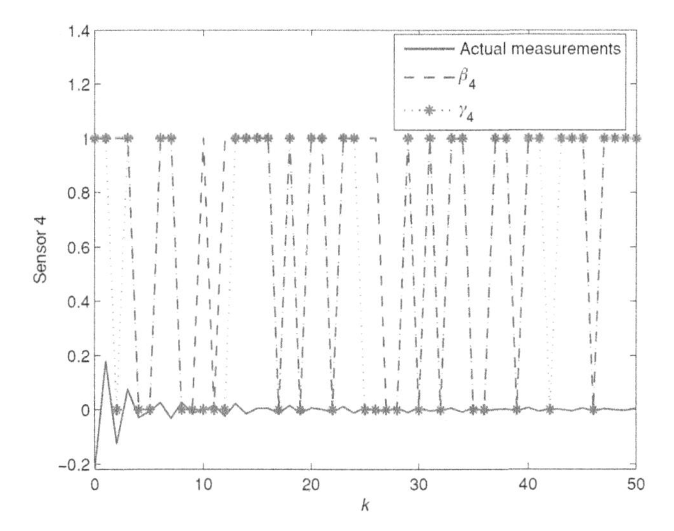

FIGURE 8.23 Measurements from Sensor 4.

Fig. 8.24 plots the filtering errors $z(k) - \hat{z}_i(k)$ $(i = 1, 2, 3, 4)$. The simulation results have confirmed the effectiveness of the distributed filtering technique presented in this chapter.

8.4 Notes

In this chapter, a new algorithm to deal with asynchronous multisensor measurements is developed. The developed algorithm, the ISKF, has been shown to be

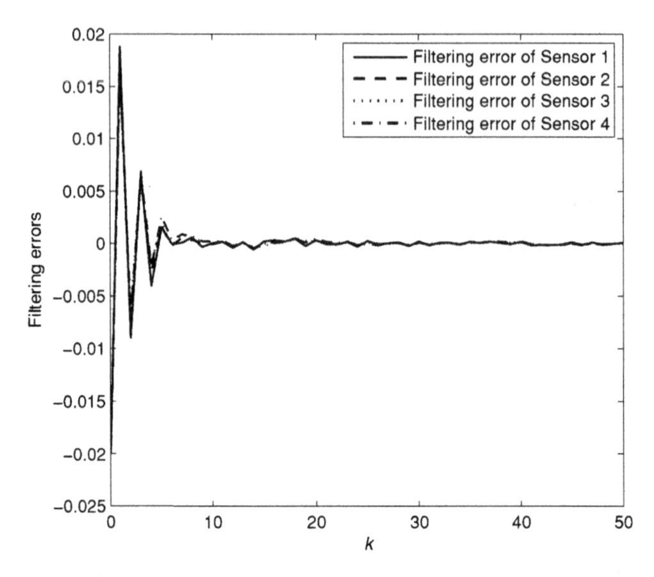

FIGURE 8.24 Filtering errors.

able to recover nondelayed estimations from delayed measurements. The ISKF has also been demonstrated to perform better compared with the regular Kalman filter if measurement loss happens. Incorporating its estimation in the fusion algorithm, the fused estimations tend to have less estimation error compared with single-sensor estimations.

Then, an improved distributed estimation method was developed based on the concept of minimum variance with a consideration of time delay appearing in the measurements of different sensor nodes. This was attributed to the time taken by physical quantities to be measured (for example, temperature, acoustic waves, to name a few) to propagate from one point in the space to another one. In this regard, it is assumed that the source of the data is located at a known position and the spatial ordinates of the sensor nodes are known. This knowledge is applied to calculate the time delay between the source of the data and any node in the network. The spectrum of the signal of interest is assumed to be known a priori. Based on this knowledge the data are first passed through a low-pass FIR filter. The signal is then jointly tracked by a sensor network, in which each node computes an estimate as a delay-adjusted weighted sum of its own and its neighbors' measurements and estimates. The filter weights are time-varying and updated locally and it has a cascade structure comprised of two loops: an inner loop producing the state estimate and an outer loop producing an estimate of the error covariance. The state estimate is thus obtained as the solution of an optimization problem with quadratic cost function and quadratic constraints.

Finally, we dealt with the distributed $\mathbb{H}_\infty$ filtering problem for systems with repeated scalar nonlinearities in the sensor network. The phenomenon of packet loss occurring between the system and each sensor has been taken into account. The distributed filters were designed for the filtering dynamics system to be stochastically stable and the estimation errors to satisfy the average $\mathbb{H}_\infty$ performance constraint. Furthermore, the explicit expression of the desired filter gains has been derived. Finally, an illustrative example highlighting the usefulness of the presented results has been discussed. It is also noted that the proposed design method for sensor networks can be easily extended to other networks, like networked control systems [36], gene regulatory networks, complex networks, and neural networks, to name a few.

Then, we dealt with the distributed $\mathbb{H}_\infty$ filtering problem for a class of nonlinear systems with randomly occurring incomplete information over sensor networks. The incomplete information considered includes both the ROSS and successive packet dropouts. The issue of ROSS has been addressed, and then the filtering dynamics have been analyzed by modeling both the ROSS and successive packet dropouts in a unified framework. The distributed filters have been designed for the filtering dynamics to be exponentially mean square stable and the filtering errors to satisfy the $\mathbb{H}_\infty$ performance constraint. Finally, an illustrative example has been provided that highlights the usefulness of the filtering approach developed.

References

[1] A. Alouani, J. Gray, D. McCabe, Theory of distributed estimation using multiple asynchronous sensors, IEEE Trans. Aerosp. Electron. Syst. 41 (2) (2005) 717–722.

[2] Y. Bar-Shalom, X. Li, T. Kirubarajan, Estimation with Applications to Tracking and Navigation, Wiley, 2001.

[3] P. Alriksson, A. Rantzer, Distributed Kalman filtering using weighted averaging, in: Proc. the 17th International Symposium on Mathematical Theory of Networks and Systems, Kyoto, Japan, 2006.

[4] D. Hall, J. Llinas, Handbook of Multisensor Data Fusion, CRC Press, 2001.

[5] Y. Bar-Shalom, Update with out-of-sequence measurements in tracking: exact solution, IEEE Trans. Aerosp. Electron. Syst. 38 (3) (2002) 769–777.

[6] S. Zhang, Y. Bar-Shalom, Optimal removal of out-of-sequence measurements from tracks using the IF-equivalent measurement, in: Proc. 49th IEEE Conference on Decision and Control (CDC), Atlanta, GA, Dec. 15–17, 2010, 2010, pp. 1312–1317.

[7] S. Sun, Z. Deng, Multi-sensor optimal information fusion Kalman filter, Automatica 40 (6) (2004) 1017–1023.

[8] Y. Xia, Li Li, G.P. Liu, P. Shi, $\mathbb{H}_\infty$ predictive control of networked control systems, Int. J. Control 84 (6) (2011) 1080–1097.

[9] Y. Xia, J. Shang, J. Chen, G.P. Liu, Networked data fusion with packet losses and variable delays, IEEE Trans. Syst. Man Cybern., Part B, Cybern. 39 (5) (2009) 1107–1120.

[10] L. Yan, D. Zhou, M. Fu, Y. Xia, State estimation for asynchronous multirate multisensor dynamic systems with missing measurements, IET Signal Process. 4 (6) (2010) 728–739.

[11] L. Yan, B. Liu, D. Zhou, Asynchronous multirate multisensor information fusion algorithm, IEEE Trans. Aerosp. Electron. Syst. 43 (3) (2007) 1135–1146.

[12] B. Sinopoli, L. Schenato, M. Franceschetti, K. Poolla, M. Jordan, S. Sastry, Kalman filtering with intermittent observations, IEEE Trans. Autom. Control 49 (9) (2004) 1453–1464.

[13] L. Yan, B. Liu, D. Zhou, The modeling and estimation of asynchronous multirate multisensor dynamic systems, Aerosp. Sci. Technol. 10 (1) (2006) 63–71.

[14] L. Yan, H. Shi, M. Du, Z. Zhu, Asynchronous multirate multisensor state fusion estimation with incomplete measurements, in: Proc. 4th Int. Conference on Wireless Communications, Networking and Mobile Computing (WiCOM'08), 2008, pp. 1–4.
[15] L. Zhang, X. Wu, Q. Pan, H. Zhang, Multiresolution modeling and estimation of multisensor data, IEEE Trans. Signal Process. 52 (11) (2004) 3170–3182.
[16] M. Mallick, S. Coraluppi, C. Carthel, Advances in asynchronous and decentralized estimation, in: Proc. IEEE Aerospace Conference, vol. 4, 2001, pp. 1873–1879.
[17] X.B. Chen, W.B. Xu, T.Y. Huang, X.Y. Ouyang, S.S. Stankovic, Pair-wise decomposition and coordinated control of complex systems, Inf. Sci. 185 (2012) 78–99.
[18] M. Saleem, G.A. Di Caro, M. Farooq, Swarm intelligence-based routing protocol for wireless sensor networks: survey and future directions, Inf. Sci. 181 (2011) 78–99.
[19] M.S. Mahmoud, Robust Control and Filtering for Time-Delay Systems, Marcek-Dekker, New York, 2000.
[20] H.K. Khalil, Nonlinear Systems, third edition, Prentice-Hall, Upper Saddle River, NJ, 2002.
[21] Y. Xiao, Y. Cao, Z. Lin, Robust filtering for discrete-time systems with saturation and its application to trans-multiplexers, IEEE Trans. Signal Process. 52 (5) (2004) 1266–1277.
[22] F. Yang, Y. Li, Set-membership filtering for systems with sensor saturation, Automatica 45 (8) (2009) 1896–1902.
[23] S. Boyd, L.E. Ghaoui, E. Feron, V. Balakrishnan, Linear Matrix Inequalities in System and Control Theory, SIAM, Philadelphia, PA, 1994.
[24] Z. Zuo, D.W.C. Ho, Y. Wang, Fault tolerant control for singular systems with actuator saturation and nonlinear perturbation, Automatica 46 (3) (2010) 569–576.
[25] Z. Wang, D.W.C. Ho, Filtering on nonlinear time-delay stochastic systems, Automatica 39 (1) (2006) 101–108.
[26] B. Shen, Z. Wang, Y.S. Hung, Distributed consensus H_∞ filtering in sensor networks with multiple missing measurements: the finite-horizon case, Automatica 46 (10) (2010) 1682–1688.
[27] M. Farina, G. Ferrari-Trecate, R. Scattolini, Distributed moving horizon estimation for sensor networks, in: Proc. the 1st IFAC Workshop on Estimation and Control of Networked Systems, Venice, Italy, 2009, pp. 126–131.
[28] P. Shi, M.S. Mahmoud, K. Nguang, A. Ismail, Robust filtering for jumping systems with mode dependent delays, Signal Process. 86 (2006) 140–152.
[29] R. Carli, A. Chiuso, L. Schenato, S. Zampieri, Distributed Kalman filtering based on consensus strategies, IEEE J. Sel. Areas Commun. 26 (2008) 622–633.
[30] L. Magni, R. Scattolini, Stabilizing decentralized model predictive control of nonlinear systems, Automatica 42 (2006) 1231–1236.
[31] K. You, L. Xie, Minimum data rate for mean square stabilizability of linear systems with Markovian packet losses, IEEE Trans. Autom. Control 56 (4) (2011) 772–785.
[32] V. Ugrinovskii, Distributed robust filtering with $\mathcal{H}_\infty$ consensus of estimations, in: Proc. the 2010 American Control Conference, 2010, pp. 1374–1379.
[33] Y. Chu, K. Glover, Bounds of the induced norm and model reduction errors for systems with repeated scalar nonlinearities, IEEE Trans. Autom. Control 44 (1999) 4215–4226.
[34] Y. Chu, Further results for systems with repeated scalar nonlinearities, IEEE Trans. Autom. Control 44 (2001) 2031–2035.
[35] H. Gao, J. Lam, C. Wang, Induced ℓ_2 and generalized $\mathcal{H}_2$ filtering for systems with repeated scalar nonlinearities, IEEE Trans. Signal Process. 53 (2005) 4215–4226.
[36] M.S. Mahmoud, Y. Xia, Networked Control Systems: Cloud Control and Secure Control, Elsevier Butterworth-Heinemann, Brazil, 2019.

Chapter 9

Advanced distributed filtering

9.1 Self-tuning Kalman filtering

As battery power and processing power of nodes in sensor networks increase, the possibility of more intelligent estimation schemes becomes more and more important. The use of sensor networks was first driven by military applications, but with cheaper technology many other areas could make use of sensor networks (see, for example, [18,19]). Advantages with wireless sensor networks typically include more robustness, because more than one unit is performing the same task, and increased flexibility. However decentralized estimation is a far more complex task than traditional centralized estimation. At least two big questions arise, i.e., what to send and how to make use of the received information. The first question has two obvious candidate answers: measurements or estimates.

In the case when measurements are transmitted the main problem is that all nodes usually do not have the possibility to communicate with all other nodes, thus the measurements have to be routed through other nodes. These nodes might experience a very high communication load, thus limiting the size of the network. This is especially evident when the number of nodes exceeds the size of the state vector to be estimated. Once the routing problem is solved, generating an estimate based on measurements is straightforward, at least if the system is linear and subject to Gaussian disturbances.

The case where estimates are communicated has been given great attention in the literature. In [20] a decentralized Kalman filter was proposed. However, this algorithm requires every node to be able to communicate with every other node, which might not be possible. An alternative approach is to only allow nodes to communicate with their neighbors. As opposed to the case where measurements are communicated, no routing is required when estimates are used as information carriers. Without direct communication between all nodes, a new problem is introduced, namely, how to combine estimates from just neighboring nodes. To optimally combine two estimates one has to know the mutual information between the estimates. Computing this quantity for a general communication graph is a difficult task that requires global knowledge of the topology. In the case of a loop-free graph the problem was solved in [21] by introduction of a channel filter. This approach was used in a coordinated search strategy application (see [22]). If the state is assumed constant, the problem can be viewed as a distributed average problem which has been studied in [23]. The problem

Discrete Networked Dynamic Systems. https://doi.org/10.1016/B978-0-12-823698-7.00017-5

was generalized to time-varying states in [24] and [25] using consensus filters. A closely related area is the combination of estimates from a number of noncommunicating estimators, so-called track-to-track fusion algorithms. This problem differs in that the combined estimate is not used to compute a new estimate in the individual nodes. The fused estimate is only communicated to a central node to be used for some task. This problem was studied in for example [26,27].

9.1.1 Multisensor data fusion

Multisensor data fusion is a synergistic approach used in industry, which involves the mechanism of fusing data from various local resources, e.g., local sensors, individual subunits, etc. The data thus collected are centralized and are used for monitoring the health of the plant, controlling various scenarios, and making instant decisions based on local conditions. One of the key issues in developing multisensor data fusion is the combination and collection of the information from various local resources and then obtaining an accurate joint estimate of the system state vector. The two basic fusion architectures are centralized and decentralized (or distributed). Both architectures can be used in specific scenarios. When the processing center can receive all measurements from the local sensors in time, the centralized Kalman filtering can be used there. This can be used in mission-critical scenarios, where every local sensor is important with its local information; this local information then transmits the processed data local state estimate to a fusion center. The main advantage of the centralized fusion method is that it involves minimum information loss. Under some regularity conditions, in particular, the assumption of cross-independent sensor noises, an optimal Kalman filtering fusion was proposed in [1], which was proved to be equivalent to the centralized Kalman filtering using all sensor measurements; therefore, such fusion is optimal. However, it may result in high computational load due to overloading of the filter with more than it can handle. On the other hand, the distributed fusion method can give the globally optimal or suboptimal state estimation by combining or weighting the local state estimators placed at ith sensors [2]. Also, then there is no feedback from the fusion center to local sensors. We propose a distributed Kalman filtering fusion formula under a mild condition, and prove that the fused state estimate is equivalent to the centralized Kalman filtering using all sensor measurements.

9.1.2 Information fusion estimation

In [3], preliminary analysis of information fusion estimation of noise statistics for multisensor systems was presented. Parameter estimation on linear time-varying systems was the subject of [4]. Some related results on fault detection and diagnosis are reported in [5–7]. Optimal filtering over linear observations with unknown parameters was treated in [8] and recent results on H_∞ filtering for systems with nonlinear perturbations are found in [9,10].

In this section, by the information matrix approach, we build on the results of [11,12] and examine carefully an approach of self-tuning Kalman filter fusion with sensor noises cross-correlated for distributed recursive state estimators of dynamic systems. In [12], the idea was implemented using an ordinary Kalman filter with no conditions of feedback from the fusion center to local sensors and the corresponding condition of fusion formula with feedback. It has been shown that the corresponding self-tuning Kalman fusion without feedback is, as the fusion with feedback, exactly identical to the corresponding centralized Kalman filtering fusion formula using all sensor measurements. The results have been evaluated by testing on a quadruple-tank system.

It has been shown that an efficient modified self-tuning Kalman filter approach has no requirement of providing a feedback loop in view of its embedded information matrix. The condition of feedback from the fusion center to the local sensors has helped in the implementation of self-tuning Kalman filters. Through lab simulation on a quadruple-tank system, the modified filter has performed without feedback in a much better way compared with the regular Kalman filter with feedback.

9.1.3 Problem formulation

Consider the following l-sensor distributed dynamic system:

$$x_{k+1} = \phi_k x_k + \upsilon_k, \; k = 0, 1, ..., \tag{9.1}$$

$$y_k^i = H_k^i x_k + w_k^i, \; i = 1, ..., \ell, \tag{9.2}$$

where ϕ_k is a matrix of order $(r \times r)$, $\widehat{x}_k, \upsilon_k \in \Re^r$, $H_i^k \in \Re^{N_i \times r}$, $y_k^i, w_i^k \in \Re^{N_i}$. The process noise υ_k and measurement noise w_i^k are both zero mean random variables independent of each other temporally, but w_i^k and w_j^k may be cross-correlated for $i \neq j$ at the same time instant k. To compare performance between the centralized and distributed filtering fusion, the stacked measurement equation is written as follows:

$$y_k = H_k x_k + w_k, \tag{9.3}$$

where

$$y_k = (y_k^{1^t}, ..., y_k^{l^t})^t, \; H_k = (H_k^{1^t}, ..., H_k^{l^t})^t,$$

$$w_k = (w_k^{1^t}, ..., w_k^{l^t})^t, \; R = diag[R_1 \; ..... \; R_L], \tag{9.4}$$

the covariance of the noise w_k is given by

$$\mathbf{Cov}(w_k) = R_k, \; \mathbf{Cov}(w_k^i) = R_k^i, \; i = 1, ..., \ell, \tag{9.5}$$

and R_k and R_k^i are both invertible for all i. According to the optimal distributed fusion Kalman filter [11,12], the local steady state optimal Kalman filter is described by

$$\widehat{x}_{k/k}^i = \widehat{\Psi}^i \widehat{x}_{k-1/k-1}^i + \widehat{K}_k^i y_k^i, \tag{9.6}$$

$$\widehat{\Psi}^i = \widehat{P}_{k/k}^i \widehat{P}_{k/k-1}^{-1} \widehat{\Phi}_k, \tag{9.7}$$

$$\widehat{K}_k^i = \widehat{P}_{k/k}^i H_k^{i^T} \widehat{R}_k^{i^{-1}}, \tag{9.8}$$

with covariance of the filtering error given by

$$\widehat{P}_{k/k}^{i^{-1}} = \widehat{P}_{k/k-1}^{i^{-1}} + H_k^{i^t} \widehat{R}_k^{i^{-1}} H_k^i, \tag{9.9}$$

where

$$\begin{aligned}
\widehat{x}_{k/k-1}^i &= \widehat{\phi}_k \widehat{x}_{k-1/k-1}^i, \\
\widehat{P}_{k/k}^i &= E[(\widehat{x}_{k/k}^i - \widehat{x}_k)(\widehat{x}_{k/k-1}^i - \widehat{x}_k)^t], \\
\widehat{P}_{k/k-1}^i &= E[(\widehat{x}_{k/k-1}^i - \widehat{x}_k)(\widehat{x}_{k/k-1}^i - \widehat{x}_k)^t].
\end{aligned}$$

From (9.1) and (9.3), we have the following steady state optimal centralized fusion Kalman filter:

$$\widehat{x}_{k/k} = \widehat{\Psi} \widehat{x}_{k-1/k-1} + P \sum_{i=1}^{L} H_k^T \widehat{R}_k^{-1} y_k. \tag{9.10}$$

Putting values, we get

$$\begin{aligned}
\widehat{x}_{k/k} &= \Psi \widehat{x}_{k-1/k-1} \\
&\quad + P \sum_{i=1}^{L} [\widehat{P}_{k/k}^{-1} \widehat{x}_{k/k} - \widehat{P}_{k/k-1}^{-1} \widehat{\Phi} \widehat{x}_{k-1/k-1}],
\end{aligned} \tag{9.11}$$

$$\widehat{K}_k = \widehat{P}_{k/k} H_k^{i^t} \widehat{R}_k^{-1}, \tag{9.12}$$

$$\widehat{P}_{k/k}^{-1} = \widehat{P}_{k/k-1}^{-1} + H_k^t \widehat{R}_k^{-1} H_k, \tag{9.13}$$

$$\widehat{\Psi} = \widehat{P}_{k/k} \widehat{P}_{k/k-1}^{-1} \Phi, \tag{9.14}$$

where

$$\begin{aligned}
\widehat{x}_{k/k-1} &= \widehat{\phi}_k \widehat{x}_{k-1/k-1}, \\
\widehat{P}_{k/k} &= E[(\widehat{x}_{k/k} - \widehat{x}_k)(\widehat{x}_{k/k-1} - \widehat{x}_k)^T], \\
\widehat{P}_{k/k-1} &= E[(\widehat{x}_{k/k-1} - \widehat{x}_k)(\widehat{x}_{k/k-1} - \widehat{x}_k)^t].
\end{aligned}$$

9.1.4 Self-tuning distributed Kalman fusion filter

Using the system identification method [13], we can obtain the consistent estimators.

Assumption 9.1. The model parameter and noise variance estimators are consistent, that is, we assume that

$$\widehat{\theta} \to \theta, \ \ \widehat{R}_i \to R_i, \ \ \text{as } k \to \infty. \tag{9.15}$$

The basic principle of the self-tuning fusion filter is that inclusion of the online estimates of the model parameters and noise variances into the optimal fusion filter will yield a self-tuning fusion filter. It can be obtained by the following three steps [3,13].

Step 1. Using the system identification method and the correlation function method, the estimators of model parameters and noise variances are obtained.
Step 2. Substituting the estimates $\widehat{\Phi}_k$ and $\widehat{R}_i$ into Eqs. (9.6)–(9.9) yields the self-tuning local Kalman filters

$$\widehat{x}_{k/k}^{S_i} = \widehat{P}_{k/k}^{S_i} \widehat{P}_{k/k-1}^{S_i^{-1}} \widehat{\Phi}_k \widehat{x}_{k-1/k-1}^{S_i} + \widehat{K}_k^{S_i} y_k^{S_i}, \tag{9.16}$$

$$\widehat{K}_k^{S_i} = \widehat{P}_{k/k}^{S_i} H_k^{i^T} \widehat{R}_k^{S_i^{-1}}, \tag{9.17}$$

$$\widehat{P}_{k/k}^{S^{-1}} = \widehat{P}_{k/k-1}^{S^{-1}} + H_k^{i^T} \widehat{R}_k^{S_i^{-1}} H_k^i. \tag{9.18}$$

Step 3. Substituting $\widehat{P}_{k/k}^{S_i}$, $\widehat{P}_{k/k-1}^{S^{-1}}$, and $\widehat{x}_{k/k}^{S}$ into the steady state optimal distributed fusion Kalman filter (9.11)–(9.14) yields the self-tuning distributed fusion Kalman filter

$$\begin{aligned} \widehat{x}_{k/k}^{S} &= \widehat{P}_{k/k}^{S} \widehat{P}_{k/k-1}^{S^{-1}} \widehat{\Phi}_k \widehat{x}_{k-1/k-1}^{S} \\ &\quad + \widehat{P}_{k/k}^{S} \sum_{i=1}^{L} [\widehat{P}_{k/k}^{S^{-1}} \widehat{x}_{k/k}^{S} - \widehat{P}_{k/k-1}^{S^{-1}} \widehat{\Phi}_k \widehat{x}_{k-1/k-1}^{S}], \end{aligned} \tag{9.19}$$

$$\widehat{P}_{k/k}^{S^{-1}} = \widehat{P}_{k/k-1}^{S^{-1}} + \sum_{i=1}^{L} [\widehat{P}_{k/k}^{S_i^{-1}} - \widehat{P}_{k/k-1}^{S_i^{-1}}], \tag{9.20}$$

with the initial value $\widehat{x}^S(0|0) = \widehat{x}_0^S$, $\widehat{P}(0|0) = \widehat{P}_0$. When sensor noises are cross-dependent, clearly

$$H_k^T \widehat{R}_k^{S^{-1}} H_k = \sum_{i=1}^{l} H_k^{i^T} \widehat{R}_k^{S_i^{-1}} H_k^i. \tag{9.21}$$

To express the centralized self-tuning filter $\widehat{x}^S_{k/k}$ in terms of the local filtering, we use (9.6)–(9.8). Then we have

$$H_k^{i^T} \widehat{R}_k^{S_i^{-1}} y_k^{S_i} = \widehat{P}_{k/k}^{S_i^{-1}} \widehat{x}_{k/k}^{S_i} - \widehat{P}_{k/k}^{S_i^{-1}} \widehat{\Phi}_k \widehat{x}_{k-1/k-1}^{S_i}. \tag{9.22}$$

Thus, substituting (9.22) into (9.19) yields

$$\begin{aligned} \widehat{P}_{k/k}^{S^{-1}} \widehat{x}_{k/k}^{S} &= \widehat{P}_{k/k}^{S^{-1}} \widehat{\Phi}_k \widehat{x}_{k-1/k-1}^{S} \\ &\quad + \sum_{i=1}^{l} (\widehat{P}_{k/k}^{S_i^{-1}} \widehat{x}_{k/k}^{S_i} - \widehat{P}_{k/k-1}^{S_i^{-1}} \widehat{\Phi}_k \widehat{x}_{k/k-1}^{S_i}), \end{aligned} \tag{9.23}$$

that is, the centralized self-tuned filtering (9.23) and error matrix (9.20) are explicitly expressed in terms of the local filtering. It was claimed in [11] that although the feedback cannot improve the performance at the fusion center, the feedback does reduce the covariance of each local tracking error, and they proceeded further from here on in proving that. Also it was claimed that when sensor noises were cross-correlated, it was still possible to express equivalently the centralized Kalman filtering in terms of the local filtering, and derive a similar performance analysis for the distributed Kalman filtering with feedback. In the following sections, it is shown that the self-tuning Kalman filter approach can perform better even without the feedback, as it gives the same results without the feedback which were given here with the feedback.

9.1.5 Distributed self-tuning Kalman filter without feedback

Assume all sensor measurement matrices H_i^k are of full row-rank. In many practical applications, for example, in the tracking problem, this assumption is fulfilled very often. Thus, we know $H_i^k (H_k^{i'})^{\dagger} = I$, where the superscript $\dagger$ stands for pseudoinverse.

Similar to the derivation of (9.20) and (9.23), the centralized Kalman filtering with cross-correlated sensor noises is given by

$$\begin{aligned} \widehat{x}_{k/k}^{S} &= \Psi \widehat{x}_{k-1/k-1}^{S} + \widehat{P}_{k/k} H_k^t \sum_{i=1}^{l} \widehat{R}_k^{-1}(*i) y_k^i, \\ K_k &= \widehat{P}_{k/k} H_k^t \widehat{R}_k^{-1}. \end{aligned} \tag{9.24}$$

Also, in (9.24),

$$\widehat{K}_k^S y_k = \widehat{P}_{k/k} H_k^t \sum_{i=1}^{l} \widehat{R}_k^{-1}(*i) y_k^i, \tag{9.25}$$

where $\widehat{R}_k^{-1}(*i)$ is the ith submatrix column of $\widehat{R}_k^{-1}$, with covariance of filtering error given by

$$\begin{aligned}\widehat{P}_{k/k}^{-1} &= \widehat{P}_{k/k-1}^{-1} + H_k^t \widehat{R}_k^{-1} H_k \\ &= \widehat{P}_{k/k-1}^{-1} + H_k^t \sum_{i=1}^{l} \widehat{R}_k^{-1}(*i)\widehat{R}_k^i (H_k^{i'}) \\ &+ H_k^{i'} \widehat{R}_k^{-i} H_k^i,\end{aligned} \tag{9.26}$$

or

$$\widehat{P}_{k/k} = (I - \widehat{K}_k H_k)\widehat{P}_{k/k-1}. \tag{9.27}$$

Using (9.9) and (9.26), the estimation error covariance of the centralized Kalman filtering is given by using the estimation error covariances of all local filters, i.e.,

$$\begin{aligned}\widehat{P}_{k/k}^{-1} &= \widehat{P}_{k/k-1}^{-1} + H_k^t \sum_{i=1}^{l} \widehat{R}_k^{-1}(*i)\widehat{R}_k^i (H_k^{i'}) \\ &+ (\widehat{P}_{k/k}^{i^{-1}} - \widehat{P}_{k/k-1}^{i^{-1}}).\end{aligned} \tag{9.28}$$

Using (9.24), (9.4), and (9.5), we have

$$\begin{aligned}K_k y_k &= \widehat{P}_{k/k} H_k^t \sum_{i=1}^{l} \widehat{R}_k^{-1}(*i) y_k^i \\ &= \widehat{P}_{k/k} H_k^t \sum_{i=1}^{l} \widehat{R}_k^{-1}(*i)\widehat{R}_k^i (H_k^{i'}) + H_k^{i'} \widehat{R}_k^{-i} H_k^i.\end{aligned} \tag{9.29}$$

To express the centralized filtering $\widehat{x}_{k/k}^S$ in terms of the local filtering, we use (9.6) and (9.29) to eliminate y_k from (9.24). From (9.6) we have

$$H_k^{i^T} \widehat{R}_k^{i^{-1}} y_k^i = [\widehat{P}_{k/k}^{i^{-1}} \widehat{x}_{k/k}^{S_i} - \widehat{P}_{k/k-1}^{i^{-1}} \Phi \widehat{x}_{k/k-1}^{S_i}]. \tag{9.30}$$

Thus, substituting (9.27), (9.29), and (9.30) into (9.24) yields

$$\begin{aligned}P_{k/k}^{-1} x_{k/k} &= P_{k/k-1}^{-1} \Phi x_{k/k-1} + H_k^t \sum_{i=1}^{l} R_k^{-1}(*i) R_k^i (H_k^{i'})^+ \\ &\times (P_{k/k}^{i^{-1}} x_{k/k}^i - P_{k/k-1}^{i^{-1}} \Phi x_{k/k-1}^i),\end{aligned} \tag{9.31}$$

that is, the centralized filtering (9.31) and error matrix (9.28) are explicitly expressed in terms of the local filtering. Hence, the performance of the distributed

Kalman filtering fusion given in (9.28) and (9.31) is the same as that of the centralized fusion.

Remark 9.1. It is not difficult to see from (9.24) that the centralized Kalman is different from the one in [12]. Moreover, in (9.31), the information vector is the main source of carrying the information.

9.1.6 Optimality of self-tuning Kalman filter with feedback

When there is feedback, the fusion center broadcasts its latest estimate to the local sensors. Thus, for all i, the following local and global one-stage predictions were modified naturally as

$$x^i_{k/k-1} = \Phi_k \hat{x}_{k-1/k-1} = x_{k/k-1}, \tag{9.32}$$

$$\hat{P}^i_{k/k-1} = \hat{P}_{k/k-1}. \tag{9.33}$$

Naturally, the following filtering fusion with feedback was suggested:

$$\hat{P}^{-1}_{k/k} = \hat{P}^{-1}_{k/k-1} + H^t_k \sum_{i=1}^{l} R^{-1}_k (*i) R^i_k (H^{i'}_k) + (\hat{P}^{i^{-1}}_{k/k} - \hat{P}^{-1}_{k/k-1}), \tag{9.34}$$

$$\hat{P}^{-1}_{k/k} x_{k/k} = \hat{P}^{-1}_{k/k-1} \Phi x_{k/k-1} + H^t_k \sum_{i=1}^{l} R^{-1}_k (*i) R^i_k (H^{i'}_k)^+ \times (\hat{P}^{i^{-1}}_{k/k} x^i_{k/k} - \hat{P}^{i^{-1}}_{k/k-1} \Phi x^i_{k/k-1}). \tag{9.35}$$

In what follows, a rigorous analysis for the feedback filtering fusion with cross-correlated sensor noises is presented:

$$\hat{x}_{k/k} = x_{k/k}, \; \hat{P}_{k/k} = P_{k/k}, \tag{9.36}$$

$$\hat{P}^i_{k/k} = E[(\hat{x}^i_{k/k})(\hat{x}^i_{k/k} - x_k)^t]. \tag{9.37}$$

The main advantage of the feedback filtering fusion is the reduction of the covariance of every local estimate error, i.e.,

$$\hat{P}^i_{k/k} \leq P^i_{k/k}, \quad i = 1, 2, ..., \ell \tag{9.38}$$

9.1.7 Global optimality of the feedback filtering fusion

We use mathematical induction to show no global performance difference exists between two types of filtering fusion with feedback and without feedback.

Clearly, it is reasonable to have the same initial values at both the local and the global filters, with or without feedback, so we have

$$\hat{x}_{0/0} = x_{0/0} = x^i_{0/0} = \hat{x}^i_{0/0},$$
$$\hat{P}_{0/0} = P_{0/0} = P^i_{0/0} = \hat{P}^i_{0/0}. \tag{9.39}$$

Suppose that

$$\hat{x}_{k-1/k-1} = x_{k-1/k-1}, \hat{x}_{k/k-1} = x_{k/k-1}, \tag{9.40}$$
$$\hat{P}_{k-1/k-1} = P_{k-1/k-1}, \hat{P}_{k/k-1} = P_{k/k-1}. \tag{9.41}$$

Therefore, at the kth stage, using the above equations and (9.34), we have

$$\begin{aligned}\hat{P}^{-1}_{k/k} = \hat{P}^{-1}_{k/k-1} + H^t_k \sum_{i=1}^{l} R^{-1}_k(*i) R^i_k (H^{i'}_k) \\ + (\hat{P}^{i^{-1}}_{k/k} - \hat{P}^{-1}_{k/k-1}).\end{aligned} \tag{9.42}$$

Note the definition of $\hat{P}^i_{k/k}$ similar to that in (9.9). We get

$$\hat{P}^{-1}_{k/k} = P^{-1}_{k/k-1} + H^{i'}_k R^{i^{-1}}_k H^i_k. \tag{9.43}$$

Substituting (9.43) into (9.42) and comparing it with (9.16)–(9.18) yields

$$\hat{P}^{-1}_{k/k} = P^{-1}_{k/kl}. \tag{9.44}$$

Then using (9.15)–(9.18) in the feedback case and (9.43), we have

$$\begin{aligned}\hat{P}^{i^{-1}}_{k/k}\hat{x}^i_{k/k} &= \hat{P}^{i^{-1}}_{k/k}\Phi x_{k/k-1} + H^{i'}_k R^{i^{-1}}_k (y^i_k - H^i_k x_{k/k-1}) \\ &= \hat{P}^{i^{-1}}_{k/k}\Phi x_{k/k-1} + H^{i^t}_k R^{i^{-1}}_k y^i_k \\ &- H^{i'}_k R^{i^{-1}}_k H^i_k x_{k/k-1} \\ &= \hat{P}^{i^{-1}}_{k/k}\Phi x_{k/k-1} + H^{i^{-1}}_k R^{i^{-1}}_k y^i_k \\ &- \hat{P}^{i^{-1}}_{k/k} x_{k/k-1} + P^{-1}_{k/k-1} x_{k/k-1} \\ &= H^{i'}_k R^{i^{-1}}_k y^i_k + P^{-1}_{k/k-1} x_{k/k-1} \\ &+ \hat{P}^{-1}_{k/k-1}(\Phi - 1) x_{k-1/k-1}.\end{aligned} \tag{9.45}$$

Remark 9.2. Note that (9.45) shows that the feedback loop is employed as followed by the steps of [12]. Obviously, it is different from the nonfeedback case. However, it will be shown below that when feedback is not applied here in

(9.45), it gives an answer similar to what was given in [12] with the application of feedback. Thus

$$\hat{P}^{i^{-1}}_{k/k}\hat{x}^{i}_{k/k} = \hat{P}^{i^{-1}}_{k/k}\Phi x_{k/k-1} + H^{i'}_{k} R^{i^{-1}}_{k}(y^{i}_{k}). \tag{9.46}$$

Similarly, from (9.19) and (9.20), we have

$$\begin{aligned} P^{-1}_{k/k}x_{k/k} &= H^{t}_{k}R^{-1}_{k}y_{k} + P^{-1}_{k/k-1}\Phi x_{k/k-1} \\ &= P^{-1}_{k/k-1}\Phi x_{k/k-1} \\ &\quad + H^{t}_{k}\sum_{i=1}^{l} R^{-1}_{k}(*i)R^{i}_{k}(H^{i'}_{k}) + H^{i'}_{k}R^{i^{-1}}_{k}y^{i}_{k}. \end{aligned} \tag{9.47}$$

Hence, using (9.35), (9.41), and (9.44)–(9.47), we have

$$\begin{aligned} \hat{x}_{k/k} &= P_{k/k}P^{-1}_{k/k-1}\Phi x_{k/k-1} \\ &\quad + P_{k/k}H^{t}_{k}\sum_{i=1}^{l} R^{-1}_{k}(*i)R^{i}_{k}(H^{i'}_{k}) + H^{i'}_{k}R^{i^{-1}}_{k}y^{i}_{k} \\ &= x_{k/k}. \end{aligned} \tag{9.48}$$

By calculating the local estimates without feedback with self-tuning Kalman filter, $\hat{P}^{i^{-1}}_{k/k}$ has been defined in (9.43). We need to verify if it is still the covariance of the local estimate error. From the last equality of (9.45) and (9.43), we know

$$\begin{aligned} \hat{x}^{i}_{k/k} &= \hat{P}^{i}_{k/k}H^{i'}_{k}R^{i^{-1}}_{k}y^{i}_{k} + \hat{P}^{i}_{k/k}\hat{P}^{i^{-1}}_{k/k}x_{k/k-1} \\ &= \hat{P}^{i}_{k/k}H^{i'}_{k}R^{i^{-1}}_{k}y^{i}_{k} + x_{k/k-1} \\ &= \hat{P}^{i}_{k/k}H^{i'}_{k}R^{i^{-1}}_{k}y^{i}_{k} + x^{i}_{k/k} - \hat{P}^{i}_{k/k}H^{i'}_{k}R^{i^{-1}}_{k}y^{i}_{k} \\ &= x_{k/k}. \end{aligned} \tag{9.49}$$

It follows that

$$\mathbb{E}[(\hat{x}^{i}_{k/k})(\hat{x}^{i}_{k/k})'] = \hat{P}^{i}_{k/k}. \tag{9.50}$$

The result shows that with the self-tuning Kalman filter, no feedback is required, which was practiced in [12]. The self-tuning Kalman filter without feedback is good enough to run and prove compatible with the ordinary Kalman filter with feedback.

9.1.8 Simulation example 9.1

In what follows, we demonstrate the effectiveness of the distributed self-tuning Kalman filter fusion algorithm as applied to a typical dynamic process. The process is a quadruple-tank system and it consists of four interconnected water tanks and two pumps. Its manipulated variables are voltages to the pumps and the controlled variables are the water levels in the two lower tanks. The quadruple-tank process is built by considering the concept of two double-tank processes. The quadruple-tank system presents a multiinput multioutput system. This system is a real-life control problem prototyped to experiment on, and we try to solve it in the most efficient way, since it deals with multiple variables, thus it gives a reflection for large systems in industry. The schematic description of the four-tank system can be visualized by Fig. 9.1. The system has two control inputs (pump throughputs) which can be manipulated to control the water level in the tanks. The two pumps are used to transfer water from a sump into four overhead tanks. By adjusting the bypass valves of the system, the proportion of the water pumped into different tanks can be changed to adjust the degree of interaction between the pump throughputs and the water levels. Thus each pump output goes to two tanks, one lower and another upper, diagonally opposite, and the ratio of the split-up is controlled by the position of the valve. Because of the large water distribution load, the pumps are supplied with 12 V each. The mathematical modeling of the quadruple-tank process can be obtained by using the Bernoulli law. Combining all the equations for the interconnected four-tank system we obtain the physical system. A fault model can then be constructed by adding extra holes to each tank. The mathematical model of the faulty quadruple-tank system can be given as

$$
\begin{aligned}
\frac{dh_1}{dt} &= -\frac{a_1}{A_1}\sqrt{2gh_1} + \frac{a_3}{A_1}\sqrt{2gh_3} + \frac{\gamma_1 k_1}{A_1}v_1 + \frac{d}{A_1} \\
&\quad - \frac{a_{leak1}}{A_1}\sqrt{2gh_1}, \\
\frac{dh_2}{dt} &= -\frac{a_2}{A_2}\sqrt{2gh_2} + \frac{a_4}{A_2}\sqrt{2gh_4} + \frac{\gamma_2 k_2}{A_2}v_2 - \frac{d}{A_2} \\
&\quad - \frac{a_{leak2}}{A_2}\sqrt{2gh_2}, \\
\frac{dh_3}{dt} &= -\frac{a_3}{A_3}\sqrt{2gh_3} + \frac{(1-\gamma_2)k_2}{A_3}v_2 \\
&\quad - \frac{a_{leak3}}{A_3}\sqrt{2gh_3}, \\
\frac{dh_4}{dt} &= -\frac{a_4}{A_4}\sqrt{2gh_4} + \frac{(1-\gamma_1)k_1}{A_4}v_1 \\
&\quad - \frac{a_{leak4}}{A_4}\sqrt{2gh_4}, \\
\frac{dv_1}{dt} &= -\frac{v_1}{\tau_1} + \frac{1}{\tau_1}u_1, \\
\frac{dv_2}{dt} &= -\frac{v_2}{\tau_2} + \frac{2}{\tau_2}u_2.
\end{aligned}
\tag{9.51}
$$

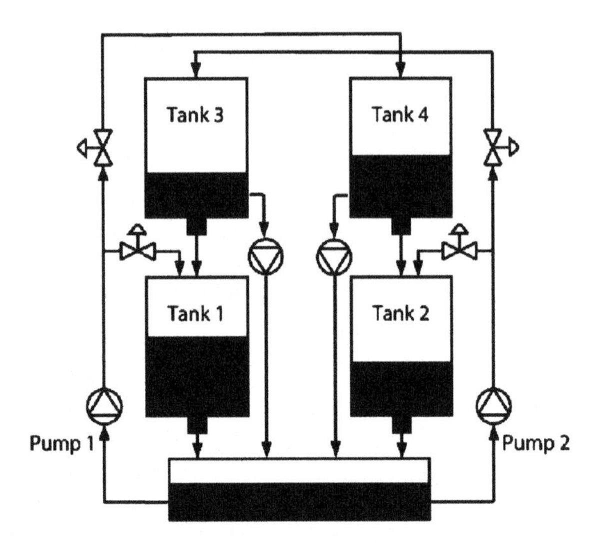

FIGURE 9.1 Schematic diagram of a quadruple-tank system.

By carefully evaluating the ensuing results of the quadruple-tank system for three scenarios,

- the regular Kalman filter fusion without feedback,
- the regular Kalman filter fusion with feedback,
- the efficient distributed self-tuning Kalman filter implementation without feedback (see Fig. 9.2),

it can be readily seen that what is achieved by [12] with feedback is also achieved by self-tuning Kalman filtering without feedback. This is due to the good performance of the structure of the self-tuning Kalman filter with information matrix.

9.2 Kalman filtering with intermittent communications

In this section, we employ the forward–backward (FB) Kalman filtering approach to tackle the distributed Kalman filtering fusion with passive packet loss or initiative intermittent communications from local estimators to a fusion center while the process noise does exist.

Under some regularity conditions, in particular, the assumption of cross-independent sensor noises, an optimal Kalman filtering fusion [14] was proved to be equivalent to the centralized Kalman filtering using all sensor measurements; therefore, such fusion is optimal. However, it may result in high computational load due to overloading of the filter with more than it can handle. For the distributed Kalman filter, let us consider the case with packet loss or intermittent communications from local sensors/estimators to the fusion center. In [15] it has been noted that the optimality of the fusion equations in reproduc-

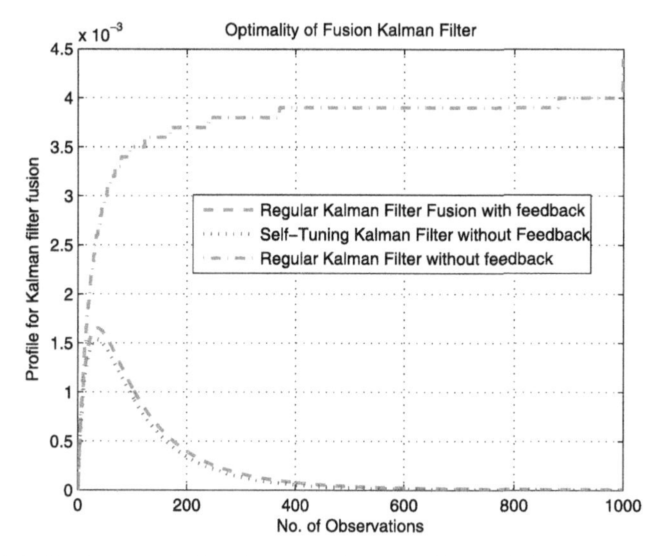

FIGURE 9.2 Results of quadruple-tank system with various Kalman fusions.

ing the centralized estimates depends on the conditional independence of the measurements given the target state.

On comparing with the original multisensor system, the centralized filtering fusion performance of the modified multisensor system may be reduced in some degree since sensor noises of the modified multisensor system are cross-correlated. Thus, as shown in [16] and [17], such fusion algorithm is suboptimal. Since the proposed suboptimal algorithm takes advantage of sensor estimates obviously more than the standard distributed Kalman filtering fusion with lost sensor estimates replaced by their predictions, it can be expected that the performance of the former should be better than that of the latter.

In this section, we develop an improved approach of FB Kalman filtering to tackle the distributed Kalman filtering fusion with passive packet loss or initiative intermittent communications from local estimators to a fusion center and in the presence of process noise.

9.2.1 Problem formulation

Consider an ℓ-sensor distributed dynamic system described by

$$x_{k+1} = \phi_k x_k + \upsilon_k,\ k = 0, 1, ..., \tag{9.52}$$

$$y_k^i = H_k^i x_k + w_k^i,\ i = 1, ..., \ell, \tag{9.53}$$

where ϕ_k is a matrix of order $r \times r$, $x_k, \upsilon_k \in \Re^r$, $H_i^k \in \Re^{N_i \times r}$, $y_k^i, w_i^k \in \Re^{N_i}$. The process noise υ_k and measurement noise w_i^k are both zero mean random variables independent of each other temporally, but w_i^k and w_j^k may be cross-correlated for $i \neq j$ at the same time instant k.

To compare the performance between the centralized and distributed filtering fusion, the stacked measurement equation is written as

$$y_k = H_k x_k + w_k, \tag{9.54}$$

where

$$\begin{aligned} y_k &= (y_k^{1^t}, ..., y_k^{\ell^t})^t, \; H_k = (H_k^{1^t}, ..., H_k^{\ell^t})^t, \\ w_k &= (w_k^{1^t}, ..., w_k^{\ell^t})^t \end{aligned} \tag{9.55}$$

and the covariance of the noise w_k is given by

$$\mathbf{Cov}(w_k) = R_k, \quad R_k^i = \mathbf{Cov}(w_k^i), \quad i = 1, ..., \ell, \tag{9.56}$$

where R_k and R_k^i are both invertible for all i. According to the standard results in FB Kalman filtering, the local Kalman filtering at the ith sensor is expressed as

$$\textbf{Forward run} : \text{for (k = 0; k < T; +k)}$$

$$\widehat{K}_k^i = \widehat{P}_{k/k}^i H_k^{i^t} \widehat{R}_k^{i^{-1}}, \tag{9.57}$$

$$\widehat{x}_{k/k}^i = \widehat{x}_{k/k-1}^i + \widehat{K}_k^i (y_k^i - H_k^i \widehat{x}_{k/k-1}^i), \tag{9.58}$$

$$\widehat{P}_{k/k}^i = \widehat{P}_{k/k-1}^i - \widehat{K}_k^i H_k \widehat{P}_{k/k-1}^i, \tag{9.59}$$

$$\textbf{Backward run} : \text{for (k = T − 1; t ≥ 0; −k)}$$

$$\begin{aligned} \widehat{\lambda}_{k-1/T}^i &= (I_{P+N} - H_k^{i^*} \widehat{K}_{f,i}^*) \widehat{\Phi}_k^i \widehat{\lambda}_{k+1/T}^i \\ &\quad + H_k^i \widehat{R}_{e,i}^{-1} (y_k^i - H_k^i \widehat{x}_{k/k-1}^i), \end{aligned} \tag{9.60}$$

$$\widehat{x}_{k-1/T}^i = \widehat{x}_{k-1/k-1}^i + \widehat{P}_{k/k-1}^i \widehat{\lambda}_{k-1/T}^i, \tag{9.61}$$

$$\widehat{P}_{k-1/T}^i = \widehat{P}_{k-1/k-1}^i, \tag{9.62}$$

where the covariance of the filtering error can be stated as

$$\widehat{P}_{k/k}^{i^{-1}} = \widehat{P}_{k/k-1}^{i^{-1}} + H_k^{i^t} \widehat{R}_k^{i^{-1}} H_k^i, \tag{9.63}$$

where

$$\begin{aligned} \widehat{x}_{k/k-1}^i &= \widehat{\Phi}_k \widehat{x}_{k-1/k-1}^i, \\ \widehat{P}_{k/k}^i &= E[(\widehat{x}_{k/k}^i - \widehat{x}_k)(\widehat{x}_{k/k-1}^i - \widehat{x}_k)^t], \\ \widehat{P}_{k/k-1}^i &= E[(\widehat{x}_{k/k-1}^i - \widehat{x}_k)(\widehat{x}_{k/k-1}^i - \widehat{x}_k)^t]. \end{aligned}$$

Similarly, the centralized FB Kalman filtering with all sensor data is given by

$$\textbf{Forward run}: \text{for } (\text{k} = 0; \text{k} < \text{T}; +\text{k})$$

$$\widehat{K}_k = \widehat{P}_{k/k} H_k^t \widehat{R}_k^{-1}, \tag{9.64}$$

$$\widehat{x}_{k/k} = \widehat{x}_{k/k-1} + \widehat{K}_k (y_k - H_k \widehat{x}_{k/k-1}), \tag{9.65}$$

$$\widehat{P}_{k/k} = \widehat{P}_{k/k-1} - \widehat{K}_k H_k \widehat{P}_{k/k-1}, \tag{9.66}$$

$$\textbf{Backward run}: \text{for } (\text{k} = \text{T} - 1; \text{t} \geq 0; -\text{k})$$

$$\widehat{\lambda}_{k-1/T} = (I_{P+N} - H_k^* \widehat{K}_{f,i}^*) \widehat{\Phi}_k \widehat{\lambda}_{k+1/T} + H_k \widehat{R}_{e,i}^{-1} (y_k^i - H_k \widehat{x}_{k/k-1}), \tag{9.67}$$

$$\widehat{x}_{k-1/T} = \widehat{x}_{k-1/k-1} + \widehat{P}_{k/k-1} \widehat{\lambda}_{k-1/T}, \tag{9.68}$$

$$\widehat{P}_{k-1/T} = \widehat{P}_{k-1/k-1}, \tag{9.69}$$

where the covariance of the filtering error can be described as

$$\widehat{P}_{k/k}^{-1} = \widehat{P}_{k/k-1}^{-1} + {H_k}^t \widehat{R}_k^{-1} H_k, \tag{9.70}$$

where

$$\widehat{x}_{k/k-1} = \widehat{\Phi}_k \widehat{x}_{k-1/k-1},$$

$$\widehat{P}_{k/k} = E[(\widehat{x}_{k/k} - \widehat{x}_k)(\widehat{x}_{k/k-1} - \widehat{x}_k)^t],$$

$$\widehat{P}_{k/k-1} = E[(\widehat{x}_{k/k-1} - \widehat{x}_k)(\widehat{x}_{k/k-1} - \widehat{x}_k)^t].$$

It is quite clear when the sensor noises are cross-dependent that

$$H_k^t \widehat{R}_k^{-1} H_k = \sum_{i=1}^{l} H_k^{i^t} \widehat{R}_k^{i^{-1}} H_k^i.$$

Likewise, it has already been proved in [1, 3, 4] that the centralized filtering and error matrix could be explicitly expressed in terms of the local filtering and error matrices as follows:

$$\widehat{P}_{k/k}^{-1} = \widehat{P}_{k/k-1}^{-1} + \sum_{i=1}^{l} (\widehat{P}_{k/k}^{i^{-1}} - \widehat{P}_{k/k-1}^{i^{-1}}) \tag{9.71}$$

and

$$\widehat{P}_{k/k}^{-1} \widehat{x}_{k/k} = \widehat{P}_{k/k-1}^{-1} + \sum_{i=1}^{l} (\widehat{P}_{k/k}^{i^{-1}} \widehat{x}_{k/k}^i - \widehat{P}_{k/k-1}^{i^{-1}} \widehat{x}_{k/k-1}^i). \tag{9.72}$$

Also,

$$H_k^{i'} \widehat{R}_k^{i^{-1}} y_k^i = \widehat{P}_{k/k}^{i^{-1}} \widehat{x}_{k/k}^i - \widehat{P}_{k/k-1}^{i^{-1}} \widehat{x}_{k/k-1}^i. \tag{9.73}$$

In what follows, we are going to deal with the practical situation in which the local sensors may fail to send their estimates to the fusion center. In this case, the measurement equation of the corresponding centralized multisensor system has to be modified, that is, the original multiple individual observations should be stacked as a modified single observation. This modification process is made by an FB Kalman filter smoother. It will be shown that these modification equations will offer improved results in cases when:

1. the process noise and sensor noise are one-step mutually correlated;
2. the sensor noises are cross-correlated, which both obviously violate the condition for the global optimality of the distributed filtering fusion.

Before proceeding further, we present a preliminary result on globally optimal distributed Kalman filtering fusion for dynamic systems with process noise one-step correlated to measurement noise and sensor noises cross-correlated.

Consider a discrete dynamic process

$$\hat{x}_{k+1} = \hat{\Phi}_k \hat{x}_k + \hat{\upsilon}_k, \; k = 0, 1, ..., \tag{9.74}$$

$$\hat{y}_k^i = \hat{H}_k^i \hat{x}_k + \hat{w}_k^i, \; i = 1, ..., \ell, \tag{9.75}$$

where the process noises $\hat{\upsilon}_k$ and $\hat{w}_k$ are both zero mean random variables independent of each other temporally. But $\hat{\upsilon}_k$ and $\hat{w}_k$ are one-step correlated here, that is,

$$\mathbb{E}(\hat{\upsilon}_k) = 0, \quad \mathbf{Cov}(\hat{\upsilon}_k) = \hat{R}\upsilon_k, \tag{9.76}$$

$$\mathbb{E}(\hat{w}_k^i) = 0, \quad \mathbf{Cov}(\hat{\upsilon}_k^i) = \hat{R}_k^i, \tag{9.77}$$

$$\mathbb{E}(\hat{\upsilon}_k \hat{w}_j^{i'}) = \hat{M}_k^i \delta_{k-j+1}, \tag{9.78}$$

where δ_{k-j+1} is the Kronecker delta function. Recall that the stacked measurement equation remains the same as (9.54) and (9.55).

Since, $\hat{\upsilon}_k$ and $\hat{w}_k$ are one-step correlated, we have the following, also with the covariance of $\hat{w}_k$:

$$\mathbb{E}(\hat{\upsilon}_k \hat{w}_j^t) = \hat{M}_k = (\hat{M}_k^1, ..., \hat{M}_k^l)\delta_{k-j+1}, \tag{9.79}$$

$$\mathbf{Cov}\hat{w}_k = \hat{R}_k. \tag{9.80}$$

However, the local measurement noises $\hat{\upsilon}_k$ and $\hat{w}_k$ are not necessarily uncorrelated here.

We are now in a position to prove the following results.

Theorem 9.1. 1. *The FB Kalman filtering of the ith sensor given by*

$$
\begin{aligned}
&\textbf{Forward run}: \text{for } (\mathrm{k}=0; \mathrm{k}<\mathrm{T}; +\mathrm{k}) \\
&\hat{K}_k^i = \hat{P}_{k/k}^i(\hat{H}_k^{i'} + \hat{P}_{k/k-1}^{i^{-1}}\hat{M}_k^i)(\hat{R}_k^{i'} - \hat{M}_k^{i'}\hat{P}_{k/k-1}^{i^{-1}}\hat{M}_k^i)^{-1}, \\
&\hat{x}_{k/k}^i = \hat{x}_{k/k-1}^i + \hat{K}_k^i(y_k^i - H_k^i\hat{x}_{k/k-1}^i), \\
&\hat{P}_{k/k}^i = \hat{P}_{k/k-1}^i - \hat{K}_k^i(\hat{H}_k^{i'} + \hat{P}_{k/k-1}^{i^{-1}}\hat{M}_k^i)\hat{P}_{k/k-1}^i, \\
&\textbf{Backward run}: \text{for } (\mathrm{k}=\mathrm{T}-1; \mathrm{t}\geq 0; -\mathrm{k}) \\
&\hat{\lambda}_{k-1/T}^i = (I_{P+N} - (\hat{H}_k^{i^*} + \hat{P}_{k/k-1}^{i^{-1}}\hat{M}_k^i)\hat{K}_{f,i}^*)\hat{\Phi}_k^i\hat{\lambda}_{k+1/T}^i \\
&\qquad + (\hat{H}_k^{i'} + \hat{P}_{k/k-1}^{i^{-1}}\hat{M}_k^i)(\hat{R}_k^{i'} - \hat{M}_k^{i'}\hat{P}_{k/k-1}^{i^{-1}}\hat{M}_k^i)^{-1}(y_k^i - H_k^i\hat{x}_{k/k-1}^i), \\
&\hat{x}_{k-1/T}^i = \hat{x}_{k-1/k-1}^i + \hat{P}_{k/k-1}^i\hat{\lambda}_{k-1/T}^i, \\
&\hat{P}_{k-1/T}^i = \hat{P}_{k-1/k-1}^i,
\end{aligned}
\tag{9.81}
$$

with covariance of the filtering error for the forward run as

$$
\begin{aligned}
\hat{P}_{k/k}^{i^{-1}} &= \hat{P}_{k/k-1}^{i^{-1}} + (\hat{H}_k^{i'} + \hat{P}_{k/k-1}^{i^{-1}}\hat{M}_k^i) \\
&\quad \times (\hat{R}_k^{i'} - \hat{M}_k^{i'}\hat{P}_{k/k-1}^{i^{-1}}\hat{M}_k^i)^{-1}(\hat{H}_k^{i'} + \hat{P}_{k/k-1}^{i^{-1}}\hat{M}_k^i)^t,
\end{aligned}
\tag{9.82}
$$

where

$$
\begin{aligned}
\hat{x}_{k/k-1}^i &= \hat{\Phi}_{k-1}\hat{x}_{k-1/k-1}^i, \\
\hat{P}_{k/k-1}^i &= E[(\hat{x}_{k/k-1}^i - \hat{x}_k)((\hat{x}_{k/k-1}^i - \hat{x}_k))^t | \hat{y}_0^i, \ldots., \hat{y}_{k-1}^t]
\end{aligned}
$$

is globally optimal in the mean square error sense.

2. *When the row dimensions of all matrices $\hat{H}_k^i + (\hat{P}_{k/k-1}^{i^{-1}}\hat{M}_k^i)^t$ are less than or equal to the dimension of the state and all of them are of full row-rank, the centralized filtering can be explicitly expressed in terms of local filtering.*

Proof. It suffices to show point **2**. The centralized FB Kalman filtering with all sensor data is given by

$$\textbf{Forward run}: \text{for } (\mathrm{k}=0; \mathrm{k}<\mathrm{T}; +\mathrm{k})$$

$$\hat{K}_k = \hat{P}_{k/k}(\hat{H}_k^t + \hat{P}_{k/k-1}^{-1}\hat{M}_k)(\hat{R}_k^t - \hat{M}_k^t\hat{P}_{k/k-1}^{-1}\hat{M}_k)^{-1}, \tag{9.83}$$

$$
\begin{aligned}
\hat{x}_{k/k} &= (I - \hat{K}_k\hat{H}_k)\hat{x}_{k/k-1} + \hat{K}_k\hat{y}_k, \\
&= (I - \hat{K}_k\hat{H}_k)\hat{x}_{k/k-1} + \hat{P}_{k/k}(\hat{H}_k^t + \hat{P}_{k/k-1}^{-1}\hat{M}_k) \\
&\quad \times \sum_{i=1}^{l}(\hat{R}_k - \hat{M}_k^t)^{-1}(\times i)\hat{y}_k,
\end{aligned}
\tag{9.84}
$$

$$\hat{P}_{k/k} = \hat{P}_{k/k-1} - \hat{K}_k(\hat{H}_k^t + \hat{P}_{k/k-1}^{-1}\hat{M}_k^i)\hat{P}_{k/k-1}, \tag{9.85}$$

$$
\begin{aligned}
&\textbf{Backward run}: \text{for } (\mathrm{k} = \mathrm{T}-1; \mathrm{t} \geq 0; -\mathrm{k}) \\
\hat{\lambda}_{k-1/T} &= (I_{P+N} - (\hat{H}_k^* + \hat{P}_{k/k-1}^{-1}\hat{M}_k)\hat{K}_{f,i}^*) \\
&\quad \times \hat{\Phi}_k\hat{\lambda}_{k+1/T} + (\hat{H}_k^t + \hat{P}_{k/k-1}^{-1}\hat{M}_k) \\
&\quad \times (\hat{R}_k^t - \hat{M}_k^t\hat{P}_{k/k-1}^{-1}\hat{M}_k)^{-1}(y_k - H_k\hat{x}_{k/k-1}), && (9.86) \\
\hat{x}_{k-1/T} &= \hat{x}_{k-1/k-1} + \hat{P}_{k/k-1}\hat{\lambda}_{k-1/T}, && (9.87) \\
\hat{P}_{k-1/T} &= \hat{P}_{k-1/k-1}, && (9.88)
\end{aligned}
$$

where the covariance of the filtering error for the forward run can also be given by

$$
\begin{aligned}
\hat{P}_{k/k}^{i^{-1}} &= \hat{P}_{k/k-1}^{i^{-1}} + (\hat{H}_k^{i'} + \hat{P}_{k/k-1}^{i^{-1}}\hat{M}_k^i) \\
&\quad \times (\hat{R}_k^{i'} - \hat{M}_k^{i'}\hat{P}_{k/k-1}^{i^{-1}}\hat{M}_k^i)^{-1}(\hat{H}_k^{i'} + \hat{P}_{k/k-1}^{i^{-1}}\hat{M}_k^i)^t. \qquad (9.89)
\end{aligned}
$$

In the sequel, we let $(\hat{R}_k^i - \hat{M}_k^{t^i}\hat{P}_{k/k-1}^{i^{-1}}\hat{M}_k^i)^{-1}$ be the ith submatrix column of $(\hat{R}_k - \hat{M}_k^{\prime}\hat{P}_{k/k-1}^{-1}\hat{M}_k)^{-1}$.

We assume the row dimensions of all matrices $\hat{H}_k^i + (\hat{P}_{k/k-1}^{i^{-1}}\hat{M}_k^i)^t$ to be less than or equal to the dimension of the state, and we assume all of them to be of full row-rank. In addition, we know $(\hat{H}_k^i + (\hat{P}_{k/k-1}^{i^{-1}}\hat{M}_k^i)^t)(\hat{H}_k^i + (\hat{P}_{k/k-1}^{i^{-1}}\hat{M}_k^i)')^+ = I$, where the superscript † stands for pseudoinverse. Using (9.83), we get

$$
\begin{aligned}
\hat{K}_k\hat{y}_k &= \hat{P}_{k/k}(\hat{H}_k^t + \hat{P}_{k/k-1}^{-1}\hat{M}_k) \\
&\quad \times \sum_{i=1}^{l}(\hat{R}_k - \hat{M}_k^t\hat{P}_{k/k-1}^{-1}\hat{M}_k)^{-1}(\times i)\hat{y}_k^i \\
&= \hat{P}_{k/k}(\hat{H}_k^t + \hat{P}_{k/k-1}^{-1}\hat{M}_k)\sum_{i=1}^{l}(\hat{R}_k - \hat{M}_k^t\hat{P}_{k/k-1}^{-1}\hat{M}_k)^{-1}(\times i) \\
&\quad \times (\hat{R}_k^i - \hat{M}_k^{i^t}\hat{P}_{k/k-1}^{-1}\hat{M}_k)(\hat{H}_k^{i^t} + \hat{P}_{k/k-1}^{i^{-1}}\hat{M}_k^i)^+ \\
&\quad \times (\hat{H}_k^{i^t} + \hat{P}_{k/k-1}^{i^{-1}})(\hat{R}_k^i - \hat{M}_k^{i^t}\hat{P}_{k/k-1}^{i^{-1}}\hat{M}_k^i)^{-1}\hat{y}_k^i. \qquad (9.90)
\end{aligned}
$$

Now, we proceed to express the FB centralized filtering $\hat{x}_{k/k}$ in terms of the local filtering, and using (9.81), we get

$$
\begin{aligned}
&(\hat{H}_k^{i'} + \hat{P}_{k/k-1}^{i^{-1}})(\hat{R}_k^i - \hat{M}_k^{i'}\hat{P}_{k/k-1}^{i^{-1}}\hat{M}_k^i)^{-1}\hat{y}_k^i \\
&= \hat{P}_{k/k}^{i^{-1}}\hat{K}_k^i\hat{y}_k^i \\
&= \hat{P}_{k/k}^{i^{-1}}[\hat{x}_{k/k}^i - (I - \hat{K}_k^i\hat{H}_k^i)\hat{x}_{k/k-1}^i]. \qquad (9.91)
\end{aligned}
$$

Using (9.90) and (9.91), we obtain

$$\begin{aligned}\hat{x}_{k/k} &= (I - \hat{K}_k \hat{H}_k)\hat{x}_{k/k-1} + \hat{P}_{k/k}(\hat{H}_k^t + \hat{P}_{k/k-1}^{-1}\hat{M}_k)\\ &\times \sum_{i=1}^{l}(\hat{R}_k - \hat{M}_k^t \hat{P}_{k/k-1}^{-1}\hat{M}_k)^{-1}(\times i)\\ &\times (\hat{R}_k^i - \hat{M}_k^{i'} \hat{P}_{k/k-1}^{i^{-1}}\hat{M}_k^i)(\hat{H}_k^{i'} + \hat{P}_{k/k-1}^{i^{-1}}\hat{M}_k^i)^+\\ &\times \hat{P}_{k/k}^{i^{-1}}[\hat{x}_{k/k}^i - (I - \hat{K}_k^i)\hat{x}_{k/k-1}^i]\end{aligned} \tag{9.92}$$

and

$$\begin{aligned}\hat{\lambda}_{k-1/T} &= I_{P+N} - (\hat{H}_k^t[\hat{P}_{k/k}(\hat{H}_k^t + \hat{P}_{k/k-1}^{-1}\hat{M}_k)\\ &\times \sum_{i=1}^{l}(\hat{R}_k^i - \hat{M}_k^{t^i} \hat{P}_{k/k-1}^{i^{-1}}\hat{M}_k^i)^{-1}\\ &\times (\hat{R}_k^i - \hat{M}_k^{i^t} \hat{P}_{k/k-1}^{-1}\hat{M}_k)(\hat{H}_k^{i^t} + \hat{P}_{k/k-1}^{i^{-1}}\hat{M}_k^i)^+\\ &\times (\hat{H}_k^{i^t} + \hat{P}_{k/k-1}^{i^{-1}})(\hat{R}_k^i - \hat{M}_k^{i'} \hat{P}_{k/k-1}^{i^{-1}}\hat{M}_k^i)^{-1}\\ &- H_k^i R_{e_i}^{-1} H_k^i \hat{x}_{k/k-1}^i])\hat{\Phi}_k \hat{\lambda}_{k+1/T} + \hat{P}_{k/k}(\hat{H}_k^t + \hat{P}_{k/k-1}^{-1}\hat{M}_k)\\ &\times \sum_{i=1}^{l}(\hat{R}_k - \hat{M}_k^t \hat{P}_{k/k-1}^{-1}\hat{M}_k)^{-1}(\times i(\hat{R}_k^i - \hat{M}_k^{t^i} \hat{P}_{k/k-1}^{i^{-1}}\hat{M}_k^i)^{-1})\\ &\times (\hat{R}_k^i - \hat{M}_k^{i'} \hat{P}_{k/k-1}^{-1}\hat{M}_k)(\hat{H}_k^{i'} + \hat{P}_{k/k-1}^{i^{-1}}\hat{M}_k^i)^+(\hat{H}_k^{i'} + \hat{P}_{k/k-1}^{i^{-1}})\\ &\times (\hat{R}_k^i - \hat{M}_k^{i'} \hat{P}_{k/k-1}^{i^{-1}}\hat{M}_k^i)^{-1}\hat{y}_k^i - H_k^i R_{e_i}^{-1} H_k^i \hat{x}_{k/k-1}^i,\\ \hat{x}_{k-1/T} &= \hat{x}_{k/k-1} + P_{k/k-1}\lambda_{k/T}.\end{aligned} \tag{9.93}$$

Simplifying, it is easy to show that the FB centralized filtering can be explicitly expressed in terms of the local filtering in the following form:

$$\begin{aligned}&\textbf{Forward run}:\\ \hat{x}_{k/k} &= (I - \hat{K}_k \hat{H}_k)\hat{x}_{k/k-1} + \hat{P}_{k/k}(\hat{H}_k^t + \hat{P}_{k/k-1}^{-1}\hat{M}_k)\\ &\times \sum_{i=1}^{\ell}(\hat{R}_k^i - \hat{M}_k^{t^i} \hat{P}_{k/k-1}^{i^{-1}}\hat{M}_k^i)^{-1}(\hat{R}_k^i - \hat{M}_k^{i^t} \hat{P}_{k/k-1}^{i^{-1}}\hat{M}_k^i)\\ &\times (\hat{H}_k^{i^t} + \hat{P}_{k/k-1}^{i^{-1}}\hat{M}_k^i)^+ \hat{P}_{k/k}^{i^{-1}}[\hat{x}_{k/k}^i - (I - \hat{K}_k^i)\hat{x}_{k/k-1}^i],\\ &\textbf{Backward run}:\\ \hat{\lambda}_{k-1/T} &= \hat{\Phi}_k \hat{\lambda}_{k+1/T} - \hat{P}_{k/k}^{i^{-1}}\hat{K}_k^i(\hat{H}_k^t \hat{P}_{k/k}\hat{\Phi}_k \hat{\lambda}_{k+1/T} - \hat{y}_k^i) - \hat{H}_k^t \hat{x}_{k/k-1}^i,\\ \hat{x}_{k-1/T} &= \hat{x}_{k/k-1} + P_{k/k-1}\lambda_{k/T}.\end{aligned}$$

We conclude that the performance of the distributed Kalman filtering fusion with process noise one-step correlated to measurement noise and sensor noises cross-correlated is better than the centralized fusion because of the smoothing effect. □

9.2.2 Impact of packet loss/intermittent communications

In the sequel, we consider the dynamic system (9.52)–(9.53) with synchronous intermittent communications from sensors to the fusion center. The packet loss from sensors to the fusion center is actually passive asynchronous intermittent communications. Then, we can easily extend the result in this section to the asynchronous intermittent communications/packet loss case.

We now assume the intermittent communication is periodical and the periods of all the local sensors are the same n in this section, i.e., the fusion center only receives $x^i_{N/N}$ $(N = kn, k = 1, 2, ...)$ from the local sensor.

9.2.3 Modified centralized multisensor system

Firstly, by stacking n-step sensor observations and some simple algebraic operations, we will convert the n-step dynamic system to the compact form with one-step correlated process and measurement noises in the following proposition.

Proposition 9.1. *Assume Φ_k to be invertible. The n-step original local dynamic system*

$$
\begin{aligned}
x_{kn+1} &= \Phi_{kn} x_{kn} + \upsilon_{kn}, \\
y^i_{kn+1} &= H^i_{kn+1} x_{kn+1} + w^i_{kn+1}, \\
&\vdots \quad \vdots \quad \vdots \\
x_{(k+1)n} &= \Phi_{(k+1)n} - 1 x_{(k+1)(n-1)} + \upsilon_{(k+1)(n-1)}, \\
y^i_{(kn+1)n} &= H^i_{(k+1)n} x_{(k+1)n} + w^i_{(k+1)n}
\end{aligned}
\tag{9.94}
$$

can be rewritten in a compact form

$$
\begin{aligned}
\tilde{x}_{k+1} &= \tilde{\Phi}_k \tilde{x}_k + \tilde{\upsilon}_k, \ k = 0, 1, ..., \\
\tilde{y}^i_{(k+1)} &= \tilde{H}^i_{(k+1)} \tilde{x}_{(k+1)} + w^i_{(k+1)}, \ i = 1, ..., l,
\end{aligned}
\tag{9.95}
$$

where

$$
\begin{aligned}
\tilde{x}_k &= \tilde{x}_{kn}, \\
\tilde{y}^i_k &= (y_{(k-1)n+1}, y^t_{(k-1)n+2}, \cdots y^t_{kn})^t, \\
\tilde{\Phi}_k &= \Pi^{n-1}_{j=0} \Phi_{kn+j},
\end{aligned}
\tag{9.96}
$$

$$\tilde{\upsilon}_k = \sum_{l=1}^{n-1} (\Pi_{j=l}^{n-1}) \upsilon_{kn+l-1} + \upsilon_{(k+l)n-1}, \tag{9.97}$$

$$\tilde{H}_{k+1}^i = \begin{bmatrix} H_{kn+1}^i (\Pi_{j=1}^{n-1} \Phi_{kn+j})^{-1} \\ \cdot \\ \cdot \\ \cdot \\ H_{(kn+1)n-1} \Phi_{(k+1)n-1}^{-1} \\ H_{(k+1)n}^i \end{bmatrix}, \tag{9.98}$$

$$\tilde{w}_{k+1}^i = \begin{bmatrix} w_{kn+1}^i \\ -H_{kn+1}^i (\Pi_{j=1}^{n-1} \Phi_{kn+j})^{-1} \sum_{l=2}^{n-1} (\Pi_{j=l}^{n-1} \Phi_{kn+j}) \upsilon_{kn+l-1} \\ -H_{kn+1}^i (\Pi_{j=1}^{n-1} \Phi_{kn+j})^{-1} \upsilon_{(k+l)n-1} \\ \cdot \\ \cdot \\ \cdot \\ w_{(k+1)n-1}^i - H_{(kn+1)n-1} \Phi_{(k+1)n-1}^{-1} \upsilon_{k+1} n - 1 \\ w_{(k+1)n}^i \end{bmatrix}. \tag{9.99}$$

Similar to (9.94) and (9.95), the n-step stacked measurement equations

$$\begin{aligned} y_{kn+1} &= H_{kn+1} x_{kn+!} + w_{(kn+1)}, \\ &\cdots \\ y_{k+1} n &= H_{(k+1)n} x_{(k+1)n} + w_{(k+1)n}, \end{aligned} \tag{9.100}$$

where $m = 1, ...n$,

$$\begin{aligned} y_{kn+m} &= (y_{kn+m}^{1^t}, ..., y_{kn+m}^{\ell^t})^t, \\ H_{kn+m} &= (H_{kn+m}^{l^t}, ..., H_{kn+m}^{\ell^t})^t, \\ w_{kn+m} &= (w_{kn+m}^{l^t}, ..., w_{kn+m}^{\ell^t})^t, \end{aligned}$$

can then be rewritten as

$$\tilde{y}_{k+1} = \tilde{H}_{k+1} \tilde{x}_{k+1} + \tilde{w}_{k+1}, \tag{9.101}$$

where

$$\begin{aligned} y_{k+1} &= (y_{k+1}^{1^t}, ..., y_{k+1}^{\ell^t})^t, \\ H_{k+1} &= (H_{k+1}^{1^t}, ..., H_{k+1}^{\ell^t})^t, \\ w_{k+1} &= (w_{k+1}^{1^t}, ..., w_{k+1}^{\ell^t})^t. \end{aligned} \tag{9.102}$$

In addition to the correlation of sensor noises as mentioned above, the process and measurement noises of the modified centralized system are also one-step correlated, that is,

$$\mathbf{Cov}(\tilde{w}_k) = \tilde{R}_k, \quad \mathbb{E}(\tilde{\upsilon}_k \tilde{w}_j^t) = \tilde{M}_k \delta_{k-j+1}. \tag{9.103}$$

In Section 9.3, we assumed that *the row dimensions of all matrices* $\hat{H}_k^i + (\hat{P}_{k/k-1}^{i^{-1}} \hat{M}_k^i)^t$ *are less than or equal to the dimension of the state, and all of them are of full row-rank.* According to whether or not this assumption is satisfied, the Kalman filtering with packet loss/intermittent communications problem can be divided into two cases to discuss.

Remark 9.3. Due to (9.97) and (9.99), since the original process noise υ_k is now involved in the new process and measurement noises, υ_k and w_k^i are one-step correlated and sensor noises w_k^i and w_k^j are cross-correlated here, i.e.,

$$\begin{aligned}
\mathbb{E}(\tilde{\upsilon}_k) = 0, \qquad & \mathbf{Cov}(\tilde{\upsilon}_k) = \tilde{R}\upsilon_k, \\
\mathbb{E}(\tilde{w}_k^i) = 0, \qquad & \mathbf{Cov}(\tilde{w}_k^i) = \tilde{R}_k^i, \\
\mathbb{E}(\tilde{w}_k^i \tilde{w}_k^{j'}) \neq 0, \; i \neq j, \; & \mathbb{E}(\tilde{\upsilon}_k \tilde{w}_j^{i'}) = \tilde{M}_k^i \delta_{k-j+1}, \; \forall k, i, j.
\end{aligned}$$

Obviously, the larger the variance of υ_k is, the larger the correlation of sensor noises $\tilde{w}_k^i$ and $\tilde{w}_k^{j'}$ is.

Two important cases are considered:

Case 1 implies that the row dimensions of all matrices $\hat{H}_k^i + (\hat{P}_{k/k-1}^{i^{-1}} \hat{M}_k^i)^t$ are less than or equal to the dimension of the state, and all of them are of full row-rank. **Case 2** treats the situation in which such an assumption does not hold.
Case 2: The assumption is satisfied. Recall we have already converted the local sensory n-step dynamic equations (9.94) to the compact form with one-step correlated process and measurement noises (9.95). According to Theorem 9.1, the local filtering of (9.95) at the ith sensor is

Forward run:

$$\begin{aligned}
\tilde{x}_{k/k-1}^i &= \tilde{\Phi}_k \tilde{x}_{k-1/k-1}^i, \\
\tilde{P}_{k/k-1}^i &= \tilde{\Phi}_k \tilde{P}_{k-1/k-1}^i \Phi_k^t + \tilde{R}\upsilon_k, \\
\tilde{P}_{k/k}^{-1} &= \tilde{P}_{k/k-1}^{i^{-1}} + (\tilde{H}_k^{i^t} + \tilde{P}_{k/k-1}^{i^{-1}} \tilde{M}_k^i) \\
&\quad \times (\tilde{R}_k^i - \tilde{M}_k^{i^t} \tilde{P}_{k/k-1}^{i^{-1}} \tilde{M}_k^i)^{-1} (\tilde{H}_k^{i'} + \tilde{P}_{k/k-1}^{i^{-1}} \tilde{M}_k^i)^t, \\
\tilde{K}_k^i &= \tilde{P}_{k/k}^i (\tilde{H}_k^{i'} + \tilde{P}_{k/k-1}^{i^{-1}} \tilde{M}_k^i) \\
&\quad \times (\tilde{R}_k^i - \tilde{M}_k^{i'} \tilde{P}_{k/k-1}^{i^{-1}} \tilde{M}_k^i)^{-1},
\end{aligned}$$

$$
\begin{aligned}
\tilde{x}^i_{k/k} &= \tilde{x}^i_{k/k-1} + \tilde{K}^i_k (y^i_k - \tilde{H}^i_k \tilde{x}^i_{k/k-1}),\\
&\textbf{Backward run}:\\
\tilde{\lambda}^i_{k-1/T} &= (I_{P+N} - (\tilde{H}^{i'}_k + \tilde{P}^{i^{-1}}_{k/k-1}\tilde{M}^i_k)K^*_{f,i})\\
&\quad \times \Phi_k \lambda_{k+1/T} + (\tilde{H}^{i'}_k + \tilde{P}^{i^{-1}}_{k/k-1}\tilde{M}^i_k)(\tilde{R}^i_k - \tilde{M}^{i'}_k \tilde{P}^{i^{-1}}_{k/k-1}\tilde{M}^i_k)^{-1}\\
&\quad \times (y_k - (\tilde{H}^{i'}_k + \tilde{P}^{i^{-1}}_{k/k-1}\tilde{M}^i_k)\tilde{x}_{k/k-1}),\\
\tilde{x}^i_{k-1/T} &= \tilde{x}^i_{k-1/k-1} + \tilde{P}^i_{k/k-1}\tilde{\lambda}^i_{k/T},\\
\tilde{P}^i_{k-1/T} &= \tilde{P}^i_{k-1/k-1}.
\end{aligned}
\tag{9.104}
$$

By Theorem 9.1, the optimality of (9.104) is the same as the standard Kalman filtering (that is, globally optimal in the mean square error sense). Combining this with the equivalence of dynamic systems (9.94) and (9.95), the state estimate $\tilde{x}^i_{k/k}$ of n-step observation Kalman filtering with one-step correlated process and measurement noises is equal to the standard Kalman filtering state estimate $x^i_{kn/kn}$, that is, $\tilde{x}^i_{k/k} = x^i_{kn/kn}$.

Since the primitive local measurement matrix $H^i_k \in R^{N_i \times r}$, $x_k \in R^r$, and the communication period of all the local sensors is the same n. If $N_i n \leq r$, the row dimension of all matrices $\tilde{H}^i_k + (\tilde{P}^{i^{-1}}_{k/k-1}\tilde{M}^i_k)^t$ equals $N_i n$ and is less than or equal to the dimension of the state. Frequently, $\tilde{H}^i_k + (\tilde{P}^{i^{-1}}_{k/k-1}\tilde{M}^i_k)^t$ is of full row-rank under the condition $N_i n \leq r$. Then the assumption in Theorem 9.1 is satisfied. Thus, we give the filtering at the fusion center as follows:

$$
\begin{aligned}
&\textbf{Forward run}:\\
\tilde{x}_{k/k-1} &= \tilde{\Phi}_k \tilde{x}_{k-1/k-1},\\
\tilde{P}_{k/k-1} &= \tilde{\Phi}_k \tilde{P}_{k-1/k-1}\Phi^t_k + \tilde{R}v_k,\\
\tilde{P}^{-1}_{k/k} &= \tilde{P}^{-1}_{k/k-1} + (\tilde{H}^t_k + \tilde{P}^{-1}_{k/k-1}\tilde{M}_k)\\
&\quad \times (\tilde{R}_k - \tilde{M}^t_k \tilde{P}^{-1}_{k/k-1}\tilde{M}_k)^{-1}(\tilde{H}^t_k + \tilde{P}^{-1}_{k/k-1}\tilde{M}_k)^t,\\
\tilde{K}_k &= \tilde{P}_{k/k}(\tilde{H}^t_k + \tilde{P}^{-1}_{k/k-1}\tilde{M}_k)(\tilde{R}_k - \tilde{M}^t_k \tilde{P}^{-1}_{k/k-1}\tilde{M}_k)^{-1},\\
\tilde{x}_{k/k} &= (I - \tilde{K}_k \tilde{H}_k)\tilde{x}_{k/k-1}\\
&\quad + \tilde{P}_{k/k}(\tilde{H}^t_k + \tilde{P}^{-1}_{k/k-1}\tilde{M}_k)\\
&\quad \times \sum_{i=1}^{l}(\tilde{R}_k - \tilde{M}^t_k \tilde{P}^{-1}_{k/k-1}\tilde{M}_k)^{-1}(\times i)\\
&\quad \times (R^i_k - \tilde{M}^{i'}_k \tilde{P}^{i^{-1}}_{k/k-1}\tilde{M}^i_k)(\tilde{H}^{it}_k + \tilde{P}^{i^{-1}}_{k/k-1}\tilde{M}^i_k)^+\\
&\quad \times \tilde{P}^{i^{-1}}_{k/k}[\tilde{x}^i_{k/k} - (I - \tilde{K}^i_k)\tilde{x}^i_{k/k-1}],
\end{aligned}
$$

Backward run :

$$\begin{aligned}
\tilde{\lambda}_{k-1/T} &= \tilde{\Phi}_k \tilde{\lambda}_{k+1/T} - \tilde{P}^{i-1}_{k/k} \tilde{K}^i_k (\tilde{H}^t_k \tilde{P}_{k/k} \tilde{\Phi}_k \tilde{\lambda}_{k+1/T} - \tilde{y}^i_k) - \tilde{H}^t_k \tilde{x}^i_{k/k-1}, \\
\tilde{x}_{k-1/T} &= \tilde{x}_{k/k-1} + P_{k/k-1} \tilde{\lambda}_{k/T}, \\
\tilde{P}_{k-1/T} &= \tilde{P}_{k-1/k-1}.
\end{aligned} \tag{9.105}$$

Remark 9.4. It has already been proved in Theorem 9.1 that under certain assumptions (roughly speaking, flat measurement matrix in local sensors and short packet loss/intermittent communications period), the centralized filtering $\tilde{x}_{k/k}$ in (9.105) can be explicitly expressed in terms of the local filtering $\tilde{x}^i_{k/k}$. Since $\tilde{x}^i_{k/k} = x^i_{kn/kn}$, the centralized filtering $\tilde{x}^i_{k/k}$ in (9.105) can be explicitly expressed in terms of $x^i_{kn/kn}$. Alternatively, the performance of the estimation at the fusion center does not change with packet loss/intermittent communications from local sensors to the fusion center even if the process noise does exist. Since there is no process noise, it is a sufficient but not a necessary condition for reconstructing global optimality of the algorithm with packet loss/intermittent communications.

Case 2: The assumption is not satisfied. It frequently occurs when the measurement matrix is not flat or the packet loss/intermittent communications period is long that the row dimension of $\tilde{H}^i_k + (\tilde{P}^{i-1}_{k/k-1} \tilde{M}^i_k)'$ is greater than the dimension of the state. In this case, our assumption is not satisfied and one has to recourse to a suboptimal distributed fusion algorithm.

The following lemma proposes a way to reduce the row dimension of $\tilde{H}^i_k + (\tilde{P}^{i-1}_{k/k-1} \tilde{M}^i_k)^t$ to the rank of $\tilde{H}^i_k + (\tilde{P}^{i-1}_{k/k-1} \tilde{M}^i_k)^t$, which is less than or equal to the dimension of the state. This dimensionality reduction does not change the estimation performance at each local sensor.

Lemma 9.1. *Suppose that a dynamic system is given by*

$$x_{k+1} = \Phi_k x_k + \upsilon_k, \; k = 0, 1, ..., \tag{9.106}$$

$$y_k = H_k x_k + w_k, \tag{9.107}$$

where $\Phi_k \in R^{r \times r}$, $x_k, \upsilon_k \in R^r$, $H_k \in R^{N \times r}$, $y_k, w_k \in R^N$, $Cov(w_k) = R_k$, $E(\upsilon_k w_j) = M_k \delta_{k-j+1}$, and $N > r$. Then one finds a dimensionality reduction matrix S_k (for example, see (9.109)), such that the dimension-reduced measurement equation

$$z_k = S_k y_k = S_k H_k x_k + S_k w_k \tag{9.108}$$

replacing (9.107) does not change the performance of the Kalman filtering.

Proof. Let $H_k + (P^{-1}_{k/k-1} M_k)^t = T_k G_k$ be a full-rank decomposition of matrix $H_k + (P^{-1}_{k/k-1} M_k)^t$, where T_k and G_k are full column-rank and full row-rank matrices, respectively, and their ranks are equal to the rank of $H_k + (P^{-1}_{k/k-1} M_k)^t$.

We will show that

$$S_k = T_k^t (R_k - M_k^t P_{k/k-1}^{-1} M_k)^{-1} \tag{9.109}$$

is a lossless dimensionality reduction matrix.

Denote the estimate error covariances using y_k and z_k by $P_{k/k}^{-1}(y_k)$ and $P_{k/k}^{-1}(z_k)$, respectively. Recalling (9.89), we have

$$\begin{aligned} P_{k/k}^{-1}(y_k) &= P_{k/k-1}^{-1} + (H_k^t + P_{k/k-1}^{-1} M_k) \\ &\quad \times (R_k - M_k^t P_{k/k-1}^{-1} M_k)^{-1} (H_k^t + P_{k/k-1}^{-1} M_k)^t. \end{aligned} \tag{9.110}$$

We denote $H_k = S_k H_k$, $\hat{w}_k = S_k w_k$. Then

$$\mathbb{E}(\upsilon_k \hat{w}_j^t) = \hat{M}_k \delta_{k-j+1}, \ \mathbf{Cov}(\hat{w}_k) = \hat{R}_k = S_k R_k S_k^t.$$

By assumption,

$$\begin{aligned} \hat{H}_k + (P_{k/k-1}^{-1} \hat{M}_k)^t &= S_k H_k + (P_{k/k-1}^{-1} M_k S_k^t)^t \\ &= S_k (H_k + (P_{k/k-1}^{-1} M_k)^t) \\ &= T_k^t (R_k - M_k^t P_{k/k-1}^{-1} M_k)^{-1} T_k G_k. \end{aligned}$$

Obviously, $T_k^t (R_k - M_k^t P_{k/k-1}^{-1} M_k)^{-1} T_k$ is an invertible matrix and then $\hat{H}_k + (P_{k/k-1}^{-1} \hat{M}_k)^t = T_k^t (R_k - M_k^t P_{k/k-1}^{-1} M_k)^{-1} T_k G_k$ is a full row-rank matrix. This leads to

$$\begin{aligned} P_{k/k}^{-1}(z_k) &= P_{k/k-1}^{-1} + (\hat{H}_k^t + P_{k/k-1}^{-1} \hat{M}_k) \\ &\quad \times (\hat{R}_k - \hat{M}_k^t P_{k/k-1}^{-1} \hat{M}_k)^{-1} (\hat{H}_k^t + P_{k/k-1}^{-1} \hat{M}_k)^t \\ &= P_{k/k-1}^{-1} + (T_k^t (R_k - M_k^t P_{k/k-1}^{-1} M_k))^{-1} \\ &\quad \times (\hat{H}_k + (P_{k/k-1}^{-1} \hat{M}_k)^t)^t (T_k^t (R_k - M_k^t P_{k/k-1}^{-1} M_k))^{-1} \\ &\quad \times (R_k - M_k^t P_{k/k-1}^{-1} M_k)(R_k - M_k^t P_{k/k-1}^{-1} M_k)^{-1} T_k^{-1} \\ &\quad \times (T_k^t (R_k - M_k^t P_{k/k-1}^{-1} M_k))^{-1} (H_k + (P_{k/k-1}^{-1} M_k)^t). \end{aligned}$$

Further algebraic manipulation yields

$$\begin{aligned} P_{k/k}^{-1}(z_k) &= P_{k/k-1}^{-1} + (T_k^t (R_k - M_k^t P_{k/k-1}^{-1} M_k))^{-1} \\ &\quad \times (\hat{H}_k + (P_{k/k-1}^{-1} \hat{M}_k)^t)^t \\ &\quad \times (T_k^t ((R_k - M_k^t P_{k/k-1}^{-1} M_k))^{-1} T_k)^{-1} \end{aligned}$$

$$
\begin{aligned}
&\times (T_k^t(R_k - M_k^t P_{k/k-1}^{-1} M_k))^{-1}(H_k + (P_{k/k-1}^{-1} M_k)^t)\\
&= P_{k/k-1}^{-1} + (H_k + (P_{k/k-1}^{-1})^t)^t\\
&\times (R_k - M_k^t P_{k/k-1}^{-1} M_k)^{-1}\\
&\times T_k(T_k^t(R_k - M_k^t P_{k/k-1}^{-1} M_k)^{-1} T_k)^{-1}\\
&\times T_k^t(R_k - M_k^t P_{k/k-1}^{-1} M_k)^{-1} T_k G_k\\
&= P_{k/k-1}^{-1} + (H_k + (P_{k/k-1}^{-1} M_k)^t)^t\\
&\times (R_k - M_k^t P_{k/k-1}^{-1} M_k)^{-1} T_k G_k\\
&= P_{k/k-1}^{-1} + (H_k + (P_{k/k-1}^{-1} M_k)^t)^t\\
&\times (R_k - M_k^t P_{k/k-1}^{-1} M_k)^{-1}(H_k + (P_{k/k-1}^{-1} M_k)^t)\\
&= P_{k/k}^{-1}(y_k),
\end{aligned}
$$

which completes the proof. □

By compressing the sensor observation, the local filtering at the ith sensor could be rewritten as

$$
\begin{aligned}
&\textbf{Forward run}:\\
\bar{x}_{k/k-1}^i &= \tilde{\Phi}_k \bar{x}_{k-1/k-1}^i,\\
\bar{P}_{k/k-1}^i &= \tilde{\Phi}_k \bar{P}_{k-1/k-1}^i \tilde{\Phi}_k^t + \tilde{R}\upsilon_k,\\
\bar{H}_k^i &= S_k^i \tilde{H}_k^i,\ \bar{M}_k^{i^t} = \tilde{M}_k^{i^t} S_k^{i^t},\ \bar{R}_k^i = S_k^i \tilde{R}_k^i S_k^{i^t},\\
\bar{P}_{k/k}^{i^{-1}} &= \bar{P}_{k/k-1}^{i^{-1}} + (\bar{H}_k^{i^t} + \bar{P}_{k/k-1}^{i^{-1}} \bar{M}_k^i)\\
&\times (\bar{R}_k^i - \bar{M}_k^{i^t} \bar{P}_{k/k-1})^{-1}(\bar{H}_k^{i^t} + \bar{P}_{k/k-1}^{i^{-1}} \bar{M}_k^i)^t,\\
\bar{K}_k^i &= \bar{P}_{k/k}^i(\bar{H}_k^{i'} + \bar{P}_{k/k-1}^{i^{-1}} \bar{M}_k^i)\\
&\times (\bar{R}_k^i - \bar{M}_k^{i^t} \bar{P}_{k/k-1})^{-1}(\bar{H}_k^{i^t} + \bar{P}_{k/k-1}^{i^{-1}}),\\
\bar{x}_{k/k}^i &= \bar{x}_{k/k-1}^i + \bar{K}_k^i(S_k^i \bar{y}_k^i - \bar{H}_k^i \bar{x}_{k/k-1}^i),\\
&\textbf{Backward run}:\\
\bar{\lambda}_{k-1/T}^i &= I_{P+N} - (\bar{H}_k^{i^*} + \bar{P}_{k/k-1}^{i^{-1}} \bar{M}_k^i \bar{K}_{f,i}^*)\bar{\Phi}_k^i \bar{\lambda}_{k+1/T}^i\\
&+ (\bar{H}_k^{i'} + \bar{P}_{k/k-1}^{i^{-1}} \bar{M}_k^i)\\
&\times (\bar{R}_k^{i'} - \bar{M}_k^{i'} \bar{P}_{k/k-1}^{i^{-1}} \bar{M}_k^i)^{-1}(S_k^i y_k^i - H_k^i \bar{x}_{k/k-1}^i),\\
\bar{x}_{k-1/T}^i &= \bar{x}_{k-1/k-1}^i + \bar{P}_{k/k-1}^i \bar{\lambda}_{k-1/T}^i,\\
\bar{P}_{k-1/T}^i &= \bar{P}_{k-1/k-1}^i,
\end{aligned}
\tag{9.111}
$$

where S_k^i is the dimensionality reduction matrix given in Lemma 9.1, that is,

$$\bar{x}_{k/k}^i = \tilde{x}_{k/k}^i = x_{kn/kn}^i.$$

Now we assume the row dimensions of all new matrices $\bar{H}_k^i + (\bar{P}_{k/k-1}^{i^{-1}} \bar{M}_k^i)$ to be less than or equal to the dimension of the state. Our assumption in Theorem 9.1 is thus satisfied. Then the proposed optimal distributed fusion algorithm above can be used for the new compressed system. Let $S = \mathbf{diag}(S_k^1, ..., S_k^l)$. The filtering at the fusion center is

$$
\begin{aligned}
&\textbf{Forward run}:\\
\bar{x}_{k/k-1} &= \tilde{\Phi}_k \bar{x}_{k-1/k-1},\\
\bar{P}_{k/k-1} &= \tilde{\Phi}_k \bar{P}_{k-1/k-1} \tilde{\Phi}_k^t + \tilde{R} \upsilon_k,\\
\bar{H}_k^i &= S_k \tilde{H}_k,\ \bar{M}_k = \tilde{M}_k S_k^t,\ \bar{R}_k = S \tilde{R}_k S^t,\\
\bar{P}_{k/k}^{-1} &= \bar{P}_{k/k-1}^{-1} + (\bar{H}_k^t + \bar{P}_{k/k-1}^{-1} \bar{M}_k)\\
&\quad \times (\bar{R}_k - \bar{M}_k^t \bar{P}_{k/k-1})^{-1} (\bar{H}_k^t + \bar{P}_{k/k-1}^{-1} \bar{M}_k)^t,\\
\bar{K}_k &= \bar{P}_{k/k} (\bar{H}_k^t + \bar{P}_{k/k-1}^{-1} \bar{M}_k)(\bar{R}_k - \bar{M}_k^t \bar{P}_{k/k-1}^{-1} \bar{M}_k)^{-1},\\
\bar{x}_{k/k}^i &= (I - \bar{K}_k \bar{H}_k) \bar{x}_{k/k-1}\\
&\quad + \bar{P}_{k/k} (\bar{H}_k^t + \bar{P}_{k/k-1}^{-1} \bar{M}_k)\\
&\quad \times \sum_{i=1}^{l} (\bar{R}_k - \bar{M}_k^t \bar{P}_{k/k-1}^{-1} \bar{M}_k)^{-1} (\times i)\\
&\quad \times (\bar{R}_k^i - \bar{M}_k^{i'} \bar{P}_{k/k-1}^{i^{-1}} \bar{M}_k^i)(\bar{H}_k^{i'} + \tilde{P}_{k/k-1}^{i^{-1}} \bar{M}_k^i)^+\\
&\quad \times \bar{P}_{k/k}^{i^{-1}} [\bar{x}_{k/k}^i - (I - \bar{K}_k^i \bar{H}_k^i) \bar{x}_{k/k-1}^i],\\
&\textbf{Backward run}:\\
\hat{\lambda}_{k-1/T} &= \hat{\Phi}_k \hat{\lambda}_{k+1/T}\\
&\quad - \hat{P}_{k/k}^{i^{-1}} \hat{K}_k^i (\hat{H}_k^t \hat{P}_{k/k} \hat{\Phi}_k \hat{\lambda}_{k+1/T} - \hat{y}_k^i) - \hat{H}_k^t \hat{x}_{k/k-1}^i,\\
\hat{x}_{k-1/T} &= \hat{x}_{k/k-1} + P_{k/k-1} \lambda_{k/T} \bar{P}_{k-1/T}^i \bar{P}_{k-1/k-1}^i.
\end{aligned} \tag{9.112}
$$

In general, however, the fusion estimate $\bar{x}_{k/k}$ in (9.112) is not globally optimal except some special case, for example, the process noise υ_k is equal to zero (therefore, the converted sensor noises are mutually uncorrelated). Thus, when the assumption for the matrices $H_k^i + (\tilde{P}_{k/k-1}^{i^{-1}} \tilde{M}_k^i)'$ is not satisfied, (9.111) and (9.112) provide a suboptimal algorithm for the packet loss or intermittent communications problem. The numerical example in the next section shows that this suboptimal algorithm still works well.

Remark 9.5. In this section, it was assumed that the packet loss or intermittent communications from local sensors to fusion center are synchronous. In fact, it is easy to extend the results to the distributed Kalman filtering fusion with asynchronous intermittent communications or packet loss. In the asynchronous communications case, we first synchronize the distributed system by regarding every possible communication instant from sensors to the fusion center as a point of the time subscript set of the whole dynamic system. Then, at each time instant, the fusion center only fuses the sensor estimates that are received by the fusion center using the proposed distributed filtering fusion algorithms above. More precisely, if, for example, $\bar{x}^i_{k-1/k-1}$ is not received by the fusion center, view $\bar{x}^i_{k/k} = \bar{x}^i_{k/k-1}$ and $\bar{K}^i_k \bar{H}^i_k = 0$ in (9.112). Since Kalman filtering is a recursive minimal mean square error estimator using all received observations up to now, such extension does not change the performance analysis in Section 9.4 at all.

9.2.4 Process data collection

The benchmarked lab-scale two-tank process control system has been used to collect data at a sampling rate of 50 ms. Process data have been generated through an experimental setup as shown in Fig. 9.3. The main objective of the benchmarked dual-tank system is to reach a reference height of 200 ml of the second tank. During this process, several faults have been introduced, such as leakage faults, sensor faults, and actuator faults. Leakage faults have been introduced through the pipe clogs of the system, knobs between the first and the second tank, etc. Sensor faults have been introduced by introducing a gain in the circuit as if there is a fault in the level sensor of the tank. Actuator faults have been introduced by introducing a gain in the setup for the actuator that is comprised of the motor and pump. A proportional integral (PI) controller works in a closed-loop configuration to reach the desired height of the second tank. Due to the inclusion of faults, the controller was finding it difficult to reach the desired level. For this reason, the power of the motor has been increased from a scale of 0 to 5 V to a scale of 5 to 18 V in order to provide it the maximum throttle to reach the desired level. In doing so, the actuator performed well in achieving its desired level but it also suppressed the faults of the system. So, it made the task of detecting the faults even more difficult. After the collection of data, techniques such as settling time, steady state value, and coherence spectra can help us to gain insight in the fault.

In this section, in particular, leakage fault has been considered. Hydraulic height and liquid output flow rate of the second tank are the inputs, while the leakage fault level on a discrete scale of 1 to 4 is the considered output. Data are collected by introducing leakage fault in the closed-loop system.

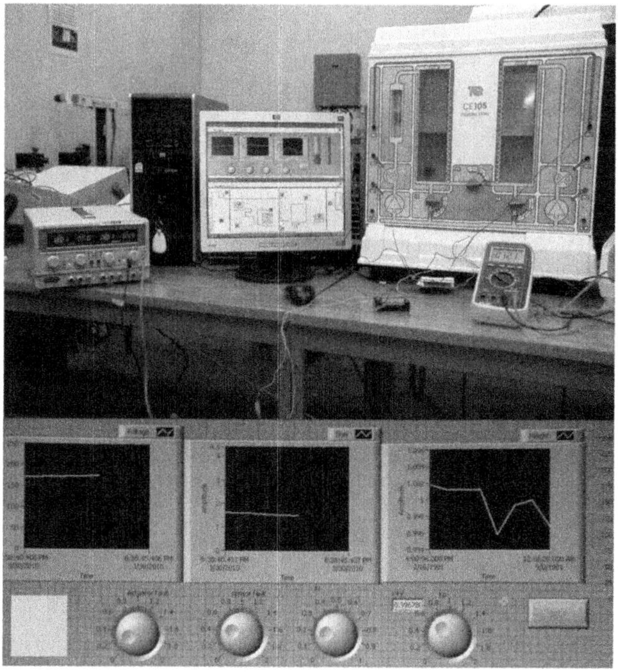

FIGURE 9.3 (Top) The two-tank system interfaced with the LabVIEW through a DAQ and the amplifier for the magnified voltage. (Bottom) The LabVIEW setup of the apparatus including the circuit window and the block diagram of the experiment.

9.2.5 Model of the coupled-tank system

The physical system under evaluation is formed of two tanks connected by a pipe. The leakage is simulated in the tank by opening the drain valve. A DC motor-driven pump supplies the fluid to the first tank and a PI controller is used to control the fluid level in the second tank by maintaining the level at a specified level, as shown in Fig. 9.4.A step input is applied to the DC motor–pump system to fill the first tank. The opening of the drainage valve introduces a leakage in the tank. Various types of leakage faults are introduced and the liquid height in the second tank, H_2, and the inflow rate, Q_i, are both measured. The National Instruments LabVIEW package is employed to collect these data. A benchmark model of a cascade connection of a DC motor and a pump relating the input to the motor, u, and the flow, Q_i, is a first-order system:

$$\dot{Q}_i = -a_m Q_i + b_m \phi(u), \tag{9.113}$$

where a_m and b_m are the parameters of the motor–pump system and $\phi(u)$ is a dead-band and saturation type of nonlinearity. It is assumed that the leakage Q_ℓ

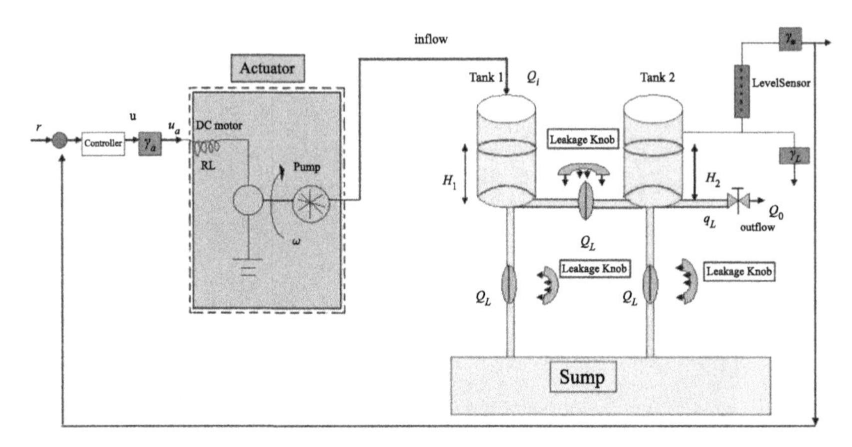

FIGURE 9.4 Process control system: A lab-scale two-tank system.

occurs in tank 1 and is given by (see Eq. (9.114))

$$Q_\ell = C_{d\ell}\sqrt{2gH_1}. \tag{9.114}$$

With the inclusion of the leakage, the liquid level system is modeled by (see Eq. (9.115))

$$A_1\frac{dH_1}{dt} = Q_i - C_{12}\varphi(H_1 - H_2) - C_\ell\varphi(H_1), \tag{9.115}$$

$$A_2\frac{dH_2}{dt} = C_{12}\varphi(H_1 - H_2) - C_O\varphi(H_2), \tag{9.116}$$

where $\varphi(.) = sign(.)\sqrt{2g(.)}$, $Q_\ell = C_\ell\varphi(H_1)$ is the leakage flow rate, $Q_0 = C_0\varphi(H_2)$ is the output flow rate, H_1 is the height of the liquid in tank 1, H_2 is the height of the liquid in tank 2, A_1 and A_2 are the cross-sectional areas of the two tanks, $g = 980$ cm/sec^2 is the gravitational constant, and C_{12} and C_o are the discharge coefficient of the intertank and output valves, respectively.

The model of the two-tank fluid control system, shown in Fig. 9.4, is of a second order and is nonlinear with a smooth square-root type of nonlinearity. For design purposes, a linearized model of the fluid system is required and is given as follows:

$$\frac{dh_1}{dt} = b_1q_i - (a_1 + \alpha)h_1 + a_1h_2, \tag{9.117}$$

$$\frac{dh_2}{dt} = a_2h_1 - (a_2 - \beta)h_2, \tag{9.118}$$

where h_1 and h_2 are the increments in the nominal (leakage-free) heights H_1^0 and H_2^0 (see Eqs. (9.119)–(9.120)):

$$b_1 = \frac{1}{A_1}, \quad a_1 = \frac{C_{db}}{2\sqrt{2g(H_1^0 - H_2^0)}}, \quad \beta = \frac{C_0}{2\sqrt{2gH_2^0}}, \tag{9.119}$$

$$a_2 = a_1 + \frac{C_{do}}{2\sqrt{2gH_2^0}}, \quad \alpha = \frac{C_{d\ell}}{2\sqrt{2gH_1^0}}, \tag{9.120}$$

and the parameter α indicates the amount of leakage.

A PI controller, with gains k_p and k_I, is used to maintain the level of tank 2 at the desired reference input r as (see Eq. (9.121))

$$\begin{aligned} \dot{x}_3 &= e = r - h_2, \\ u &= k_p e + k_I x_3. \end{aligned} \tag{9.121}$$

The linearized model of the entire system formed by the motor, the pump, and the tanks is given by

$$\dot{x} = Ax + Br, \quad y = Cx, \tag{9.122}$$

where

$$x = \begin{bmatrix} h_1 \\ h_2 \\ x_3 \\ q_i \end{bmatrix}, \quad A = \begin{bmatrix} -a_1 - \alpha & a_1 & 0 & b_1 \\ a_2 & -a_2 - \beta & 0 & 0 \\ -1 & 0 & 0 & 0 \\ -b_m k_p & 0 & b_m k_I & -a_m \end{bmatrix}, \tag{9.123}$$

$$B = \begin{bmatrix} 0 & 0 & 1 & b_m k_p \end{bmatrix}^T, \quad C = [1 \quad 0 \quad 0 \quad 0],$$

where q_i, q_ℓ, q_0, h_1, and h_2 are the increments in Q_i, Q_ℓ, Q_0, H_1^0, and H_2^0, respectively, the parameters a_1 and a_2 are associated with linearization, and the parameters α and β are respectively associated with the leakage and the output flow rate, i.e., $q_\ell = \alpha h_1$, $q_o = \beta h_2$.

9.2.6 Evaluation of results

The distributed FB Kalman filtering fusion with intermittent observation has been employed here on a coupled-tank system. The intermittent observations have been created in the height profile of the water in the coupled-tank system. Two profiles are shown in Figs. 9.5 and 9.6, first, with the ordinary hydraulic height without the intermittent observations, second, with the fusion center Kalman filter with intermittent observation, and third, with the FB Kalman fusion center. It has been shown that the FB Kalman fusion performs well with its

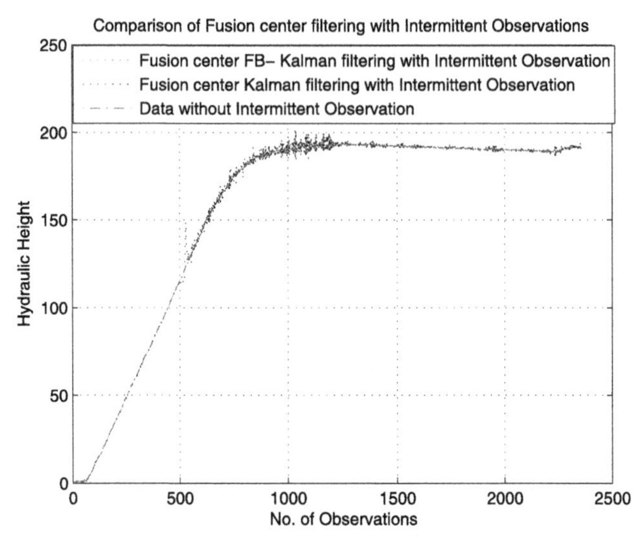

FIGURE 9.5 Implementation results of fusion center-based Kalman filtering and FB Kalman filtering on the intermittent observations.

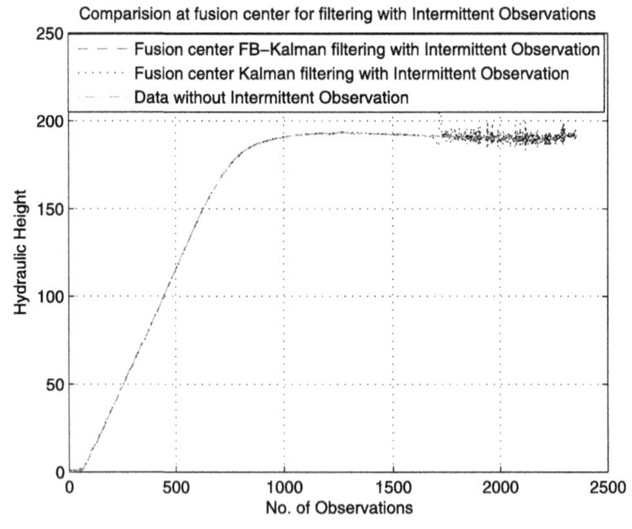

FIGURE 9.6 Implementation results of fusion center-based Kalman filtering and FB Kalman filtering on the intermittent observations.

habitual instincts of smoothing and providing better results compared with the ordinary Kalman filter fusion.

Also, the mean square error at the fusion center with ordinary Kalman filter for the first scenario in Fig. 9.5 is 0.0017, whereas the mean square error at the fusion center with FB Kalman filter is 0.00019144. This shows that by calculating the values in the backward loop, we are able to minimize the error. Likewise, in Fig. 9.6, it is 0.0027 for the ordinary Kalman filter, whereas it is 0.00012147 for the FB Kalman filter.

9.3 Information-based algorithms

As we learned, the process of state estimation yields the estimate in the form of a mean and covariance matrix or in the form of a distribution function. In this way, it is effectively used in providing a strict surveillance system for an appropriate supervision. One of the methods to achieve such sort of estimation often requires a group of distributed sensors which provide information of the local targets.

In [28], the distributed state estimation problem is investigated for a class of sensor networks described by uncertain discrete-time dynamical systems with Markovian jumping parameters and distributed time delays. In [29], an FB Kalman filter-based estimation in a fault diagnosis scheme is provided with a wide coverage of multitarget tracking architectures that are applied to complex applications such as condition and health monitoring of aircraft, industrial plants, and electrical infrastructures [30].

9.3.1 Introduction

The problem of multitarget tracking utilizing information from multiple sensors employed has been in focus since many years [31], [32]. Tackling this problem, many fusion algorithms and filters were derived to combine local estimates [33], [34] to prove better efficiency and effectiveness. For example, the state vectors can be fused using weighted covariance [31], information matrices [35], and covariance intersection [36]. The algorithms differ with respect to the method by which they treat the covariance. As for the performance of different algorithms, [37] shows that the performance of the weighted covariance algorithm is consistently worse compared with the measurement fusion method. Covariance intersection avoids cross-covariance computation and its fusion result will be a consistent estimate, but its conservative estimates reduce performance [36]. However, covariance intersection is also used for simultaneous localization and mapping to maintain the full correlation structure.

In [38] the identification of distributed two-dimensional systems and/or network-connected systems is described in the framework of distributed optimization methods such as the alternating direction method of multipliers. Distributed systems are expressed as interconnections of simple systems, and we introduce artificial signals in order to make the resulting optimization problem have a separable objective function. An enhanced method for distributed estimation based on prior information is developed in [39], where a Bayesian-based FB Kalman filter is employed together with a prescribed estimation fusion procedure.

9.3.2 Problem formulation

Consider a distributed tracking system in which $N \geq 2$ sensors are tracking the same target. The mathematical model describing object dynamics is assumed to be linear time-invariant and of the form

$$x_{k+1} = Fx_k + Gv_k,\ k = 0, 1, 2, ..., \tag{9.124}$$

where $x_k \in \Re^{n_1}$ is the state vector of the target at time k, F is the state transition matrix, $v_k \in \Re^{n_2}$ is zero mean white Gaussian process noise with known covariance Q, and G is the input matrix. The target is tracked by N sensors, where the measurement model of sensor $j = 1, ..., N$ is described by

$$z_k^j = H^j x_k + w_k^j, \tag{9.125}$$

where $H^j \in \Re^{n_3 \times n_1}$ is the associated measurement matrix and $w_k^j \in \Re^{n_3}$ is zero mean white Gaussian measurement noise with covariance $\mathbf{R}_k^j$. It is assumed that local estimates $\hat{x}_{k|k}^j$ and $P_{k|k}^j$, where $j = 1, ..., N$, are obtained by each sensor's information-based filter based on a measurement sequence

$$\mathbf{Z}_k^j = \{z_i^j,\ i = 1, 2, ..., k\}$$

and are optimal in the sense of minimum variance. At the end of each n sampling interval, each sensor transmits its local estimate to the fusion center, where track association and fusion are performed. For the fused estimate, there are two choices: either be sent back to the sensor to improve local estimation performance or to be stored on the fusion center. For the sake of simplicity, the dimensions of the fused track and all local tracks are assumed to be the same. The distributed track fusion problem is to generate an optimal estimate $\hat{x}_{k|k}$ from all local track information, i.e., $\hat{x}_{k|k}^j$ and $P_{k|k}^j$, and prior information about local and fused estimation if possible [32].

9.4 Covariance intersection

In what follows, we consider a track-to-track system in which two sensors are tracking the same target and where the intersection is characterized by the convex combination of the covariances [41], and therefore the covariance intersection algorithm is expressed as

$$\hat{x}_{k|k} = P_{k|k}(\omega P_{k|k}^{-i} \hat{x}_{k|k}^i + (1 - \omega) P_{k|k}^{-i} \hat{x}_{k|k}^j), \tag{9.126}$$

$$K_i = \omega P_{k|k} P_{k|k}^{-i}, \tag{9.127}$$

$$K_j = (1 - \omega) P_{k|k} P_{k|k}^{-j}, \tag{9.128}$$

where K_i, K_j are the gains at sensors i, j, respectively, and $\omega \in [0, 1]$ and it manipulates the weights which are assigned to $\hat{x}^i_{k|k}$ and $\hat{x}^j_{k|k}$. The covariance of the filtering error is given by

$$P_{k|k} = (\omega P^{-i}_{k|k} + (1-\omega)P^{-j}_{k|k})^{-1}. \tag{9.129}$$

Equivalently,

$$P^{-1}_{k|k} = \omega P^{-i}_{k|k} + (1-\omega)P^{-j}_{k|k}, \tag{9.130}$$

where $\omega = (K_i/P_{k|k}).P^i_{k|k}$ and $1-\omega = (K_j/P_{k|k}).P^j_{k|k}$, where $P^{-i}_{k|k}$ and $P^{-j}_{k|k}$ are the error covariance matrices in the Kalman filter equations for each sensor.

Thus substituting (9.127), (9.128), and (9.130) into (9.126) yields

$$P^{-1}_{k|k}\hat{x}_{k|k} = \omega P^{-i}_{k|k}\hat{x}^i_{k|k} + (1-\omega)P^{-j}_{k|k}\hat{x}^j_{k|k}. \tag{9.131}$$

Remark 9.6. Different choices of ω can be used to optimize the update with respect to different performance criteria such as minimizing the trace or determinant of $P_{k|k}$. Note that the covariance matrix $P^{-i}_{k|k}$ is the true covariance at sensor i and the mean square error matrix $P_{k|k}$ is the estimated covariance matrix used by the filter/fusion processing.

9.5 Information-based covariance intersection filter

To discuss the information-based covariance intersection filter and provide the relative features, we start with the underlying algorithm.

9.5.1 Algorithm

Consider the target dynamic model of (9.124) and (9.125) of the form [41]

$$x_{k+1} = F^i x^i_k + F^j x^j_k + Gv_k, \tag{9.132}$$

$$z^j_k = K_1 x^i_k + K_2 x^j_k + w_k, \tag{9.133}$$

where matrices F^i, K_1 and F^j, K_j are associated with sensors i and j, respectively. The key idea of the information matrix filter is to identify the common information shared by estimates that are to be fused, and then the information is removed or decorrelation is implemented. It will take into consideration the common information but not the common process noise.

Under the assumption of no feedback, the estimation using information-based filtering in the case of covariance intersection is as follows:

$$\begin{aligned} P^{-1}_{k|k}\hat{x}_{k|k} = {} & P^{-1}_{k|k-n}\hat{x}_{k|k-n} + \omega P^{-i}_{k|k}\hat{x}^i_{k|k} - \omega P^{-i}_{k|k-n}\hat{x}^i_{k|k-n} + (1-\omega)P^{-j}_{k|k}\hat{x}^j_{k|k} \\ & - (1-\omega)P^{-j}_{k|k-n}\hat{x}^j_{k|k-n}, \end{aligned} \tag{9.134}$$

$$P_{k|k}^{-1} = P_{k|k-n}^{-1} + \omega P_{k|k}^{-i} - \omega P_{k|k-n}^{-i} + (1-\omega) P_{k|k}^{-j} - (1-\omega) P_{k|k-n}^{-j}, \tag{9.135}$$

where the n step fusion state prediction is

$$x_{k|k-n} = F^i x_k^i + F^j x_k^j, \tag{9.136}$$

where matrices F^i, F^j, K_j are associated with sensors i and j. The associated covariance is computed in the sequel. Following [42], since $v_k \in \Re^m$ is assumed to be a zero mean white noise process and $x_k \in \Re^n$ is a state vector, it can be easily seen from $x_{k+1} = F x_k^i + F x_k^j + G v_k$ that the covariance matrix of x_k obeys the recursion

$$\Pi_{i+1} = F_k \Pi_k^i F_k^t + F_k \Pi_k^j F_k^t + G_i Q_i G_i^t, \tag{9.137}$$

where $\Pi_k^i = \mathbb{E}[x_k^i x_k^{i^t}]$ and $\Pi_k^j = \mathbb{E} x_k^j x_k^{j^t}$.

Likewise, since $\hat{x}_{k|k-n} = F^i x_k^i + F^j x_k^j$, it satisfies the recursion

$$\Sigma_{i+1} = F^i \Sigma_k^i F^{i^t} + F^j \Sigma_k^j F^{j^t}, \tag{9.138}$$

where $\Sigma_k^i = \mathbb{E}[\hat{x}_{k|k-1}^i \hat{x}_{k|k-1}^{i^t}]$ and $\Sigma_k^j = \mathbb{E}[\hat{x}_{k|k-1}^j \hat{x}_{k|k-1}^{j^t}]$ with initial condition $\Sigma_0 = 0$. Now the orthogonal decomposition $x_i = \hat{x}_{k|k-1} + \hat{x}_{i|i-1}$ shows that $\Pi_i = \Sigma_k^i + \Sigma_k^j + P_{k|k-1}$. It is then immediate to conclude that $P_{k+1|k} = \Sigma_{k+1} - \Sigma_{k+1}^i + \Sigma_{k+1}^j$ satisfies the recursion

$$P_{k+1|k} = F^i P_{k|k-1} F^{i^t} + G_i Q_i G_i^t. \tag{9.139}$$

For the distributed tracking system, the communication network is considered to be large; therefore, the fused state estimate and associated covariance depend upon the local estimates as

$$\hat{x}_{k|k-n}^i + \hat{x}_{k|k-n}^j = \hat{x}_{k|k-n}, \tag{9.140}$$

$$P_{k|k-n}^i + P_{k|k-n}^j = P_{k|k-n}. \tag{9.141}$$

9.5.2 Complete feedback case

For the case of complete feedback, a closed-form analytical solution of steady fused covariance of information-based covariance intersection filtering with N sensors is derived below. From (9.132) and (9.133), it is easy to show that the following two equations hold:

$$\begin{aligned} P_{k|k}^{-1} \hat{x}_{k|k} = {} & P_{k|k-n}^{-1} \hat{x}_{k|k-n} + \omega P_{k|k}^{-i} \hat{x}_{k|k}^i - \omega P_{k|k-n}^{-i} \hat{x}_{k|k-n}^i + (1-\omega) P_{k|k}^{-j} \hat{x}_{k|k}^j \\ & - (1-\omega) P_{k|k-n}^{-j} \hat{x}_{k|k-n}^j, \end{aligned} \tag{9.142}$$

$$P_{k|k}^{-1} = P_{k|k-n}^{-1} + \omega P_{k|k}^{-i} - \omega P_{k|k-n}^{-i} + (1-\omega) P_{k|k}^{-j} - (1-\omega) P_{k|k-n}^{-j}. \tag{9.143}$$

For the two local sensors in covariance intersection, i.e., i and j, it is possible to write

$$\hat{x}_{k|k} = \omega P_{k|k} {P_{k|k}^{i}}^{-1} F^{i} x_{k|k}^{i} + (1-\omega) P_{k|k} {P_{k|k}^{j}}^{-1} F^{j} x_{k|k}^{j}. \tag{9.144}$$

Using (9.144) and (9.140), we have

$$\hat{x}_{k|k} = A_n x_{k|k}^{i} + B_i x_{k|k}^{j}, \tag{9.145}$$

where, $\forall\, i = 1, ..., n$, we have

$$A_0 = I,\ A_i = \omega A_{i-1} P_{k|k} (P_{k|k}^{i})^{-1} F^{i},\ B_j = (1-\omega) A_{j-1} P_{k|k} (P_{k|k}^{j})^{-1} F^{j}.$$

Under the assumption of complete feedback, (9.142) and (9.143) can be rewritten as

$$\begin{aligned} P_{k|k}^{-1} \hat{x}_{k|k} &= -(N-1) P_{k|k-n}^{-1} \hat{x}_{k|k-n} + \omega P_{k|k}^{-i} \hat{x}_{k|k}^{i} \\ &\quad + (1-\omega) P_{k|k}^{-j} \hat{x}_{k|k}^{j}, \end{aligned} \tag{9.146}$$

$$P_{k|k}^{-1} = -(N-1) P_{k|k-n}^{-1} + \omega P_{k|k}^{-i} + (1-\omega) P_{k|k}^{-j}. \tag{9.147}$$

To compute the steady state error covariance of the fused state estimate, subtracting $P_{k|k}^{-1} x_k$ from both sides of (9.146) and substituting (9.145) yields

$$\begin{aligned} P_{k|k}^{-1} (\hat{x}_{k|k} - x_k) &= -P_{k|k}^{-1} x_k - (N-1) P_{k|k-n}^{-1} \hat{x}_{k|k-n} \\ &\quad + \omega {P_{k|k}^{i}}^{-1} \hat{x}_{k|k}^{i} + (1-\omega) P_{k|k}^{-j} \hat{x}_{k|k}^{j} \\ &= -(N-1) P_{k|k-n}^{-1} F^{n} (\hat{x}_{k|k-n} - x_{k-n}) \\ &\quad - P_{k|k}^{-1} x_k - (N-1) P_{k|k-n}^{-1} F^{n} x_{k-n} \\ &\quad + P_{k|k}^{-1} [A_n x_{k|k}^{i} + B_i x_{k|k}^{j}]. \end{aligned} \tag{9.148}$$

Algebraic manipulation and substituting (9.147) into (9.148) gives

$$\begin{aligned} P_{k|k}^{-1} (\hat{x}_{k|k} - x_k) &= \{-(N-1) P_{k|k-n}^{-1} F^{n} + P_{k|k}^{-1} A_n\} (\hat{x}_{k-n|k-n} - x_{k-n}) \\ &\quad + P_{k|k}^{-1} A_n \hat{x}_{k-n} - P_{k|k}^{-1} x_k \\ &\quad - (N-1) P_{k|k-n}^{-1} F^{n} x_{k-n} + P_{k|k}^{-1} B_i x_{k|k}^{j} \\ &= \{-(N-1) P_{k|k-n}^{-1} F^{n} + P_{k|k}^{-1} A_n\} (\hat{x}_{k-n|k-n} - x_{k-n}) \\ &\quad + P_{k|k}^{-1} A_n \hat{x}_{k-n} - (N-1) P_{k|k-n}^{-1} F^{n} x_{k-n} \\ &\quad + P_{k|k}^{-1} B_i w_{k-n+i} - P_{k|k}^{-1} x_k \end{aligned}$$

$$+ P_{k|k}^{-1} B_i (K_1 F^i x_{k-n}^i + K_2 F^j x_{k-n}^j)$$

$$+ P_{k|k}^{-1} B_j \sum_{h=1}^{i} (K_1 + K_2) F^{i-h} G v_{k-n+h}. \tag{9.149}$$

It has been shown that A_n satisfies the following identity:

$$A_n = -\sum_{i=1}^{n} B_i K F^i + F^n. \tag{9.150}$$

Substituting (9.150) and (9.147) into (9.149), we have

$$\begin{aligned}
P_{k|k}^{-1}(\hat{x}_{k|k} - x_k) &= \{-(N-1) P_{k|k-n}^{-1} F^n + P_{k|k}^{-1} A_n\}(\hat{x}_{k-n|k-n} - x_{k-n}) \\
&+ P_{k|k}^{-1} A_n x_{k-n} - (N-1) P_{k|k-n}^{-1} F^n x_{k-n} \\
&+ P_{k|k}^{-1} B_j w_{k-n+i} - P_{k|k}^{-1} x_k \\
&+ P_{k|k}^{-1} (F^n - A_n) x_{k-n} \\
&+ P_{k|k}^{-1} B_j \sum_{h=1}^{i} F^{i-h} G v_{k-n+h} \\
&= \{-(N-1) P_{k|k-n}^{-1} F^n + P_{k|k}^{-1} A_n\}(\hat{x}_{k-n|k-n} - x_{k-n}) \\
&+ P_{k|k}^{-1} B_j w_{k-n+i} + (P_{k|k}^{-1} B_j \sum_{h=i}^{n} (K_1 \\
&+ K_2) F^{h-i} - P_{k|k}^{-1} F^{n-i}) G v_{k-n+i}.
\end{aligned} \tag{9.151}$$

Using (9.151), we show the following Lyapunov form:

$$\Omega_x = C_f \Omega_x C_f^t + \Omega_f, \tag{9.152}$$

where

$$\begin{aligned}
C_f &= \lim_{k \to \infty} P_{k|k}(-(N-1) P_{k|k-n}^{-1} F^n + P_{k|k}^{i^{-1}} A_n^i + P_{k|k}^{j^{-1}} A_n^j), \\
\Omega_f &= W_s(k) R W_s(k)' + V_s(k) G Q G' V_s(k), \\
W_s(k) &= \lim_{k \to \infty} P_{k|k} P_{k|k}^{-1} B_i, \\
V_s(k) &= \lim_{k \to \infty} P_{k|k} P_{k|k}^{-1} B_i \sum_{h=1}^{n} (K_1 + K_2) F^{h-i} - P_{k|k} P_{k|k}^{-1} F^{n-i}.
\end{aligned} \tag{9.153}$$

9.5.3 Partial feedback case

In the case of partial feedback, (9.142) and (9.143) can be formulated as follows:

$$\begin{aligned} P_{k|k}^{-1}\hat{x}_{k|k} &= P_{k|k-n}^{-1}\hat{x}_{k|k-n} + \omega(P_{k|k}^{i})^{-1}\hat{x}_{k|k}^{i} \\ &\quad - \omega(P_{k|k-n}^{i})^{-1}\hat{x}_{k|k-n} + (1-\omega)(P_{k|k}^{j})^{-1}\hat{x}_{k|k}^{j} \\ &\quad - (1-\omega)(P_{k|k-n}^{j})^{-1}\hat{x}_{k|k-n}, \end{aligned} \tag{9.154}$$

$$\begin{aligned} P_{k|k}^{-1} &= P_{k|k-n}^{-1} + \omega(P_{k|k}^{i})^{-1} - \omega(P_{k|k-n}^{i})^{-1} + (1-\omega)(P_{k|k}^{j})^{-1} \\ &\quad - (1-\omega)(P_{k|k-n}^{j})^{-1}. \end{aligned} \tag{9.155}$$

Note that changing the value of N does not alter the forms of (9.154) and (9.155) and only the length of the summation item needs to be adjusted. Like the case of complete feedback, there is also a discrete Lyapunov equation,

$$\Omega_x = C_p \Omega_x C_p' + \Omega_p, \tag{9.156}$$

$$\begin{aligned} C_p &= \lim_{k\to\infty} P_{k|k}[P_{k|k}^{i^{-1}} A_n^i + P_{k|k}^{j^{-1}} A_n^j - P_{k|k-n}^{i^{-1}} F^n \\ &\quad - P_{k|k-n}^{j^{-1}} F^n + P_{k|k-n}^{-1} F^n], \end{aligned} \tag{9.157}$$

where Ω_p has the same definition as Ω_f in (9.153).

9.5.4 Weighted covariance

Given the system formulation in (9.132)–(9.133) and following up on the standard results of covariance intersection in [41], the weighted covariance at the sensor is

$$\hat{x}_{k|k} = A_k^i \hat{x}_{k|k}^i + A_k^j \hat{x}_{k|k}^j, \tag{9.158}$$

where the weighted matrices of two local estimates are calculated as

$$A_k^i = (P_{k|k}^j - \Sigma_{k|k}^{j,i})(P_{k|k}^i + P_{k|k}^j - \Sigma_{k|k}^{i,j} - \Sigma_{k|k}^{j,i})^{-1}, \tag{9.159}$$

$$A_k^j = (P_{k|k}^i - \Sigma_{k|k}^{i,j})(P_{k|k}^i + P_{k|k}^j - \Sigma_{k|k}^{i,j} - \Sigma_{k|k}^{j,i})^{-1} \tag{9.160}$$

and the covariance of the fused estimate is computed as follows:

$$P_{k|k} = P_{k|k}^j - (P_{k|k}^j - \Sigma_{k|k}^{j,i})(P_{k|k}^i + P_{k|k}^j - \Sigma_{k|k}^{i,j} - \Sigma_{k|k}^{j,i})^{-1}(P_{k|k}^j - \Sigma_{k|k}^{j,i})^t. \tag{9.161}$$

Alternatively,

$$\begin{aligned} P_{k|k}^{-1} &= (P_{k|k}^j - (P_{k|k}^j - \Sigma_{k|k}^{j,i})(P_{k|k}^i + P_{k|k}^j - \Sigma_{k|k}^{ij} - \Sigma_{k|k}^{ji})^{-1} \\ &\quad \times (P_{k|k}^j - \Sigma_{k|k}^{ji})^t)^{-1}, \end{aligned} \tag{9.162}$$

where

$$\Sigma^{i,j}_{1|1} = (I - K^i_1 H^i_1) Q_0 (I - K^i_1 H^i_1)^t,$$
$$\Sigma^{i,j}_{k|k} = (I - K^i_k H^i_k) F_{k-1} \Sigma^{i,j}_{k-1|k-1} F^t_{k-1} (I - K^i_k H^i_k)^t$$
$$+ (I - K^i_k H^i_k) Q_{k-1} (I - K^i_k H^i_k)^t,$$
$$\Sigma^{j,i}_{k|k} = (\Sigma^{i,j}_{k|k})^t.$$

Multiplying (9.162) with (9.158) gives

$$P^{-1}_{k|k} \hat{x}_{k|k} = (P^j_{k|k} - (P^j_{k|k} - \Sigma^{j,i}_{k|k})(P^i_{k|k} + P^j_{k|k} - \Sigma^{ij}_{k|k} - \Sigma^{ji}_{k|k})^{-1} (P^j_{k|k} - \Sigma^{ji}_{k|k})^T)^{-1} (A^i_k \hat{x}^i_{k|k} + A^j_k \hat{x}^j_{k|k}). \tag{9.163}$$

9.5.5 Weighted covariance filter algorithm

For the case of deriving an information-based weighted covariance filter, the target dynamic model of (9.124) and (9.125) will be of the form

$$x_{k+1} = F x_k + G w_k, \tag{9.164}$$
$$z_k = H^i x_k + H^j x_k + v^i + v^j. \tag{9.165}$$

The key idea of the information matrix filter is to identify the common information shared by estimates that are to be fused, and then the information is removed or decorrelation is implemented. It will take into consideration the common information but not the common process noise. Under the assumption of no feedback, the estimation using information-based filtering in the case of weighted covariance is as follows:

$$\begin{aligned} P^{-1}_{k|k} \hat{x}_{k|k} &= P^{-1}_{k|k-n} \hat{x}_{k|k-n} + (P^j_{k|k} - (P^j_{k|k} - \Sigma^{j,i}_{k|k}) \\ &\quad \times (P^i_{k|k} + P^j_{k|k} - \Sigma^{i,j}_{k|k} - \Sigma^{j,i}_{k|k})^{-1} (P^j_{k|k} - \Sigma^{j,i}_{k|k})^t)^{-1} \\ &\quad \times (A^i_k \hat{x}^i_{k|k} + A^j_k \hat{x}^j_{k|k}) - (P^j_{k|k-n} - (P^j_{k|k-n} - \Sigma^{j,i}_{k|k-n}) \\ &\quad \times (P^i_{k|k-n} + P^j_{k|k-n} - \Sigma^{i,j}_{k|k-n} - \Sigma^{j,i}_{k|k-n})^{-1} \\ &\quad \times (P^j_{k|k-n} - \Sigma^{j,i}_{k|k-n})^t)^{-1} (A^i_k \hat{x}^i_{k|k-n} + A^j_k \hat{x}^j_{k|k-n}), \end{aligned} \tag{9.166}$$

$$\begin{aligned} P^{-1}_{k|k} &= P^{-1}_{k|k-n} + (P^j_{k|k} - (P^j_{k|k} - \Sigma^{j,i}_{k|k}) \times (P^i_{k|k} + P^j_{k|k} \\ &\quad - \Sigma^{i,j}_{k|k} - \Sigma^{j,i}_{k|k})^{-1} (P^j_{k|k} - \Sigma^{j,i}_{k|k})^t)^{-1} - (P^j_{k|k-n} \\ &\quad - (P^j_{k|k-n} - \Sigma^{j,i}_{k|k-n}) \times (P^i_{k|k-n} + P^j_{k|k-n} - \Sigma^{i,j}_{k|k-n} \\ &\quad - \Sigma^{j,i}_{k|k-n})^{-1} \times (P^j_{k|k-n} - \Sigma^{j,i}_{k|k-n})^t)^{-1}. \end{aligned} \tag{9.167}$$

The n-step fusion state prediction and associated covariance are given by

$$\hat{x}_{k|k-n} = F^i \hat{x}_{k-n|k-n} + F^j \hat{x}_{k-n|k-n}, \tag{9.168}$$

$$P_{k+1|k} = F^i P_{k|k-1} F^{i^t} + G_i Q_i G_i^t. \tag{9.169}$$

The fused state estimate and associated covariance depend upon the local estimates as

$$\hat{x}^i_{k|k-n} + \hat{x}^j_{k|k-n} = \hat{x}_{k|k-n}, \tag{9.170}$$

$$P^i_{k|k-n} + P^j_{k|k-n} = P_{k|k-n}. \tag{9.171}$$

9.5.6 Weighted covariance filter: complete feedback case

For the case of complete feedback, a closed-form analytical solution of the steady fused covariance of an information-based covariance intersection filter with N sensors is derived below. From (9.164) and (9.165), it is easy to show that the following two equations hold:

$$x_k = F^i x_{k-n} + F^j x_{k-n} + \sum_{i=1}^{n} F^{n-i} G v_{k-n+i}, \tag{9.172}$$

$$\begin{aligned} z_k = {} & H^i F^i x_{k-n} + H^j F^j x_{k-n} + w^i_{k-n+i} + w^j_{k-n+j} \\ & + H^i F^i G v_{k-n+i} + H^j F^j G v_{k-n+j}. \end{aligned} \tag{9.173}$$

For the local sensors, it is possible to write weighted covariance as

$$\begin{aligned} \hat{x}_{k|k} = {} & P_{k|k}(P^j_{k|k} F \hat{x}_{k|k-n} + (P^j_{k|k} - \Sigma^{ji}_{k|k})(P^i_{k|k} + P^j_{k|k} \\ & - \Sigma^{ij}_{k|k} - \Sigma^{ji}_{k|k})^{-1} (P^j_{k|k} - \Sigma^{ji}_{k|k})^T)^{-1} P_{k|k} \\ & \times (A^i_k F \hat{x}^i_{k|k} + A^j_k F \hat{x}^j_{k|k}). \end{aligned} \tag{9.174}$$

Using (9.174) and (9.172), we have

$$\hat{x}_{k|k} = A_n P_{k|k} A^i_k F^i x^i_{k|k} + A_n P_{k|k} A^j_k F^j x^j_{k|k}, \tag{9.175}$$

where, $\forall\, i = 1, \ldots, n$, we have $A_0 = I$, $A_i = A_{i-1} P_{k|k}(P^j_{k|k} F^j \hat{x}_{k|k-n} + (P^j_{k|k} - \Sigma^{j,i}_{k|k})(P^i_{k|k} + P^j_{k|k} - \Sigma^{i,j}_{k|k} - \Sigma^{j,i}_{k|k})^{-1}(P^j_{k|k} - \Sigma^{j,i}_{k|k})^t)^{-1}$.

Under the assumption of complete feedback, (9.166) and (9.167) can be rewritten as

$$\begin{aligned} P^{-1}_{k|k} \hat{x}_{k|k} = {} & -(N-1) P^{-1}_{k|k-n} \hat{x}_{k|k-n} + (P^j_{k|k} - (P^j_{k|k} - \Sigma^{j,i}_{k|k}) \\ & \times (P^i_{k|k} + P^j_{k|k} - \Sigma^{i,j}_{k|k} - \Sigma^{j,i}_{k|k})^{-1} \times (P^j_{k|k} - \Sigma^{j,i}_{k|k})^T)^{-1} \\ & \times (A^i_k \hat{x}^i_{k|k} + A^j_k \hat{x}^j_{k|k}), \end{aligned} \tag{9.176}$$

$$\begin{aligned}P_{k|k}^{-1} &= -(N-1)P_{k|k-n}^{-1} + (P_{k|k}^{j} - (P_{k|k}^{j} - \Sigma_{k|k}^{j,i})\\ &\times (P_{k|k}^{i} + P_{k|k}^{j} - \Sigma_{k|k}^{i,j} - \Sigma_{k|k}^{j,i})^{-1}\\ &\times (P_{k|k}^{j} - \Sigma_{k|k}^{j,i})^{t})^{-1}.\end{aligned} \tag{9.177}$$

To compute the steady state error covariance of the fused state estimate, subtracting $P_{k|k}^{-1}x_k$ from both sides of (9.176) and substituting (9.175) yields

$$\begin{aligned}P_{k|k}^{-1}(\hat{x}_{k|k} - x_k) &= -P_{k|k}^{-1}x_k - (N-1)P_{k|k-n}^{-1}\hat{x}_{k|k-n}\\ &\quad - (N-1)P_{k|k-n}^{-1}\hat{x}_{k|k-n} + (P_{k|k}^{j} - (P_{k|k}^{j}\\ &\quad - \Sigma_{k|k}^{j,i}).(P_{k|k}^{i} + P_{k|k}^{j} - \Sigma_{k|k}^{ij} - \Sigma_{k|k}^{ji})^{-1}\\ &\quad \times (P_{k|k}^{j} - \Sigma_{k|k}^{ji})^{T})^{-1}.(A_k^i\hat{x}_{k|k}^i + A_k^j\hat{x}_{k|k}^j)\\ &= -(N-1)P_{k|k-n}^{-1}F^n(\hat{x}_{k-n|k-n} - x_{k-n})\\ &\quad - P_{k|k}^{-1}x_k - (N-1)P_{k|k-n}^{-1}F^n x_{k-n}\\ &\quad + P_{k|k}^{-1}(A_n P_{k|k}A_k^i F^i x_{k|k}^i\\ &\quad + A_n P_{k|k}A_k^j F^j \hat{x}_{k|k}^j).\end{aligned} \tag{9.178}$$

Through simple algebraic manipulations and substituting (9.173), we can rewrite (9.178) as

$$\begin{aligned}P_{k|k}^{-1}(\hat{x}_{k|k} - x_k) &= (-(N-1)P_{k|k-n}^{-1}F^n + P_{k|k}^{-1}A_n P_{k|k}A_k^i F\\ &\quad + P_{k|k}^{-1}A_n P_{k|k}A_k^j F).(\hat{x}_{k-n|k-n} - \hat{x}_{k|k}^i\\ &\quad - \hat{x}_{k|k}^j) + P_{k|k}^{-1}A_n P_{k|k}A_k^i F\hat{x}_{k-n}^i + P_{k|k}^{-1}\\ &\quad \times A_n P_{k|k}A_k^j F\hat{x}_{k|k-n}^j - P_{k|k}^{-1}x_k\\ &\quad - (N-1)P_{k|k-n}^{-1}F^n x_{k-n}.\end{aligned} \tag{9.179}$$

Using (9.179), we show the following Lyapunov form:

$$\Omega_x = C_f\Omega_x C_f^t + \Omega_f, \tag{9.180}$$

$$\begin{aligned}C_f &= \lim_{k\to\infty} P_{k|k}(-(N-1)P_{k|k-n}^{-1}F^n + P_{k|k}^{-1}A_n\\ &\quad \times P_{k|k}A_k^i F^i + P_{k|k}^{-1}A_n P_{k|k}A_k^j F^j),\\ \Omega_f &= W_s(k)RW_s(k)',\\ W_s(k) &= \lim_{k\to\infty} P_{k|k}P_{k|k}^{-1}A_n P_{k|k}(A_k^i + A_k^j).\end{aligned} \tag{9.181}$$

9.5.7 Partial feedback case

In the case of partial feedback, (9.166) and (9.167) can be formulated as follows:

$$
\begin{aligned}
P_{k|k}^{-1}\hat{x}_{k|k} &= P_{k|k-n}^{-1}\hat{x}_{k|k-n} + (P_{k|k}^{j} - (P_{k|k}^{j} - \Sigma_{k|k}^{j,i}) \\
&\times (P_{k|k}^{i} + P_{k|k}^{j} - \Sigma_{k|k}^{ij} - \Sigma_{k|k}^{ji})^{-1}(P_{k|k}^{j} \\
&- \Sigma_{k|k}^{ji})^{t})^{-1}.(A_{k}^{i}\hat{x}_{k|k}^{i} + A_{k}^{j}\hat{x}_{k|k}^{j}) - (P_{k|k-n}^{j} \\
&- (P_{k|k-n}^{j} - \Sigma_{k|k-n}^{j,i}).(P_{k|k-n}^{i} + P_{k|k-n}^{j} - \Sigma_{k|k-n}^{ij} \\
&- \Sigma_{k|k-n}^{ji})^{-1}.(P_{k|k-n}^{j} - \Sigma_{k|k-n}^{ji})^{t})^{-1}(A_{k}^{i}\hat{x}_{k|k-n} \\
&+ A_{k}^{j}\hat{x}_{k|k-n}),
\end{aligned}
\tag{9.182}
$$

$$
\begin{aligned}
P_{k|k}^{-1} &= P_{k|k-n}^{-1} + (P_{k|k}^{j} - (P_{k|k}^{j} - \Sigma_{k|k}^{j,i}).(P_{k|k}^{i} + P_{k|k}^{j} \\
&- \Sigma_{k|k}^{ij} - \Sigma_{k|k}^{ji})^{-1}(P_{k|k}^{j} - \Sigma_{k|k}^{ji})^{T})^{-1} \\
&- (P_{k|k-n}^{j} - (P_{k|k-n}^{j} - \Sigma_{k|k-n}^{ij} \\
&- \Sigma_{k|k-n}^{ji})^{-1}.(P_{k|k-n}^{j} - \Sigma_{k|k-n}^{ji})^{t})^{-1}.
\end{aligned}
\tag{9.183}
$$

Note that changing the value of N does not alter the forms of (9.182) and (9.183) and only the length of the summation item needs to be adjusted. Like the case of complete feedback, there is also a discrete Lyapunov equation,

$$
\Omega_{x} = C_{p}\Omega_{x}C_{p}^{t} + \Omega_{p}, \tag{9.184}
$$

$$
\begin{aligned}
C_{p} = \lim_{k\to\infty} P_{k|k}[&P_{k|k}^{i^{-1}} A_{n}^{i} P_{k|k} A_{k}^{i} F + P_{k|k}^{j^{-1}} A_{n}^{j} P_{k|k} A_{k}^{j} F \\
&- P_{k|k-n}^{i^{-1}} F^{n} - P_{k|k-n}^{j^{-1}} F^{n} + P_{k|k-n}^{-1} F^{n}],
\end{aligned}
\tag{9.185}
$$

where Ω_p has the same definition as Ω_f in (9.181).

9.5.8 Kalman-like particle filter

In this section, we will derive an information-based Kalman-like particle filter, where the simple Kalman-like particle filter is expressed in [43]. The question arises here why the Kalman-like particle filter has been preferred over a basic Kalman filter. The justification for the approach with respect to filter is given in [43]; moreover, it is preferred here over the basic Kalman filter because of the following (see Fig. 9.7 for the comparison of estimates of a basic Kalman filter and a Kalman-like particle filter; in Fig. 9.8 it can be seen how the mean square error is reduced in a lower number of iterations for the particle filter compared with a regular Kalman filter).

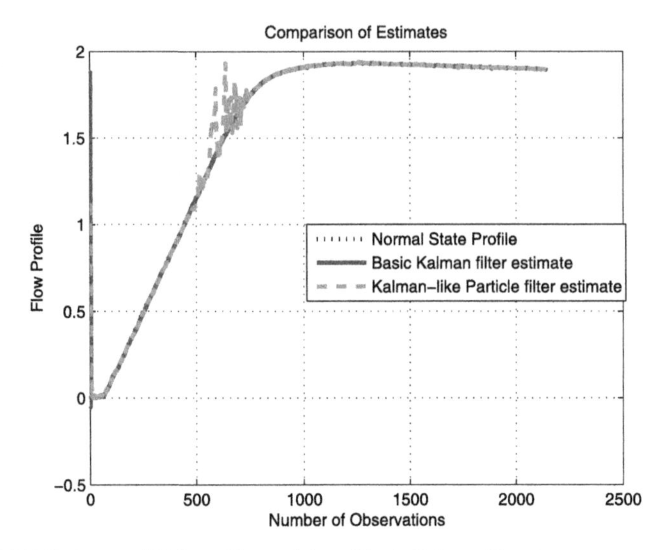

FIGURE 9.7 Estimates of Kalman-like particle and basic Kalman filters.

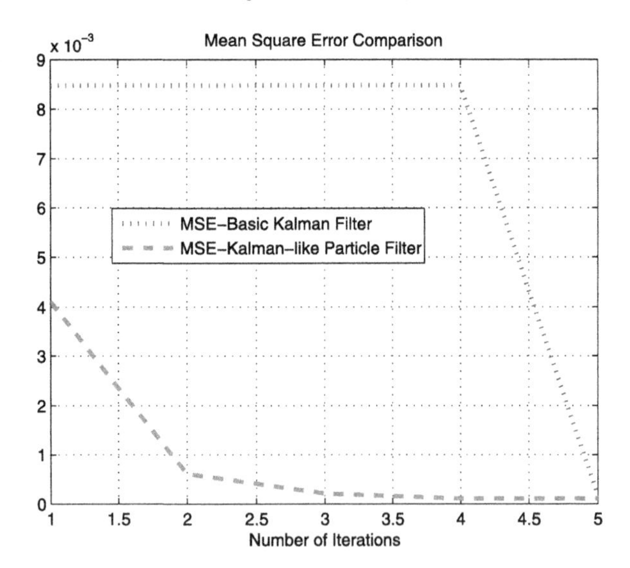

FIGURE 9.8 Mean square error: Kalman-like particle and basic Kalman filters.

According to the standard results of the Kalman-like particle filter in [43], the Kalman-like particle filter at sensor is

$$\begin{aligned}\hat{x}_{k|k} &= \hat{x}_{k|k-1} + \frac{P_k H_k^t}{H_k P_k H_k^t + \sigma_v^2}(y_k - H_k \hat{x}_{k|k-1}) \\ &= (I - \frac{P_k H_k^t}{H_k P_k H_k^t + \sigma_v^2} H_k)\hat{x}_{k|k-1} + \frac{P_k H_k^t}{H_k P_k H_k^t + \sigma_v^2} y_k, \end{aligned} \tag{9.186}$$

with the covariance of the filtering error given by

$$P_{k|k} = (I - \frac{P_k H_k^t}{H_k P_k H_k^t + \sigma_v^2} H_k) P_{k|k-1},$$

$$P_{k|k-1}^{-1} = P_{k|k}^{-1} (I - \frac{P_k H_k^t}{H_k P_k H_k^t + \sigma_v^2} H_k), \tag{9.187}$$

or

$$P_{k|k}^{-1} = P_{k|k-1}^{-1} + P_{k|k}^{-1} \frac{P_k H_k^t H_k}{H_k P_k H_k^t + \sigma_v^2}. \tag{9.188}$$

Thus substituting (9.187) into (9.186) yields

$$P_{k|k}^{-1} \hat{x}_{k|k} = P_{k|k-1}^{-1} \hat{x}_{k|k-1} + P_{k|k}^{-1} (\frac{P_k H_k^T H_k}{H_k P_k H_k^T + \sigma_v^2}) \hat{x}_{k|k}. \tag{9.189}$$

9.5.9 Kalman-like particle filter algorithm

The key idea of the information matrix filter is to identify the common information shared by estimates that are to be fused, and then the information is removed or decorrelation is implemented. It will take into consideration the common information but not the common process noise. Under the assumption of no feedback, the estimation using an information-based filter in the case of Kalman-like particle filtering is as follows:

$$\begin{aligned} P_{k|k}^{-1} \hat{x}_{k|k} = P_{k|k-1}^{-1} \hat{x}_{k|k-1} &+ P_{k|k}^{j^{-1}} (\frac{P_k^j H_k^{j^T} H_k^j}{H_k^j P_k^j H_k^{j^T} + \sigma_v^2}) \hat{x}_{k|k}^j \\ &- P_{k|k-n}^{j^{-1}} (\frac{P_k^j H_k^{j^T} H_k^j}{H_k^j P_k^j H_k^{j^T} + \sigma_v^2}) \hat{x}_{k|k-n}^j, \end{aligned} \tag{9.190}$$

$$\begin{aligned} P_{k|k}^{-1} = P_{k|k-1}^{-1} &+ P_{k|k}^{j^{-1}} (\frac{P_k^j H_k^{j^T} H_k^j}{H_k^j P_k^j H_k^{j^T} + \sigma_v^2}) \\ &- P_{k|k-n}^{j^{-1}} (\frac{P_k^j H_k^{j^T} H_k^j}{H_k^j P_k^j H_k^{j^T} + \sigma_v^2}). \end{aligned} \tag{9.191}$$

The n-step fusion state prediction and associated covariance are given by

$$x_{k|k-n} = F^n \hat{x}_{k-n|k-n}, \tag{9.192}$$

$$P_{k|k-n} = F^n P_{k-n|k-n} F^{nt} + F^{n-i} G Q G^t F^{n-i^t}, \tag{9.193}$$

where the n-step fusion state prediction and associated covariance is written as

$$\hat{x}^j_{k|k-n} = \hat{x}_{k|k-n}, \tag{9.194}$$

$$P^j_{k|k-n} = P_{k|k-n}. \tag{9.195}$$

9.5.10 Complete feedback case

For the case of complete feedback, a closed-form analytical solution of the steady fused covariance of the information-based Kalman-like particle filter with N sensors is derived below. From (9.124) and (9.125), it is easy to show that the following two equations hold:

$$x_k = F^i x_{k-n} + F^{n-i} G v_{k-n+i}, \tag{9.196}$$

$$z^j_{k-n+i} = H^j F^j x_{k-n} + w^j_{k-n+i} \sum_{h=1}^{i} H^j F^{i-h} G v_{k-n+h}. \tag{9.197}$$

For the two local sensors in the Kalman-like particle filter, it is possible to write as

$$\hat{x}^j_{k|k} = P_{k|k} P^{j^{-1}}_{k|k} F^j \hat{x}^j_{k|k-1} + P^j_{k|k} P^{j^{-1}}_{k|k} \frac{P^j_k H^{j^t}_k H^j_k}{H^j_k P^j_k H^j_k + \sigma^2_v} \hat{x}_{k|k}. \tag{9.198}$$

Utilizing (9.194) and (9.198), we have

$$\hat{x}^j_{k|k} = A^j_n \hat{x}_{k-n|k-n} + \sum_{i=1}^{n} B^j_i \hat{x}_{k|k}, \tag{9.199}$$

where, $\forall\, i = 1, \ldots, n$, we have $A^j_0 = I$, $A^j_i = P_{k-i+1|k-i+1} P^{j^{-1}}_{k-i+1|k-i+1} F$, $B^j = A^j_{i-1} P^j_{k-i+1|k-i+1} P^{j^{-1}}_{k-i+1|k-i+1} (P^j_k H^{j^T}_k H^j_k / (H^j_k P^j_k H^j_k + \sigma^2_v))F$.

Under the assumption of complete feedback, (9.190) and (9.191) can be re-written as

$$\begin{aligned} P^{-1}_{k|k} \hat{x}_{k|k} &= -(N-1) P^{-1}_{k|k-n} \hat{x}_{k|k-n} \\ &\quad + \sum_{j=1}^{N} P^{j^{-1}}_{k|k} \frac{P_k H^T_k H_k}{H_k P_k H^T_k + \sigma^2_v} \hat{x}^j_{k|k}, \end{aligned} \tag{9.200}$$

$$\begin{aligned} P^{-1}_{k|k} &= -(N-1) P^{-1}_{k|k-n} \\ &\quad + \sum_{j=1}^{N} P^{j^{-1}}_{k|k} \frac{P_k H^T_k H_k}{H_k P_k H^T_k + \sigma^2_v}. \end{aligned} \tag{9.201}$$

To compute the steady state error covariance of the fused state estimate, subtracting $P_{k|k}^{-1}x_k$ from both sides of (9.201) and substituting (9.199) yields

$$\begin{aligned}
P_{k|k}^{-1}(\hat{x}_{k|k}-x_k) &= -P_{k|k}^{-1}x_k-(N-1)P_{k|k-n}^{-1}\hat{x}_{k|k-n} \\
&+\sum_{j=1}^{N}P_{k|k}^{j^{-1}}\frac{P_kH_k^TH_k}{H_kP_kH_k^T+\sigma_v^2}\hat{x}_{k|k}^j \\
&= -(N-1)P_{k|k-n}^{-1}F^n(\hat{x}_{k|k-n}-x_{k-n}) \\
&- P_{k|k}^{-1}x_k-(N-1)P_{k|k-n}^{-1}F^nx_{k-n} \\
&+\sum_{j=1}^{N}P_{k|k}^{j^{-1}}\frac{P_kH_k^TH_k}{H_kP_kH_k^T+\sigma_v^2} \\
&\times[A_n^j\hat{x}_{k-n|k-n}+\sum_{i=1}^{n}B_i^jx_{k|k}].
\end{aligned} \tag{9.202}$$

By algebraic manipulations and substituting (9.198), we can rewrite (9.202) as

$$\begin{aligned}
P_{k|k}^{-1}(\hat{x}_{k|k}-x_k) &= (-(N-1)P_{k|k-n}^{-1}F^n \\
&+\sum_{j=1}^{N}P_{k|k}^{j^{-1}}(\frac{P_kH_k^TH_k}{H_kP_kH_k^T+\sigma_v^2})A_n^j) \\
&\times(\hat{x}_{k-n|k-n}-x_{k-n})+\sum_{j=1}^{N}P_{k|k}^{j^{-1}} \\
&\times(\frac{P_kH_k^TH_k}{H_kP_kH_k^T+\sigma_v^2})A_n^jx_{k-n}-P_{k|k}^{-1}x_k \\
&-(N-1)P_{k|k-n}^{-1}F^nx_{k-n} \\
&+(\sum_{j=1}^{N}P_{k|k}^{j^{-1}}\sum_{i=1}^{n}B_i^jx_{k|k}).
\end{aligned} \tag{9.203}$$

Using (9.203), we show the following Lyapunov form:

$$\Omega_x = C_f\Omega_xC_f'+\Omega_f, \tag{9.204}$$

where

$$\begin{aligned}
C_f = \lim_{k\to\infty}P_{k|k}\big(&-(N-1)P_{k|k-n}^{-1}F^n \\
&+\sum_{j=1}^{n}P_{k|k}^{-1}\frac{P_kH_k^TH_k}{H_kP_kH_k^T+\sigma_v^2}A_n^j\big),
\end{aligned}$$

$$\Omega_f = \sum_{j=1}^{N}\sum_{k=1}^{n} W_s^j(k) R^j W_s^j(k)',$$

$$W_s^j(k) = \lim_{k\to\infty} P_{k|k} P_{k|k}^{j^{-1}} B_i^j. \tag{9.205}$$

9.5.11 Kalman-like particle filter: partial feedback case

In the case of partial feedback, (9.190) and (9.191) can be formulated as follows:

$$\begin{aligned} P_{k|k}^{-1}\hat{x}_{k|k} &= P_{k|k-n}^{-1}\hat{x}_{k|k-n} \\ &+ \sum_{j=1}^{N} P_{k|k}^{j^{-1}} \frac{P_k H_k^T H_k}{H_k P_k H_k^T + \sigma_v^2}\hat{x}_{k|k}^j \\ &- P_{k|k-n}^{j^{-1}} \frac{P_k^j H_k^{j^T} H_k^j}{H_k^j P_k^j H_k^{j^T} + \sigma_v^2}\hat{x}_{k|k-n}^j, \end{aligned} \tag{9.206}$$

$$\begin{aligned} P_{k|k}^{-1} &= P_{k|k-n}^{-1} + \sum_{j=1}^{N} P_{k|k}^{j^{-1}} (\frac{P_k H_k^T H_k}{H_k P_k H_k^T + \sigma_v^2}) \\ &- P_{k|k-n}^{j^{-1}} (\frac{P_k^j H_k^{j^T} H_k^j}{H_k^j P_k^j H_k^{j^T} + \sigma_v^2}). \end{aligned} \tag{9.207}$$

Note that changing the value of N does not alter the forms of (9.206) and (9.207) and only the length of the summation item needs to be adjusted. Like the case of complete feedback, there is also a discrete Lyapunov equation,

$$\Omega_x = C_p \Omega_x C_p' + \Omega_p, \tag{9.208}$$

where

$$C_p = \lim_{k\to\infty} P_{k|k}[\sum_{j=1}^{n}(P_{k|k}^{-1} \cdot \frac{P_k H_k^T H_k}{H_k P_k H_k^T + \sigma_v^2} A_n^j - P_{k|k-n}^{j^{-1}} F^n) + P_{k|k-n}^{-1} F^n], \tag{9.209}$$

where Ω_p has the same definition as Ω_f in (9.205).

9.5.12 Measurement fusion algorithm

The information captured in each of the information-based filter cases is designed for a distributed structure. The idea is taken from the fusion methods in [40].

Suppose there are X sensors. For every measurement coming from these sensors that is received in the fusion center, there is a corresponding estimation based solely on these individual sensors. The information can be structured as estimated information or prior estimated information in the following two ways: the measurement fusion method and the state vector fusion method, as shown in Figs. 9.9 and 9.10, respectively.

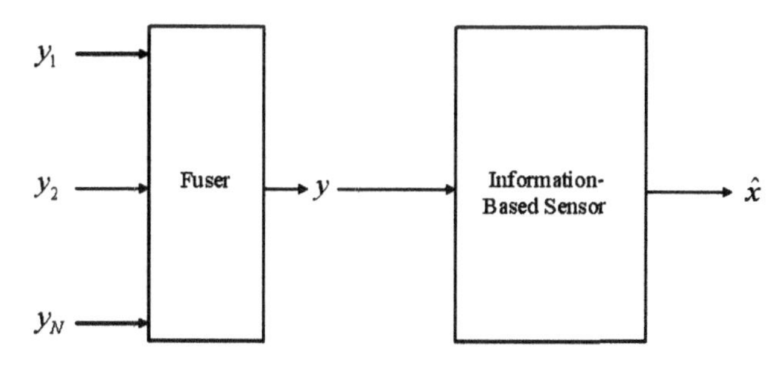

FIGURE 9.9 Measurement fusion employed for information-based sensors.

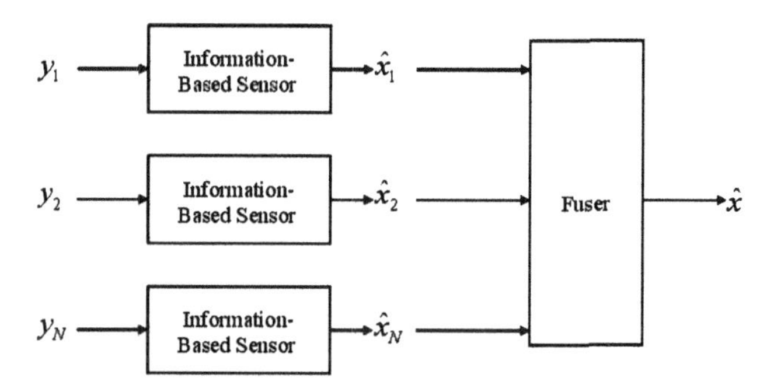

FIGURE 9.10 State vector fusion employed for information-based sensors.

9.5.13 Measurement fusion method

The measurement fusion method integrates the sensor measurement information by augmenting the observation vector as follows:

$$y(k) = y^{(mf)}(k) = [y_1(k) \; \dots \; y_N(k)]^T, \tag{9.210}$$

$$C(k) = C^{(mf)}(k) = [C_1(k) \; \dots \; C_N(k)]^T, \tag{9.211}$$

$$R(k) = R^{(mf)}(k) = diag[R_1(k) \; \dots \; R_N(k)], \tag{9.212}$$

where the superscript mf stands for the measurement fusion.

9.5.14 State vector fusion method

The state vector fusion method obtains the fused measurement information by weighted observation as follows:

$$y(k) = y^{(sf)}(k) = [\sum_{j=1}^{N} R_j^{-1}(k)]^{-1} \sum_{j=1}^{N} R_j^{-1}(k) y_j(k), \tag{9.213}$$

$$C(k) = C^{(sf)}(k) = [\sum_{j=1}^{N} R_j^{-1}(k)]^{-1} \sum_{j=1}^{N} R_j^{-1}(k) C_j(k), \tag{9.214}$$

$$R(k) = R^{(sf)}(k) = [\sum_{j=1}^{N} R_j^{-1}(k)]^{-1}, \tag{9.215}$$

where the superscript sf stands for state vector fusion.

9.5.15 Functional equivalence results

Comparing (9.210)–(9.212) with (9.213)–(9.215), we note that the treatment in the measurement fusion schemes is quite different. With reference to [40], we will show here that there exists a functional equivalence between the two methods.

Theorem 9.2. *If the N sensors used for data fusion with different and independent noise characteristics have identical measurement matrices, that is, $C_1(k) = C_2(k) = \ldots = C_N(k)$, then the measurement fusion method is functionally equivalent to the state vector fusion.*

Proof. If the information-based covariance intersection filter is used to demonstrate the functional equivalence of the two measurement fusion methods, then we only need to check whether the terms $(K_1 + K_2)C_k$ and $(K_1 + K_2)(k)y(k)$ in the measurement fusion method are functionally equivalent to those in the state vector fusion method. Alternatively, if the information filter is used, then we need to check the functional equivalence between terms $C^T(k)R^{-1}(k)C(k)$ and $C^T(k)R^{-1}(k)y(k)$ in both methods.

Consider the case when the information-based covariance intersection filter is applied, and $(K_1 + K_2)^{(mf)}$ is given by

$$\begin{aligned}
(K_1 + K_2)^{(mf)}(k) &= \omega P^{(mf)}(k|k-1)(C^{(sf)})^T \Xi_i^{(mf)} \\
&\quad + (1-\omega) P^{(mf)}(k|k-1)(C^{(sf)})^T \Xi_j^{(mf)}, \\
\Xi_i^{(mf)} &= [C(k)P^i(k|k-1)C(k) + R(k)]^{-1}, \\
\Xi_j^{(mf)} &= [C(k)P^j(k|k-1)C(k) + R(k)]^{-1}.
\end{aligned} \tag{9.216}$$

Proceeding further, we get

$$(K_1+K_2)^{(mf)}(k)=\omega P^{(mf)}(k|k-1)(C^{(sf)})^T$$
$$\times\begin{bmatrix} R_1+\Xi_i^{(mf)} & \Xi_i^{(mf)} \\ \Xi_i^{(mf)} & R_2+\Xi_i^{(mf)} \end{bmatrix}^{-1}$$
$$+(1-\omega)P^{mf}(k|k-1)(C^{(sf)})^t$$
$$\times\begin{bmatrix} R_1+\Xi_j^{(mf)} & \Xi_j^{(mf)} \\ \Xi_j^{(mf)} & R_2+\Xi_j^{(mf)} \end{bmatrix}^{-1}. \quad (9.217)$$

Algebraic manipulations using matrix inversion and block matrix inversion lemmas (see Chapter 1) yield

$$(K_1+K_2)^{(mf)}(k)=\omega P^{(mf)}(k|k-1)(C)^T[(R_2+\Xi_i^{(mf)})^{-1}$$
$$\times R_2[R_1+\Xi_i^{(mf)}-\Xi_i^{mf}(R_2+\Xi_i^{(mf)})^{(mf)}]^{-1}(R_2+\Xi_i^{(mf)})^{-1}$$
$$-(R_2+\Xi_i^{(mf)})^{-1}R_2[R_1+\Xi_i^{(mf)}-\Xi_i^{(mf)}(R_2+\Xi_i^{(mf)})^{-1}\Xi_i^{(mf)}]^{-1}$$
$$\times\Xi_i^{(mf)}(R_2+\Xi_i^{(mf)})^{-1}$$
$$+(1-\omega)P^{(mf)}(k|k-1)C^T[(R_2+\Xi_j^{(mf)})^{-1}R_2[R_1+\Xi_j^{(mf)}$$
$$-\Xi_j^{(mf)}(R_2\Xi_j^{(mf)})^{-1}\Xi_j^{(mf)}]^{-1}(R_2+\Xi_j^{(mf)})^{-1}$$
$$-(R_2+\Xi_j^{(mf)})^{-1}R_2[R_1+\Xi_j^{(mf)}-\Xi_i^{(mf)}(R_2+\Xi_j^{(mf)})^{-1}\Xi_j^{(mf)}]^{-1}$$
$$\times\Xi_j^{(mf)}(R_2+\Xi_j^{(mf)})^{-1}. \quad (9.218)$$

Following [40], it can be established that

$$(R_2+\Xi_i^{(mf)})^{-1}R_2[R_1+\Xi^{(mf)}-\Xi_i^{(mf)}(R_2+\Xi_i^{(mf)})^{-1}\Xi_i^{(mf)}]^{-1}$$
$$=[\Xi_i^{(mf)}+R_1(R_1+R_2)^{-1}R_2]^{-1}R_2(R_1+R_2)^{-1} \quad (9.219)$$

and

$$(R_2+\Xi_i^{(mf)})^{-1}-(R_2+\Xi_i^{(mf)})^{-1}R_2[R_1+\Xi_i^{(mf)}-\Xi^{(mf)}$$
$$\times(R_2+\Xi_i^{(mf)})^{-1}\Xi_i^{(mf)}]^{-1}\Xi_i^{(mf)}(R_2+\Xi_i^{(mf)})^{-1}$$
$$=[\Xi_i^{(mf)}+R_1(R_1+R_2)^{-1}R_2]^{-1}R_1(R_1+R_2)^{-1}. \quad (9.220)$$

Similar expressions like (9.219) and (9.220) can be derived for $\Xi_j^{(mf)}$.

Based on (9.218)–(9.220), we have

$$(K_1+K_2)^{(mf)}(k)=$$
$$\omega P^{(mf)}(k|k-1)C^t[CP^{i^{(mf)}}(k|k-1)C^t+R_1(R_1+R_2)^{-1}R_2]^{-1}$$

$$\times [R_2(R_1+R_2)^{-1}R_1(R_1+R_2)^{-1}]$$
$$+(1-\omega)P^{(mf)}(k|k-1)C^t[CP^{j^{(mf)}}(k|k-1)C^t$$
$$+R_1(R_1+R_2)^{-1}R_2]^{-1}\times[R_2(R_1+R_2)^{-1}R_1(R_1+R_2)^{-1}]. \tag{9.221}$$

It follows from (9.221) that

$$(K_1+K_2)^{(mf)}(k)C^{(mf)}(k)=\omega P^{(mf)}(k|k-1)C^t$$
$$\times [CP^{i^{(mf)}}(k|k-1)C^t+R_1(R_1+R_2)^{-1}R_2]^{-1}C$$
$$+(1-\omega)P^{(mf)}(k|k-1)C^t$$
$$\times [CP^{j^{(mf)}}(k|k-1)C^t+R_1(R_1+R_2)^{-1}R_2]^{-1}C, \tag{9.222}$$
$$(K_1+K_2)^{(mf)}(k)y^{(mf)}(k)=\omega P^{(mf)}(k|k-1)C^T$$
$$\times [CP^{i^{(mf)}}(k|k-1)C^t+R_1(R_1+R_2)^{-1}R_2]^{-1}$$
$$\times [R_2(R_1+R_2)^{-1}y_1(k+1)$$
$$+R_1(R_1+R_2)^{-1}y_2(k+1)]$$
$$+(1-\omega)P^{(mf)}(k|k-1)C^t\times[CP^{j^{(mf)}}(k|k-1)C^t$$
$$+R_1(R_1+R_2)^{-1}R_2]^{-1}\times[R_2(R_1+R_2)^{-1}y_1(k+1)$$
$$+R_1(R_1+R_2)^{-1}y_2(k+1)]. \tag{9.223}$$

If $C_1=C_2=C$, then we obtain the Kalman gain in the state vector method as follows:

$$(K_1+K_2)^{(sf)}(k)=\omega P^{(sf)}(k|k-1)C^T$$
$$\times [CP^{i^{(sf)}}(k|k-1)C^T R_1(R_1+R_2)^{-1}R_2]^{-1}$$
$$+(1-\omega)P^{(sf)}(k|k-1)C^T[CP^{j^{(sf)}}(k|k-1)C^T$$
$$+R_1(R_1+R_2)^{-1}R_2]^{-1}. \tag{9.224}$$

Similarly, we can derive the terms $K^{(sf)}(k)C^{(sf)}(k)$ and $K^{(sf)}(k)y^{(sf)}(k)$ as follows:

$$\begin{aligned}(K_1+K_2)^{(sf)}(k)C^{(sf)}(k) &= \omega P^{(sf)}(k|k-1)C^T\times[CP^{i^{(sf)}}(k|k-1)\\ &\quad\times C^T+R_1(R_1+R_2)^{-1}R_2]^{-1}C+(1-\omega)\\ &\quad\times P^{(sf)}(k|k-1)C^T\times[CP^{j^{(sf)}}(k|k-1)C^T\\ &\quad+R_1(R_1+R_2)^{-1}R_2]^{-1}C, \end{aligned}\tag{9.225}$$
$$\begin{aligned}(K_1+K_2)^{(sf)}(k)y^{(sf)}(k) &= \omega P^{(sf)}(k|k-1)C^T\times[CP^{i^{(sf)}}(k|k-1)C^T\\ &\quad+R_1(R_1+R_2)^{-1}R_2]^{-1}\times[R_2(R_1+R_2)^{-1}\end{aligned}$$

$$\begin{aligned}&\times y_1(k+1)+R_1(R_1+R_2)^{-1}y_2(k+1)]\\&+(1-\omega)P^{(sf)}(k|k-1)C^T\\&\times[CP^{j^{(sf)}}(k|k-1)C^T+R_1(R_1+R_2)^{-1}R_2]^{-1}\\&\times[R_2(R_1+R_2)^{-1}y_1(k+1)\\&+R_1(R_1+R_2)^{-1}y_2(k+1)].\end{aligned}\tag{9.226}$$

It is important to note that (9.222) and (9.225) have the same form and that (9.223) and (9.226) are also in the same form. Therefore, with the same initial conditions, i.e., $P^{(mf)}(0|0) = P^{(sf)}(0|0)$ and $\hat{x}^{(mf)}(0|0) = \hat{x}^{(sf)}(0|0)$, the Kalman filters based on the observation information generated by (9.210)–(9.212) and (9.213)–(9.215), respectively, will result in the same state estimate $\hat{x}(k|k)$. This means that the two measurement fusion methods are functionally equivalent in the sensor-to-sensor case. □

Now, we consider the case when the information filter is applied. From (9.210)–(9.215), it is easy to prove the following equalities:

$$[C^{(mf)}(k)]^T[R^{(mf)}(k)]^{-1}C^{(mf)}(k)=\sum_{j=1}^{N}C_j^TR_j^{-1}C_j,\tag{9.227}$$

$$(C^{(mf)}(k))^T[R^{(mf)}(k)]^{-1}y^{(mf)}(k)=\sum_{j=1}^{N}C_j^TR_j^{-1}y_j,\tag{9.228}$$

and

$$\begin{aligned}&(C^{(sf)}(k))^T[R^{(sf)}(k)]^{-1}C^{(sf)}(k)=\\&[(\sum_{j=1}^{N}R_j^{-1})^{-1}\sum_{j=1}^{N}R_j^{-1}C_j]^T\sum_{j=1}^{N}R_j^{-1}C_j,\end{aligned}\tag{9.229}$$

$$\begin{aligned}&(C^{(sf)}(k))^T[R^{(sf)}(k)]^{-1}y^{(sf)}(k)=\\&[(\sum_{j=1}^{N}R_j^{-1})^{-1}\sum_{j=1}^{N}R_j^{-1}C_j]^T\sum_{j=1}^{N}R_j^{-1}y_j.\end{aligned}\tag{9.230}$$

If $C_j = C$, $j = 1, 2, \ldots, N$, then we have

$$(C^{(mf)}(k))^T[R^{(mf)}(k)]^{-1}C^{(mf)}(k)=[C^{(sf)}(k)]^T[R^{(sf)}(k)]^{-1}C^{(sf)}(k),\tag{9.231}$$

$$(C^{(mf)}(k))^T[R^{(mf)}(k)]^{-1}y^{(mf)}(k)=[C^{(sf)}(k)]^T[R^{(sf)}(k)]^{-1}y^{(sf)}(k).\tag{9.232}$$

Remark 9.7. The functional equivalence proved here before has considered the gain K as the center of existence for all the calculations, which can be the case for information-based weighted covariance filters too, but not for information-based Kalman-like particle filters where the gain K is not present.

9.6 Simulation example 9.2

In this section, simulation studies will be carried out to demonstrate the effectiveness of the developed distributed estimation.

An industrial utility boiler [44] is utilized to test and evaluate the developed algorithms. In the boiler system, the principal input variables are u_1, feed-water flow rate (kg/s); u_2, fuel flow rate (kg/s); and u_3, temperature spray flow rate (kg/s). The states are x_1, fluid density, x_2, drum pressure, x_3, water flow input, x_4, fuel flow input, and x_5, spray flow input. The principal output variables are y_1, drum level (m); y_2, drum pressure (kPa); and y_3, steam temperature (°C). The schematic diagram of the utility boiler can be seen in Fig. 9.11.

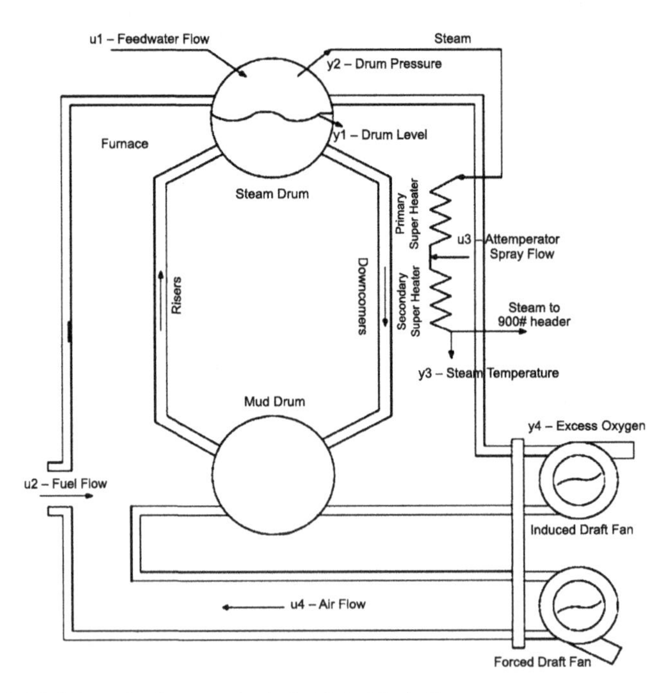

FIGURE 9.11 Schematic diagram of an industrial utility boiler.

The mathematical model of the faulty utility boiler can be described where faults occurring at steam pressure are reflected in state 4 (fuel flow input) and state 5 (spray flow input), respectively.

In the normal operation of the utility boiler, the steam temperature must be kept at a certain level to avoid overheating of the superheaters. By applying a step to the water flow input (state 3), the steam temperature increases and the steam temperature dynamics behave like a first-order system. Applying a step to the fuel flow input (state 4), the steam temperature increases and the system behaves like a second-order system. Applying a step to the spray flow input (state 5), the steam temperature decreases and the system behaves like a first-order system. Then, a third-order system is selected for the steam temperature model. Steam pressure is added there in states 4 and 5, resulting in a more uncontrolled nonlinear system.

Following [44] and the proposed fault scenarios, a scaled and dimensionless discrete-time fault model of the system can be described as

$$x_1(k+1) = -0.002044x_1(k) - 0.0001933x_2(k) + 0.00645u_1(k), \quad (9.233)$$

$$x_2(k+1) = -0.0024x_2(k) - 0.0404u_1(k) + 3.025u_2(k), \quad (9.234)$$

$$\begin{aligned} x_3(k+1) = &-0.0211x_2(k) + x_4(k) - 0.0010967u_1(k) + 0.0475u_2(k) \\ &+ 3.1846u_3(k), \end{aligned} \quad (9.235)$$

$$\begin{aligned} x_4(k+1) = &\ 0.0015x_2(k) + x_5(k) - 0.001u_1(k) \\ &+ 0.32u_2(k) - 2.9461u_3(k) + (a_{st\ pr})x_2(k), \end{aligned} \quad (9.236)$$

$$\begin{aligned} x_5(k+1) = &-1.278 \times 10^{-3}x_2(k) - 0.00025831\, x_3(k) - 0.29747\, x_4(k) \\ &- 0.8787621548\, x_5(k) - 0.00082\, u_1(k) - 0.2652778\, u_2(k) \\ &+ 2.491\, u_3(k) + (a_{st\ pr})x_2(k). \end{aligned} \quad (9.237)$$

9.6.1 Evaluation of results

In what follows, we present simulation results for the proposed information-based versions of filters. The simulations have been performed on the utility boiler system where the faults due to steam pressure have been introduced in states 4 and 5, respectively. In all subsequent simulation studies, the computational accuracy was set to 10^{-4}. Firstly, the data generated from the simulation of the plant have been initialized and the parameters have been optimized, which involves the preprocessing and normalization of the data. The comparison of results for the distributed estimation, and normal estimation with different feedback generated from faults, and the basic profile of that particular state has been carried out. Moreover, the same pattern of comparison has been followed for all versions of information-based filters.

9.6.2 Covariance intersection filter

The information-based covariance intersection filter has been simulated here for the utility boiler steam pressure fault of state 4. Simulations have been made for the estimate of each case using the state vector fusion method. In the simulation, comparisons of various profiles have been made, that is,

- the profile of the normal fault-free state,
- the estimate of the normal fault-free state,
- the estimate of the faulty state, and
- the distributed estimate based on state vector fusion for different feedback strategies.

The comparison of profiles mentioned above for the cases of complete feedback, partial feedback, and no feedback profiles can be readily seen in Figs. 9.12–9.14, respectively. Moreover, a comparison for all feedback strategies is depicted in Fig. 9.15.

9.6.3 Weighted covariance filter

The information-based weighted covariance filter has been simulated here for the utility boiler steam pressure fault of state 4. Simulations have been made for the estimate of each case using the state vector fusion method. In the simulation, comparisons of various profiles have been made similar to the foregoing case of the information-based covariance intersection filter. The comparison of profiles with complete feedback and partial feedback profiles can be readily seen in Figs. 9.16 and 9.17, respectively. Moreover, a full comparison for all feedback strategies can be found in Fig. 9.18.

9.6.4 Kalman-like particle filter

The information-based Kalman-like particle filter has been simulated here for the utility boiler steam pressure fault of state 4. Simulation has been made for the estimate of each case using the state vector fusion method. In the simulation, comparisons of various profiles have been made, i.e., the profile of the normal fault-free state, the estimate of the normal fault-free state, the estimate of the faulty state, and the distributed estimate based on state vector fusion for different feedback strategies. The comparison of profiles mentioned above for complete feedback and partial feedback profiles can be seen in Figs. 9.19 and 9.20, respectively. Moreover, the one-on-one full comparison for all feedback strategies can be seen in Fig. 9.21. A profile comparison for the measurement fusion method can be seen in Fig. 9.22 for a complete feedback case.

It can be readily concluded that:

- *In the case of information-based covariance intersection, the complete feedback case performs better than the partial and no feedback cases.*
- *In the case of information-based weighted covariance, the no feedback case performs better than the partial feedback and complete feedback cases. The complete feedback strategy has the lowest performance.* One interpretation for this result is that the structure of the information-based weighted covariance filter does not require additional information for processing fused estimates and hence the feedback action does not add any value.

- *In the case of the information-based Kalman-like particle filter, the partial feedback case performs slightly better than the complete feedback and no feedback cases. The no feedback strategy has the lowest performance.* One interpretation of this result is that the information afforded by the partial feedback strategy is sufficient for distributed fusion of estimates and hence complete feedback does not give any additional value.

9.6.5 Mean square error comparison

In this section, we make a comparison of all versions of information-based filters with complete, partial, and no feedback, respectively. It can be seen from Table 9.1 that the feedback versions perform differently for a particular case of information-based filter. The mean square error value of complete feedback is minimum in the case of the information-based covariance intersection filter and Kalman-like particle filter, respectively, whereas partial feedback performs well in the case of the information-based weighted covariance filter.

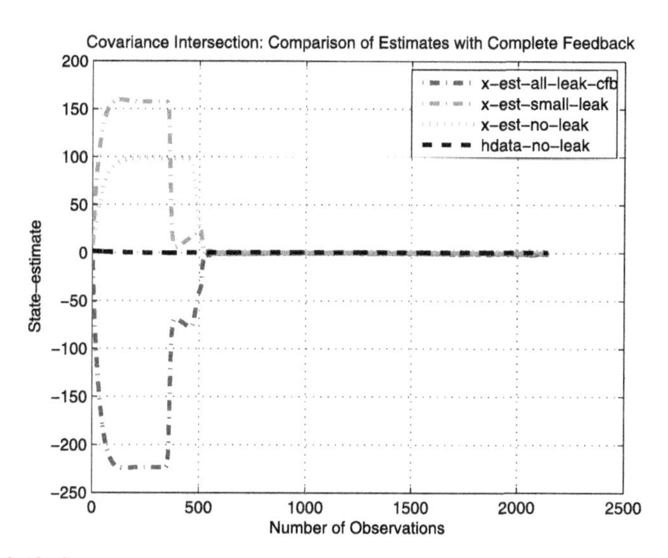

FIGURE 9.12 Covariance intersection: Complete feedback comparison.

Note in Fig. 9.12 that $x - est - all - leak - cft$ means the estimate of x with full leakage using complete feedback, $x - est - small - leak$ means the estimate of x with small leakage using complete feedback, $x - est - no - leak$ means the estimate of x without leakage using complete feedback, and $hdata - no - leak$ means historical data without leakage.

Note in Fig. 9.13 that $x - est - all - leak - pft$ means the estimate of x. Note in Fig. 9.14 that $x - est - all - leak - nft$ means the estimate of x with full leakage without feedback.

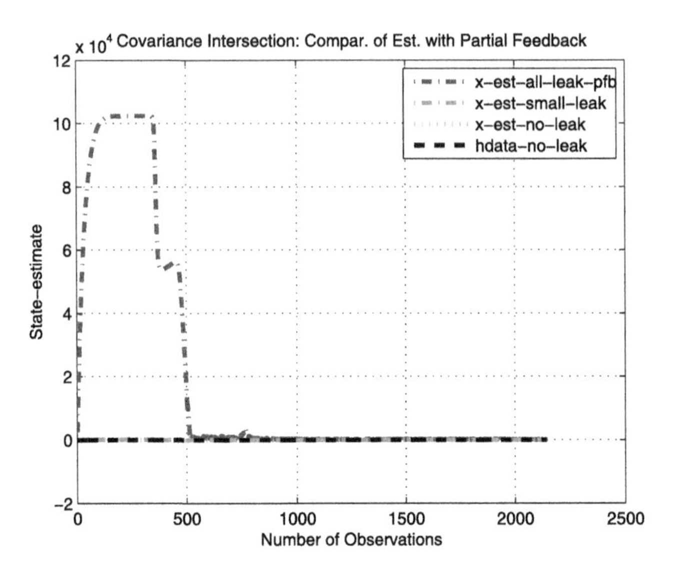

FIGURE 9.13 Covariance intersection: Partial feedback comparison.

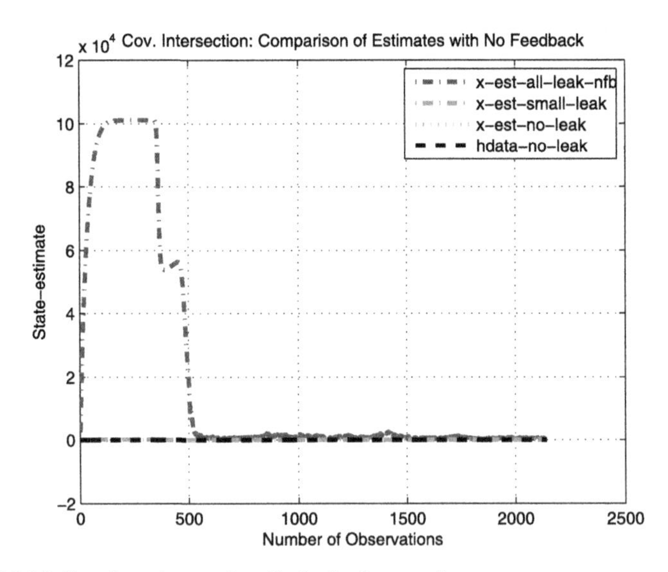

FIGURE 9.14 Covariance intersection: No feedback comparison.

Note in Fig. 9.16 that $x - est - all - leak - cft$ means the estimate of x with full leakage using complete feedback, $x - est - small - leak$ means the estimate of x with small leakage using complete feedback, $x - est - no - leak$ means the estimate of x without leakage using complete feedback, and $hdata - no - leak$ means historical data without leakage.

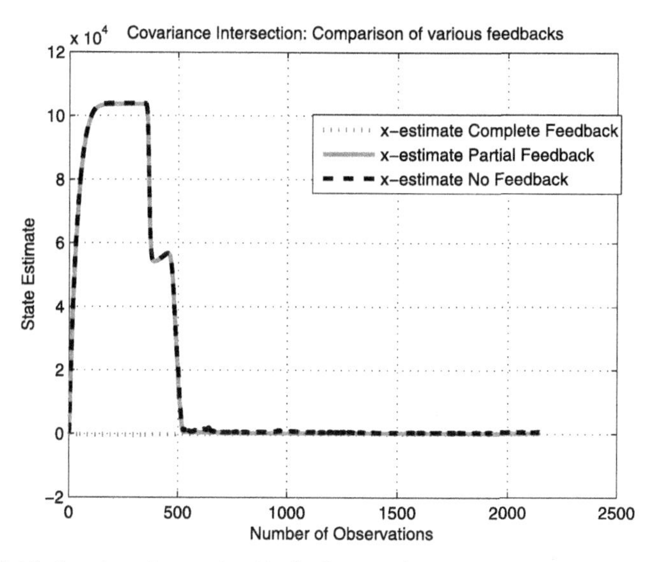

FIGURE 9.15 Covariance intersection: Feedback comparison.

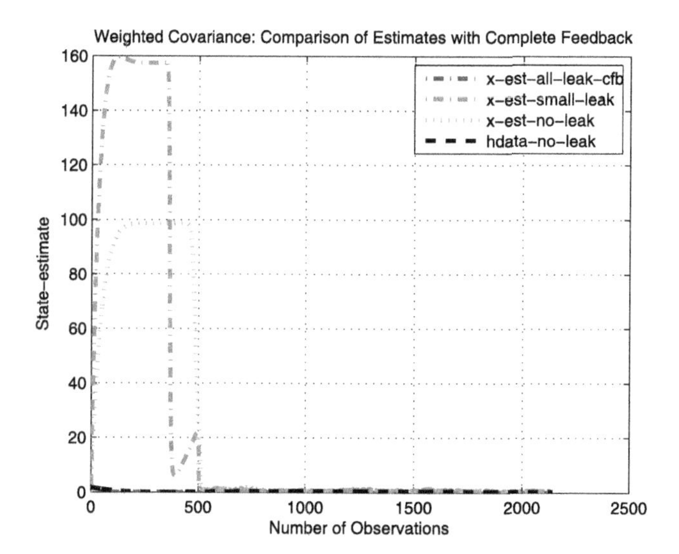

FIGURE 9.16 Weighted covariance: Complete feedback comparison.

Note in Fig. 9.17 that $x - est - all - leak - pft$ means the estimate of x with full leakage using partial feedback.

Note in Fig. 9.19 that $x - est - all - leak - cft$ means the estimate of x with full leakage using complete feedback, $x - est - small - leak$ means the estimate of x with small leakage using complete feedback, $x - est - no - leak$ means the

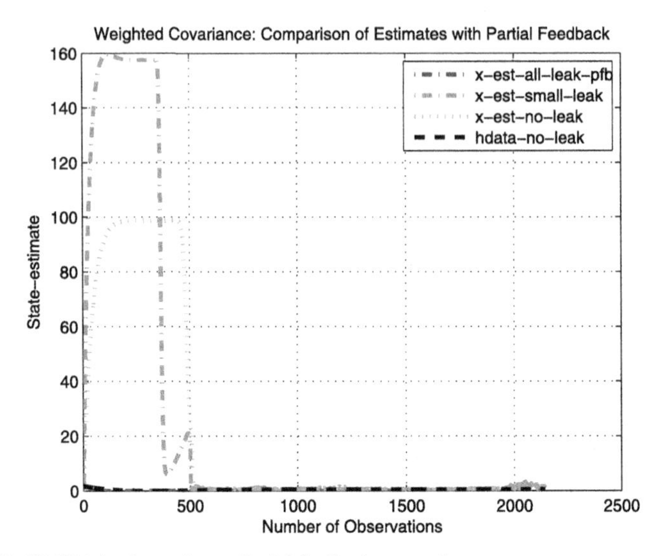

FIGURE 9.17 Weighted covariance: Partial feedback comparison.

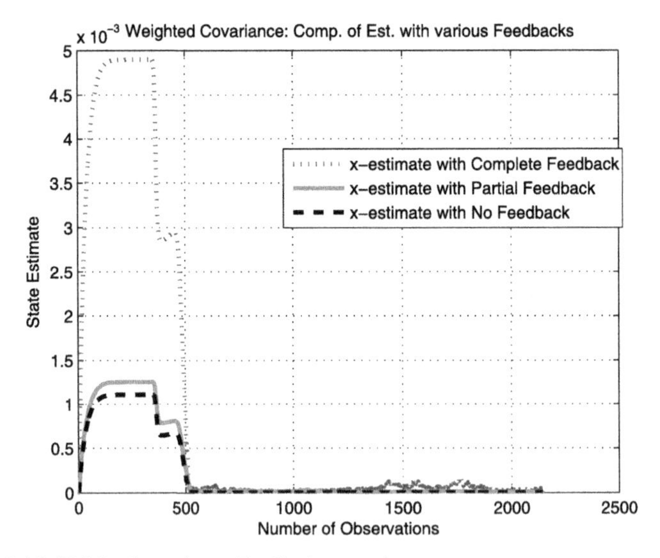

FIGURE 9.18 Weighted covariance: Feedback comparison.

estimate of x without leakage using complete feedback, and $hdata - no - leak$ means historical data without leakage.

Note in Fig. 9.19 that $x - est - all - leak - pft$ means the estimate of x with full leakage using partial feedback.

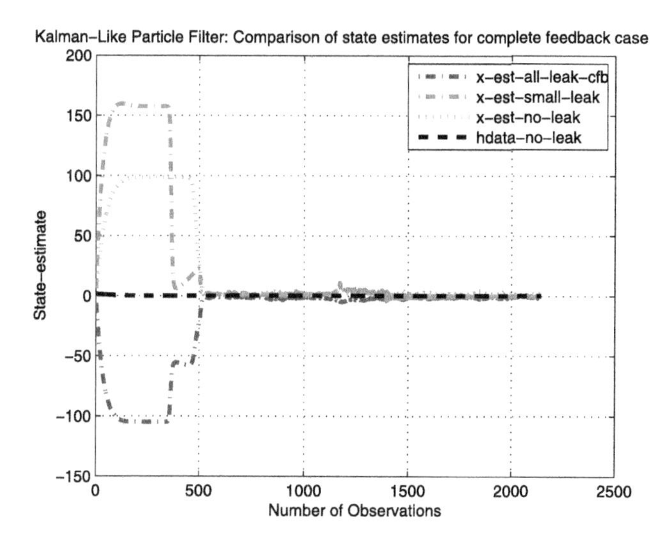

FIGURE 9.19 Kalman-like particle filter: Complete feedback comparison.

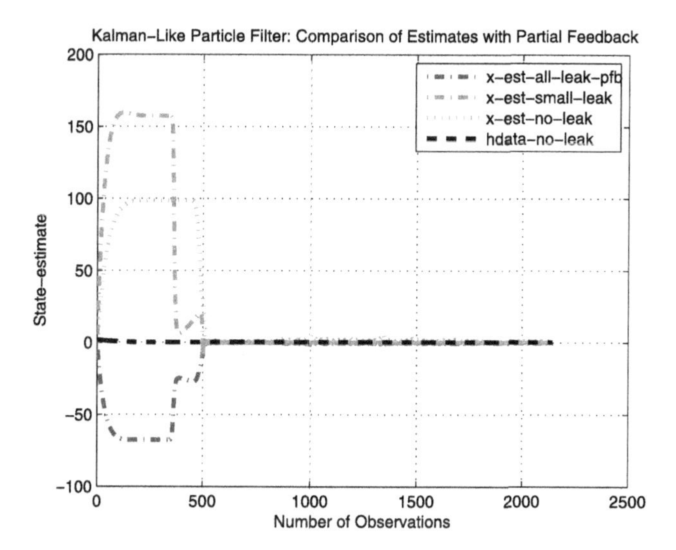

FIGURE 9.20 Kalman-like particle filter: Partial feedback comparison.

Note in Fig. 9.22 that $x - est - all - leak - pft - measurement - fusion$ means the estimate of x with full leakage using complete feedback with measurement fusion.

Remark 9.8. Table 9.1 shows the comparison of all the versions of information-based filters, where FB stands for feedback, CI stands for covariance intersection, WC stands for weighted covariance, and KLPF stands for Kalman-like particle filter.

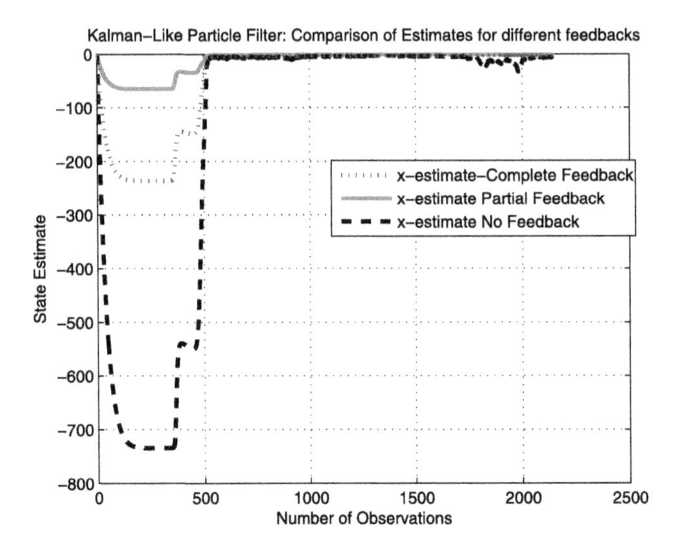

FIGURE 9.21 Kalman-like particle filter: Feedback comparison.

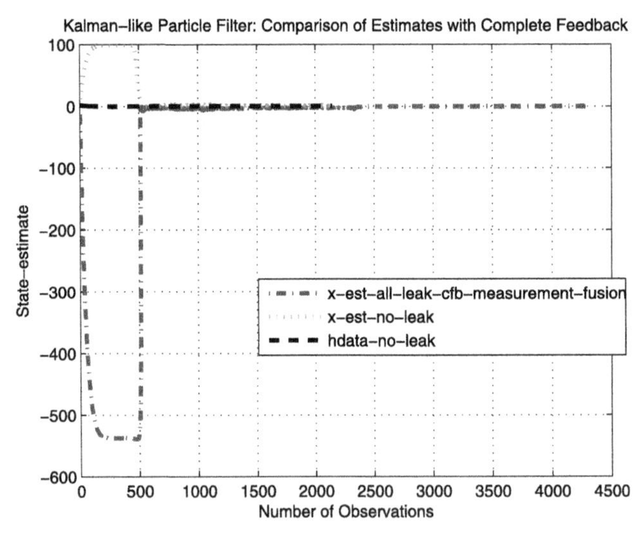

FIGURE 9.22 Kalman-like particle filter: Complete feedback with the measurement fusion method.

9.7 Notes

This chapter has presented a state filtering for linear stochastic discrete-time systems subject to deception attacks and data losses on the control signals transmitted by the controller to the plant. A bias state-dependent intermittent unknown input disabled at the occurrence time of data losses has been used to

TABLE 9.1 Mean square error comparison for all information-based filters.

Filter	Complete FB	Partial FB	No FB
CI	6.424	8.2759	8.411
WC	1.031×10^{-3}	1.0273×10^{-3}	1.0275×10^{-3}
KLPF	0.565	0.703	0.6223

derive a fixed-dimensional augmented state model of the plant, allowing a direct application of the intermittent unknown input Kalman filter.

It has been shown that the self-tuning Kalman filter approach has no requirement of providing a feedback loop in view of its embedded information matrix. The condition of feedback from fusion center to the local sensors has helped in the implementation of the self-tuning Kalman filter. The developed scheme has been successfully evaluated on a quadruple-tank system to demonstrate its effectiveness in practical implementations.

Then, an FB Kalman filter fusion approach was implemented on the developed scheme on the regular Kalman filter. It has been shown that the FB Kalman filter has a smoothing property which provides better results than the regular Kalman filter. Moreover, the conditions with intermittent observations are handled in a satisfactory manner identical to the communication losses.

A distributed estimation has been proposed using various versions of information matrix filters. Different feedback strategies were evaluated and the focal point is the relation of the performance and the number of sensors. It is shown that for algorithms, the feedback strategies perform differently, that is, information-based covariance intersection and the Kalman-like particle filter perform better with complete feedback, whereas information-based weighted covariance performs better with partial feedback.

References

[1] H.R. Hashmipour, S. Roy, A.J. Laub, Decentralized structures for parallel Kalman filtering, IEEE Trans. Autom. Control 33 (1) (1988) 88–93.

[2] Y.M. Zhu, Z. You, J. Zhao, K.S. Zhang, X.K. Li, The optimality for the distributed Kalman filtering fusion with feedback, Automatica 37 (2001) 1489–1493.

[3] Y. Gao, W.L. Wang, Z.L. Deng, Information fusion estimation of noise statistics for multisensor systems, in: Chinese Control and Decision Conference, Guilin, 2009, pp. 1127–1131.

[4] L. Claudio, A. Souza, R.M. Palhares, Parameter estimation on linear time-varying systems, J. Franklin Inst. 348 (4) (May 2011) 777–789.

[5] T. Li, Y. Zhang, Fault detection and diagnosis for stochastic systems via output PDFs, J. Franklin Inst. 348 (6) (August 2011) 1140–1152.

[6] N. Orani, A. Pisano, E. Usai, Fault diagnosis for the vertical three-tank system via high-order sliding-mode observation, J. Franklin Inst. 347 (6) (August 2010) 923–939.

[7] H.R. Karimi, M. Zapateiro, M. Luo, A linear matrix inequality approach to robust fault detection filter design of linear systems with mixed time-varying delays and Nonlinear perturbations, J. Franklin Inst. 347 (6) (August 2010) 957–973.

[8] M. Basin, D. Calderon-Alvarez, Optimal filtering over linear observations with unknown parameters, J. Franklin Inst. 347 (6) (August 2010) 988–1000.

[9] D. Zhang, L. Yu, $\mathcal{H}_\infty$ filtering for linear neutral systems with mixed time-varying delays and nonlinear perturbations, J. Franklin Inst. 347 (7) (September 2010) 1374–1390.
[10] M. Basin, P. Shi, P. Soto, Central suboptimal $\mathcal{H}_\infty$ filtering for nonlinear polynomial systems with multiplicative noise, J. Franklin Inst. 347 (9) (November 2010) 1740–1754.
[11] G.L. Tao, X.H. Guan, Z.L. Deng, Self-tuning distributed fusion Kalman filter with asymptotic global optimality, in: Proc. the 29th Chinese Control Conference, July 29–31, 2010, pp. 1268–1272.
[12] E. Song, Y. Zhu, J. Zhou, The optimality of Kalman filtering fusion with cross-correlated sensor noises, in: Proceedings of the 43rd IEEE Conference on Decision and Control, Atlantis, Paradise Island, Bahamas, December 14–17, 2004, pp. 4637–4642.
[13] L. Ljung, System Identification: Theory for the User, second edition, Prentice Hall, NY, 1999.
[14] M.S. Mahmoud, H.M. Khalid, Distributed Kalman filtering: a bibliographic review, IET Control Theory Appl. 7 (4) (April 2013) 483–501.
[15] M. Liggins, C.Y. Chong, I. Kadar, M.G. Alford, V. Vannicola, S. Thomopoulos, Distributed fusion architectures and algorithms for target tracking, Proc. IEEE 85 (1) (1997) 95–107.
[16] D.I. Schizas, G.B. Giannakis, Z.Q. Luo, Distributed estimation using reduced-dimensionality sensor observations, IEEE Trans. Signal Process. 55 (8) (2007) 4285–4299.
[17] E.B. Song, Y.M. Zhu, J. Zhou, Sensors' optimal dimensionality compression matrix in estimation fusion, Automatica 41 (12) (2005) 2131–2139.
[18] D. Hall, J. Llinas, An Introduction to multi-sensor data fusion, Proc. IEEE 85 (1) (1997) 6–23.
[19] R. Viswanathan, P. Varshney, Distributed detection with multiple sensors Part I: fundamentals, Proc. IEEE 85 (1) (1997) 54–63.
[20] H. Durrant-Whyte, B. Rao, H. Hu, Toward a fully decentralized architecture for multi-sensor data fusion, in: Proc. IEEE Int. Conference on Robotics and Automation, vol. 2, 1990, pp. 1331–1336.
[21] S. Grime, H.F. Durrant-Whyte, P. Ho, Communication in decentralized data-fusion systems, in: Proc. American Control Conference, 1992, pp. 3299–3303.
[22] F. Bourgault, H.F. Durrant-Whyte, Communication in general decentralized filters and the coordinated search strategy, in: Proc. 7th Int. Conference on Information Fusion, July 2004.
[23] L. Xiao, S. Boyd, Fast linear iterations for distributed averaging, Syst. Control Lett. 53 (1) (2004) 65–78.
[24] D.P. Spanos, R. Olfati-Saber, R.M. Murray, Distributed sensor fusion using dynamic consensus, in: Proc. 16th IFAC World Congress, July 2005.
[25] R. Olfati-Saber, Distributed Kalman filter with embedded consensus filters, in: Proc. 44th IEEE Conference on Decision and Control and European Control Conference, December 2005.
[26] K. Chang, T. Zhi, R. Saha, Performance evaluation of track fusion with information matrix filter, IEEE Trans. Aerosp. Electron. Syst. 38 (2) (2002) 455–466.
[27] S. Sun, Multi-sensor optimal information fusion Kalman filter for discrete multi-channel ARMA signals, in: IEEE Int. Symposium on Intelligent Control, 2003, pp. 377–382.
[28] J. Liang, Z. Wang, X. Liu, Distributed state estimation for uncertain Markov-type sensor networks with mode-dependent distributed delays, Int. J. Robust Nonlinear Control 22 (3) (February 2012) 331–346.
[29] M.S. Mahmoud, H.M. Khalid, Expectation maximization approach to data-based fault diagnostics, Inf. Sci. 235 (2013) 80–96.
[30] M.S. Mahmoud, H.M. Khalid, Bibliographical review on distributed Kalman filtering, IET Control Theory Appl. 7 (4) (2013) 483–501.
[31] Y. Shalom, On the track-to-track cross-covariance problem, IEEE Trans. Autom. Control 26 (1981) 571–572.
[32] C.Y. Chong, S. Mori, Convex combination and covariance intersection algorithm in distributed fusion, in: Proc. the 4th Int. Information Fusion Conference, Montreal, Canada: ISIF, 2001.
[33] S.L. Sun, Z.L. Deng, Multi-sensor optimal information fusion Kalman filter, Automatica 40 (2004) 1017–1023.
[34] B.F. Scala, A. Farina, Choosing a track association method, Inf. Fusion 3 (2001) 119–133.

[35] C.Y. Chong, K.C. Chang, S. Mori, Distributed tracking in distributed sensor networks, in: Proc. the American Control Conference, Seattle, 1986, pp. 1863–1868.
[36] S.J. Julier, J.K. Uhlman, A non-divergent estimation algorithm in the presence of unknown correlations, in: Proc. the American Control Conference, 1997, pp. 2369–2373.
[37] J.A. Roecker, C.D. McGillem, Comparison of two-sensor tracking methods based on state vector fusion and measurement fusion, IEEE Trans. Aerosp. Electron. Syst. 24 (1988) 447–449.
[38] A. Hansson, M. Verhaegen, Distributed system identification with ADMM, in: Proc. the 53rd IEEE Conference on Decision and Control, Los Angeles, California, USA, December 15–17, 2014, pp. 290–295.
[39] H.M. Khalid, J.C. Peng, M.S. Mahmoud, Enhanced distributed estimation based on prior information, IET Signal Process. 9 (1) (2015) 60–72.
[40] Q. Gan, C.J. Harris, Comparison of two measurement fusion methods for Kalman-filter-based multi-sensor data fusion, IEEE Trans. Aerosp. Electron. Syst. 37 (1) (2001) 273–279.
[41] Z. Liu, M. Wang, J. Huang, An evaluation of several fusion algorithms multi-sensor tracking system, J. Inf. Comput. Sci. 7 (10) (2010) 2101–2109.
[42] A.H. Sayed, Fundamentals of Adaptive Filtering, Wiley, NJ, 2003.
[43] R.T. Sukhavasi, B. Hassibi, The Kalman-like particle filter: optimal estimation with quantized innovations/measurements, in: Proc. Conference on Decision and Control, 2009, pp. 4446–4451.
[44] H. Marquez, M. Riaz, Robust state observer design with application to an industrial boiler system, Control Eng. Pract. 13 (2005) 713–728.

Index

K

L

M

N

O

P

T

Printed in the United States
By Bookmasters